LEEDS BE
1700026792

The Geography of Urban Transportation

Edited by
Susan Hanson
Clark University

The Guilford Press
New York London

To Perry, Kristin, and Erik—and our favorite form of transportation, the bicycle.

A Division of Guilford Publications, Inc.
200 Park Avenue South, New York, N.Y. 10003

Printed in the United States of America

Library of Congress Cataloging-in-Publication Data

The Geography of Urban transportation.

Includes bibliographical references and index.
1. Urban transportation. I. Hanson, Susan.
HE305.G46 1986 388.4 86-7620
ISBN 0-89862-775-3

Contributors

Gerald Barber, Ph.D.
University of Victoria, British Columbia, Canada

Gordon J. Fielding, Ph.D.
Institute for Transportation Studies, University of California, Irvine

Genevieve Giuliano, Ph.D.
Institute for Transportation Studies, University of California, Irvine

Susan Hanson, Ph.D.
Graduate School of Geography, Clark University, Worcester, Massachusetts

Geoffrey J. D. Hewings, Ph.D.
Department of Geography, University of Illinois, Urbana-Champaign

David Hodge, Ph.D.
Department of Geography, University of Washington, Seattle

Joel L. Horowitz, Ph.D.
Departments of Geography and Economics, University of Iowa, Iowa City

Donald G. Janelle, Ph.D.
Department of Geography, University of Western Ontario, London, Canada

Ronald L. Mitchelson, Ph.D.
Department of Geography, University of Georgia, Athens

Peter O. Muller, Ph.D.
Department of Geography, University of Miami, Coral Gables, Florida

Eric I. Pas, Ph.D.
Department of Civil and Environmental Engineering, Duke University, Durham, North Carolina

John S. Pipkin, Ph.D.
Department of Geography and Regional Planning, State University of New York at Albany

David A. Plane, Ph.D.
Department of Geography and Regional Development, University of Arizona, Tucson

Margo Schwab
Graduate School of Geography, Clark University, Worcester, Massachusetts

Eric Sheppard, Ph.D.
Department of Geography, University of Minnesota, Minneapolis

Frederick P. Stutz, Ph.D.
Department of Geography, San Diego State University, California

James O. Wheeler, Ph.D.
Department of Geography, University of Georgia, Athens

Preface

On the inside flap of a box of Celestial Seasoning Cinnamon Rose tea I recently encountered the following tidbit of wisdom: we find our individual freedom by choosing not a destination but a direction.* Seeing words like destination and direction, I knew I had stumbled on a gem that held special significance for a transportation geographer. Was not individual freedom after all one of the fundamental reasons for studying urban movement patterns? But then, if directions are the only route to freedom, why have transportation geographers devoted untold years of research time to the issue of destination choice? Should we now shift our efforts to modeling direction choice? Or was the hidden message that transportation geographers had long ago chosen a direction (or perspective) from which to view the urban transportation problem that was somewhat different from the direction selected by other transportation professionals trained as civil engineers, economists, or planners? This geographic perspective, one might argue (in line with the tea flap), is one that is less concerned with arriving at a particular destination (a transportation system of a given configuration) than with understanding movement in cities and the role of transportation in urban life and urban systems.

*Attributed to Marilyn Ferguson.

Enough musing on a tea box! Understanding urban transportation—movement in cities—is critical to understanding how cities work. As Russell Baker noted some time ago, a good chunk of one's life is spent moving about: "The year of the suburban housewife and mother includes 400 hours of driving from home to school to supermarket to school to supermarket to home. Doesn't sound like much, maybe, but it's 64 hours longer than a two-week vacation."* His figures might be apocryphal, but what is not disputable is the central role of transportation to urban life and hence to urban analysis.

Despite the importance of urban transportation to the geography of the city and despite the fact that many of us teach courses in urban transportation geography, we have not had a book that tackles urban transportation from a geographic perspective. This

**New York Times Magazine*, 31 December 1978.

book is very much a collective effort at filling this long-felt need. I would not have taken on the job of trying to put together a book like this had not the lack of an urban transportation geography text been a major topic of discussion (and despair) at two successive annual meetings of the AAG Transportation Specialty Group (TSG). That discussion (and subsequent prodding) mobilized the contributors to develop original chapters for a text aimed at upper-level undergraduates and beginning graduate students.

We have tried to structure a book that lays out both theoretical and practical concerns and shows how social scientists and planners have gone about analyzing these problems. After three introductory chapters that describe the urban transportation context, the role of transportation in urban form, and the urban transportation planning process, respectively, six chapters are devoted to outlining analytical approaches to transportation problems. Here a key distinction is in the level, or scale, of analysis: the first three chapters of this second section deal with aggregate approaches and the second three chapters focus on disaggregate approaches. Within each section there is first a chapter devoted to describing flows at the scale in question, then one on modeling these flows, and finally a chapter that provides examples of the procedures outlined in the modeling chapter. The problem orientation of the book comes to the fore in the third section, in which each chapter spotlights a particular transportation-related policy concern: public transportation, land use impacts, energy issues, social and environmental impacts, substituting communications for travel, and finally an overview of policy options. Some of the analytical techniques described in part 2 are used in specific policy contexts in this last section. Throughout the book we have tried to minimize the amount of mathematics a student must have in order to comprehend the material presented here.

Beginning as mere gleams in our eyes at AAG Transportation Specialty Group meetings, our idea for a book like this would never have been transformed into this book without the hard work and cooperation of all the contributors. To them—and especially to Jim Wheeler and Pete Fielding, the past and present chairs of the TSG respectively, who gave special impetus to the project—I owe many thanks. Janet Crane, geographer and Guilford Press editor, also played a leading role in the genesis and realization of this book. I appreciate her vision, her patience, and her persistence. Thanks to all of these people, we now have an urban transportation book with a geographic direction.

Susan Hanson
Worcester, Massachusetts
February 1986

Contents

POLICY CONCERNS

SECTION I
INTRODUCTION

1 DIMENSIONS OF THE URBAN TRANSPORTATION PROBLEM

SUSAN HANSON
Clark University

Urban transportation. What do these words conjure up? Sleek new buses? Graffiti-covered, putty-colored New York subway cars? Quaint cable cars rattling up and down the hills of San Francisco? An automobile-clogged "free"way? The clean efficiency of a modern rapid rail system such as Washington, D.C.'s, Metro? For many people the phrase "urban transportation" brings to mind images like these—pictures of all sorts of transportation rolling stock, the hardware of mobility. But is this hardware the essence of urban transportation? Think for a moment what urban life would be like without the ease of movement that we have come to take for granted. The blizzards that periodically envelop major cities have given urban populations a fleeting taste of what it is like to be held captive (quite literally) in one's own home for several days. With roads buried in six feet of packed snow, how can one obtain food, visit friends, get medical care for a sick child—not to mention get to work? Urban life is paralyzed when transportation arteries are impassable and mobility is impossible.

Thanks to my research assistants, Ibipo Johnston and Margo Schwab, for their help in assembling some of the materials for this chapter.

MOBILITY AND ACCESSIBILITY

At the core of transportation are *access to activity sites*—proximity to places of work, recreation, socializing, shopping, medical care—and *mobility*, the ability to move between these activity sites. The spatial separation of activity sites demands mobility for a city to function. The urban landscape consists of spatially distinct, highly specialized land uses—food stores, laundromats, hardware stores, banks, drug stores, hospitals, schools, post offices, and so on—which people must get to if they want to obtain necessary goods and services. Moreover, home and work are now in the same location for only a very few people (about 2.3 percent of the American work force in 1979), so that to earn an income as well as to spend it, one must travel. As a result people select a place to live not only for the characteristics of a housing unit itself (for example, a two-bedroom apartment in an old three-story brownstone) or for the characteristics of the neighborhood in which the apartment is located (such as a good school district, the presence of many other families with young children), but also for the set of linkages to activity sites that a particular location affords. Although the principles of mobility and accessibility discussed here apply to the movement of goods and information as well as to people, the focus of this book is on the movement of people in cities. The book's first part furnishes a background against which to understand the analytical issues tackled in Part II and the policy issues addressed in Part III. After this introductory chapter, there is a chapter providing a historical overview of the movement of people in cities, followed by a chapter outlining the urban transportation planning process.

Only occasionally do people engage in travel for its own sake, as in taking a Sunday drive or a family bike ride. Most urban travel occurs as a by-product of trying to accomplish some other (nontravel) activity such as work or shopping. In this sense the demand for urban transportation is a *derived demand*, derived as it is from the need or desire to do something (other than travel) at some place other than home. There is always a trade-off between doing an activity at home (such as cutting hair, watching a film on TV, doing laundry) or adding the cost of movement to accomplish that activity or a similar one somewhere else.

Transportation and Land Use

All movement incurs a cost of some sort, which is usually measured in terms of time or money. Some kinds of travel, such as that by auto, bus, or train, incur both time and monetary costs; other trips, such as those made on foot, involve an outlay primarily of time. In deciding which mode(s) to use on a given trip, travelers often trade off time versus money costs, as the more costly travel modes are usually the faster ones. There is also a trade-off involved in the decision to make a trip; the traveler weighs the expected benefits to be gained at the destination against the expected costs of getting there. Each trip represents a triumph of such anticipated benefits over costs, although for the many trips that are made out of habit this intricate weighing of costs and benefits does not occur before each and every trip. Because people generally place a negative value on travel time, they do not want to be too far away from the places they need to visit regularly. (The desire for proximity, however, does not usually hold in the extreme: people want to be close to schools and shops and work, but not *too* close.) The desire not to be too far away from key activity sites—and entrepreneurs' awareness of this preference—is an important determinant of land use patterns in the city.

Thus transportation can be seen as the *consequence* of the fact that different types of land uses in the city are spatially separated, but enhanced mobility can also be seen as *contributing* to increased separation of land uses. Because improved transportation facilities enable people to travel farther in a given amount of time than they could previously, transportation improvements contribute to the growing spatial separation between activity sites (especially between home and work) in urban areas.

One reason geographers are interested in urban transportation is this symbiotic relationship between transportation and land use. One could never hope to understand the spatial structure of the metropolis or to grasp how it is changing without a knowledge of movement patterns. The accessibility of places has a major impact upon their land values (and hence the use to which the land is put), and the location of a place within the transportation network determines its accessibility. Thus in the long run the transportation system (and the travel on it) shapes the land use pattern. In the next chapter Peter Muller provides numerous historical examples of this interaction between transportation innovation and urban land use patterns. In the short run, however, the existing land use configuration helps to shape travel patterns. The intimate relationship between transportation and land use is explicitly acknowledged by the fact that at the heart of every city's master plan is a long-run transportation plan.

Measuring Accessibility

We can talk about the accessibility of *places* (as above: how easily certain places can be reached) or of *people* (how easily a person or a group of people can reach activity sites). An individual's level of accessibility will depend largely on where activity sites are located vis-à-vis the person's home and the transportation network, but it will also be affected by when such sites are open and even how much time the person can spare for making trips. Many scholars have argued that the ease with which people can reach employment locations, retail and service outlets, and recreational opportunities should be considered in any assessment of the health of a city. They have implied that accessibility should be a central part of any measure of the quality of life (see, for example, Koenig, 1980; Pirie, 1979; Wachs & Kumagi, 1973).

Personal accessibility is usually measured by counting the number of activity sites (also called "opportunities") available at a given distance from the person's home and "discounting" that number by the intervening distance. Often accessibility measures are calculated for specific types of opportunities, such as shops, employment places, or medical facilities.

$$A_i = \sum_j O_j \, d_{ij}^{-b} \quad (1)$$

where A_i is the accessibility of person i, O_j is the number of opportunities at distance j from person i's home, and d_{ij} is some measure of the separation between i and j (this could be travel time, travel cost, or simple distance). Such an accessibility index is a measure of the number of potential destinations available to a person and how easily they can be reached. Accessibility is usually assessed in relation to the person's home because that is the base from which most trips are made; there are, however, other important bases (such as the workplace) around which personal accessibility indices could (and perhaps should) be computed.

The accessibility of a place to other places in the city can be measured in a similar way:

$$A_i = \sum_j O_j \, d_{ij}^{-b} \quad (2)$$

where A_i is the accessibility of zone i, O_j is the number of opportunities in zone j, and d_{ij} is, as before, a measure of the separation between i and j.

Although measures 1 and 2 are structurally alike, the difference between the two is important. The first measures the accessibility of individuals, and the second indicates the accessibility of places (or zones) within the city. In the second instance every one living in zone i is assumed to have the same level of accessibility to activity sites in the city, and there is no way of distinguishing among different types of people within a zone, such as those with or without a car.

In Figure 1-1 an accessibility index has been computed for each person in a sample of people living throughout a city (Uppsala, Sweden), and the resulting individual accessibility scores have been mapped. The sample consisted of women who were not in the paid labor force and who lacked access to a car. You can see that accessibility levels vary fairly systematically according to a person's location within the city; in this case, people living nearest to the center of the city, the place with the densest pattern of land uses, have the highest accessibility levels. As the city in Figure 1-1 was a classic single-center city at the time these data were collected, this circular accessibility pattern is to be expected. If we were to construct such a map for one of the multinucleated cities described in the next chapter, a more complicated pattern would no doubt emerge. The point to note from Figure 1-1 is how dependent people's accessibility levels are upon the spatial configuration of activity sites.

A person's ability to reach places depends

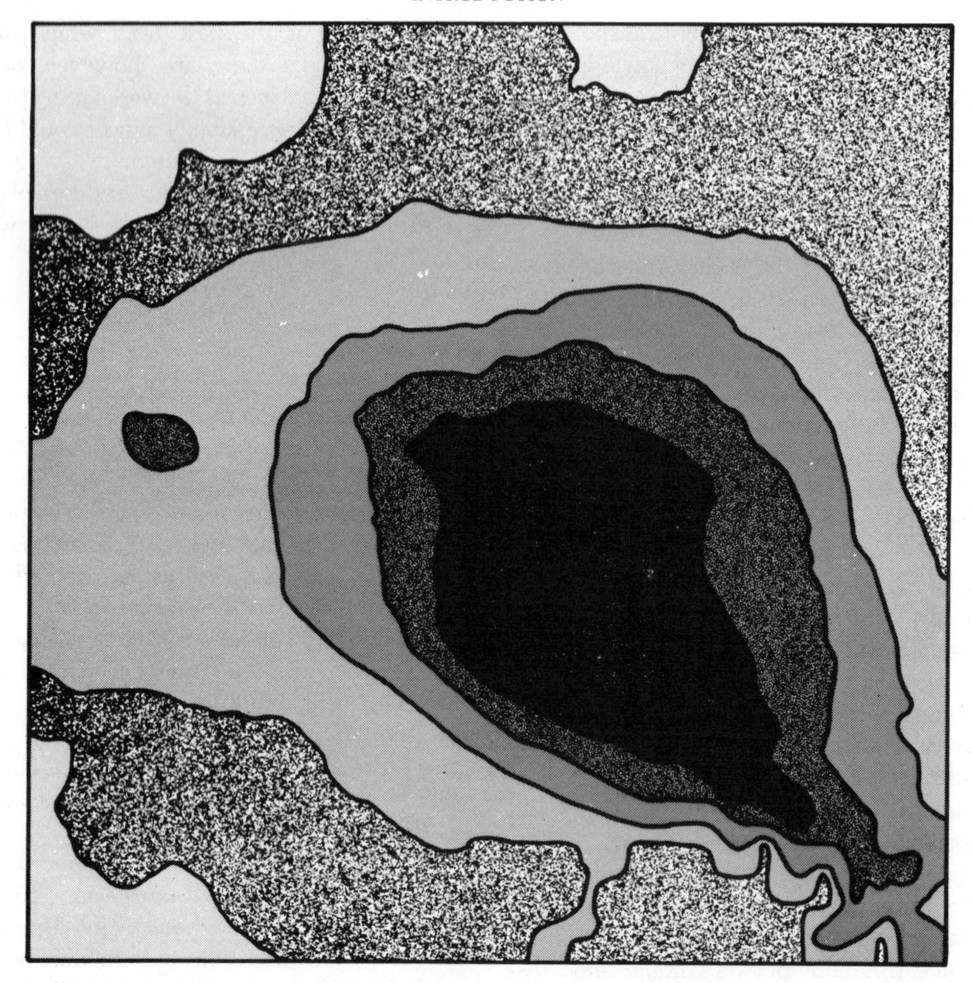

Figure 1-1. Accessiblity of unemployed women without cars to retail and service establishments in Uppsala, Sweden, 1971, darker areas indicate higher access.

only in part on the relative location of those places; it also depends on mobility, the ability to move to activity sites. We have seen how the spatial organization of contemporary society demands—indeed assumes—mobility; yet not all urban residents enjoy the high level of mobility that the contemporary city requires for the conduct of daily life. Those without access to cars are especially likely to lack the mobility necessary to reach job locations or other activity sites. This lack of mobility among certain groups of people is one part of the urban transportation problem. As transportation is essential to the very fabric of urban life, transportation issues and problems are inextricably bound up with other urban issues and problems such as social inequities and environmental pollution. Before we consider other aspects of the urban transportation problem, the next section of the chapter provides some background on the context within which travel takes place in the North American city.

THE URBAN CONTEXT

Because the primary concern of transportation planners is (or should be) to provide a system that enables people to move to activity sites, it is necessary, in order to get an overview of the urban transportation system (and urban transportation problem), to see where people live vis-à-vis activity sites, especially employment opportunities. Because both residential and employment distributions shift over time and because certain groups' accessibility to employment can be altered as a result, it is also important to see how these patterns have been changing in recent decades. Instead of making generalizations about American cities as a whole, I shall focus much of the discussion on one city in particular—Worcester, Massachusetts—because the trends revealed by a close look at Worcester are also found in many other large North American cities.

Overview of Residential Patterns

The census figures for Worcester shown in Table 1-1 disclose a number of trends from 1960–1980 that hold important implications for travel patterns and for urban transportation planning. First, while the population of the metropolitan area (SMSA) as a whole has grown modestly, the number of households and the number of single-person households has increased dramatically. From 1970 to 1980 the SMSA population grew by only 8%, but the number of households increased 25%. The greater increase in households relative to population has implications for trip making because the number of trips made per person per day generally declines as household size increases. The trend to more households and more single-person households contributes, then, to an overall growth in travel within the SMSA. A second trend is that despite the increase in population and in households, the proportion of the population residing in the central city has declined. A larger proportion of the metropolitan population now lives in the lower-density suburbs, which are more difficult to serve efficiently with public transportation.

Third, the proportion of households having no vehicle has dropped since 1960 both for the SMSA and for the city of Worcester, and the percentage of households without a car has remained noticeably higher in the city than in the suburbs. This latter point is to be expected, given the higher incidence of low-income households in the central city and given the greater availability of public transportation there. Nevertheless, despite the fact that a smaller proportion of households are now carless than was the case in 1960, there are still many people who must rely for mobility upon the bus, the taxi, a bicycle, their own feet, or rides from other people. Fourth, while the proportion of carless households has declined, the proportion of households with more than one vehicle has grown in both city and suburbs; by 1980 more than half of Worcester's suburban households had more than one car.

A fifth trend that is evident from examining Table 1-1 is that since 1960 there has been an increase in the number (and proportion) of households that are likely to have special transportation needs—the elderly, the poor, and female-headed households. The proportion of the population that is over 65 years of age has grown in both the SMSA and the central city, with a higher proportion of the population in the city than in the SMSA being elderly. An important point that is not revealed in Table 1-1, however, is that the proportion of the SMSA's elderly who live in the central city has actually been decreasing—from 66.2% of the elderly in 1960 to 52.6% in 1980; this means that the number of elderly people living in the suburbs has increased markedly. Because many older people do not drive, their presence in the suburbs—where bus service is often infrequent or nonexistent—raises questions about how the mobility needs of this group can be met. The travel problems of single-parent households, most of which are headed by women, stem from the difficulty of running a household single-handedly; earning an income, shopping, obtaining medical care and child care all must be done by the one adult in the household, often without the aid of an automobile.

Journey-to-Work Patterns

In addition to changing demographic patterns, Table 1-1 discloses some important

Table 1-1. Changes in population and travel patterns, Worcester, Massachusetts, 1960–1980

	1960	1970	1980
Population of SMSA[a]	323,306	344,320	372,940
Number of households in SMSA	94,680	104,694	130,785
Percent of single-person households in SMSA	13.6	17.6	23.2
Percent of SMSA population living in central city	57.7	51.3	43.4
Percent of households with no vehicle			
SMSA	22.0	17.7	14.5
City	29.3	26.2	23.0
Suburbs	no data	7.6	7.5
Percent of households with more than one vehicle			
SMSA	15.2	28.6	43.2
City	11.1	19.4	29.1
Suburbs	no data	39.3	53.0
Percent of population over 65 years of age			
SMSA	11.9	12.0	13.4
City	13.6	14.7	16.3
Suburbs	9.5	9.2	11.3
Percent of families below the poverty level			
SMSA	no data	5.4	7.5
City		7.1	11.2
Suburbs		3.7	4.7
Percent of families headed by women			
SMSA	no data	11.3	15.1
City		15.2	21.1
Suburbs		7.2	10.9
Percent of suburban workers who work in central city	42.0	39.6	36.5
Percent of workers using transit on worktrip			
SMSA	11.7	7.6	3.6
City	16.7	12.2	7.0
Suburbs	4.5	2.6	1.2

Notes: [a]SMSA = Standard Metropolitan Statistical Area.

Adapted from *U.S. Censuses of Population and Housing*, 1960, 1970, and 1980, by the U.S. Bureau of the Census, Washington, D.C.

trends regarding the journey to work. Of those employed people who lived in the suburbs, fewer than half have worked in the central city since 1960. The notion that the suburbs serve as bedrooms for the central-city work force clearly has little basis in fact. By 1980 only little more than a third (36.5%) of workers who lived in the suburbs commuted to work in the central city. As the suburb-to-city work trip is more easily made by public transportation than is the suburb-to-suburb or the city-to-suburb work trip, we should not be surprised to see that the decline in suburb-to-city commuters is accompanied by a decline in the proportion of workers using public transit to get to work. By 1980 not even 4% of the metropolitan work force commuted by bus, although the proportion of bus users was considerably higher in the central city (7%) than it was for the SMSA as a whole.

In a recent study David Plane (1981) demonstrates just how wrong is the idea that commuting patterns are dominated by suburban-to-central-city flows and how complex are the spatial patterns that contemporary commuters actually etch. He identifies five different types of work trips defined in terms of their origins and destinations (see Figure 1-2) and shows the overwhelming importance of lateral (suburb-to-suburb) flows in the commuting patterns of 28 New England urban fields (an urban field—or a daily urban system—is delimited by a line beyond which there are no workers commuting inward on a daily basis to the central city). Not only are inward flows a very small proportion of total work trips in these New England urban fields, but cross flows (whereby a worker lives in one urban field and works in another) comprise a substantial proportion of the work trips (see Table 1-2). Plane points out that in New England, where cities are relatively close together, urban fields overlap each other, and as a result most people live within more than one urban field; as much as 65% of the work force lives within three or four *different* daily urban systems. Because of this overlap, some work trips are counted more than once; the flow in Figure 1-3, for example, can be counted both as an inward flow for Urban Field 1 and as a cross flow for Urban Field 2. Plane points to the difficulties that these complex patterns pose for interpreting commuting data:

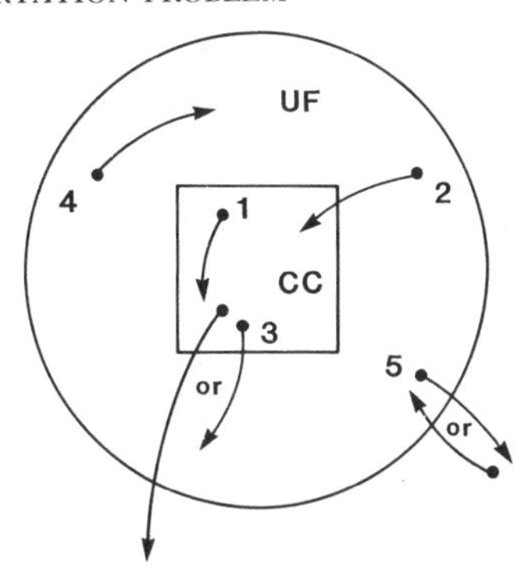

Figure 1-2. Typology of commuting flows: (1) within central city, (2) inward, (3) reverse, (4) lateral, (5) cross. From "The Geography of Urban Commuting Fields: Some Empirical Evidence from New England" by D. A. Plane, 1981, *Professional Geographer*, 33, p. 184. Copyright 1981 by the Association of American Geographers. Reprinted by permission.

Table 1-2. Commuting flows, by type, for selected urban fields, 1970

Flow type	Percent of all flows: Worcester, Mass.[a]	Percent of all flows: Median for 28 New England urban fields
Within central city	4.5	4.8
Inward: suburb to city	2.7	3.4
Reverse: city to suburb	0.8	1.7
Lateral: suburb to suburb	60.5	61.4
Cross: between urban fields	31.5	25.4
Total	100.0	100.0

Notes: [a]Based on a total of 1,226,000 workers.

From *Journey to Work* by U.S. Department of Commerce, Bureau of the Census, 1973, Washington, D.C.: U.S. Government Printing Office. See also Plane, 1981, p. 185.

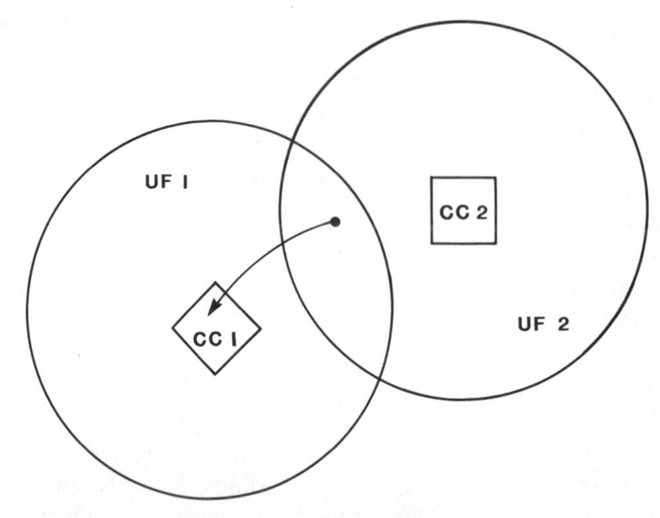

Figure 1-3. Overlapping urban commuting fields. From "The Geography of Urban Commuting Fields: Some Empirical Evidence from New England" by D. A. Plane, 1981, *Professional Geographer*, 33, p. 186. Copyright 1981 by the Association of American Geographers. Reprinted by permission.

> Because of such multiple categorization, the percentages of "within" and "inward" flows for individual central cities understate the overall share of workplaces located within the 28 central cities. Overall, 38.2 percent of total tabulated commuting trips are to jobs within these cities. Although this is substantially greater than the 8.2 percent median for the 28 urban fields considered separately, nearly two-thirds of all work trips are missed if the focus is placed exclusively on nodal ["within" and "inward"] flows. (1981, p. 186)

This example shows the impact that the decentralization of workplaces has had on journey-to-work patterns; it also demonstrates the necessity of testing long-held ideas against data and the complexities involved in doing so.

The decentralization of employment locations has been accompanied by the suburbanization of residences, but because not all social groups have decentralized to the same degree, the accessibility of different groups to employment has been altered over time. In a carefully designed study Ottensmann (1980) demonstrates how changes in the patterns of residential and work locations between 1927 and 1963 altered accessibility to employment in Milwaukee County, Wisconsin. By comparing the 1927 and 1963 maps of accessibility to employment, using equation 3, he shows that it was the region's central areas that experienced a decline in access to employment sites, whereas outlying areas to the north and west had increases exceeding 40%, so that the ratio of 1963 employment accessibility to 1927 employment accessibility was more than 1.4 (see Figure 1-4).

$$A_i = \frac{\sum_{j=1}^{110} E_j / D_{ij}^2}{\sum_{j=1}^{110} E_j} \tag{3}$$

where A_i is accessibility of district i to employment, E_j is employment in district j measured by number of work-trip destinations, and D_{ij} is distance in miles between the centers of districts i and j (there were 110 districts).

Ottensmann goes on to show that the areas with the greatest decrease in employment accessibility coincided with the areas having the highest incidence of poverty (see Figure 1-5). As low-income households have remained concentrated in central-city areas since the 1960s and as economic activity (and hence employment opportunities) has continued to decentralize in recent years, there is no reason to believe that the employment accessibility of low-income areas has improved relative to high-income areas since the 1960s.

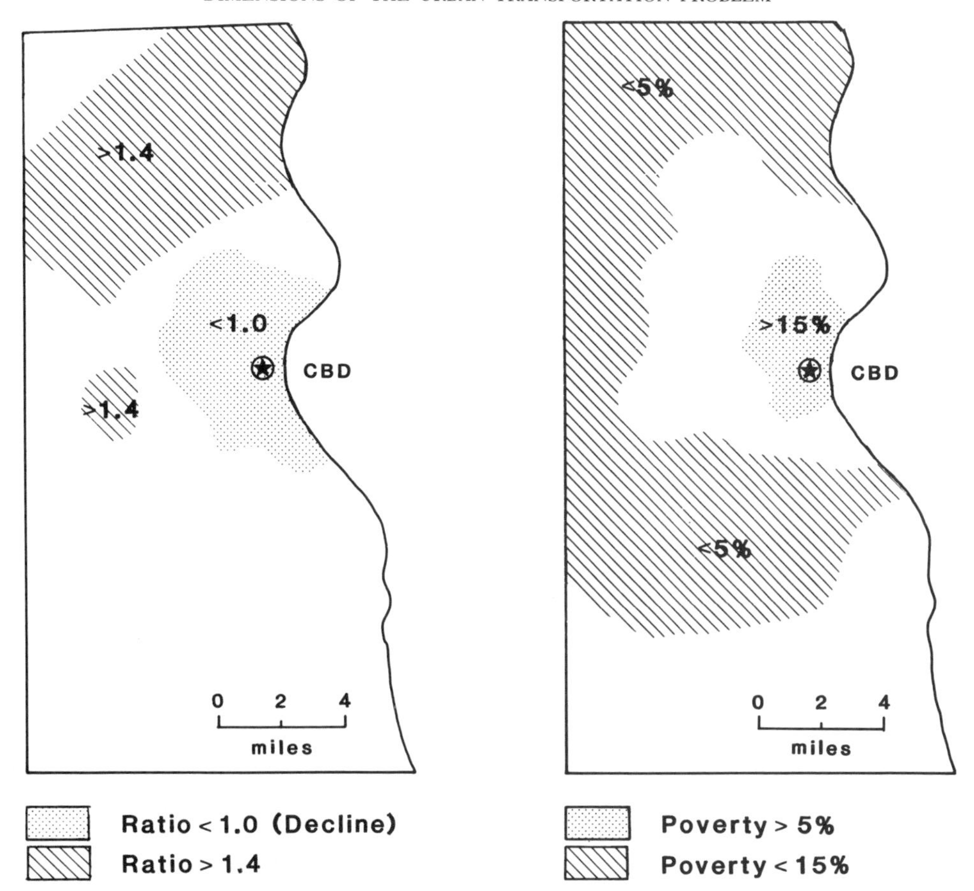

Figure 1-4. Ratio of 1963 employment accessibility to 1927 employment accessibility as a measure of accessibility change in Milwaukee County from 1927 to 1963. From "Changes in Accessibility to Employment in an Urban Area: Milwaukee, 1927–1963" by J. R. Ottensmann, 1980, *The Professional Geographer, 32,* p. 427. Copyright 1980 by the Association of American Geographers. Reprinted by permission.

Figure 1-5. Poverty levels as indicated by percentage of families with incomes under $3,000 in Milwaukee County, 1960. From "Changes in Accessibility to Employment in an Urban Area: Milwaukee, 1927–1963" by J. R. Ottensmann, 1980, *The Professional Geographer, 32,* p. 427. Copyright 1980 by the Association of American Geographers. Reprinted by permission.

The persistence of residential segregation in American cities has important implications, then, for access to employment.

The interest in changing patterns of employment accessibility is rooted not only in concern about the deteriorating accessibility of jobs to the poor but also in the important role that journey-to-work patterns play in the overall urban transportation picture. First, in sheer volume of travel involved, the journey to work is a force to be reckoned with. Of all the purposes for which people travel (including work, socializing, recreation, shopping, and personal business), work is the purpose for which the largest proportion of trips is made; somewhere between a third and a half of all intraurban trips are work trips (Brunso & Hartgen, 1983, p. 24; Domenich & McFadden, 1975, pp. 186–190). Second, people tend to travel longer distances for work than for nonwork purposes. Also, although national patterns of changes in work-trip length (in minutes and miles) over time are difficult to establish because such information was not collected as part of the decennial census until 1980, there is some indication that work trips have recently increased in terms of travel distance but not in terms of travel time. In 1979 the average work trip in the United States was about 11 miles and the average travel time was about 23 minutes (U.S. Department of Commerce, 1982). This represented an increase in both distance and time from 1975, when average travel distance was about 9 miles and average travel time was approximately 20 minutes (U.S. Department of Commerce, 1979), but whereas the increase in *distance* was large enough to be statistically significant, the increase in *time* was not (U.S. Department of Commerce, 1982). These statistics fit the more general pattern Donald Janelle reports in chapter 15 of this book, namely that travel distances have been increasing while travel times have remained relatively unchanged. The reason for this is not hard to discern; improvements in the transportation system (especially in highways) make it easy to travel farther within a given amount of time. The upshot is an increase in both vehicle miles and person miles traveled.

A third reason for focusing on the journey to work is that work trips, unlike trips for other purposes, are extremely concentrated in time. Because most people have to be at work between 7 and 9 A.M. and leave 8 hours later, the journey to and from work creates a "peaking problem"; it is this peak load that places the greatest demands on the transportation system and that planners therefore seek to accommodate. As you will see in the chapter explaining the urban transportation planning process (chapter 3) and later in a chapter describing the use of aggregate models in planning (chapter 6), urban transportation planners have traditionally aimed to provide a transportation system with enough capacity to handle the work trip, under the assumption that such a system can then easily accommodate travel for other purposes.

The Issue of Scale

The census-derived figures in Table 1-1 together with the studies reviewed above provide a useful snapshot of some important urban processes that have transportation implications—decentralization of population and employment and increasing concentrations of low-income, carless, elderly, and female-headed households in the central city. But the spatial resolution of the information discussed thus far is quite gross in that no finer distinction has been made than that between the whole SMSA and the central city. Of course maps at a finer scale, for example at the census-tract level, will show considerable variability *within* each of these general areas in the proportion of the population that is elderly, in the proportion of households with no vehicles, or in the proportion of households below the poverty line. If transportation policies and facilities are to be tailored to the specific needs of different kinds of people such as the elderly or the carless, then it is important to know where, within the SMSA and within the central city, these target groups live. By mapping census data, we can get a general idea of the relative densities of different groups of interest and an overview of how the spatial distributions of selected groups have been changing over time.

Maps of several census variables for Worcester, Massachusetts, at the census-tract level reveal the stark intracity contrasts that are masked by the city-level statistics presented in Table 1-1 (see Figures 1-6 to 1-12). Two sets of maps show trends from 1960 to 1980 in the proportion of the population in each tract that is over 65 years of age (Figures 1-6 to 1-8) and in the proportion of households in each tract without a vehicle (Figures 1-9 to 1-11). A look at the three maps showing the distribution of the elderly reveals that in the two decades prior to 1980 this group became more evenly spread throughout the city; by 1980 only 6 of the city's 41 census tracts were below the SMSA median in percentage of elderly, whereas 17 tracts were below the median in 1960. (Note that the map categories are keyed to the overall mean of the distribution for the SMSA, which has been increasing since 1960.)

By contrast the three maps showing the

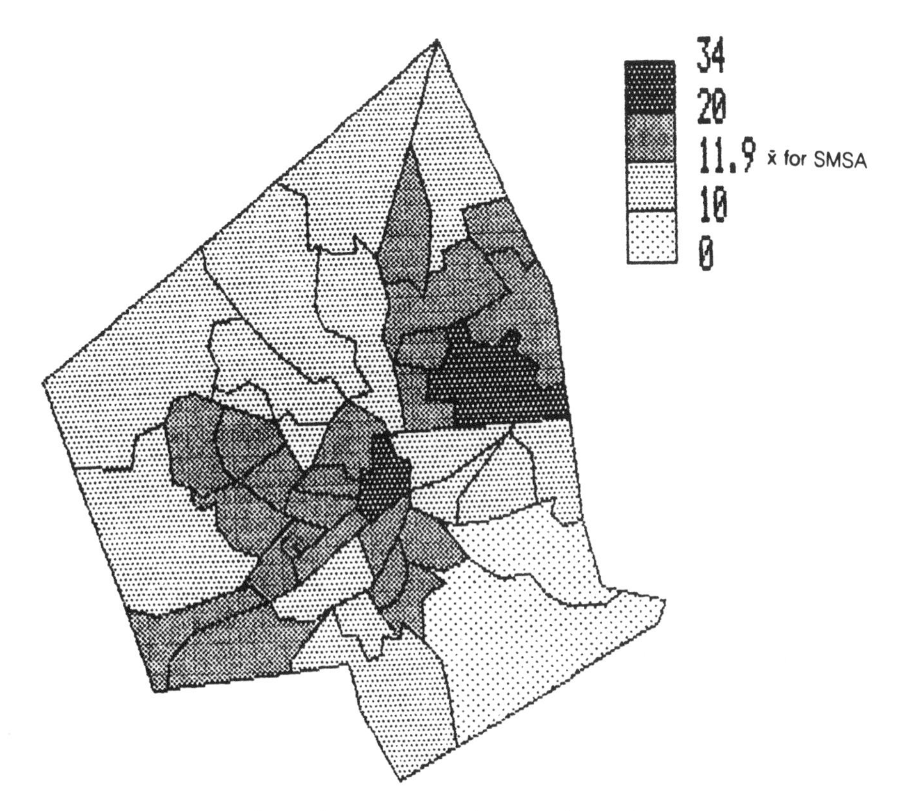

Figure 1-6. Percent of population over 65 years of age, Worcester, Mass., 1960.

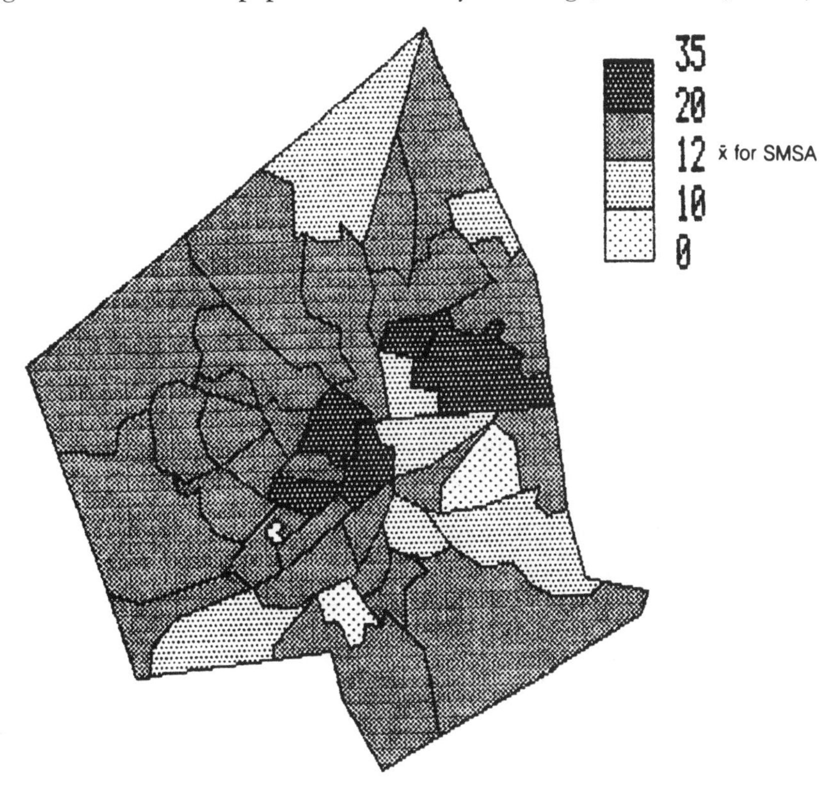

Figure 1-7. Percent of population over 65 years of age, Worcester, Mass., 1970.

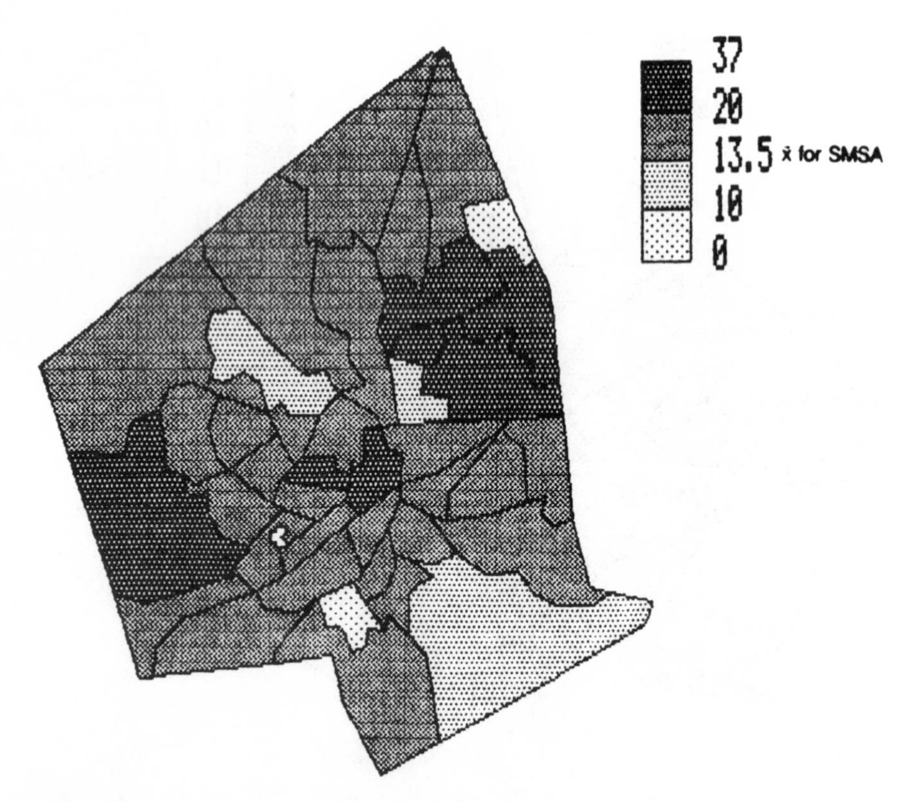

Figure 1-8. Percent of population over 65 years of age, Worcester, Mass., 1980.

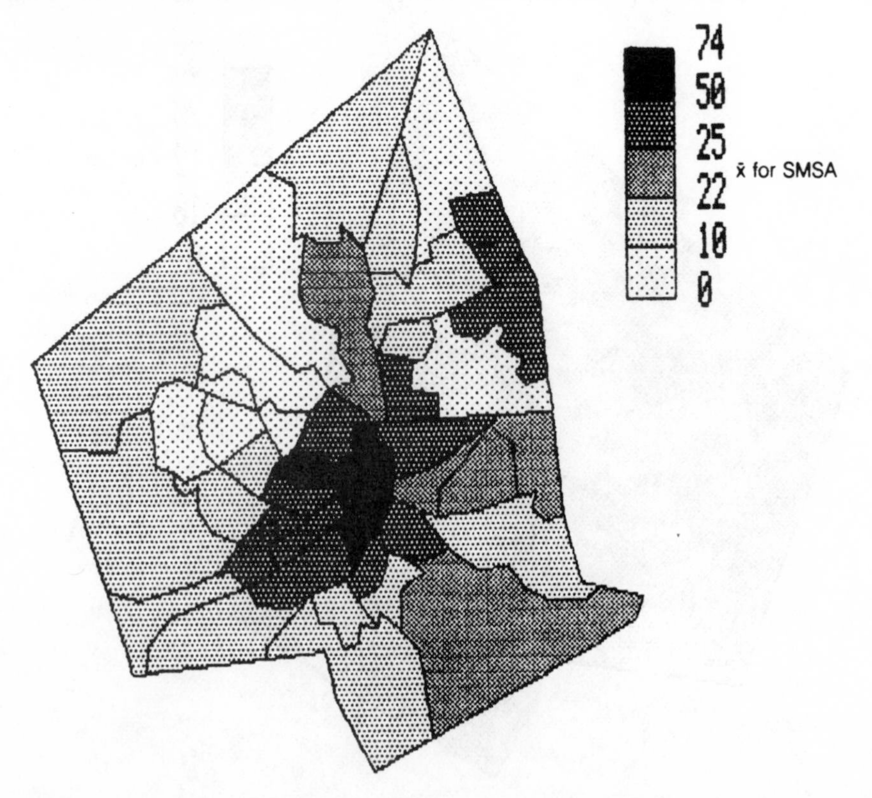

Figure 1-9. Proportion of households with no vehicle, Worcester, Mass., 1960.

Figure 1-10. Proportion of households with no vehicle, Worcester, Mass., 1970.

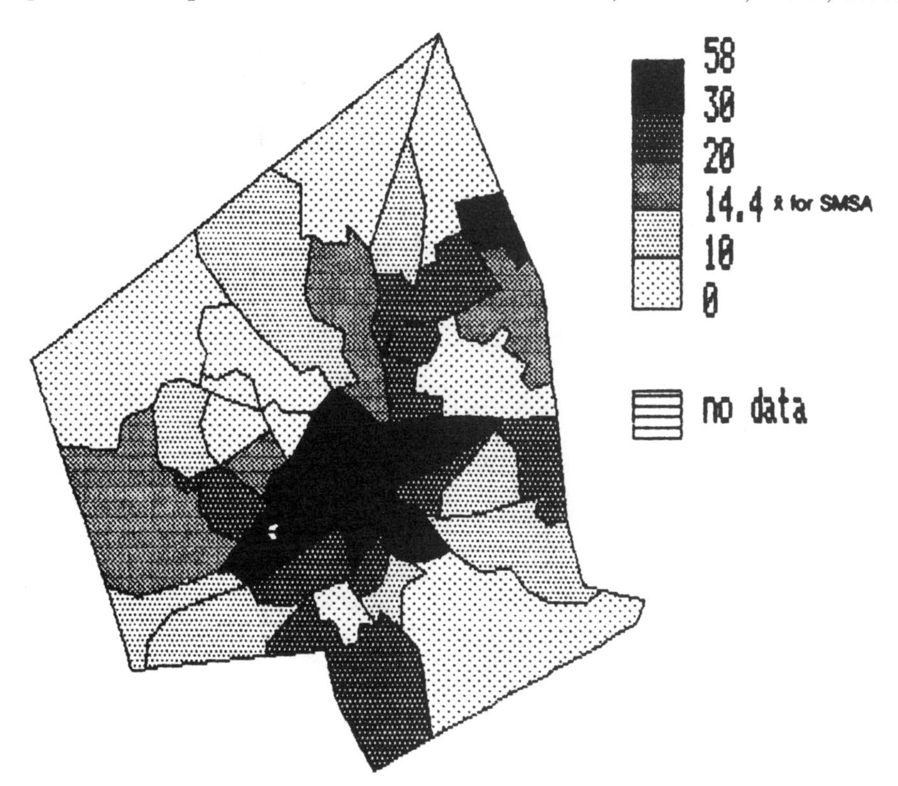

Figure 1-11. Proportion of households with no vehicle, Worcester, Mass., 1980.

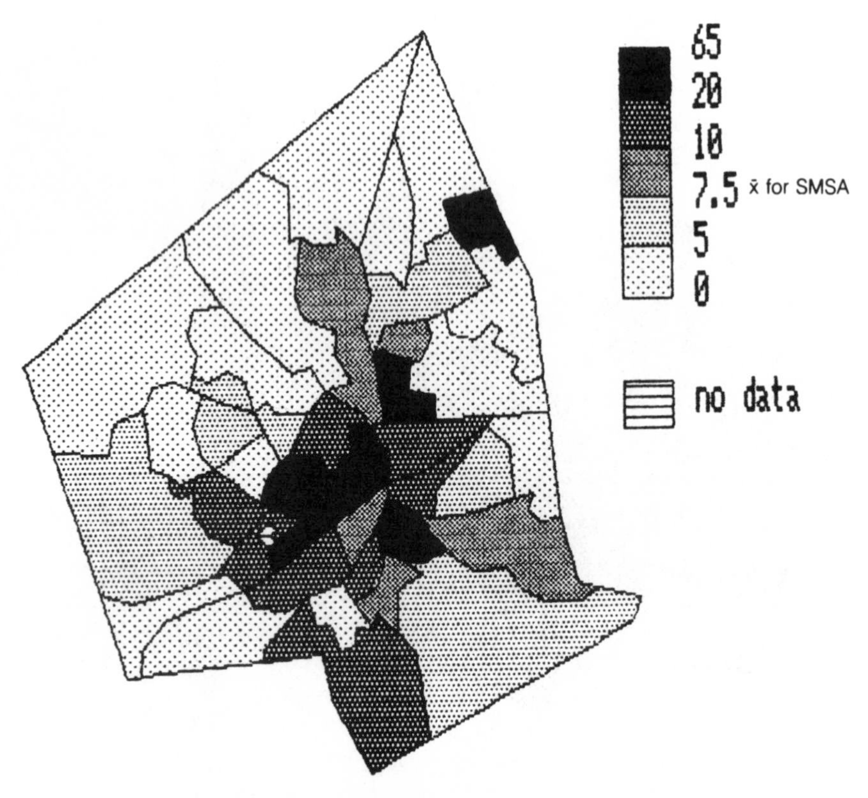

Figure 1-12. Proportion of households below poverty level, Worcester, Mass., 1980.

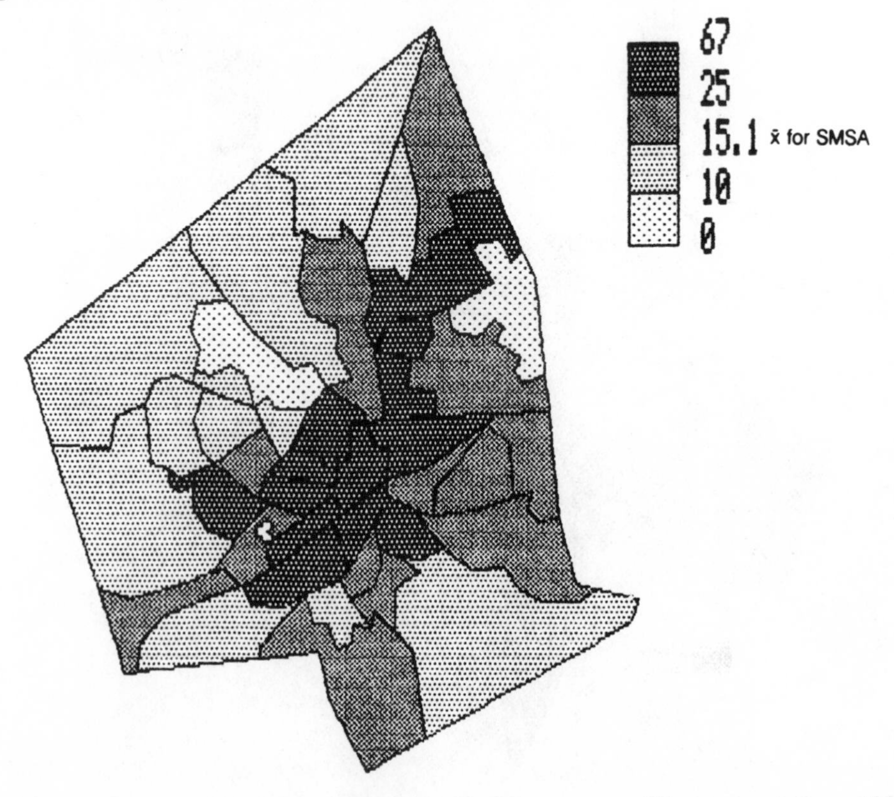

Figure 1-13. Proportion of households headed by women, Worcester, Mass., 1980.

distributions of households without vehicles disclose little change between 1960 and 1980. In each case carless households are concentrated near the center of the city, and the affluent northern and western parts of the city stand out as having very low proportions of households without vehicles. A look at maps showing the proportion of households below the poverty level for 1960, 1970, and 1980 would reveal an uncanny resemblance to the maps (Figures 1-9 to 1-11) of the carless. As is the case with the distribution of carless households, the distribution of the poor within Worcester has remained remarkably stable from 1960 to 1980 although the overall percentage of the city's households falling below the poverty line has increased.

Maps of poverty (Figure 1-12) and of female-headed households (Figure 1-13) only for 1980 are included here for comparison with the other 1980 distributions shown. The 1980 maps show a spatial coincidence of female-headed households, the carless, and the poor (Figures 1-11, 1-12, and 1-13); that is, the same areas tend to have a high poverty rate, a high percentage of female-headed households, and a high proportion of carless households. But what do these maps tell us about the transportation needs of *people* living in various parts of Worcester? Policies aimed at providing mobility for carless people could of course target the census tracts that have large percentages of carless households. Such an areally targeted policy would, however, miss the many carless households that do *not* live in the target census tracts. Also, there are numerous *people* (rather than households) who are carless for much of the day—people, for example, who remain at home while someone else takes the household's one car to work. The census-tract maps are little help in locating these people.

What these maps show is the familiar phenomenon of similar people clustering together in space. What they do not show is the extent to which heterogeneity exists within each census tract or the extent to which certain variables that covary at the areal (tract) level also covary at the individual level. What percentage of female-headed households within a tract are carless, for instance?

Consider the three hypothetical census tracts in Figure 1-14; all have an average household income of $15,000, and in this purely fictitious example we have information not only for the tract but also for households within the tract. In the tract in Figure 1-14a every household's income is identical, exactly $15,000, so the zonal average income is an accurate measure of the individual household incomes within the zone. In Figure 1-14b the zonal average masks two distinct subareas within the zone. In one part every household's income is $18,000 and in the other every household's income is $13,000. In Figure 1-14c the $13,000 households are interspersed with the $18,000 households. The point is that the complete zonal homo-

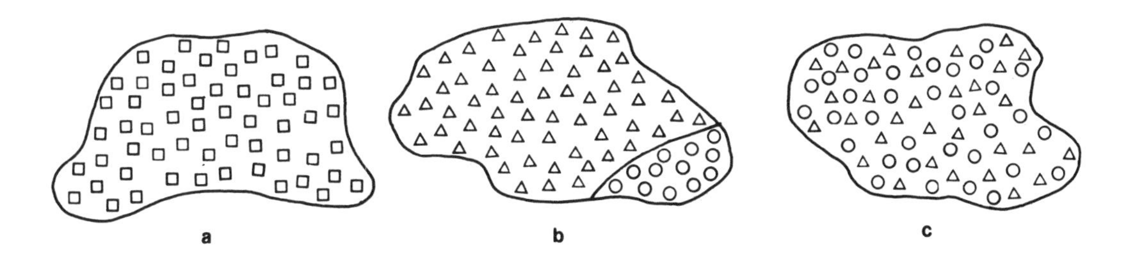

Figure 1-14. Hypothetical distributions of households with different income levels.

geneity depicted in Figure 1-14a simply does not occur in the real world; data for areas (or zones) smooth whatever internal heterogeneity exists and may create a fictitious average: people in $18,000 households are likely to have quite different travel patterns from people in the $13,000 households, but the zonal data will portray only an "average" behavior for the people of that zone. The more homogeneous an area is, the closer the zonal data will come to approximating the characteristics of the individuals living within that zone. Census-tract boundaries or the boundaries of traffic zones (areal units often used in transportation studies) sometimes split relatively homogeneous areas, adding heterogeneity to the resulting zones.

The kind of areal (or zonal) data displayed in Figures 1-6 to 1-13 are useful for providing an overview of population distributions and employment locations within the SMSA, for showing where certain population characteristics coincide in space, and for suggesting where certain transportation policies might best be deployed. They are not particularly useful for indicating what characteristics occur together at the individual level or for investigating how and why people make travel decisions or how they might respond to a particular transportation policy such as increased headways (a longer time between buses) on a bus route. Such questions require data for individuals rather than for areas.

Transportation analysts use both areal (aggregate) and individual-level (disaggregate) data in studying movement patterns in cities. Until about 1970, transportation studies were conducted with data for areal units called traffic zones. In this aggregate approach separate trips are grouped together according to their zone of origin and their zone of destination (see Figure 1-15), and the focus is on these flows between zones: how many trips does a particular zone "produce" (in other words, how many trips leave zone i)? How many of those trips leaving zone i end in zone j? What are the characteristics of zone i and zone j that might account for the volume of flows leaving i and arriving at j? Can you explain the size of the flow from i to j in terms of the attractiveness of j and its distance from i? These are some of the questions the aggregate approach seeks to answer. In recent years, transportation analysts have begun using individual- and household-level data in addition to areal data. The questions posed in a disaggregate context are aimed at individuals or households rather than at zones: what sociodemographic characteristics are related to a person's level of trip making? What factors affect why a person selects one destination or mode rather than another? What proportion of those who live in the suburbs and work in the central city will shift from commuting by drive-alone auto to carpooling if a high-speed, high-occupancy ve-

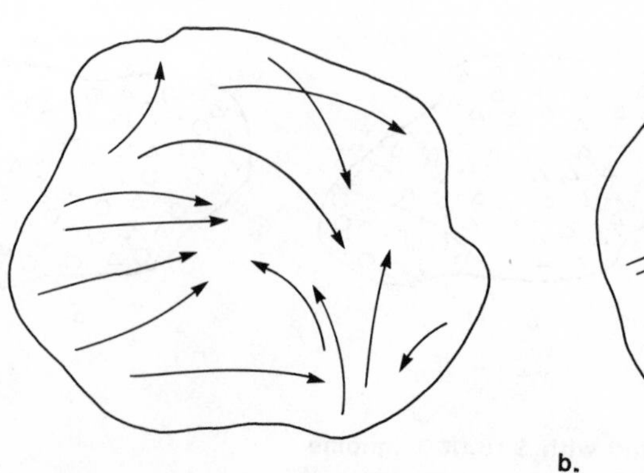

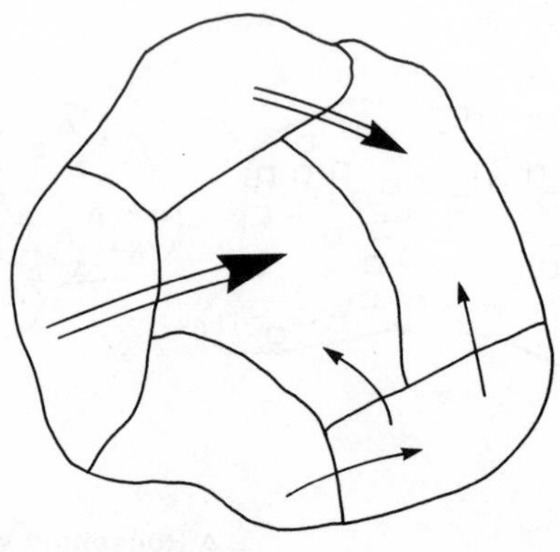

Figure 1-15. (a) Separate trips between points within the city. (b) Individual trips have been aggregated by origin and destination zones. Thickness of arrow indicates volume by flow between zones.

hicle (HOV) lane is installed on their journey-to-work route?

This scale distinction—between aggregate and disaggregate approaches—is the organizing principle behind this book's second part, which delves into the ways in which movement patterns can be analyzed. In each subsection of Part II, there is first a chapter examining how the flows at that particular scale can be *described;* there is then a chapter focusing on the ways in which *models* can help to simplify and make more comprehensible some of the overwhelming complexity that is evident in the descriptions of flow patterns; finally, for each scale level, there is a chapter that provides an *example* of how the models are used in a transportation planning context.

POLICY CONCERNS

What are the issues, the problems, the questions that transportation analysts seek to understand and to remedy with the approaches and tools outlined in Part II? Some are evident from the above general discussion of the contemporary urban context within which travel and transportation planning take place. The increasing separation between home and work and between activity sites in general—together with the growth in population, in households, in the civilian labor force, and in consumption—mean not only that more travel is required for each individual to carry out the round of daily activities but also that more and more people are traveling more and more miles. Congestion was long seen as the main urban transportation problem that transportation planners had to grapple with; they sought to solve the congestion problem by constructing more roads with greater capacity. However, since the 1950s we have learned the ironic lesson that increased highway capacity usually brings only greater congestion.

There are now a great many concerns other than traffic congestion that demand the attention of transportation policy makers, and a number of them are addressed in Part III of this book. Not every major transportation-related problem is accorded a separate chapter in Part III. One example is health. The growing distance between activity sites along with the overwhelming automobile orientation of American society makes travel on foot or by bicycle difficult and often dangerous. One might argue, therefore, that part of the urban transportation problem is the threat to health posed by the fact that the bulk of urban travel is accomplished in motorized vehicles. Air pollution, traffic accidents, and lack of exercise are all health problems that stem from urban transportation. Another example of a problem that affects transportation policy but that is not addressed specifically in Part III is the plethora of political jurisdictions and government agencies that in one way or another have a hand in decisions affecting transportation. Planners are trying to alleviate congestion on the arteries leading to Boston's Logan Airport by providing a bus service that would use designated express bus lanes to the airport. But setting aside these lanes will involve the Massachusetts Turnpike Authority, the Department of Public Works, the State Police, the City of Boston, the Environmental Protection Agency, the Federal Highway Administration, and even the Public Utilities Commission ("Boston's Traffic Mess," 1985, p. 22). Often decisions are hamstrung by the gaggle of authorities involved in the planning and decision-making process. Precisely because travel integrates different parts of an urban area and impinges on so many aspects of urban life, a single regionwide agency needs to be empowered to make transportation planning decisions.

The policy concerns that are addressed in Part III reflect the range of questions that transportation geographers and planners are grappling with—transit, land use change, energy consumption, social and environmental impacts, and the telecommunications revolution. Certainly one of the major current transportation issues is identifying the appropriate role for public transport in American cities. In the 1960s and the early 1970s planners (and the public) looked to transit to reduce air pollution, energy consumption, and congestion as well as to revitalize downtown areas and to promote mobility for the carless. It is now clear that public transportation is not a panacea for a host of urban problems but that, if carefully managed, transit could fill a more narrowly defined role quite successfully. Transit deficits have been rising dramatically in recent years; what are the reasons behind the precarious finances of transit companies in American cities? What is an appropriate role for transit in a country as devoted to the private automobile as is the United States?

The intimate relationship between transportation and land use was acknowledged at the outset of this chapter, but what are the policy implications of this close relationship? To what extent are transportation projects responsible for increasing urban land values and for generating urban development? Can urban sprawl be attributed to large-scale transportation improvements? Can certain transportation investments such as a light-rail rapid transit line, be used as a means of changing urban land uses (for example, intensifying land use and revitalizing certain portions of the city)?

Transportation is a major consumer of energy, especially energy from petroleum. In 1980 transportation of all types consumed more than one quarter of the energy used in the United States but more than 60% of the petroleum consumed (Holcomb & Kosky, 1984). Although the United States has only 45% of the world's population, it consumes 5% of the world's gasoline (Yearbook of World Energy Statistics, 1980). In the 1970s the price of energy rose substantially, and the threat of curtailed petroleum supplies forced its way into the American consciousness. What impact have the changes in energy price and availability had upon American travel patterns? To what extent have land use patterns been altered by the higher costs of travel? How effective can transportation policies be in reducing the consumption of energy?

Because social status in the American city is closely related to location, the placement of different transportation projects will affect various social groups differently. One dimension of the urban transportation problem is, then, who pays for and who benefits from any given transportation investment. Are public transportation deficits, in particular, distributed evenly among transit users? How can transportation services be provided in an equitable manner?

Because most travel is conducted in motor vehicles, another dimension of the urban transportation problem is the set of environmental impacts stemming from facility construction and from the use of motor vehicles. For example, 60% (by weight) of air pollutants in the United States comes from transportation; large proportions of carbon monoxide, hydrocarbons, and nitrous oxide can be attributed to transportation sources. "In 1975 highway vehicles alone accounted for 76 percent of nationwide NO_x emissions" (Horowitz, 1982, p. 22). What role can transportation analysts play in reducing air pollution? How can transportation investments be made so as to minimize adverse environmental impacts?

Recent advances in telecommunications technology have spurred speculation about dramatic increases in the substitution of communication for travel and about the likelihood of massive restructuring of urban systems. Can improved technology result in substitution of electronic communications for personal movement? Will telecommunications succeed in reducing the number of vehicle miles traveled (VMT), eliminating congestion, reducing pollution, and reducing the need for proximity? Will, in fact, travel become obsolete?

The chapters in the third section of this book examine the evidence that bears upon the questions raised above. An interesting theme that emerges from these chapters is that careful empirical analysis often yields results that challenge long-held ideas. Some of these established, accepted notions emerged from microeconomic theory; others came from earlier, less carefully controlled empirical work. But the message comes through again and again in Part III that we cannot complacently assume that an assertion is true simply because it has been accepted and unquestioned for a long time.

An additional point to remember in the context of transportation policy issues is that policy makers do not have the ability to manipulate all the variables that impinge upon urban travel. Even if transportation analysts understood perfectly how individuals and households make travel decisions, they could not design policies to alter certain aspects of travel. For example, we know that travel decisions (such as which mode to use) are affected not only by the characteristics of the different modes, but also by the characteristics of the traveler. Transportation policies can change the characteristics of the modes to increase the attractiveness of one relative to another (as by reducing the cost of transit relative to the cost of a drive-alone auto), but, at least in the United States, transportation policy makers have a more difficult time making decisions that affect the age, sex, or the income level of the traveler—characteristics that are known to be significantly related to mode use. If transportation analysis is to be useful in a policy sense, it must identify and include those variables that are measurable and policy-relevant, that is, susceptible to manipulation by policy makers.

MOBILITY RECONSIDERED

As is evident from this overview of Part III, the goals of transportation policy are legion. But prime among these must be this chapter's starting point: mobility and accessibility. In order to assess the impact of policies on mobility, we need a measure of this concept. The two measures of accessibility that were introduced earlier were both highly simplified representations; neither addressed mobility per se or dimensions of the concept such as the ability to visit places at different times of day.

An alternative measure—that of space–time autonomy—takes both accessibility and mobility into consideration; it is far more satisfying conceptually but far more difficult operationally. The concept of space–time autonomy has been developed in the context of time geography and focuses on the constraints impinging on a person's freedom of movement (Burns, 1979; Hägerstrand, 1970; Pred, 1981). These include capability constraints (the limited ability to perform certain tasks within a given transportation technology and the fact that we can be only one place a time), coupling constraints (the need to undertake certain activities at certain places with other people), and authority constraints (the social, political, and legal restrictions on access). Clearly the number, type, and location of the activities one can engage in over the course of the day are restricted by these constraints.

For a given individual, one can measure the level of space–time autonomy by examining the space–time prism, which shows the possibilities in space and time that are open to an individual. In Figure 1-16a the maximum speed at which this hypothetical person can travel is v and the prism shows the places this person can reach between times t_1 and t_2. Furthermore, the impact of transportation policies on space–time autonomy can be gauged by changes in the prism resulting from the policy. For example, flextime work schedules, longer store hours, and purchasing a second car all enhance space–time autonomy by adding margins to the space–time prism as shown in Figure 1-16b. Speeding laws, rigid school hours, traffic congestion, and new stop lights all constrain choice. Large families impose coupling constraints, which often affect women more than men. Babysitters, day-care centers, and children's growing up all reintroduce space–time autonomy.

Certainly increased space–time autonomy seems desirable in that it implies a greater accessibility to places and more discretion in spending one's time. We might question, however, the need for ever-increasing space–time autonomy, ever-increasing personal mobility. One question that transportation

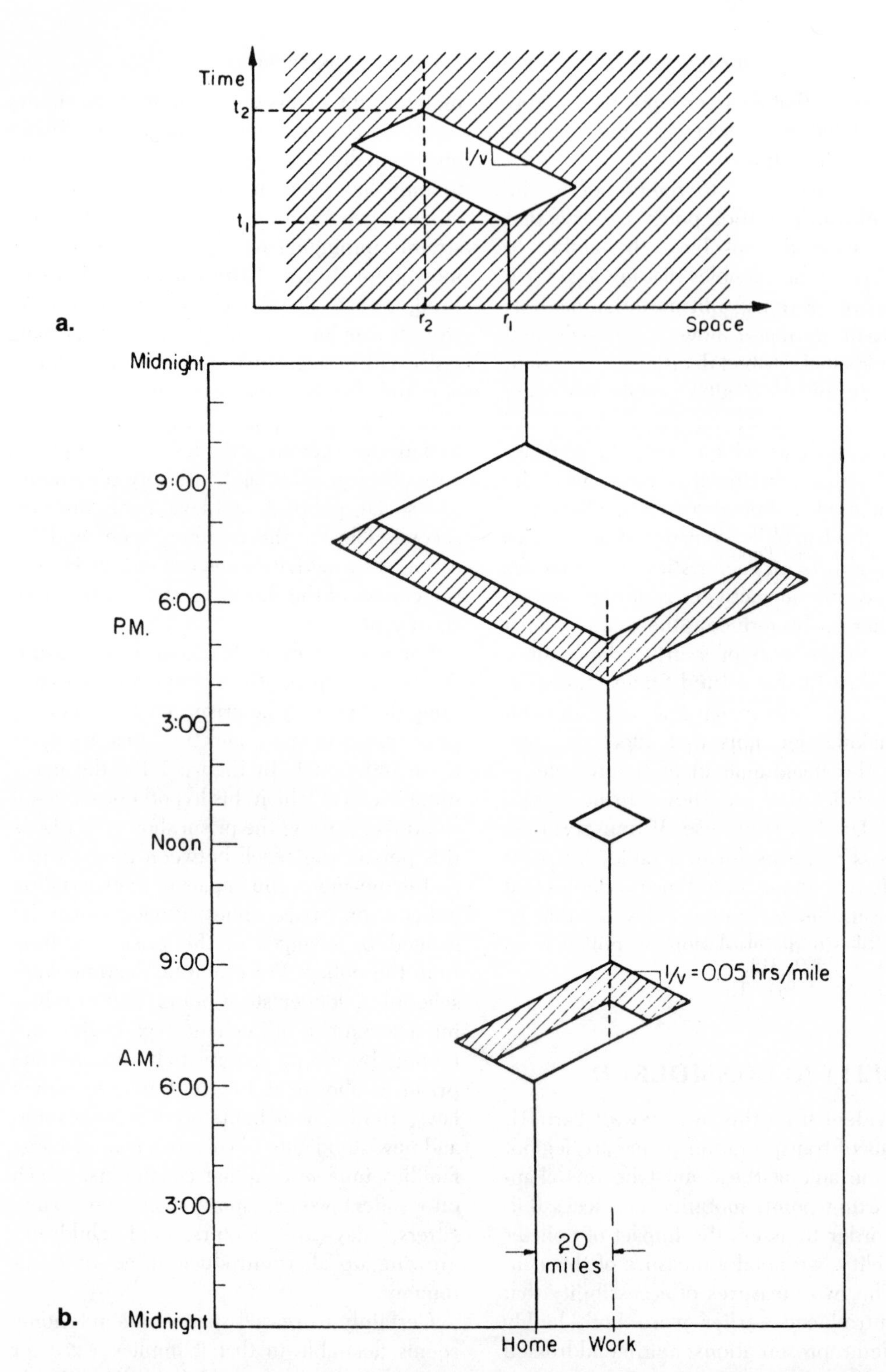

Figure 1-16. Space–time prisms. (a) Individual with origin and destination coupling constraints; prism shows options available when individual starting from place r_1 at time t_1 must arrive at r_2 by t_2. (b) Effects of flextime on space–time autonomy; flextime increases the individual's space–time autonomy by enlarging the prisms that define possibilities. Shading represents the increase in space–time autonomy gained from flextime. From *Transportation, Temporal, and Spatial Components of Accessibility* by L. Burns, 1979, pp. 13 and 19, Lexington, Mass.: Lexington Books. Copyright 1979 by D. C. Heath. Reprinted by permission.

geographers should probably ponder is whether or not there is such a thing as too much mobility!

References

"Boston's Traffic Mess." (1985). *Boston Globe*. December 3, p. 22.

Brunso, J. M., & Hartgen, D. (1983). *An update on households' reported trip generation rates*. Transportation Analysis Report #31. Albany: New York State Department of Transportation.

Burns, L. (1979). *Transportation, temporal, and spatial components of accessibility*. Lexington, Mass.: Lexington Books.

Domenich, T. A., & McFadden, D. (1975). *Urban travel demand*. Amsterdam: North Holland.

Hägerstrand, T. (1970). What about people in regional science? *Papers, Regional Science Association, 24*, 7–21.

Holcomb, M. C., & Kosky, S. (1984). *Transportation Energy Data Book*, 7th ed. Oak Ridge, Tenn.: Oak Ridge National Laboratory.

Horowitz, J. (1982). *Air quality analysis for urban transportation planning*. Cambridge: MIT Press.

Koenig, J. G. (1980). Indicators of urban accessibility: Theory and application. *Transportation, 9*, 145–172.

Ottensmann, J. R. (1980). Changes in assessibility to employment in an urban area: Milwaukee, 1927–1963. *The Professional Geographer, 32*, 421–430.

Pirie, G. H. (1979). Measuring accessibility: A review and proposal. *Environment and Planning A, 11*, 299–312.

Plane, D. A. (1981). The geography of urban commuting fields: Some empirical evidence from New England. *The Professional Geographer, 33*, 182–188.

Pred, A. (1981). Social reproduction and the time geography of everyday life. *Geografiska Annaler B, 63*, 5–22.

United Nations. (1981). 1980 Yearbook of World Energy Statistics. Department of International Economic and Social Affairs, Statistical Office. New York.

U.S. Bureau of Census. (1960). *U.S. Censuses of Population and Housing: 1960*. PHC(1)-175. Washington, D.C.: U.S. Government Printing Office.

U.S. Bureau of Census. (1972). *U.S. Censuses of Population and Housing: March 1970*. PHC(1)-236. Washington, D.C.: U.S. Government Printing Office.

U.S. Bureau of Census. (1983). *U.S. Censuses of Population and Housing: March 1980*. PHC(80-2)-376. Washington, D.C.: U.S. Government Printing Office.

U.S. Department of Commerce, Bureau of the Census. (1973). *Journey to Work*. (Census of Population 1970 Subject Reports. Final Report PC (2)-6D.) Washington, D.C.: U.S. Government Printing Office.

U.S. Department of Commerce, Bureau of the Census. (1979). *The Journey to Work in the United States: 1975*. Washington, D.C.: U.S. Government Printing Office.

U.S. Department of Commerce, Bureau of the Census. (1982). *The Journey to Work in the United States: 1979*. Washington, D.C.: U.S. Government Printing Office.

Wachs, M., & Kumagi, T. G. (1973). Physical accessibility as a social indicator. *Socioeconomic Planning Sciences, 7*, 437–456.

2 TRANSPORTATION AND URBAN FORM: *STAGES IN THE SPATIAL EVOLUTION OF THE AMERICAN METROPOLIS*

PETER O. MULLER
University of Miami

As the opening chapter demonstrated, the movement of people, goods, and information within the local metropolitan area is critically important to the functioning of cities. In this chapter, I review the American urban experience of the past two centuries and trace a persistently strong relationship between the intraurban transportation system and the spatial form and organization of the metropolis. Following an overview of the cultural foundations of urbanism in the United States, a four-stage model of intrametropolitan transport eras and associated growth patterns is introduced. Within that framework it will become clear that a distinctive spatial structure dominated each stage of urban transportation development and that geographical reorganization swiftly followed the breakthrough in movement technology that launched the next era of metropolitan expansion.[1] Finally, the contemporary scene is briefly considered, both as an evolutionary composite of the past and a dynamic arena in which new forces may already be forging a decidedly different future.

CULTURAL FOUNDATIONS OF THE AMERICAN URBAN EXPERIENCE

Americans, by and large, were not urban dwellers by design: the emergence of large cities between the Civil War and World War I was an unintended by-product of the nation's rapid industrialization. Berry (1975), recalling the observations of the eighteenth-century French traveler Hector St. Jean de Crèvecoeur, succinctly summarized the cultural values that have shaped attitudes toward

[1]This approach, while emphasizing the key role of transportation, does not mean to suggest that movement processes are the only forces shaping intraurban growth and spatial organization. As will be demonstrated throughout this chapter, urban geographical patterns are also very much the products of social values, land resources, investment-capital availability, the actions of private markets, and other infrastructural technologies. For a further discussion of the balancing of these forces, see Tobin (1976).

urban living in the United States for the past two centuries:

> Foremost . . . was a *love of newness*. Second was the overwhelming desire to be *near to nature*. *Freedom to move* was essential if goals were to be realized, and *individualism* was basic to the self-made man's pursuit of his goals, yet *violence* was the accompaniment if not the condition of success—the competitive urge, the struggle to succeed, the fight to win. Finally, [there is] a great *melting pot* of peoples, and a manifest *sense of destiny*. (p. 175)

As the indigenous culture of the emergent nation took root, its popular Jeffersonian view of democracy nurtured a powerful rural ideal that regarded cities as centers of corruption, social inequalities, and disorder (see White & White, 1962). When mass urbanization became unavoidable as the Industrial Revolution blossomed after 1850, Americans brought their agrarian ideal with them and sought to make their new manufacturing centers non-cities. For the affluent this process began almost as soon as the railroad reached the city in the 1830s; by mid-century, numerous railside residential clusters had materialized just outside the built-up urban area. But middle-income city dwellers could not afford this living pattern because of the extra time and commuting costs it demanded. With cities increasingly unlivable as industrialization intensified, pressures mounted after 1850 to improve the urban transportation system to permit the burgeoning middle class to have access to the high-amenity environment of the peri-urban zone.

The necessary technological breakthrough—in the form of the electric (streetcar) traction motor—was finally achieved in the late 1880s. By the opening of the final decade of the nineteenth century, the city began to spill over into the much-desired surrounding countryside. By 1900, the decentralization of the middle-income masses was no longer a trickle but a widening migration stream (which has yet to cease its flow) that rapidly spawned the emergence of the full-fledged metropolis, wherein a steadily increasing multitude of urban dwellers shunned the residential life of the industrial city altogether. Hardly had this initial transformation of the American city been completed when the automobile introduced mass private transportation in the 1920s for all but the poorest urban dwellers. As the metropolitan highway network expanded in the interwar period, new rounds of peripheral residential development were launched, and the urban perimeter was pushed ever farther from the downtown core. But these centrifugal forces still operated at a rather leisurely pace, undoubtedly slowed after 1930 by a decade and a half of economic depression and global war.

After the conclusion of World War II, however, all constraints were removed, and a massive new wave of deconcentration was spawned. Spurred by a reviving economy, widespread housing demands, federal home-loan policies that favored new urban development, copious highway construction, and more efficient cars, the exodus from the nation's cities reached unprecedented proportions between 1945 and 1970. The proliferation of urban freeways (introduced in southern California in the late 1930s) heightened the centrifugal drift. With the completion of these high-speed, limited-access networks in the 1960s and 1970s came the elimination of the core-city central business district's (CBD's) regionwide centrality advantages, as superior intrametropolitan accessibility became a spatial good available at any expressway-interchange location. As entrepreneurs swiftly realized the consequences of this structural reorganization of the metropolis, nonresidential activities of every sort began their own massive wave of intraurban deconcentration. Manufacturing and retailing led the way. By 1980 the erstwhile ring of bedroom communities that girdled the aging central city had become transformed into a diversified, expanding *outer city* that was increasingly home to more than half of the metropolitan area's industrial, service, and office-based business employers.

Moreover, major new multipurpose activity centers have been emerging in the outer city since 1970, attracting so many high-order urban functions that residents of surrounding areas have reorganized their entire lives around them. Thus, the compact industrial city of the recent past is today rapidly turning inside out; the rise of downtown-type centers in the increasingly independent outer city is also forging a decidedly polycentric metropolis, the product of both the cumulative spatial processes outlined above and the emerging forces of a postindustrial society that appear to be shaping new urbanization patterns that represent a clean break with the past.

The legacy of more than two centuries of intraurban transportation innovations and the development patterns they helped stamp on the landscape of metropolitan America is *suburbanization*—the growth of the edges of the urbanized area at a rate faster than that in the already developed interior. Since the spatial extent of the continuously built-up urban area has, throughout history, exhibited a fairly constant time–distance radius of about 45 minutes' travel from the center, each breakthrough in higher-speed transport technology extended that radius into a new outer zone of suburban residential opportunity. In the nineteenth century, commuter railroads, horse-drawn trolleys, and electric streetcars each created their own suburbs—*and thereby also created the large industrial city,* which could not have been formed without incorporating these new suburbs into the preexisting compact urban center. But the suburbs that materialized in the early twentieth century began to assert their independence from the ever more undesirably perceived central cities. Few significant municipal consolidations occurred after the 1920s, except in postwar Texas and certain other Sunbelt locales. As the automobile greatly reinforced the dispersal trend of the metropolitan population, the distinction between central city and suburban ring grew as well. And as freeways eventually eliminated the friction effects of intrametropolitan distance for most urban functions, nonresidential activities deconcentrated to such an extent that by the 1970s the emerging outer city had become at least a coequal alongside the central city that spawned it—and relegated the word *sub*urb into obsolescence.

This urban experience of the United States over the past two centuries is the product of uniquely American cultural values and contrasts sharply with modern urbanization trends in Europe. The European metropolis has retained a far more tightly agglomerated spatial structure, and the historic central city continues to dominate its immediate urban region. Sommers (1983) has succinctly described the concentrative forces that have shaped the cities of postwar Europe:

> Age is a principal factor, but ethnic and environmental differences also play major roles in the appearance of the European city. Politics, war, fire, religion, culture, and economics also have played a role. Land is expensive due to its scarcity, and capital for private enterprise development has been insufficient, so government-built housing is quite common. Land ownership has been fragmented over the years due to inheritance systems that often split land among sons. Prices for real estate and rent have been government controlled in many countries. Planning and zoning codes as well as the development of utilities are determined by government policies. These are characteristics of a region with a long history, dense population, scarce land, and strong government control of urban land development. (p. 97)

Further evidence of the persistent dominance of the European central city is shown in Figure 2-1. The density gradient pattern of the North American metropolis (graph A) is marked by progressive deconcentration whereas the counterpart European pattern (graph B) exhibits sustained intraurban centralization. The newest metropolitan trends of the last decade in Europe do show accelerating suburbanization, but the central city decidedly remains the intraurban social and economic core.

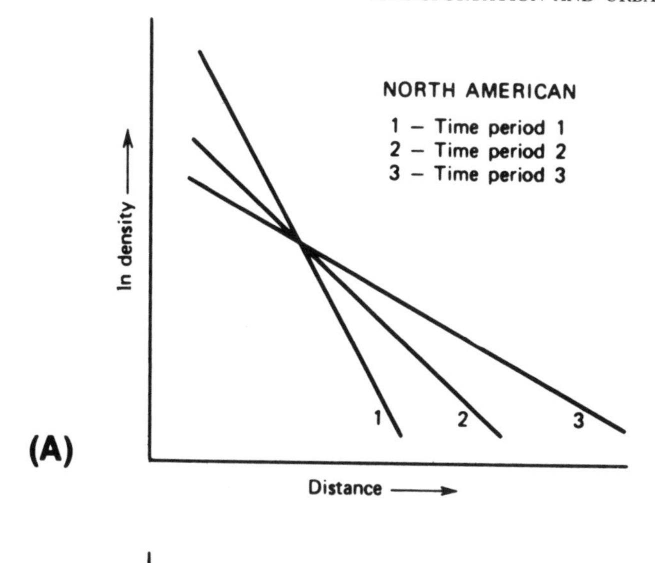

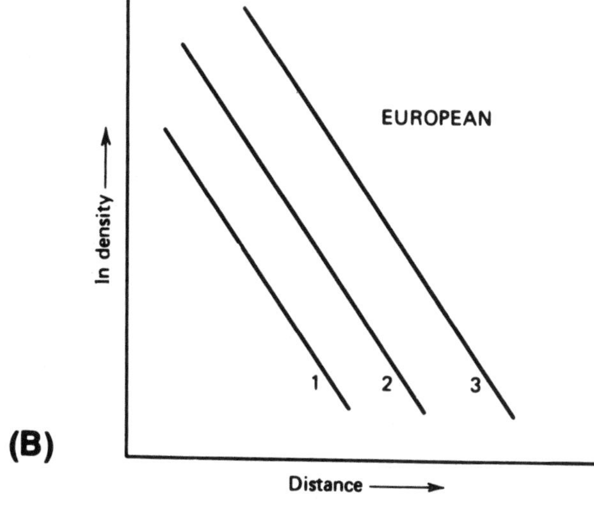

Figure 2-1. Density gradients over time in the North American and European metropolis. From *Interpreting the City: An Urban Geography* (p. 218) by T. A. Hartshorn, 1980, New York: John Wiley & Sons. Copyright 1980 by John Wiley & Sons. Reprinted by permission.

THE FOUR ERAS OF INTRAMETROPOLITAN GROWTH AND TRANSPORT DEVELOPMENT

The evolving form and structure of the American metropolis, briefly outlined in the previous section, may be traced within the framework of four transportation-related eras identified by Adams (1970). Each growth stage is dominated by a particular movement technology and network expansion process that shaped a distinctive pattern of intraurban spatial organization:

1. Walking–Horsecar Era (1800–1890)
2. Electric Streetcar Era (1890–1920)
3. Recreational Automobile Era (1920–1945)
4. Freeway Era (1945–)

This model is diagrammed in Figure 2-2 and reveals two sharply different morphological properties over time. During Eras 1 and 3 uniform transport surface conditions prevailed (as much of the urban region was similarly accessible), permitting directional freedom of movement and a decidedly compact overall development pattern. During Eras 2 and 4 pronounced network biases were dominant, producing an irregularly

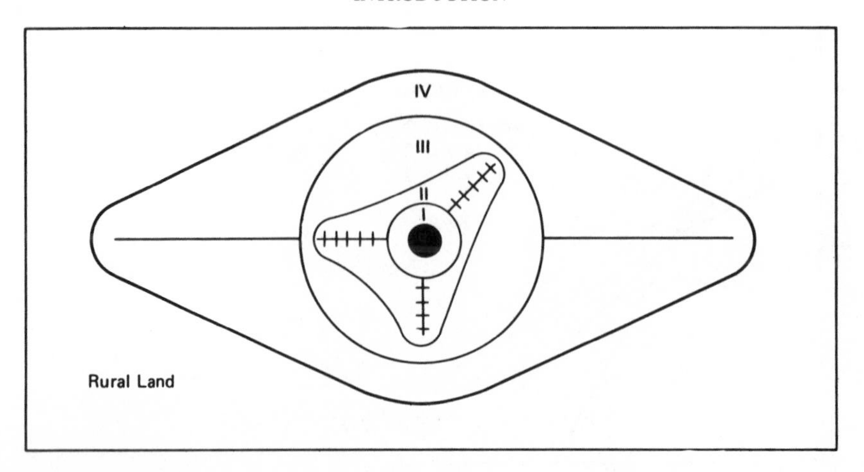

Figure 2-2. Intraurban transport eras and metropolitan growth patterns: (I) Walking–Horsecar Era, (II) Electric Streetcar Era, (III) Recreational Auto Era, and (IV) Freeway Era. From "Residential Structures of Midwestern Cities" by J. S. Adams, 1970, *Annals of the Association of American Geographers, 60*, p. 56. Copyright 1970 by the Association of American Geographers. Adapted by permission.

shaped metropolis in which axial development along radial transport routes overshadowed growth in the relatively inaccessible interstices. A generalized model of this kind, while organizationally convenient, risks oversimplification because the building processes of several simultaneously developing cities do not fall into neat time-space compartments (Tarr, 1984, pp. 5–6). An examination of Figure 2-3, which maps Chicago's growth for the past 150 years, reveals numerous empirical irregularities, suggesting that the overall urban growth pattern is somewhat more complex than a simple continuous, centrifugal thrust. Yet, when developmental ebb-and-flow pulsations, leapfrogging, backfilling, and other departures from the normative scheme are considered, there still remains a reasonably good correspondence between the model and historical–geographical reality. With that in mind, each of the four eras is now treated in detail.

Walking–Horsecar Era (1800–1890)

Prior to the middle of the nineteenth century, the American city was a highly agglomerated urban settlement in which the dominant means of getting about was on foot (Figure 2-2, Era I). Thus, people and activities were required to cluster within close proximity of one another. Initially, this meant less than a 30-minute walk from the center, later extended to about 45 minutes when the pressures of industrial growth intensified after 1830. Any attempt to break these mobility constraints courted urban failure: Washington, D.C., struggled enormously for much of its first century on L'Enfant's 1791 plan that dispersed blocks and facilities too widely for a pedestrian city, prompting Charles Dickens to observe during his 1842 visit that buildings were located "anywhere, but the more entirely out of everyone's way the better" (Schaeffer & Sclar, 1975, p. 12). Within the walking city, there were recognizable concentrations of activities as well as the beginnings of income-based residential congregations. The latter behavior was clearly evinced by the wealthy who walled themselves off in their larger homes near the city center; they also favored the privacy of horse-drawn carriages to move about town—undoubtedly the earliest American form of wheeled intraurban transportation. The rest of the population resided in tiny overcrowded quarters, of which Philadelphia's

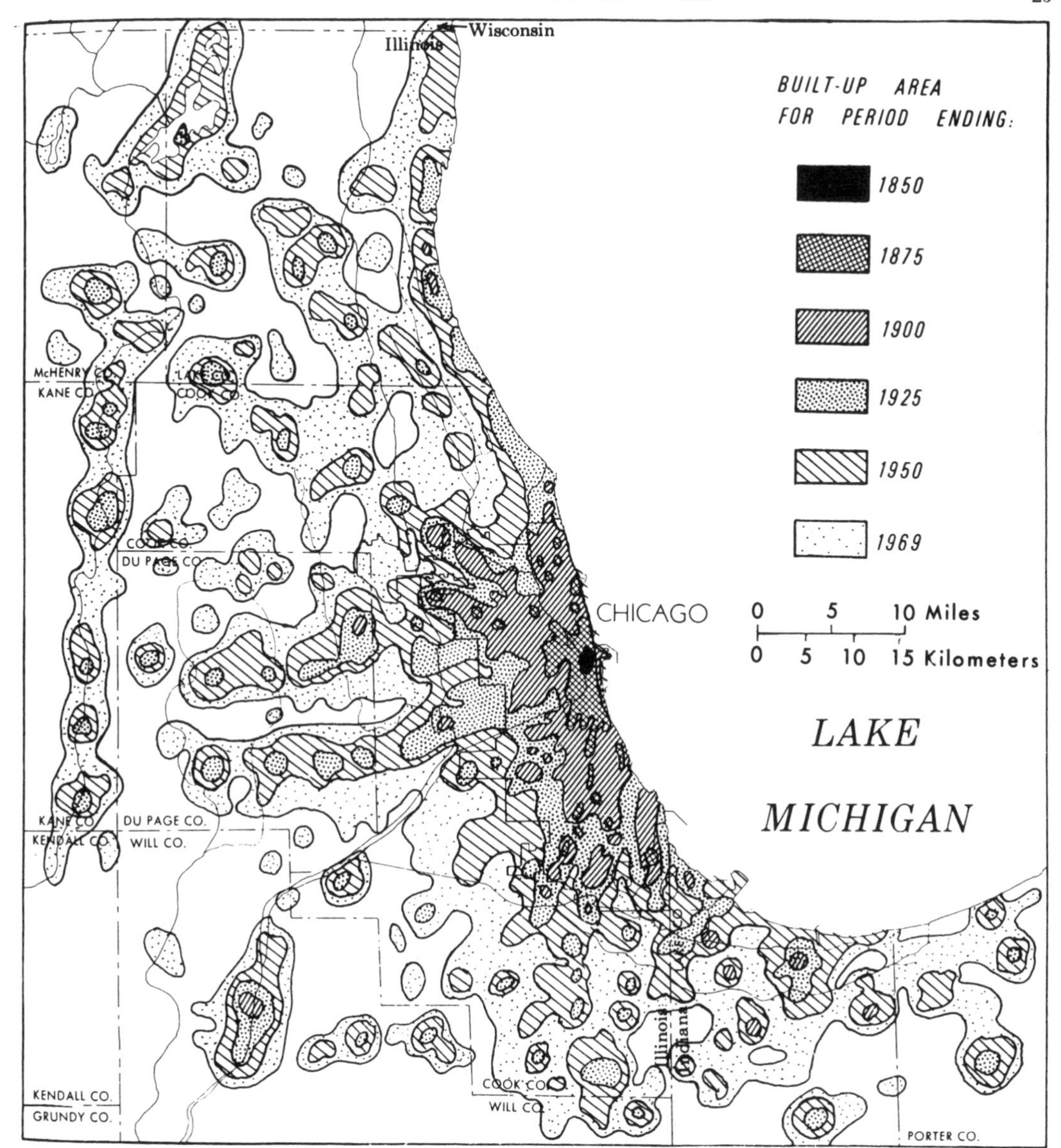

Figure 2-3. The suburban expansion of metropolitan Chicago since the mid-nineteenth century. From *Chicago: Transformations of an Urban System* (p. 9) by B. J. L. Berry et al., 1976, Cambridge, Mass.: Ballinger Publishing Company. Copyright 1976 by Brian J. L. Berry. Reprinted by permission.

now-restored Elfreth's Alley was typical (see Figure 2-4).

The rather crude environment of the compact preindustrial city prompted those of means to seek an escape from its noise as well as the frequent epidemics that resulted from the unsanitary conditions. Horse-and-carriage transportation permitted the wealthy to reside in the nearby countryside for the disease-prone summer months. The arrival of the railroad in the early 1830s soon provided the opportunity for year-round daily travel to and from posh new trackside suburbs. By 1840 hundreds of affluent businessmen in Boston, New York, and Philadelphia were making these round trips every week-

Figure 2-4. Elfreth's Alley in downtown Philadelphia, whose restoration provides a good feel for the lack of spaciousness in the revolutionary-era city. Photo courtesy of Roman A. Cybriwsky.

day. The "commutation" of their fares to lower prices when purchasing tickets in monthly quantities introduced a new word to describe the journey to work—*commuting*. A few years later, these privileges extended to the *nouveau riche* professional class (well over 100 trains a day ran between Boston and its suburbs in 1850), and a spate of planned rail suburbs, such as Riverside near Chicago, soon materialized.

As industrialization and its teeming concentrations of modest working-class housing increasingly engulfed the mid-nineteenth-century city, the worsening physical and social environment heightened the desires of middle-income residents to suburbanize as well. Unable to afford the cost and time of commuting and with the pedestrian city stretched to its morphological limit, these desires intensified the pressures to improve intraurban transport technology. As early as the 1820s, New York, Philadelphia, and Baltimore had established *omnibus* lines. These intracity adaptations of the stagecoach eventually developed dense networks in and around downtown (see Figure 2-5); other cities experimented with cable systems and even the steam railroad, but most efforts proved impractical. With omnibuses unable to carry more than a dozen or so passengers or attain speeds of people on foot, the first meaningful breakthrough toward establishing intracity "mass" transit was finally introduced in New York City in 1852 in the form of the horse-drawn streetcar (see Figure 2-6). Lighter street rails were easy to install, overcame the problems of muddy unpaved roadways, and enabled horsecars to be hauled along them at speeds slightly faster (5 miles per hour) than those of pedestrians. This modest improvement in mobility allowed a narrow band of land at the city's edge to be opened for new home construction. Middle-income urbanites flocked to these *horsecar suburbs*, which proliferated rapidly after 1860. Radial routes were usually the first to spawn such peripheral development, but the steady demand for housing necessitated the construction of cross-town horsecar lines, thereby filling in the interstices and preserving the generally circular shape of the city.

The nonaffluent remainder of the urban population was confined to the old pedestrian city and its bleak, high-density industrial appendages. With the massive influx of unskilled laborers, increasingly of European origin after the Civil War, huge blue-collar neighborhoods surrounded the factories, often built by the mill owners themselves. Since factory shifts ran 10 or more hours 6 days a week, their modestly paid workers could not afford to commute and were forced to reside within walking distance of the plant. Newcomers to the city, however, were accommodated in this nearby housing quite literally in the order in which they arrived, thereby denying immigrant factory workers even the small luxury of living in the immediate com-

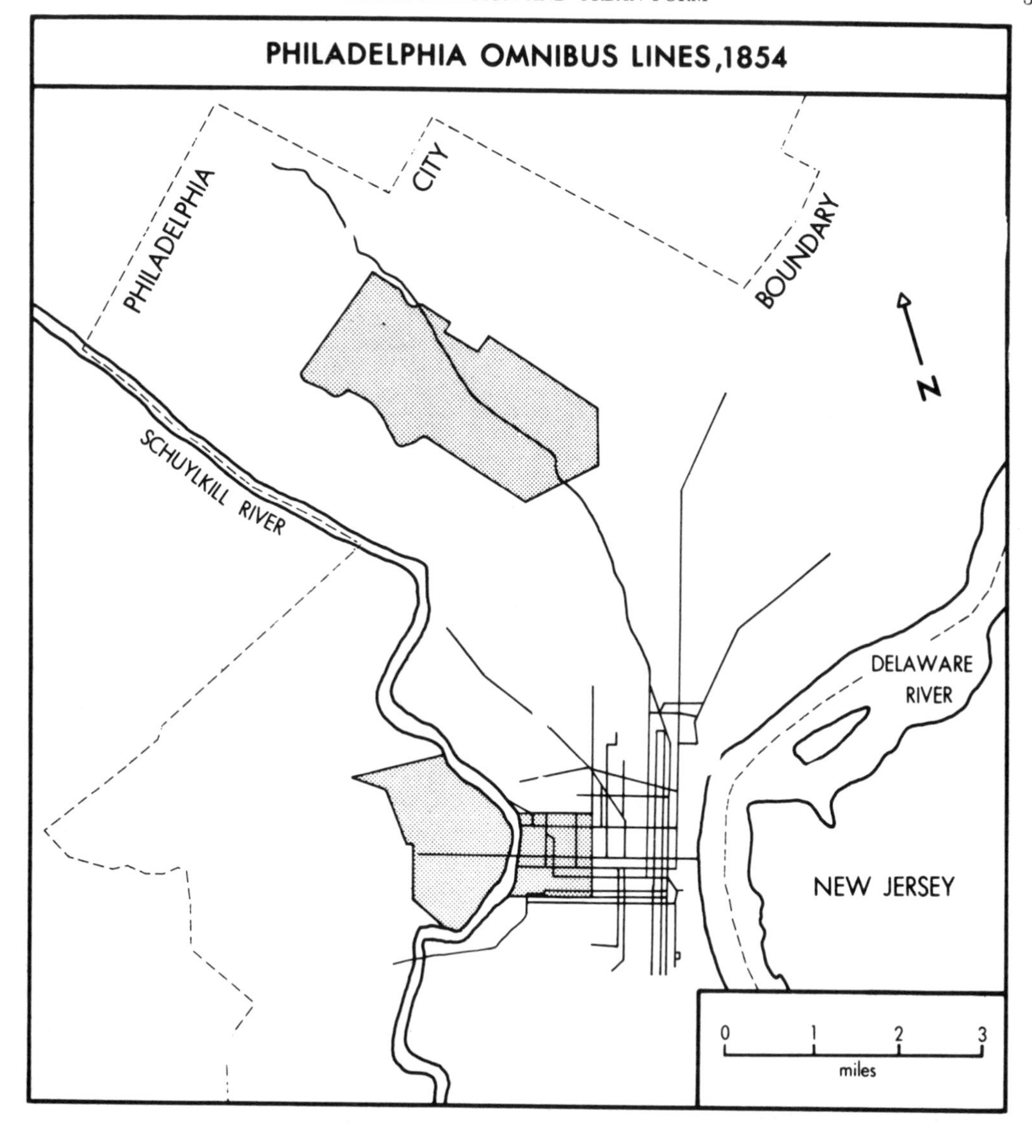

Figure 2-5. Philadelphia omnibus lines, 1854. From "Household Activity Patterns in Nineteenth-Century Suburbs: A Time–Geographic Exploration" by R. Miller, 1982, *Annals of the Association of American Geographers*, 72, p. 364. Copyright 1982 by the Association of American Geographers. Reprinted by permission.

pany of their fellow ethnics. Not surprisingly, such heterogeneous residential patterning almost immediately engendered social stresses and episodic conflicts that persisted until the end of the century, when the electric trolley would at last enable the formation of modern ethnic neighborhood communities.

Toward the end of the Walking–Horsecar Era, the scale of the city was slowly but inexorably expanding. One by-product was the emergence of the downtown central business district (CBD). As needs intensified for specialized commercial, retailing, and other services, it was quickly realized that they could best be provided from a single center at the most accessible urban location. With immigrants continuing to pour into the all-but-bursting industrial city in the late nineteenth century, pressures redoubled to improve intraurban transit and open up more

Figure 2-6. The horsecar introduced mass transportation to the American city. This dictograph was taken in downtown Minneapolis in the late 1880s. Source: Minnesota Historical Society.

of the adjacent countryside. In retrospect, horsecars had only been a stopgap measure—relieving overcrowding temporarily but incapable of bringing enough new residential space within the effective commuting range of the burgeoning middle class. The hazards of relying on horses for motive power were also becoming unacceptable; apart from high costs and the sanitation problem, disease was an ever-present threat: for example, thousands of horses succumbed in New York and Philadelphia in 1872 when respiratory illness swept through the municipal stables (Schaeffer & Sclar, 1975, p. 22). By the late 1880s, that desperately needed transit revolution was at long last in the making. When it came, it swiftly transformed both city and suburban periphery into the modern metropolis.

The Electric Streetcar Era (1890–1920)

The key to the first urban transport revolution was the invention of the electric traction motor by one of Thomas Edison's technicians, Frank Sprague. This innovation must rank among the most important in American history. The first electric trolley line opened in Richmond in 1888, was adopted by two dozen other major cities within a year, and by the early 1890s was the dominant mode of intraurban transit. The rapidity of the diffusion of this innovation was enhanced by the immediate recognition of its ability to mitigate the urban transportation problems of the day: motors could be attached to existing horsecars to convert them into self-propelled vehicles, powered via easily constructed overhead wires. Accordingly, the tripling of average speeds (to over 15 miles per hour) brought a large band of open land beyond the city's perimeter into trolley-commuting range.

The most dramatic impact of the Electric Streetcar Era was the swift residential development of those urban fringes, which extended the emerging metropolis into a decidedly star-shaped spatial entity (Figure 2-2, Era II). This morphological pattern was produced by radial trolley corridors extending several miles beyond the compact city's limits; with so much new space available for home building within easy walking distance of these trolley lines, there was no need to extend trackage laterally. Consequently, the interstices remained undeveloped. The typical streetcar suburb of the turn of the century was a continuous corridor whose backbone was the road carrying the trolley tracks (usually lined with stores and other local commercial facilities), from which gridded residential streets fanned out for several blocks on both sides of the tracks. This spatial framework is illustrated in Figure 2-7, whose map reconstructs street and property subdivisioning in a portion of streetcar-era Cambridge, Massachusetts, just outside Boston. By 1900 most of the open spaces between these streets were themselves subdivided into small rectangular lots that contained modest single-family houses. In general, the quality of housing and prosperity of streetcar suburbs increased with distance from the central-city line. However, as Warner (1962) pointed out in his classic study, these continuous developments were home to a highly mobile middle-class population, finely stratified according to a plethora of minor income and status differences. With frequent upward (and local spatial) mobility the norm, community formation became an elusive goal, a process further retarded by the relentless grid-settlement morphology and the heavy dependence on distant downtown for employment and most shopping. As Warner (1962) put it so aptly, this kind of a society generated "not integrated communities arranged about common centers, but a historical and accidental traffic pattern" (p. 158). The desire to exclude the working class also shaped the social transportation geography of suburban streetcar corridors. "Definitional" conflicts usually revolved around the entry of saloons, with middle-income areas voting to remain "dry" while the blue-collar mill towns of lower-status trolley and intercity-freight-rail corridors chose to go "wet" (Schwartz, 1976, pp. 13–18).

Within the city, too, the streetcar sparked a spatial transformation. The ubiquity and low fare of the electric trolley now provided every resident with access to the intracity circulatory system, thereby introducing truly mass transit to urban America in the closing years of the nineteenth century. For nonresidential activities, this ease of movement among the city's various parts quickly triggered the emergence of specialized land use districts for commerce, industry, and transportation as well as the continued growth of the multipurpose CBD—now abetted by the elevator which allowed the construction of much taller buildings. But the widest impact of the streetcar was on the central city's social geography, because it made possible the congregation of ethnic groups in their own neighborhoods. No longer were these moderate-income masses forced to reside in the heterogeneous jumble of rowhouses and tenements that ringed the factories. The trolley brought them the opportunity to "live with their own kind," enabling the sorting of discrete groups into their own inner-city social territories within convenient and inexpensive traveling distance of the workplace.

The latter years of the Electric Streetcar Era also witnessed additional breakthroughs in public urban rail transportation. The faster electric commuter train superseded steam locomotives in the wealthiest suburban corridors, which had resisted the middle-class incursions of the streetcar sectors because the rich always seek to preserve their social distance from those of lesser status. In some of the newer metropolises that lacked the street-rail legacy, heavier electric railways became the cornerstone of the movement system; Los Angeles is the outstanding example, with the interurban routes of the Pacific Electric network (Figure 2-8) launching a dispersed settlement fabric in preautomobile days, and many lines forging rights-of-way that were later upgraded into major boulevards and even freeways (see Banham, 1971, pp. 32–36). Finally, within the city proper elevated and underground rapid transit lines made their appearances, the "el" in New York as early as 1868 (using steam engines—the electric elevated was born in Chicago in 1892) and the subway in Boston in 1898. Such rapid transit was always enormously expensive to build and could be justified only in the largest cities that generated the highest traffic volumes. Therefore, els

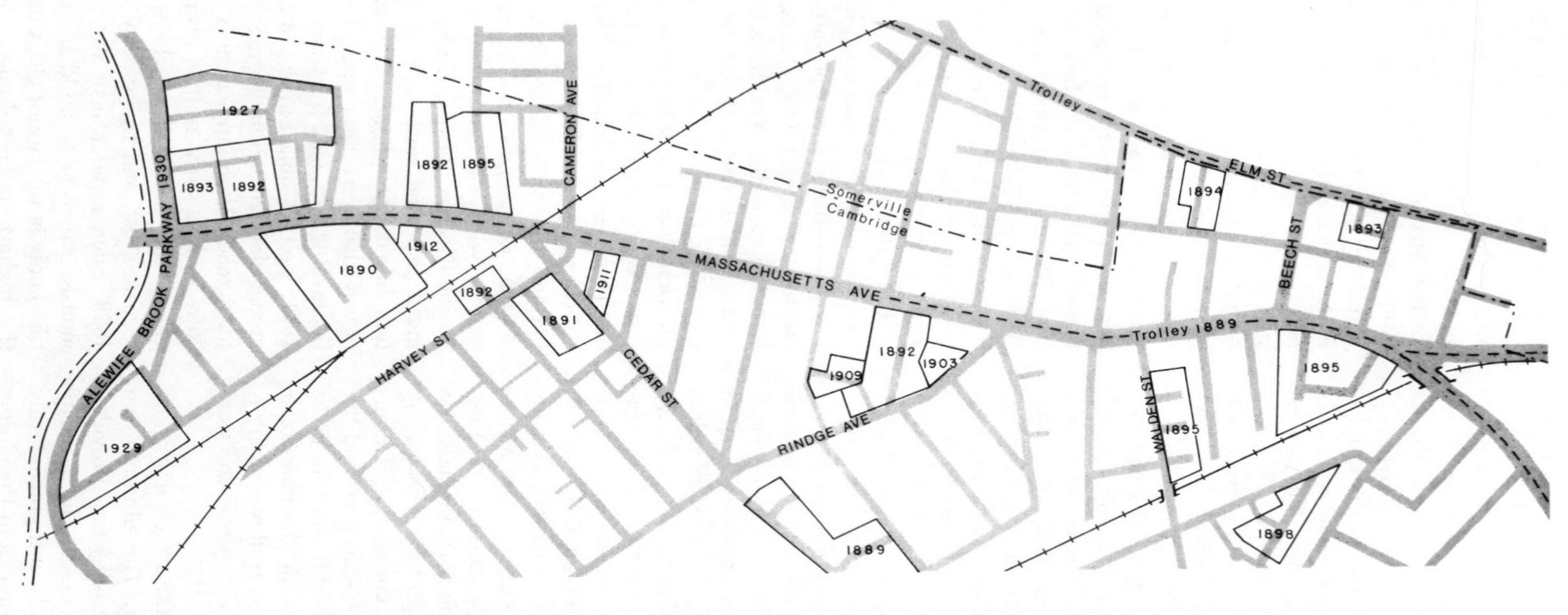

Figure 2-7. North Cambridge, Massachusetts, streetcar subdivisions, 1890–1930. From *Northwest Cambridge: Report Five, Survey of Architectural History in Cambridge* (p. 44) by A. J. Krim et al., 1977. Cambridge, Mass.: Cambridge Historical Commission and MIT Press. Copyright 1977 by Cambridge Historical Commission. Reprinted by permission.

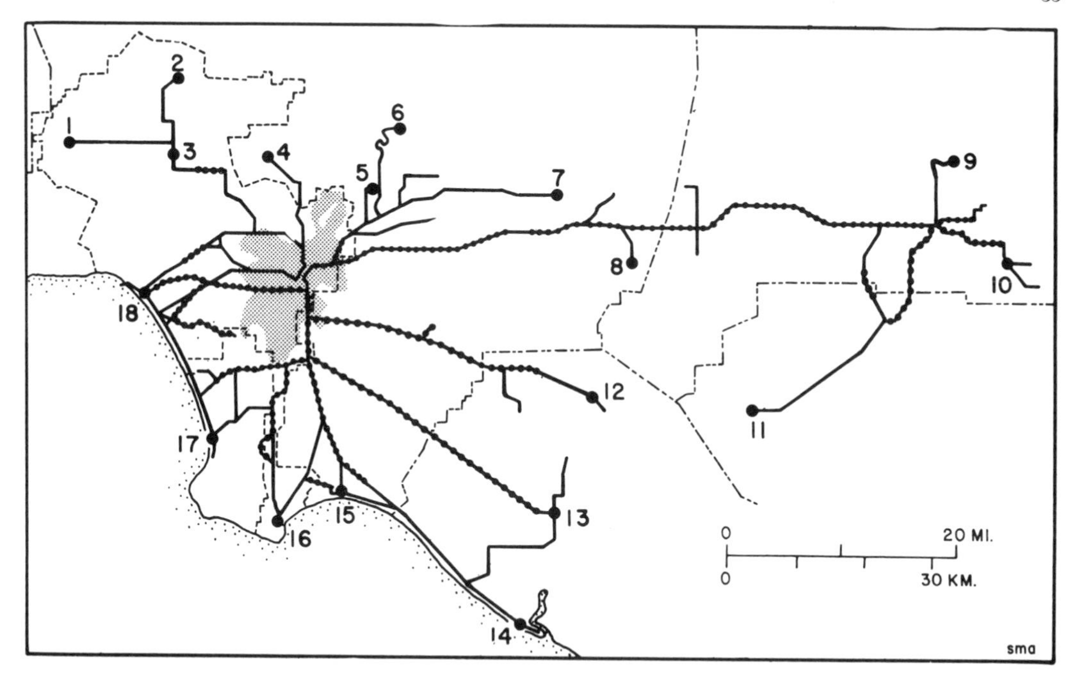

Figure 2-8. Interurban railway routes of the Pacific Electric system at their greatest cumulative extent as of the mid-1920s. Shading denotes portions of central Los Angeles that were situated within a half mile of narrow-gauge streetcar lines in the early 1920s. Selected interurban destinations are numbered in clockwise sequence as follows: 1, Canoga Park; 2, San Fernando; 3, Van Nuys; 4, Burbank; 5, Pasadena; 6, Mount Lowe; 7, Glendora; 8, Pomona; 9, Arrowhead Springs; 10, Redlands; 11, Corona; 12, Yorba Linda; 13, Santa Ana; 14, Newport Beach; 15, Long Beach; 16, San Pedro; 17, Redondo Beach; 18, Santa Monica. From *Los Angeles: The Centrifugal City* (p. 95) by R. Steiner, 1981, Dubuque, Iowa: Kendall/Hunt. Copyright 1981 by Kendall/Hunt. Reprinted by permission.

and subways were restricted to New York, Boston, Philadelphia, and Chicago, and most construction concluded by the 1920s. Rapid-transit-system building did not resume until the 1960s with the San Francisco Bay Area's Bay Area Rapid Transit (BART) network, followed in the 1970s and 1980s by projects in Cleveland, Washington, Atlanta, Baltimore, and Miami.

The Recreational Automobile Era (1920–1945)

By 1920, the electric trolleys, trains, interurbans, and subways had transformed the tracked city into a full-fledged metropolis whose streetcar suburbs and mill-town intercity rail corridors, in the largest cases, spread out to encompass an urban complex more than 20 miles in diameter. It was at this point in time, many geographers and planners would agree, that intrametropolitan transportation had achieved its greatest level of efficiency—that the burgeoning city truly "worked." How much closer the American metropolis might have approached optimal workability for all its millions of residents, however, will never be known because the second urban transportation revolution was already beginning to assert itself through the increasingly popular automobile. While many scholars have vilified the automobile as the destroyer of the city, Americans took to cars as completely and wholeheartedly as they did to anything in the nation's long cultural history. More balanced assessments of the role of the automobile (see, for instance, Bruce-Briggs, 1977) recognize its overwhelming acceptance for what it was—the long-hoped-for attainment of private mass transportation that offered users almost total freedom to travel whenever and wherever they chose. Cars came to the metropolis in ever greater numbers throughout the interwar period, a union culminating in accelerated deconcentration, through the development of the bypassed streetcar-era interstices and the pushing of the suburban frontier farther into the countryside, to produce again a compact, regular-shaped urban entity (Figure 2-2, Era III).

Although it came to produce a dramatic impact on the urban fabric by the eve of World War II, the entrance of the automobile into the American city in the 1920s and 1930s was a leisurely paced one. The first cars had appeared in both Western Europe and the United States in the 1890s, and the wealthy on both sides of the Atlantic quickly took to this innovation because it offered a better means of personal transport. It was Henry Ford, however, with his revolutionary assembly-line manufacturing techniques, who first mass-produced cars to lower selling prices to a level that soon converted them from the playthings of the rich into a transport mode available to a majority of Americans. By 1916 over 2 million autos were on the road, a total that quadrupled by 1920 despite wartime constraints. During the 1920s the total tripled to 23 million and increased another 4½ million by the end of the depression-plagued 1930s. Passenger car registrations paralleled these increases, as is shown in Table 2-1. The earliest flurry of auto adoptions had been in rural areas, where farmers badly needed better access to local service centers; thus, much of the early paved-road construction effort was concentrated in rural America. In the cities, cars were initially used for weekend outings—hence the *Recreational* Auto Era—and some of the first paved roadways built were landscaped parkways that followed scenic waterways (New York's Bronx River Parkway, Chicago's Lake Shore Drive, and the East and West River Drives along the Schuylkill in Philadelphia's Fairmount Park).

Table 2-1. Passenger car registrations in the United States (in millions), 1910–1983

Year	Registered vehicles
1910	0.5
1920	8.1
1930	23.0
1940	27.5
1950	40.4
1960	61.7
1970	89.2
1980	121.6
1983	125.4

Note: From *Statistical Abstract of the United States*, 1974, 1985, Washington, D.C.: U.S. Government Printing Office.

However, in the suburbs, where the overall growth rate now for the first time exceeded that of the central cities, cars were making a decisive penetration throughout the economically prosperous 1920s. Flink (1975, p. 14) reported that, as early as 1922, 135,000 suburban dwellings in 60 metropolises were completely dependent on motor vehicles. In fact, the subsequent rapid expansion of automobile suburbia by 1930 so adversely affected the metropolitan public transportation system that, through significant diversions of streetcar and commuter-rail passengers, the large cities started to feel the negative effects of the car years before accommodating to its actual arrival. By abetting the opening of unbuilt areas lying between suburban rail axes, the automobile effectively lured residential developers away from densely populated traction-line corridors into the now-accessible interstices. Thus, the suburban home-building industry no longer found it necessary to subsidize privately owned streetcar companies to provide cheap access to their trolley-line housing tracts.

Without this financial underpinning, the modern urban transit crisis soon began to surface. Traction companies, obliged under their charters to provide good-quality service, could not raise fares to the prohibitive levels that would have been necessary to earn profits high enough to attract new capital in the highly competitive money markets. As this economic squeeze intensified, particularly during the Great Depression of the 1930s, local governments were forced to intervene with subsidies from public funds and eventually, as ridership continued to decline in the postwar period, outright takeovers when lines could not be closed down without harming communities. Several additional factors also combined to accelerate the interwar deterioration of the superlative trolley-era metropolitan transit network: the growing intrasuburban dispersal of population that no longer generated passenger volumes great enough to support new fixed-route public transportation facilities; dispersion of employment sites within the central city, thereby diffusing commuter destinations as well as origins; shortening of the workweek from 6 to 5 days; worsening street congestion where trolleys and auto traffic increasingly mixed; and the pronounced distaste for commuting to the city by bus, a more flexibly routed new transit mode that never caught on in the suburbs.

Ironically, recreational motorways helped to intensify the decentralization of the urban population. Most were radial highways that penetrated deeply into the suburban ring; those connecting to major new bridges and tunnels—such as the Golden Gate and Bay Bridges in San Francisco, the George Washington Bridge and Holland and Lincoln Tunnels in New York—so served to open empty outer metropolitan sectors. Sunday motorists, therefore, had easy access to this urban countryside and obviously were captivated with what they saw. They responded in steadily increasing numbers to the home-sales pitches of developers who had shrewdly located their new tract housing subdivisions beside the suburban highways. As more and more city dwellers relocated to these automobile suburbs, by the end of the interwar era many recreational parkways were turning into heavily traveled commuter thoroughfares—especially near New York City, where the suburban parkway network devised by planner Robert Moses reached far into Westchester County and Long Island, and in the Los Angeles Basin, where the first freeway (the Arroyo Seco or Pasadena) was opened in 1940.

The residential development of automobile suburbia followed a simple formula that was devised in the prewar years and perfected and greatly magnified in scale in the decade after 1945. The leading motivation was developer profit from the quick turnover of land, which was acquired in large parcels, subdivided, and auctioned off. Accordingly, developers much preferred open areas at the metropolitan fringe where large packages of cheap land could be readily assembled. As the process became more sophisticated in the 1940s, developer–builders came to the forefront and produced huge complexes of inexpensive housing—with William J. Levitt and his Levittowns in the vanguard. Of course, silently approving and underwriting this uncontrolled spread of residential suburbia were public policies at all levels of government that included the financing of highway construction, obligating lending institutions to invest in new home building, insuring individual mortgages, and providing low-interest loans to Federal Housing Administration (FHA) and Veterans Administration (VA) clients.

Although the conventional-wisdom view of U.S. suburbanization is that most of it occurred after World War II, longitudinal demographic data indicate that intrametropolitan population decentralization achieved sizable proportions during the interwar era. Table 2-2 reveals that suburban growth rates began to surpass those of the central cities as early as the 1920s, and that after 1930 the outer ring took a commanding lead (which has not ceased widening to this day). With an ever-larger segment of the urban population residing in automobile suburbs, their spatial organization was already forming the framework of contemporary metropolitan society. Because automobility removed most of the preexisting movement constraints, suburban social geography now became dominated by locally homogeneous income-group

Table 2-2. Intrametropolitan population growth trends, 1910–1960

Decade	Central city growth rate	Suburban growth rate	Percent total SMSA growth in suburbs[a]	Suburban growth per 100 increase in central city population
1910–1920	27.7	20.0	28.4	39.6
1920–1930	24.3	32.3	40.7	68.5
1930–1940	5.6	14.6	59.0	144.0
1940–1950	14.7	35.9	59.3	145.9
1950–1960	10.7	48.5	76.2	320.3

Notes: U.S. Census of Population.
[a]SMSA = Standard Metropolitan Statistical Area, comprised of the central city and the county-level political units of the surrounding suburban ring.

clusters that isolated themselves from dissimilar neighbors. Gone was the highly localized stratification of streetcar suburbia; in its place arose a far more dispersed, increasingly fragmented residential mosaic that builders were only too happy to cater to, helping shape this kaleidoscopic settlement pattern by shrewdly constructing the most expensive houses that could be sold in each locality.

The long-standing partitioning of suburban society was further legitimized by the widespread adoption of zoning (legalized in 1916). This legal device gave municipalities control of lot and building standards, which, in turn, assured dwelling prices that would only attract newcomers whose incomes at least equaled those of the existing local population. For the middle class, especially, such exclusionary economic practices were enthusiastically supported because it now extended to them the capability that upper-income groups had enjoyed to maintain their social and geographic distance from people of lower socioeconomic status.

Nonresidential activities were also suburbanizing at a steadily increasing rate during the Recreational Auto Era. Indeed, many large-scale manufacturers had decentralized during the previous streetcar era, choosing suburban freight-rail locations that rapidly spawned surrounding working-class towns. These industrial suburbs became important satellites of the central city in the emerging metropolitan constellation (see Taylor, 1915/1970). The economic geography of the interwar era reflected an intensification of this trend, as shown in the curves of activity gradients in Figure 2-9. Industrial employers accelerated their intraurban deconcentration in this period as more efficient horizontal fabrication methods were replacing older techniques requiring multistoried plants—thereby generating greater space needs that were too expensive to satisfy in the high-density inner central city. Newly suburbanizing manufacturers, however, continued their spatial affiliation with intercity rail corridors, because motor trucks were not yet able to operate with their present-day efficiencies and the highway network of the outer ring remained inadequate until the 1950s.

The other major nonresidential activity of interwar suburbia was retailing. Clusters of automobile-oriented stores had first appeared in the urban fringes before World War I. By the early 1920s, the roadside commercial strip had become a common sight in many southern California suburbs. Retail activities were also featured in dozens of planned automobile suburbs that sprang up in the 1920s, most notably in outer Kansas City's Country Club District where builder Jesse Clyde Nichols opened the nation's first complete shopping center in 1922. But these diversified retail centers spread rather slowly before the 1950s; however, such chains as Sears, Roebuck and Montgomery Ward quickly discovered that stores situated alongside main suburban highways could be very successful, a harbinger of things to come in post–World War II metropolitan America.

The central city's growth in the interwar

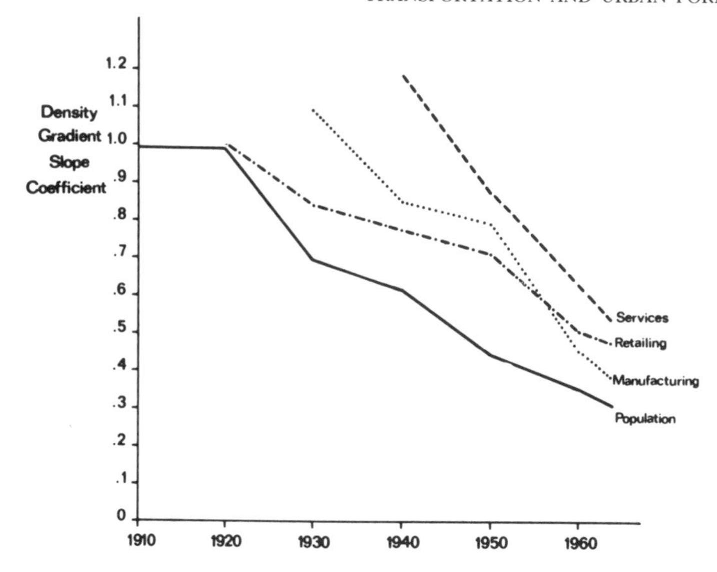

Figure 2-9. Intrametropolitan population and activity density gradients, 1910–1963. The lower the slope coefficient, the more dispersed the spatial distribution of an activity. From "Urban Density Functions" by E. S. Mills, 1970, *Urban Studies*, 7, p. 14. Copyright 1970 by Urban Studies. Reprinted by permission.

era reached its zenith and began to level off (Table 2-2) as metropolitan development after 1925 increasingly concentrated in the urban fringe zone that now widely resisted political unification with the city. Whereas the transit infrastructure of the streetcar era remained dominant in the industrial city (see Figure 2-10), the late-arriving automobile was adapted to this high-density urban environment as much as possible, but not without greatly

Figure 2-10. The automobile did not become a major force in the central city until the postwar era, but its presence can already be detected by the 1930s. This photograph was taken in St. Paul in 1932. Source: Minnesota Historical Society.

aggravating existing traffic congestion. The structure of the American city during the second quarter of this century was best summarized in the well-known *concentric-ring, sector,* and *multiple nuclei* models (reviewed in Harris & Ullman, 1945), which together described the generalized spatial organization of urban land usage. The social geography of the core city was also beginning to undergo significant change at this time as the suburban exodus of the middle class was accompanied by the arrival of southern blacks. These parallel migration streams would achieve massive proportions after World War II. The southern newcomers were attracted to the northern city by declining agricultural opportunities in the rural South and by offers of employment in the factories as industrial entrepreneurs sought a new source of cheap labor to replace the European immigrants, whose numbers were sharply curtailed by restrictive new legislation after the mid-1920s. Urban whites, however, refused to share residential space with in-migrating blacks, and racial segregation of the metropolitan population swiftly intensified as citywide dual housing markets dictated the formation of ghettos of nonwhites within their own distinctive social territories.

Freeway Era (1945–Present)

Unlike the two preceding eras, the postwar Freeway Era was not sparked by a revolution in urban transportation. Rather, it represented the coming of age of the automobile culture, which coincided with a historic watershed as a new nation emerged from 15 years of depression and war. Suddenly, the automobile was no longer a luxury or a recreational diversion: it quickly became a necessity for commuting, shopping, and socializing—essential to the successful exploitation of personal opportunities for a rapidly expanding majority of the metropolitan population. People snapped up cars as fast as the reviving peacetime automobile industry could roll them off the assembly lines, and a prodigious highway-building effort was launched, spearheaded by high-speed, limited-access expressways. Given impetus by the 1956 Interstate Highway Act, these new freeways would soon reshape every corner of urban America as the new suburbs they engendered represented nothing less than the turning inside out of the historic metropolitan city. In retrospect, this massive acceleration of the deconcentration process

> cannot be considered a break in long-standing trends, but rather the later, perhaps more dynamic, evolutionary stages of a transformation which was based on a pyramiding of small scale innovations and underlying social desires. (Sternlieb & Hughes, 1975, p. 12)

The snowballing effect of these changes is expressed spatially in the much-expanded metropolis of the postwar era (Figure 2-2, Era IV), whose expressway-dominated infrastructure again produced a network-biased development pattern reminiscent of the streetcar period (Figure 2-2, Era II). A more detailed representation of contemporary intraurban morphology is observed in Figure 2-11, showing the culmination of six decades of automobile suburbanization. Most striking is the enormous band of growth that was added between 1950 and 1980, with freeway sectors pushing the metropolitan frontier deeply into the surrounding zone of exurbia. The huge curvilinear outer city that arose within this new suburban ring was most heavily shaped by the circumferential freeway segments that girdled the central city—a universal feature of the metropolitan expressway system, originally designed to allow long-distance interstate highways to bypass the congested urban core. Today, nearly 100 of these expressways form complete *beltways* that are the most heavily traveled roadways in their regions. The prototype high-speed circumferential was suburban Boston's Route 128, completed in the early 1950s; by the 1980s, such freeways as Houston's Loop, Atlanta's Perimeter, Chicago's Tri-State Tollway, New York–New Jersey's Garden State Parkway, Miami's Palmetto Expressway, and the Beltways ringing Washington and Balti-

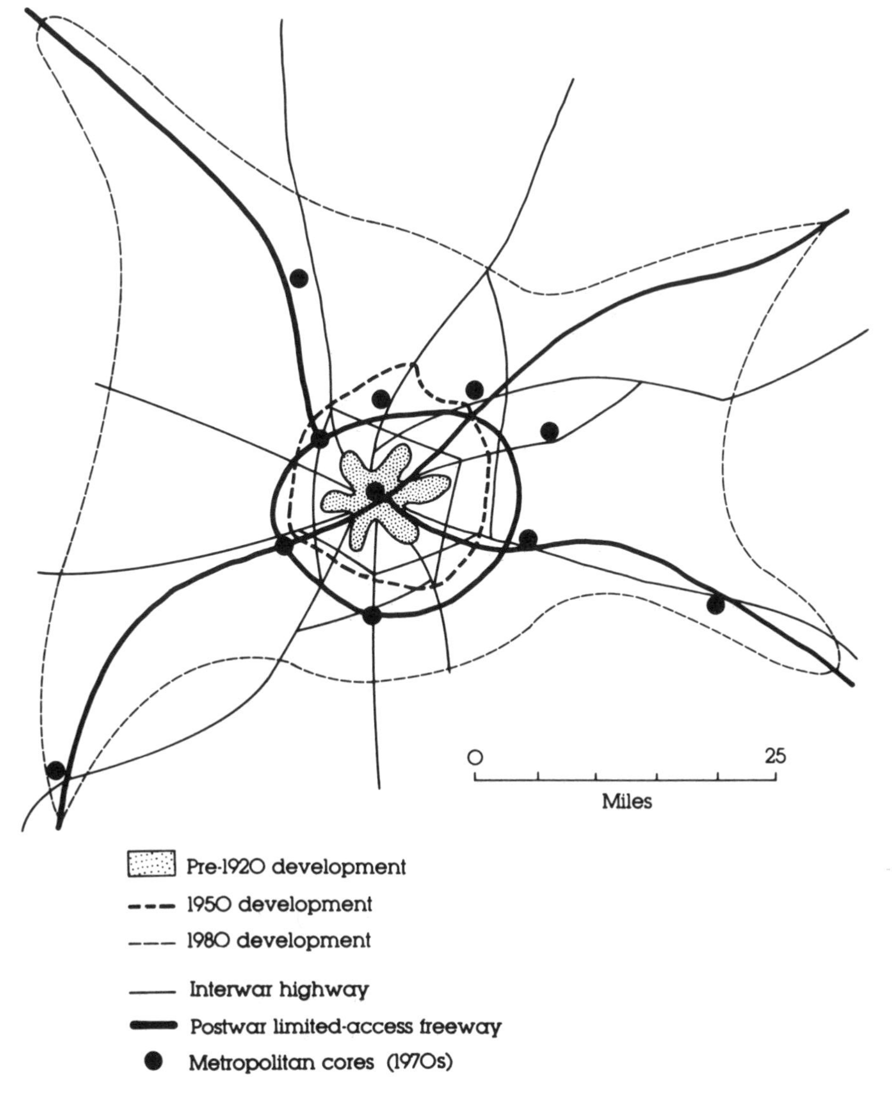

Figure 2-11. The spatial pattern of growth in automobile suburbia since 1920. From "The Role of the Suburbs in the Contemporary Metropolitan Systems" by P. O. Muller, 1982, in C. M. Christian and R. A. Harper (eds.), *Modern Metropolitan Systems* (p. 257), Columbus, Ohio: Charles E. Merrill. Copyright 1982 by Charles E. Merrill. Reprinted by permission.

more had become some of the best-known urban arteries in the nation.

The maturing freeway system was the primary force that turned the metropolis inside out after the mid-1960s, because it eliminated the regionwide centrality advantage of the central city's CBD. Now *any* location on that expressway network could easily be reached by motor vehicles, and intraurban accessibility had suddenly become an all but ubiquitous spatial good. Ironically, large cities had encouraged the construction of radial expressways in the 1950s and 1960s because they appeared to enable downtown to remain accessible to the swiftly dispersing suburban population. However, as one economic activity after another discovered its new locational footlooseness in the freeway metropolis, nonresidential deconcentration greatly accelerated. Much of this suburban growth has grav-

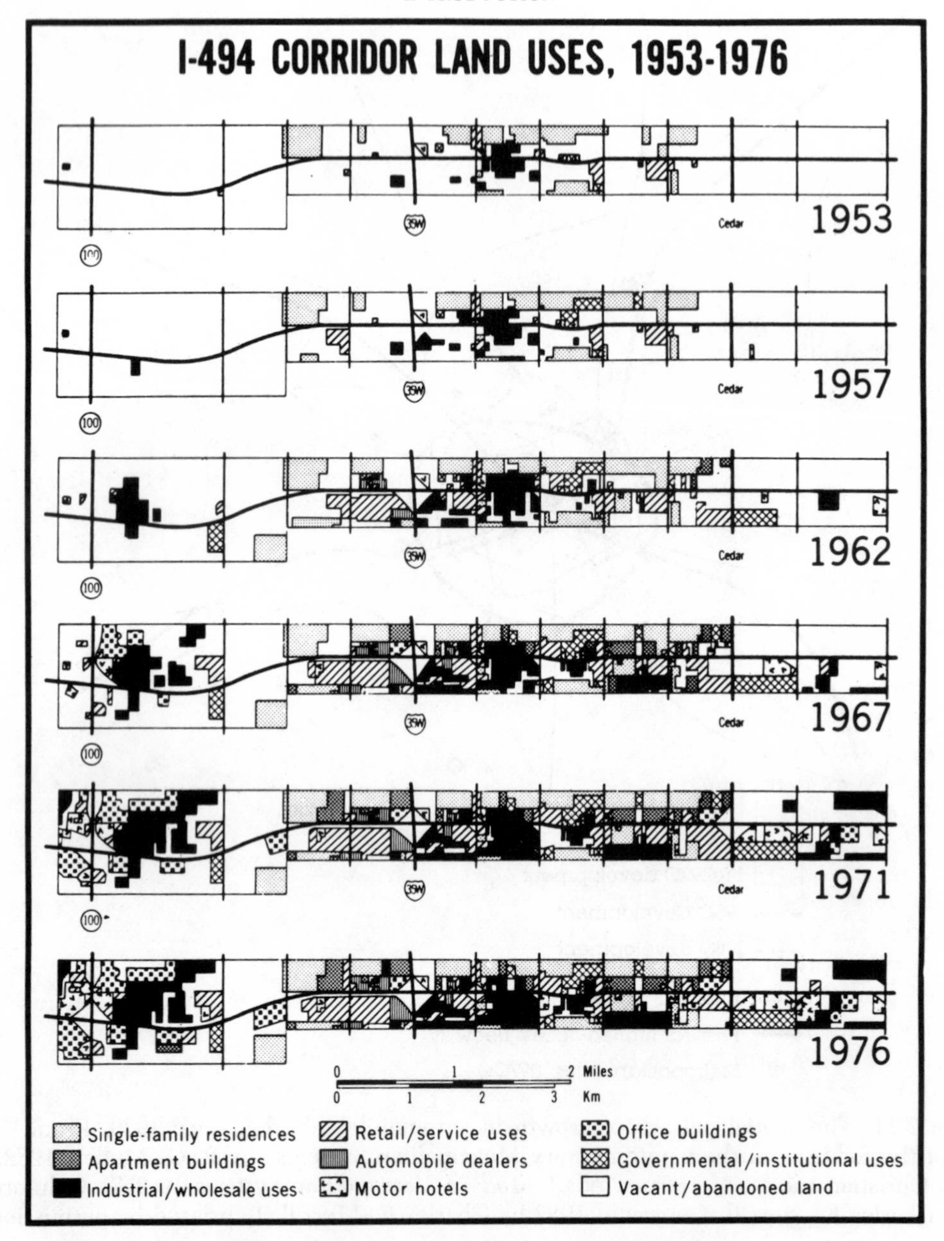

Figure 2-12. I-494 corridor land uses south of Minneapolis, 1953–1976. From "The Emergence of a New 'Downtown' " by T. J. Baerwald, 1978, *Geographical Review*, 68, p. 312. Copyright 1978 by the American Geographical Society. Reprinted by permission.

itated toward beltway corridors, and Figure 2-12 displays the typical sequence of land use development along a segment of circumferential I-494 just south of Minneapolis.

As high-speed expressways expanded the radius of commuting to encompass the entire dispersed metropolis, residential location constraints were relaxed as well. No longer were most urbanites required to live within a short distance of their job. Quite to the contrary, the workplace had now become a locus of opportunity offering access to the best possible residence that an individual could afford anywhere within the urbanized area. Thus, the kaleidoscopic patterning of uniform sociospatial clusters that had emerged in prewar automobile suburbia was writ ever larger in the Freeway Era, giving rise to a *mosaic culture* stratified not only along class lines but now also according to age, occupational status, and a host of minor life-style differences (Berry, 1981, pp. 64–66). This has fostered a great deal of local separatism, intensifying the balkanization of metropolitan society as a whole:

> With massive auto transportation, people have found a way to isolate themselves; . . . a way to privacy among their peer group. . . . they have stratified the urban landscape like a checker board, here a piece for the young married, there one for health care, here one for shopping, there one for the swinging jet set, here one for industry, there one for the aged. . . . When people move from square to square, they move purposefully, determinedly. . . . They see nothing except what they are determined to see. Everything else is shut out from their experience. (Schaeffer & Sclar, 1975, p. 119)

As the most frenetic period of change introduced by the freeway era comes to an end in the 1980s, certain structural transformations are emerging from what was, in retrospect, one of the most tumultuous upheavals in American urban history. Figure 2-11 reveals the existence of several new outlying metropolitan-level cores. These downtown-like suburban concentrations of retailing, business, and light industry have become common features near the major highway intersections of the outer city that now encircles every large central city. A representative *minicity* of this genre is mapped in Figure 2-13, revealing the array of high-order activities that have agglomerated around the King of Prussia Plaza shopping center at the most important expressway junction in Philadelphia's northern and western suburbs.[2] Dozens of such diversified centers from coast to coast have matured since 1970, and minicities such as Houston's Post Oak Galleria, Los Angeles' South Coast Metro, Chicago's Schaumburg, Washington's Tyson's Corner, New York's White Plains and Stamford, and Miami's Dadeland have also achieved national reputations (see Kowinski, 1985).

As the suburban downtowns of the outer city achieve economic–geographical parity with each other (as well as the CBD of the nearby aging central city), they provide the totality of urban goods and services to their surrounding populations and thereby make each sector of the metropolis an increasingly self-sufficient functional entity. The transition to a polycentric metropolis of *realms*—the term coined by Vance (1964) to describe the increasingly independent areas served by new downtown-like activity centers—requires the use of more up-to-date generalizations than are provided by the "classical" concentric-zone, sector, and multiple nuclei models. Such an alternative to these obsolete core-periphery models of the interwar metropolis is seen in Figure 2-14, including an application to the New York metropolitan area that shows the central city and no less than eight suburban realms (each organized along a radial freeway spine).[3] Another useful contemporary model is Erickson's (1983) generalization of the evolution of the suburban space–economy (Figure 2-15). He divides the Freeway Era into two segments (stages *b* and *c*), the pre-1960 phase consisting of rapid activity dispersal and diversification of suburban municipal economies, and the post-1960 phase dominated by infilling and especially by the emergence of high-order nucleations.

[2] My research under way in 1985 indicates substantial intensification of development in this minicity since 1975, the base year for the map in Figure 2-13. Spearheading recent growth was the completion in 1983 of the Court at King of Prussia, a new adjoining mall that doubled the size of the original superregional shopping center and made the entire complex the largest suburban mall in the United States as of the end of 1984.

[3] Numerous minicities are located within these outer realms: I (Uniondale), II (Stamford), III (White Plains), IV (Nyack-Nanuet), V (Paramus, Wayne), VI (Piscataway), VII (Woodbridge, East Brunswick), and VIII (Neptune).

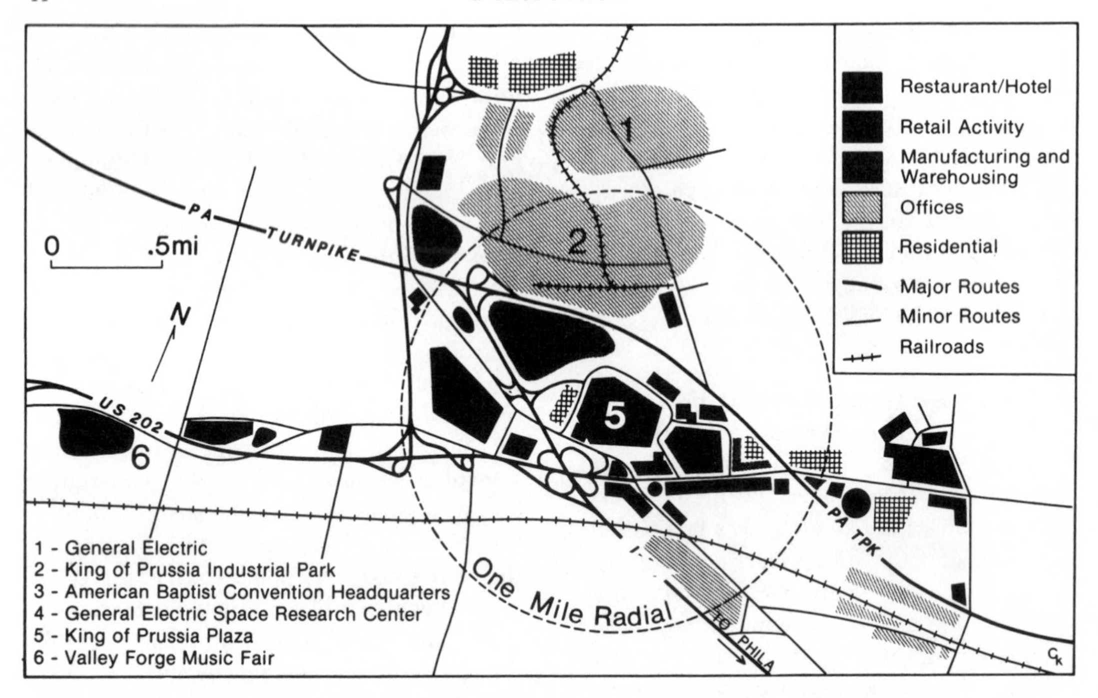

Figure 2-13. The internal activity structure of the typical suburban minicity: King of Prussia, Pennsylvania. From *The Outer City: Geographical Consequences of Urbanization of the Suburbs* (p. 41) by P. O. Muller, 1976, Washington, D.C.: Association of American Geographers, Resource Publication No. 75-2. Copyright 1976 by the Association of American Geographers. Adapted by permission.

URBAN TRANSPORTATION IN THE POSTINDUSTRIAL METROPOLIS

As the nation completes its transition to a postindustrial economy and society in the closing years of the twentieth century, intraurban movement patterns will continue to adjust to new geographical circumstances. The quaternary (information-related) and quinary (managerial–decision-making-based) economic activities that will increasingly dominate the American labor force have already demonstrated locational preferences that assume that the outer metropolitan ring will continue to be the essence of the contemporary American city for a long time to come. Above all, these activities seek the most prestigious metropolitan sites. Therefore, it is hardly a coincidence that the major concentrations of the pace-setting electronics–computer industry—from California's Silicon Valley to North Carolina's Research Triangle Park, from outer Boston's Route 128 to outer Dallas's Silicon Prairie—are localized in high-amenity suburbs. With intraurban location costs fairly well equalized, *noneconomic forces* now shape the distribution of high-tech (and most other) activities (see Berry & Cohen, 1973; Clark, 1985; and Saxenian, 1985). A recent survey of the locational decisions of such employers (Zanarini, 1983) revealed the following influences: a nearby major research university, close proximity to a cosmopolitan urban center, a large local pool of skilled labor, 300 days of annual sunshine, recreational water within a short drive, high-quality housing, and a relaxed business environment.

With the huge and still-growing commitment to the new suburban business complexes of postindustrial America, it is not surprising that two major gasoline shortages

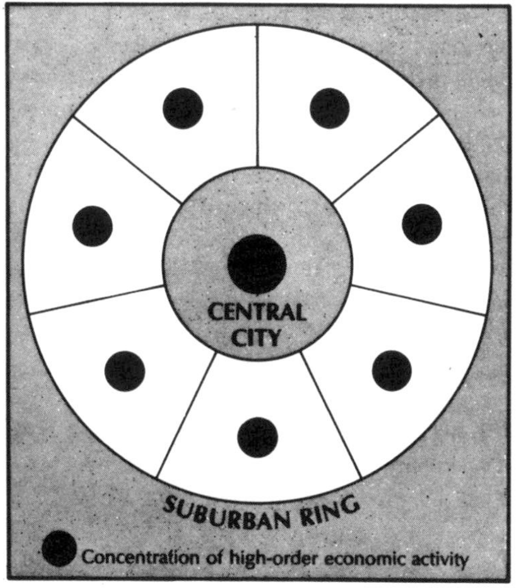

a.

Figure 2-14. Metropolitan realms. (a) Model. From *Economic Geography* (p. 169) by J. O. Wheeler and P. O. Muller, 1981, New York: John Wiley & Sons. Copyright 1981 by John Wiley & Sons. Reprinted by permission. (b) Urban realms of metropolitan New York. From "The Role of Suburbs in the Contemporary Metropolitan System" by P. O. Muller, 1982, in C. M. Christian and R. A. Harper (eds.), *Modern Metropolitan Systems* (p. 253), Columbus, Ohio: Charles E. Merrill. Copyright 1982 by Charles E. Merrill. Reprinted by permission.

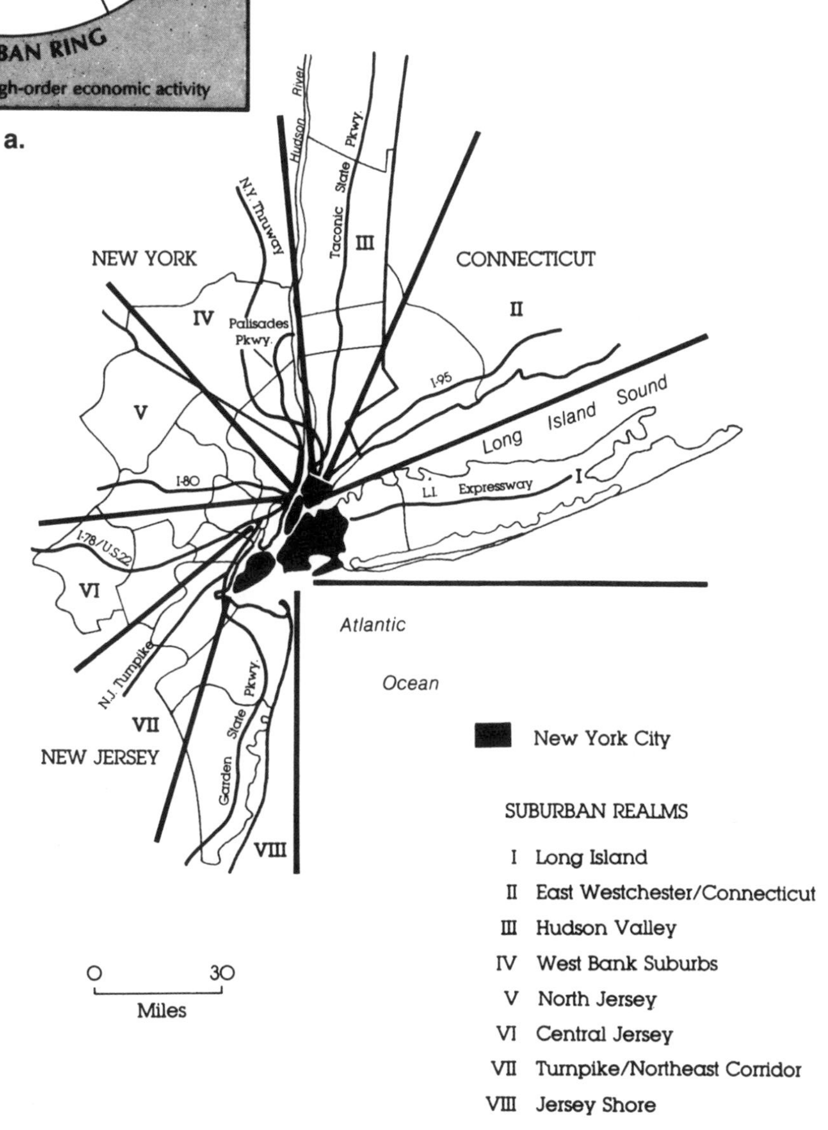

b.

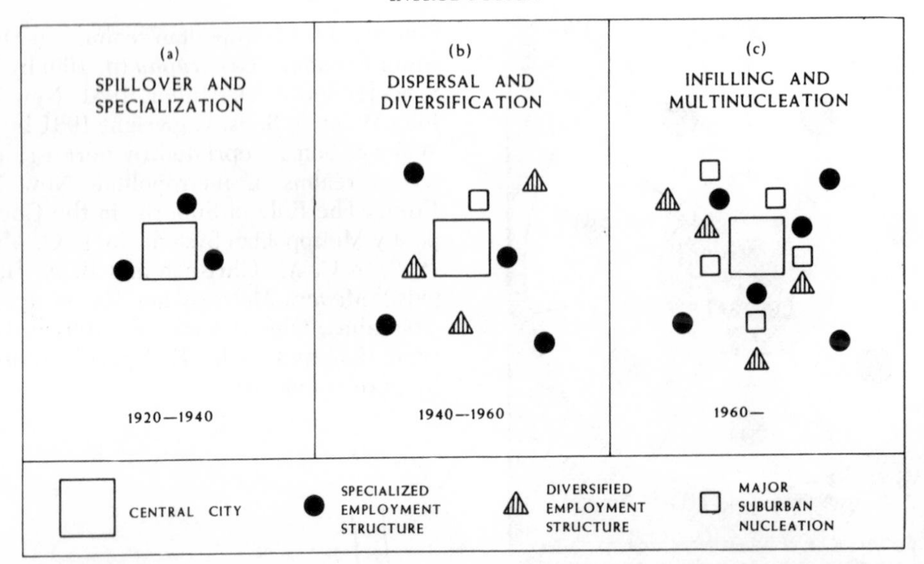

Figure 2-15. The evolution of the suburban space-economy. From "The Evolution of the Suburban Space Economy" by R. A. Erickson, 1983, *Urban Geography, 4*, p. 96. Copyright 1983 by V. H. Winston. Reprinted by permission.

in the 1970s, which resulted in permanently higher fuel prices, have not deterred deconcentration in the least. Thus, suburban contraction and metropolitan recentralization vis-à-vis the central city has become an unrealistic future scenario no matter how dire an energy-rationing forecast one subscribes to (chapter 12 outlines some of the reasons for this). On the contrary, the same decade that witnessed two serious gasoline supply shortages saw the revival of growth in nonmetropolitan areas for the first time in nearly a century as the city spilled into exurbia and beyond (a trend that may have been short lived because early-1980s population data appear to herald another turnabout). Improved highway transportation was obviously a primary force in this episode of "counterurbanization," and Lewis (1983) even proposed the fascinating hypothesis of the *galactic metropolis* in which urban America is viewed as a continuous archipelago of development clusters that follow interstate expressways through the rural zones that separate metropolises from one another. The leading urban transportation concerns of the 1980s, however, focus on the efficiencies of moving people about the dispersed, polycentric city of realms.

Although urban freeways spawned the contemporary multinodal metropolis, it is unlikely that many more will be built in the foreseeable future. Grass-roots resistance is intensifying, and governments at all levels today are increasingly unable to afford the massive construction costs that are often heightened by the need to conform to stricter environmental regulations (see chapter 14). Besides, there is considerable evidence that building new expressways does not improve the flow of traffic: metropolitan Los Angeles and Houston, for example, each possess extensive modern superhighway networks, yet each new road link seems to create more traffic jams rather than easing existing congestion (some of the reasons for this are outlined in chapter 5). Moreover, metropolises with inferior freeway systems—such as Phoenix, Denver, San Francisco, and Philadelphia—still have managed to achieve the overall movement efficiencies (and transformed spatial structures) of better-equipped urban areas.

As an alternative to additional highways, the construction of new public mass transit systems is currently being pursued as a possible solution to urban transportation problems (a subject treated fully in chapter 10). Since the 1960s, heavy-rail electric-train systems have been started in metropolitan San Francisco, Washington, Cleveland, Atlanta, Miami, and Baltimore. Moreover, less expensive light-rail trolley lines are planned for San Diego, Buffalo, Dallas, and at least a dozen other cities; major bus-system improvements are scheduled for Detroit, Indianapolis, and Dallas. Yet, the fact remains that ridership levels have been declining in all these metropolises over the past two decades and that transit lines are incapable of serving even a significant minority of travel demands in the outer suburban city. Too many newly planned transit systems also fail to recognize that the traditional CBD-focused, hub-and-spoke network has become increasingly irrelevant to urban travel patterns since the 1920s. The (San Francisco) Bay Area Rapid Transit (BART) system taught planners that lesson in the 1970s when it failed to capture its hoped-for ridership and was unable to exert any influence on development patterns around its stations (see Webber, 1976). Small wonder, then, that when a similar heavy-rail system which avoided the airport and many intermediate suburban activity centers was completed in Miami in 1985, ridership averaged only about 27,000 daily (under 25% of the level predicted).

As the postindustrial metropolis matures, its transport problems may also be mitigated by the increasing substitution of communication for the physical movement of people (a matter explored in chapter 15). Already more people are working at home than ever before, telecommuting to their offices by various information-transmission means; moreover, cable-television systems increasingly facilitate at-home shopping, and the proliferation of video cassette recorders (VCRs) and other home entertainment equipment may well reduce social and recreational travel in the future. Thus, the city of the longer-term future may well become the *wired metropolis,* whose spatial expressions are as yet unknown. Almost surely these developments portend further deconcentration, with activity centers potentially capable of locating at almost any site offering access to global computer, cable, and satellite networks.

Acknowledgments

The comments and suggestions of the editor, her students, and the following are gratefully acknowledged: Raymond A. Mohl of the Department of History at Florida Atlantic University, Thomas J. Baerwald of the Science Museum of Minnesota, and my colleague Ira M. Sheskin of the Department of Geography at the University of Miami. Roman A. Cybriwsky of Temple University, and Kristine Bradof and Tom Baerwald of the Science Museum of Minnesota, kindly assisted in the obtaining of the photographs.

References

Abbott, C. (1981). *The new urban America: Growth and politics in sunbelt cities.* Chapel Hill: University of North Carolina Press.

Adams, J. S. (1970). Residential structure of Midwestern cities. *Annals of the Association of American Geographers, 60,* 37–62.

Baerwald, T. J. (1978). The emergence of a new "downtown." *Geographical Review, 68,* 308–318.

Banham, R. (1971). *Los Angeles: The architecture of four ecologies.* New York: Penguin Books.

Berry, B. J. L. (1981). *Comparative Urbanization: Divergent paths in the twentieth-century,* 2nd ed. New York: St. Martin's Press.

Berry, B. J. L. (1975). The decline of the aging metropolis: Cultural bases and social process. In G. Sternlieb & J. W. Hughes (Eds.), *Post-industrial America: Metropolitan decline and inter-regional job shifts* (pp. 175–185). New Brunswick, N.J.: Rutgers University, Center for Urban Policy Research.

Berry, B. J. L., & Cohen, Y. S. (1973). Decentralization of commerce and industry. In L. H. Masotti & J. K. Hadden (Eds.), *The urbanization of the suburbs* (pp. 431–455). Beverly Hills, Calif.: Sage Publications.

Bruce-Briggs, B. (1977). *The war against the automobile.* New York: E. P. Dutton.

Clark, D. (1985). *Post-industrial America: A geographical perspective.* New York & London: Methuen.

Erickson, R. A. (1983). The evolution of the suburban space economy. *Urban Geography, 4,* 95–121.

Flink, J. J. (1975). *The car culture*. Cambridge: MIT Press.

Harris, C. D., & Ullman, E. L. (1945). The nature of cities. *Annals of the American Academy of Political and Social Science, 242* (November), 7–17.

Hartshorn, T. A. (1980). *Interpreting the city: An urban geography*. New York: John Wiley.

Jackson, K. T. (1985). *Crabgrass frontier: The suburbanization of the United States*. New York: Oxford University Press.

Kowinski, W. S. (1985). *The malling of America: An inside look at the great consumer paradise*. New York: William Morrow.

Krim, A. J., et al. (1977). *Northwest Cambridge: Report five, Survey of architectural history in Cambridge*. Cambridge: Cambridge Historical Commission and MIT Press.

Lewis, P. F. (1983). The galactic metropolis. In R. Platt & G. Macinko (Eds.), *Beyond the urban fringe: Land-use issues of nonmetropolitan America* (pp. 23–49). Minneapolis: University of Minnesota Press.

Miller, R. (1982). Household activity patterns in nineteenty-century suburbs: A time–geographic exploration. *Annals of the Association of American Geographers, 72*, 355–371.

Mills, E. S. (1970). Urban density functions. *Urban Studies, 7*, 5–20.

Muller, P. O. (1981). *Contemporary Suburban America*. Englewood Cliffs, N.J.: Prentice-Hall.

Muller, P. O. (1982). The role of the suburbs in contemporary metropolitan systems. In C. M. Christian & R. A. Harper (Eds.), *Modern Metropolitan systems* (pp. 251–276). Columbus, Ohio: Charles E. Merrill.

Saxenian, A. (1985). The genesis of Silicon Valley. In P. Hall & A. Markusen (Eds.), *Silicon landscapes* (pp. 20–34). Winchester, Mass.: Allen & Unwin.

Schaeffer, K. H., & Sclar, E. (1975). *Access for all: Transportation and urban growth*. Baltimore: Penguin Books.

Schwartz, J. (1976). The evolution of the suburbs. In P. Dolce (Ed.), *Suburbia: The American dream and dilemma* (pp. 1–36). Garden City, N.Y.: Anchor Press/Doubleday.

Sommers, L. M. (1983). Cities of Western Europe. In S. D. Brunn & J. F. Williams (Eds.), *Cities of the world: World regional urban development* (pp. 84–121). New York: Harper & Row.

Steiner, R. (1981). *Los Angeles: The centrifugal city*. Dubuque, Iowa: Kendall/Hunt.

Sternlieb, G., & Hughes, J. W. (Eds.). (1975). *Post-industrial America: Metropolitan decline and inter-regional job shifts*. New Brunswick, N.J.: Rutgers University, Center for Urban Policy Research.

Tarr, J. A. (1984). The evolution of the urban infrastructure in the nineteenth and twentieth centuries. In R. Hanson (Ed.), *Perspectives on urban infrastructure*, (pp. 4–66). Washington, D.C.: National Academy Press.

Taylor, G. R. (1970). *Satellite cities: A study of industrial suburbs*. New York: Arno Press. (Original work published 1915)

Tobin, G. A. (1976). Suburbanization and the development of motor transportation: Transportation technology and the suburbanization process. In B. Schwartz (Ed.), *The changing face of the suburbs* (pp. 95–111). Chicago: University of Chicago Press.

Vance, J. E., Jr. (1964). *Geography and urban evolution in the San Francisco Bay Area*. Berkeley: University of California, Berkeley, Institute of Governmental Studies.

Warner, S. B., Jr. (1962). *Streetcar suburbs: The process of growth in Boston, 1870–1900*. Cambridge: Harvard University and MIT Press.

Webber, M. M. (1976). The BART experience—what have we learned? *Public Interest*, Fall, 79–108.

White, M., & White, L. (1962). *The intellectual versus the city: From Thomas Jefferson to Frank Lloyd Wright*. Cambridge. Harvard University & MIT Press.

Zanarini, R. (1983). ULI session examines high tech locations. *AAG Urban Geography Specialty Group Newsletter, 3*, 3.

3 THE URBAN TRANSPORTATION PLANNING PROCESS

ERIC I. PAS
Duke University

INTRODUCTION

The title of this chapter implies that there exists a single, definitive urban transportation planning process. This is not the case; transportation planning in urban areas is conducted at different temporal and geographic scales by various organizations having diverse responsibilities. Furthermore, urban transportation planning is constantly evolving and has undergone considerable change during its lifetime of approximately 30 years. Nevertheless, there is a general understanding of what is meant by "the urban transportation planning process," and this process is described in this chapter.

The traditional view of planning is that its purpose is to guide future decisions; that is, to develop a master plan to be used as a framework within which specific decisions can be made in a rational manner. This view of planning, often referred to as Bolan's (1967) "classical model," underlies the urban transportation planning process that developed during the 1950s and 1960s. This type of planning emphasizes long-range forecasts of the performance of regionwide systems; the transportation plan is the backbone of any urban master plan.

By the end of the 1960s, societal changes began pushing the urban transportation planning process in new directions, and urban transportation planning began to be increasingly recognized as an activity that provides information to decision makers on the consequences of alternative courses of action. DeBettencourt et al. (1982) note that planning today is also considered to serve a number of other potential purposes, including (1) educating planners, decision makers, and the general public, (2) responding to regulations, and (3) supporting the image or position of the planning agency or important decision maker.

The remainder of this chapter has five sections. The second section provides a brief historical review of urban transportation planning. This review shows how the process of urban transportation planning has been shaped by, and has responded to, broader societal changes. The third section provides an introduction to the urban transportation planning process, which is viewed here as comprising three phases; namely, preanalysis, modeling and technical analysis, and postanalysis. Each of these phases is described in some detail in the middle sections of this chapter. The final section summarizes the chapter and concludes with a brief examination of the outlook for urban transportation planning.

This chapter describes the urban transportation planning process as it has evolved in the United States. Generally similar planning processes have developed and been applied in most of the Western industrial nations. However, there are some differences in the processes and technical tools employed by planners in different countries. These differences reflect, to some extent, variations in transportation decision making resulting from diverse political ideologies (Colcord, 1979).

HISTORICAL REVIEW

The Early Days

Prior to the mid-1940s, urban transportation was dominated by traffic engineers who were concerned with specific problems such as a congested bridge or intersection. In 1944 the Bureau of Public Roads conducted the first "origin–destination" survey. This was the first attempt to collect data that would contribute to an understanding of observed traffic volumes, in contrast to earlier traffic-count studies where no effort was made to understand the underlying process that was generating the observed traffic (Oi & Shuldiner, 1962).

The urban transportation planning process began developing in the United States in the early 1950s to address a problem that had existed in one form or another for many years, even centuries (Stopher & Meyburg, 1975). An important factor in the development of the analytical tools that formed the basis of early urban transportation planning studies was the availability of digital computers capable of manipulating large quantities of data. Computers allowed planners to examine urban travel patterns on a regionwide basis and encouraged efforts to develop mathematical equations describing these patterns. A large number of other factors contributed to the development of the urban transportation planning process in the United States during the 1950s. They include (1) rapid population growth (particularly in urban areas), (2) rapid growth in car ownership, (3) increasing movement of population to suburban areas, and (4) increasing federal involvement in funding urban development while requiring comprehensive urban planning.

Pioneering urban transportation studies that used the emerging analytical techniques were undertaken in San Juan, Detroit, and Chicago. The Detroit Study was the first to employ a process that included data collection and goal formulation, development of forecasting procedures, and testing and evaluation of alternatives. These studies were followed by a number of others in the late 1950s, including studies in Washington, D.C., Baltimore, Pittsburgh, and Philadelphia. All these studies were regionwide efforts undertaken by a large full-time staff. Most employed the computerized procedures developed during the Chicago Area Transportation Study (CATS) and had the objective of forecasting future trip-making patterns and producing a long-range regionwide transportation plan. The emphasis in these studies was on planning a highway system that would cater to the expected large increases in automobile travel in urban areas.

The Institutionalization of Urban Transportation Planning

A major turning point in the development of the urban transportation planning process occurred with the passage of the 1962 Federal Aid Highway Act. This act required that urban areas employ a continuing, comprehensive, and cooperative transportation planning process (dubbed the 3C process) in order to qualify for federal matching funds for construction of urban transportation facilities. By referring to urban areas rather than to cities, the act ensured that transportation planning would take place at the metropolitan or regional level, rather than the city level. The act also allocated funds specifically for planning and research.

The Bureau of Public Roads soon began implementing the planning requirements of the act and issued memoranda for carrying out all aspects of the 3C planning process. Between 1963 and 1967 the bureau published a large number of manuals dealing with the technical aspects of the planning process. The procedures developed in the 1950s and early 1960s were codified and institutionalized and became the technical standards for the coming decade. The advantage of this development was that forecasting tools were made available to a larger community of planners. On the other hand, once a particular set of procedures became readily available and was institutionalized, there was a great resistance to change. This is a particularly important observation because the early forecasting procedures were developed during ongoing studies by planners who needed practical tools that could be employed immediately. Furthermore, the early travel forecasting models were developed at a time when highway planning was the major concern, car ownership was increasing rapidly, and there was an apparent abundance of both natural and monetary resources. Thus, the procedures that became institutionalized as part of the urban transportation planning process were oriented almost exclusively toward long-term, capital-intensive expansions of the transportation system, primarily in the form of highways.

Adaptation to Change

The decade beginning in the early 1960s saw substantial changes in American society. The major new thrusts were the struggle for civil rights and increasing concern for the environment; both had important impacts on urban transportation planning. The civil rights movement highlighted the question of the distribution of the impacts brought about by transportation system changes and raised awareness concerning the transportation needs of the elderly, the handicapped, and the poor. The increasing concern for the environment during the 1960s is reflected in the National Environmental Policy Act (NEPA) of 1969, which had direct effects on transportation planning and policy. (These effects are described in detail in Chapter 14 of this book.)

In September 1975, the Federal Highway Administration (FHWA) and the Urban Mass Transportation Administration (UMTA) issued joint regulations to guide urban transportation planning (U.S. Department of Transportation, 1975). These new regulations may be seen, at least in part, as a response to the societal changes described above. A major change in the planning regulations required a transportation plan to contain both the traditional long-range element as well as a short-range Transportation System Management (TSM) element. In particular, the regulations published in 1975 specified that certain options be considered in the development of the TSM element. These options can be characterized as being low-capital-cost alternatives that can be implemented in a relatively short time frame and that aim to make better use of existing facilities, either by operational changes or by better management of travel demand. In addition, these options address concerns regarding energy, the environment, and the provision of transportation services for the elderly and handicapped. Options entailing operational changes include implementation of high-occupancy-vehicle (HOV) lanes on freeways and arterials and bus priority schemes at signalized intersections. Options designed to manage the demand for travel include schemes for flexible and staggered work hours and automobile-free zones. The 1975 regulations also specified, for the first time, that urban transportation changes be programmed for implementation, by requiring the formulation of a multiyear Transportation Improvement Program (TIP) consistent with the transportation plan.

The 1975 regulations represent a major change in the philosophy of urban transportation planning. The regulations required planners to consider options that would either

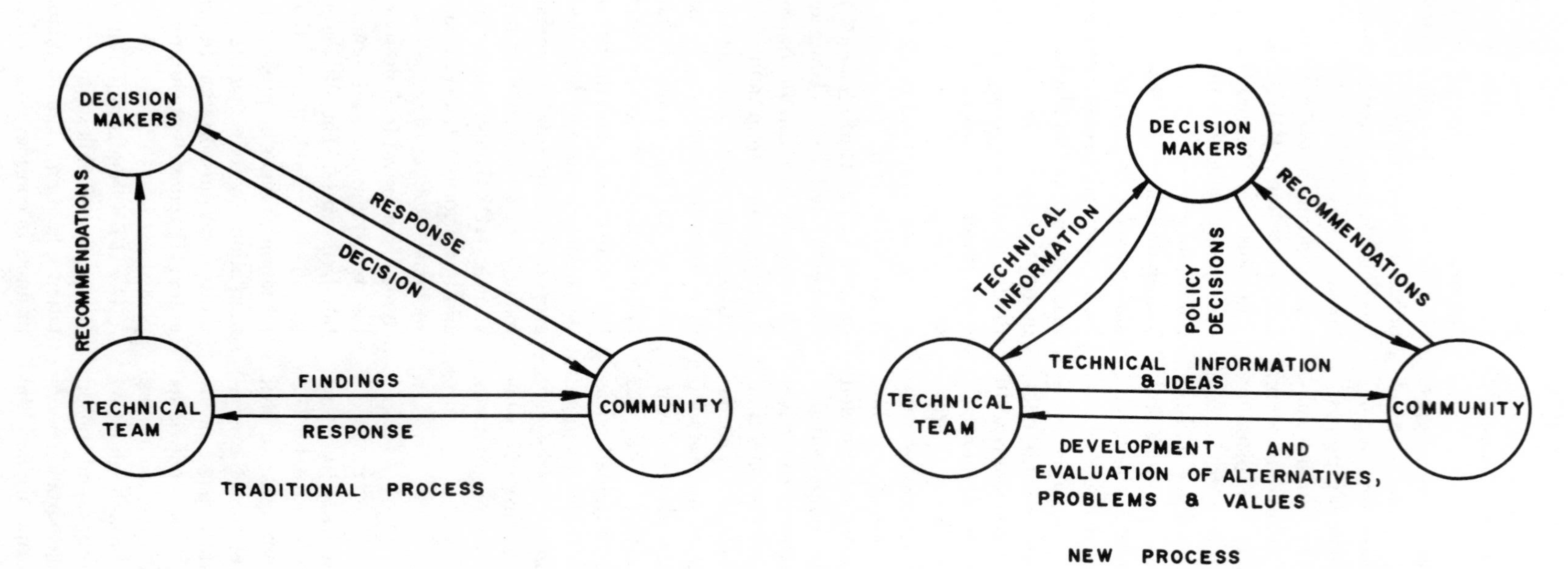

Figure 3-1. Major actors in the urban transportation planning process and their interrelationships. From "Metropolitan Transportation Planning: Process Reform" by W. G. Hansen & S. C. Lockwood, 1976, *Transportation Research Record*, No. 582, p. 4. Used with permission.

reduce the peak demand for travel or that would increase the person-carrying capacity (as opposed to vehicle-carrying capacity) of existing systems, rather than catering to the anticipated future demand for travel by increasing the supply of transportation services. The shift in emphasis away from highway systems and toward multimodal transportation systems is reflected by institutional name changes. For example, the Highway Research Board (HRB) became the Transportation Research Board (TRB), and the American Association of State Highway Officials (AASHO) became the American Association of State Highway and Transportation Officials (AASHTO).

Traditionally, urban transportation planning had been considered a technical process in which quantitative tools are used to perform objective analyses to arrive at objectively chosen solutions to technological problems. However, by the late 1960s the political nature of urban transportation decisions began to be recognized together with the need to "open" the planning process to allow meaningful citizen participation. The roles of the technical team, the decision makers (usually elected officials), and the community required modification in this changed environment.

Figure 3-1 illustrates the roles of these major actors in the traditional and revised planning processes (Hansen & Lockwood, 1976). In the revised planning process, the planner works *with* the community, rather than planning *for* the community. The planner also provides information and ideas to the community and the decision makers, rather than making recommendations to the decision makers. Two major reviews of urban transportation planning studies, undertaken in Boston and Toronto during the mid-1970s, are prototypical examples of the "open" urban transportation planning process (Gakenheimer, 1976; Pill, 1978). Allen (1985) refers to these as "postclassical" transportation studies.

The changes in urban transportation planning issues and concerns also had important implications for the methods used in the planning process to forecast travel on alternative transportation systems. First, the travel forecasting methods had to be able to deal with a much wider range of options, including so-called policy options (as opposed to physical facility options). Second, the forecasting tools had to be able to produce both long- and short-range forecasts. In particular, the short-range forecasting tools had to be able to produce results in a much shorter time frame and at much less cost than the traditional long-range forecasting tools. Third, the new planning process required forecasting methods that allowed the assessment of the impacts on specific groups of the population. Fortunately, during the late 1960s transportation researchers had begun developing travel analysis tools that respond to these needs.

Federal Disengagement

Recent and ongoing changes in the urban transportation planning process can be traced to the Reagan administration's desire to reduce federal involvement in what the administration considers local and state government responsibilities. The revised regulations issued jointly by FHWA and UMTA in June 1983 (U.S. Department of Transportation, 1983) mark another major turning point in the evolution of urban transportation planning in that they reduce, for the first time, federal requirements and responsibilities. In particular, while the federal government remains committed to an urban transportation planning process of the 3C variety, it no longer specifies how the process is to be performed (Weiner, 1983). Furthermore, many of the elements are to be self-certified. It is too early to assess the impacts of the regulations. However, it is likely that the coming years will see a continuation of the trend toward less federal involvement in urban transportation matters and a growing private sector role in the provision of urban transportation services (Weiner, 1984). The final section of this chapter looks at possible future developments in urban transportation planning.

OVERVIEW OF THE URBAN TRANSPORTATION PLANNING PROCESS

Transportation planning is undertaken at many levels (from strategic planning to project-level planning) and geographic scales in any urban area. Furthermore, the systemwide regional studies that became synonymous with urban transportation planning in the 1950s and 1960s have undergone substantial changes over the years. Nevertheless, it is possible to identify a planning process and associated technical analyses that are commonly considered to be the urban transportation planning process. In this section we provide a very brief overview of this planning process. We do not present this as a recommended process for urban transportation planning, nor do we suggest this as being the way in which urban transportation planning is always undertaken. Rather, we intend this to be a general framework within which our discussion of specific aspects can proceed.

The urban transportation planning process is viewed here as comprising three major, interrelated components (Figure 3-2). These components are referred to as the preanalysis phase, the technical analysis phase, and the postanalysis phase. A major reason for describing the process in this manner relates to the roles of the various actors at various stages of the process. These actors include the technical team, the decision makers, and the citizens (Figure 3-1). The activities in the technical analysis phase are conducted al-

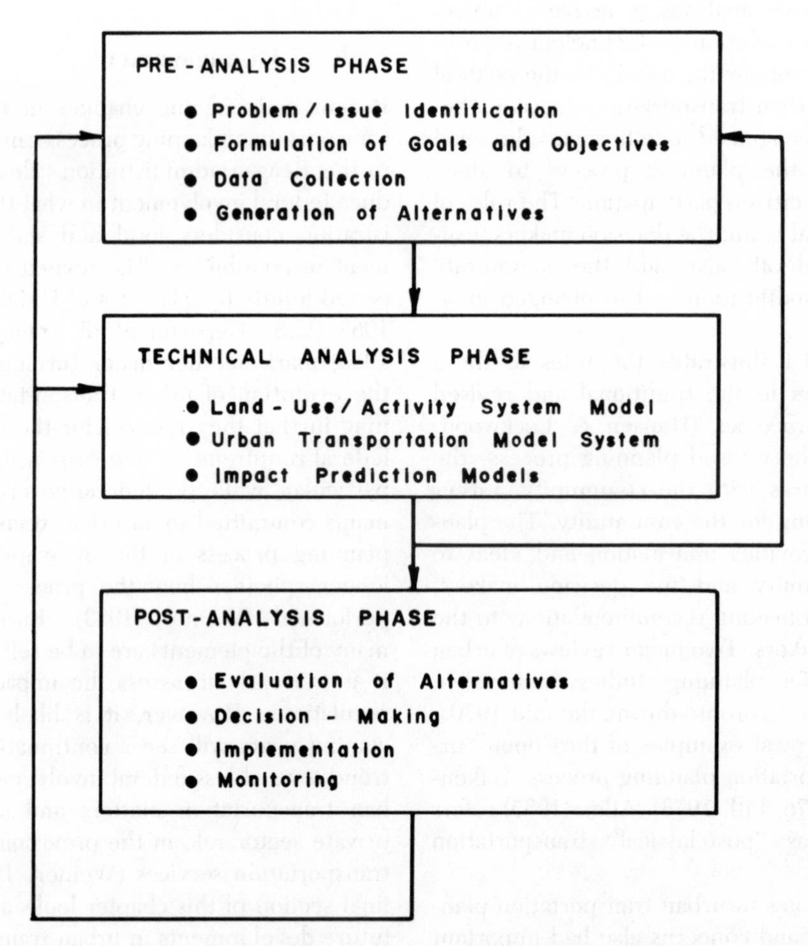

Figure 3-2. A general representation of the urban transportation planning process.

most exclusively by the technical team, while the decision makers and the citizens should be involved in the pre- and postanalysis phases.

The technical analysis phase of the urban transportation planning process is concerned with predicting the impacts of alternative courses of action. In this phase of the planning process, mathematical models (see Mathematical Models of Travel Behavior, below) are used to predict the transportation and related impacts (consequences) of alternative plans and policies. These impacts include capital and operating costs, energy usage, land requirements, air quality and noise levels, and accident rates, in addition to the quantity and quality of traffic flow on the transportation network.

The central component of the technical analysis phase is concerned with predicting the quantity and quality of traffic flow on the distinct pieces of a specified transportation network. The forecasting techniques used for this purpose are often referred to generically as the Urban Transportation Model System (UTMS). An important input to this model system is the distribution of employment, housing, and other activities in the urban area. These distributions are predicted by land use (or activity system) models. The output from the UTMS is input to a set of impact models that are used to predict the range of impacts described above.

A number of activities provide necessary inputs to the technical analysis phase, and Manheim (1979) refers to the preanalysis phase as "setup." The validity of the technical analysis phase depends on the success of the preanalysis phase. The preanalysis phase includes problem–issue identification and formulation, the development of study area goals and objectives, the collection of data concerning the existing transportation and related systems as well as existing travel patterns, and the identification of the alternative transportation systems to be analyzed. That is, the preanalysis phase has two components. The first concerns defining the current situation and problems and specifying the desired characteristics of improvements. The second aspect of the preanalysis phase concerns providing the data to be used in the technical analyses and formulating the alternative plans to be tested.

The postanalysis phase is concerned with assessing the impacts of the alternative plans and policies, selecting the preferred alternative, implementing the preferred alternative, and monitoring the performance of the implemented plans. The monitoring activity emphasizes the continuing nature of the urban transportation planning process. Each of the phases of the urban transportation planning process is described in greater detail below.

THE PREANALYSIS PHASE

The preanalysis phase plays a vital role in the urban transportation planning process in a variety of ways. These are discussed in the descriptions below of each of the components of the preanalysis phase.

Problem–Issue Identification

The objective of this stage of the planning process is the identification and definition of the problems and issues to be addressed. Meyer and Miller (1984) stress the importance of including in this step the identification of opportunities as well as problems. The identified problems and issues should be defined as broadly as possible so as not to constrain the set of possible solutions.

No amount of sophisticated analysis and assessment can overcome problem definitions that are too narrow and that result in a very narrow set of possible solutions. For example, if the problem in a particular area is defined as being "limited highway capacity in corridors *X* and *Y*," all public-transit-oriented alternatives are ruled out of consideration. In addition, policies for reducing the peak-period highway traffic loads in corridors *X* and *Y* are also eliminated from consideration. In fact, the only alternatives consistent with this problem definition are ways of increasing the highway capacity. On the other hand, the problem definition would encourage a considerably broader set of possible solutions if it read "high traffic volumes, relative to highway capacity, during the peak periods, in corridors *X* and *Y*."

Formulation of Goals and Objectives

A second aspect of the preanalysis phase is the definition of the desired states toward which the planning process should assist in moving the urban area. These broad general statements are termed "goals" and they are derived from consideration of the "values" of the society. Values are the basic desires and drives governing behavior (Wachs & Schofer, 1969). The goals are operationalized by a set of more specific "objectives" against which the performance of the alternative courses of action may be evaluated. The specific measures of objective attainment are termed "criteria" (or "measures of effectiveness") while the minimum (or maximum) acceptable level of performance on a criterion is termed a "standard." The hierarchical relationship among values, goals, objectives, and criteria is illustrated by the example in Figure 3-3.

Values are observed, or assumed, to be shared by groups of similar people; thus it is possible to speak of societal or cultural values (Wachs & Schofer, 1969). However, the diverse groups present in most urban areas would not necessarily share the same values, and they would certainly have a diversity of goals. Therefore, the planner must be careful to recognize exactly whose goals are being employed to guide the planning process. Furthermore, Wachs and Schofer (1969) describe the difficulty of formulating goals and objectives in the absence of specific proposals. Thus, while it is important to formulate representative goals and objectives to guide the planning process, this is a difficult task.

Data Collection

The urban transportation planning process and the urban transportation model system (see Figure 3-2) generally require collection of substantial quantities of data. Historically, a very large percentage of the budget for urban transportation studies has been spent on data collection and related activities, at the expense of other phases of the planning process. Hillegass (1969) notes that urban transportation planning studies typically spent

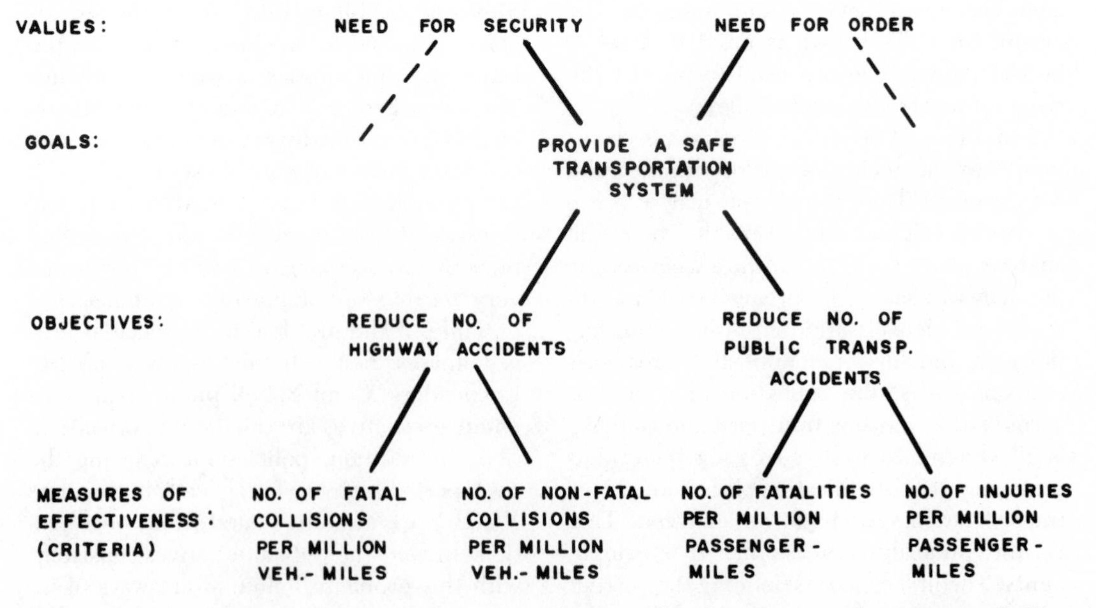

Figure 3-3. Hierarchical relationships among values, goals, objectives, and measures of effectiveness.

approximately 30% of the study budget on data collection. In general, a conventional urban transportation planning study requires an inventory of the existing transportation system, an inventory of present land use, a description of current travel patterns, and data on population growth, economic activity, employment, income, car ownership, housing, travel preference, and other related factors.

The transportation system inventory gathers data concerning the existing transportation systems in the study area. For the highway system these data include the physical attributes (number of lanes, grades, and so on), and the quantity (volumes) and quality (travel speeds) of traffic flow on the transportation system. For the public transport system, the data include service area, route structure, passenger volumes, cost and revenue data, and system operation data. The transportation system inventory is compiled from maps, public transportation operator records, and field surveys.

The travel inventory data are obtained from household travel surveys, commercial vehicle surveys, on-board surveys, and roadside surveys. The most important of these is the household travel survey, which is used to collect travel and sociodemographic data from a sample of the residents of the study area. The early urban transportation studies used simple random samples of between 4 and 20% of the households in the study area, depending upon the population of the area. More recent studies employ considerably smaller samples and use alternative sampling procedures.

A number of methods can be used to conduct the household travel survey, including the home interview survey, telephone survey, and mail survey. In a home interview survey, each sample household is visited by an interviewer who requests information concerning the household (such as number of people, number of cars, and so on) and the trips made by its members on the previous day. Mail and telephone surveys are generally considered cheaper to conduct than home interview surveys, although the latter yield higher response rates. Careful attention to survey administration details can improve the results obtained with the cheaper data-collection methods, and they are increasingly being used in urban transportation planning studies (Stopher, 1983).

A number of other methods are used to gather travel and/or traveler information. These include roadside origin–destination surveys and public transit on-board surveys. In a roadside origin–destination survey, a sample of the vehicles crossing preselected points is intercepted and the occupants are interviewed briefly. In a public transit on-board survey, interviewers board a sample of vehicles and interview passengers on board or ask passengers to complete and return a questionnaire. Both these methods employ what is termed choice-based sampling (Lerman & Manski, 1979). That is, the population from which the sample is drawn is dependent, in these cases, on the particular mode of travel chosen. The roadside origin–destination survey and the public transit on-board survey may be used to supplement the data gathered in a home interview survey. A commercial vehicle survey is also used to determine the current pattern of truck and taxi travel within the study area. Interviews are conducted with the owners or operators of a sample of commercial vehicles garaged in the study area.

Generation of Alternatives

We have described the modern view of urban transportation planning as the provision of information to decision makers on the consequences of alternative courses of action. A crucial element in this activity is the identification and specification of the alternative plans and policies to be examined. Clearly, the quality of the information provided to the decision makers is constrained by the set of alternatives that are analyzed. In particular, the set of alternatives analyzed should be as broad as possible and should facilitate identification of the trade-offs necessary in making a decision.

Herald has undertaken a major state-of-the-art review of the way in which urban transportation alternatives are generated and defined. He concludes that in both theory and practice the generation of alternatives is most commonly "a loosely structured creative trial-and-error method" (Herald, 1980). Some structured techniques, including mathematical programming, have been proposed. However, such techniques have seen little application in practice.

THE TECHNICAL ANALYSIS PHASE

During this phase of the urban transportation planning process, mathematical descriptions of travel and related behavior are used to predict the consequences of each alternative transportation plan that is to be evaluated. This phase consists of three major components: land use or activity system models, the urban transportation model system, and impact prediction (or resource consumption) models. Each of these components is described below. First, however, we introduce the notion of mathematical models of travel and related behavior.

Mathematical Models of Travel Behavior

A general model of travel behavior may be represented as follows:

$$y = f(\mathbf{x}, \mathbf{b}) \qquad (1)$$

where y is the response (dependent) variable; $\mathbf{x}$ is a vector of explanatory (independent) variables; $\mathbf{b}$ is a vector of model parameters; and f is some function.

For example, consider a model in which y is the number of daily trips made by a household. A typical model of this type is given by

$$y = b_0 + b_1 x_1 + b_2 x_2 \qquad (2)$$

where y is the number of trips per household per day; x_1 and x_2 are the explanatory variables (such as household size and income); and b_0, b_1, and b_2 are the model parameters (or coefficients).

An important question which immediately comes to mind is "how are the parameters ($\mathbf{b}$) in a model such as Equation (2) obtained?" Essentially, the parameters are obtained by collecting data on the response and explanatory variables (see Data Collection, above) and using one of a number of statistical procedures to *estimate* the parameters. This process is referred to as *model calibration* or *model estimation*. The two most commonly used estimation procedures are *least squares regression* and *maximum likelihood estimation*.

Estimation procedures are mathematical–statistical tools, and the analyst is responsible for proper specification of the model. Model specification includes identifying those explanatory variables likely to be causally related to the response variable and identifying the functional form of the relationship.

Mathematical models of travel behavior are estimated using basically three levels of analysis: the person, household, and zonal levels of analysis. For example, the model of Equation (2) is a household level model. As noted in chapter 1, a basic distinction is made between (1) models in which the response variable describes the behavior of a single household or person, and (2) models in which the response variable describes the (average or total) behavior of a geographic grouping of such households or people; we refer to the former as a *disaggregate model* and the latter as an *aggregate model*. In addition to theoretical distinctions between aggregate and disaggregate models, there are major differences in the types of data used.

A model is generally estimated with a set of data collected in one context and is used to *predict* travel in a context different from the estimation context either in space or in time, or both. Thus, data collected in City X at time t_1, may be used to estimate a model that is used to predict (1) travel in City X at time t_2, (2) travel in City Y at time t_1, or (3) travel in City Y at time t_2. In each case, we assume that the parameters estimated in the original estimation context provide useful information about travel behavior in the application environment. The question of model transfer from the estimation context to an application context has received considerable research attention (see, for example, Atherton & Ben-Akiva, 1976; Koppelman & Wilmot, 1982).

Land Use–Activity System Models

The spatial distribution of people and activities within an urban area has an important impact on travel in the region, because, as noted in chapter 1, travel is a derived demand that arises when people (or goods) are spatially separated from the locations at which they wish (or need) to be. In the short term, the activity system can be considered fixed; however, in the long run the activity system changes and affects travel patterns, while the activity system itself is affected by travel patterns and by changes in the transportation system. That is, the land use and transportation systems are interdependent, but this interdependence is only observable in the longer term because major changes in either system take considerable time.

In practice, the planning and analysis of urban activity and transportation systems are undertaken essentially separately in spite of the important interaction between these systems. The primary reason is that different agencies are generally responsible for land use and transportation planning. Thus, a land use forecast is generally used as an exogenous input to the traffic forecasting model, with the land use forecast itself predicated upon some assumed future transportation system.

Activity system models generally use regional population and employment projections as input, and they distribute these totals spatially over the region. A large variety of techniques for predicting urban activity patterns have been developed and applied, and no single technique is commonly included in available computer packages. Land use forecasting techniques may be classified into judgmental approaches and mathematical forecasting models. The latter may be further classified into heuristic models, simulation models, and econometric models (Meyer & Miller, 1984).

Heuristic models employ a set of ad hoc rules to allocate activities to the various parts of the region. The best-known heuristic land use model is the Lowry model (Lowry, 1968). Simulation models attempt to simulate explicitly the decisions of the key actors in land development. Example simulation models are the NBER model (Ingram, Kain, & Ginn, 1972) and the CAM model (Birch, Atkinson, Sandstrom, & Stack, 1974).

Econometric models employ a set of simultaneous structural equations. These equations are specified a priori, and the model is calibrated econometrically using data collected at two points in time. The EMPIRIC model (Hill, Brand, & Hansen, 1966), for example, predicts the distribution of white-collar population, blue-collar population, retail and wholesale employment, manufacturing employment, and all other employment. Changes in each of these variables, termed located variables, are predicted on the basis of changes in the other four, the base-year distribution, and changes in a set of locator variables. The latter include descriptions of zonal land use development, transportation system variables, and municipal service variables. An application of the EMPIRIC model is described in chapter 6.

The Urban Transportation Model System (UTMS)

The models used to predict the flows on the links of a particular transportation network as a function of the socioeconomic system (which generates travel) and the characteristics of the transportation system (which serves the generated travel) is generally known as the Urban Transportation Model System (UTMS). The urban transportation model system that evolved during the early studies in Detroit and Chicago is often referred to as the four-step sequential model because it comprises four submodels that are employed in a sequential process. These submodels are

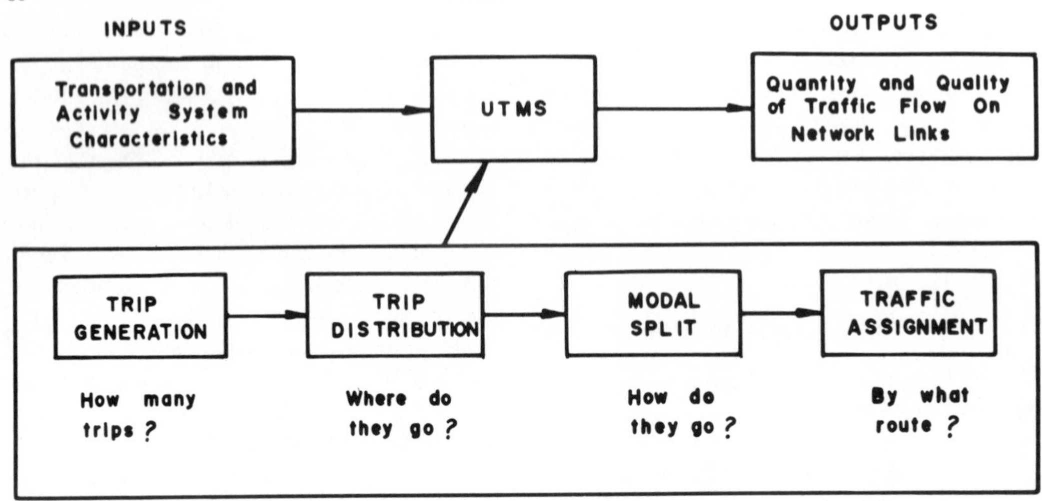

Figure 3-4. The Urban Transportation Model System (UTMS).

generally referred to as trip generation, trip distribution, modal split (or mode choice), and trip assignment (see Figure 3-4).

- *Trip generation* is concerned with predicting the number of trips produced by and attracted to each zone. That is, trip generation models address the question of how many trips are made to and from each zone of the study area.
- *Trip distribution* is concerned with predicting where the trips go. Thus, the trip distribution model links the origin and destination ends of the trips generated by the trip generation model.
- *Modal split* (or mode choice) addresses the question of how the various trips are made. That is, modal split models predict the proportion of trips by each mode of travel, between each origin and destination zone.
- *Trip assignment* is concerned with predicting the route(s) used by the trips from a given origin to a given destination by a particular mode.

These travel forecasting models are implemented in the Urban Transportation Planning System (UTPS) computer package distributed by the Urban Mass Transportation Administration (U.S. Department of Transportation, 1977).

The input to the urban transportation model system comprises the characteristics of the transportation and activity systems (Figure 3-4). In particular, UTMS requires estimates of the residential population and employment levels in each traffic analysis zone. These estimates are obtained from the land use (or activity system) models (see Land Use–Activity System Models, above). Each of the submodels of the four-step traffic forecasting model is introduced below.

Trip Generation.

Trip generation is the first submodel of the conventional four-step model sequence. This submodel is concerned with the total quantity of travel, while the other models allocate the total trips to alternative destinations, modes, and routes. Thus, trip generation is an extremely important part of the urban transportation model system.

Trip generation analysis is concerned with predicting the quantity of traffic flow to and from a given piece of land. That is, trip generation addresses the question of how many trips will be made to and from each zone within the study area. The basic idea is that the number of trips produced by or attracted to a given piece of land over some time period depends upon the characteristics of that piece of land, including land use type and intensity and the socioeconomic characteristics of the activities using the land.

Thus, the concept of a trip generation model may be described as follows:

$$t_i = f(L_i, I_i, S_i) \quad (3)$$

where t_i is the number of trips generated by a given parcel of land (or zone), designated i; L_i is the type of land use; I_i is the intensity of land use; S_i is the socioeconomic description of the activities using the land; and f is some function.

Two basic approaches have been used to develop trip generation models: least squares regression and category (or cross-classification) analysis. These methods are similar, and neither is tied to a particular unit of analysis. In fact, each method has been applied at the zonal, household, and individual levels of analysis. The fundamental difference is that least squares regression models are generally linear, while a category analysis model assumes no particular functional form.

The majority of trips in an urban area are home-based trips; that is, they begin or end at the home of the person making the trip. Thus, models of residential trip generation have received considerable attention in the past and they are introduced below.

The earliest urban transportation studies employed least squares regression analysis at the zonal level to model residential trip generation. Thus, although the data were collected from households (see Data Collection, above), each traffic analysis zone was treated as a single observation. A zonal level trip generation model attempts to explain the between-zone variations in trip making as a function of zonal characteristics, while ignoring the within-zone (between-household) variations in trip making.

Most recent trip generation modeling efforts have been undertaken at the household level. When trip generation analysis is conducted at the household level, each household is treated as a separate observation. Hence, the analysis attempts to explain the between-household variations in trip making, as a function of household characteristics. A household-level least squares regression trip generation model is given by:

$$t = b_0 + b_1X_1 + b_2X_2 + \ldots + b_KX_K \quad (4)$$

where t is the number of home-based trips per household per day; $X_1, X_2, \ldots, X_K$ are the explanatory variables (number of people in the household, household income, and so on); and $b_0, b_1, \ldots, b_K$ are the model parameters.

A household-level category analysis trip generation model is developed by classifying each household into one of a set of mutually exclusive classes, based upon the explanatory variables used in the model. For example, we might have the following household categories: 0 cars, 1 person; 1 car, 1 person; 0 cars, 2 persons; and so on. The model consists of representing each class by the mean trip-making rate of the sample households in that class. Thus, we compute

$$\bar{t}_c = \frac{1}{N_c} \sum_{i=1}^{N_c} t_i \quad (5)$$

where $\bar{t}_c$ is the mean trip-making rate in category c; for example, the average number of trips per household per day, for 2-person, 1-car households; t_i is the measure of trip-making for household i of category c; and N_c is the number of households in category c. The appeal of the category analysis model is its intuitive simplicity. Sample trip-generation models are presented in chapters 5 and 8.

Trip Distribution.

A trip-distribution model links trip ends (that is, origins and destinations). That is, a trip-distribution model predicts how many of the trips originating in Zone i will terminate in each of Zones $1, 2, \ldots, J$, and how many of the trips terminating in Zone j originate in each of Zones $1, 2, \ldots, I$. In other words, trip distribution is concerned with the question "Where do the trips go?"

A variety of trip-distribution models have been proposed. The discussion here will focus on the most commonly applied trip distribution model, the gravity model. The interaction (number of trips) from i to $j(I_{ij})$ is considered to be dependent on (1) the number of trips leaving i, (2) the attractiveness of j, and (3) the difficulty of traveling from i to j. This idea may be written in equation form as follows:

$$I_{ij} = f(O_i, A_j, d_{ij}) \tag{6}$$

where I_{ij} is the interaction (number of trips) from i to j; O_i is the number of trips produced by zone i (the origin zone); A_j is a measure of the attractiveness of zone j (the destination zone); d_{ij} is a measure of the spatial separation of zones i and j; and f denotes some function.

We expect I_{ij} to be directly proportional to O_i and A_j and inversely related to d_{ij}. The most commonly used form of the gravity model reflects these relationships, as shown by the following equation:

$$I_{ij} = O_i \frac{A_j f(d_{ij})}{\sum_k A_k f(d_{ik})} \tag{7}$$

where $f(d_{ij})$ is known as the friction factor; and f is some decreasing function of d_{ij}, known as the friction factor function. The measure of spatial separation (d_{ij}) is generally either travel time, travel cost, or a combination of time and cost.

The calibration of a gravity model involves determining the friction factor function. This procedure is described in chapters 5 and 6.

Modal Split (Mode Choice)

Modal split (or mode choice) is concerned with predicting the number of trips from each origin to each destination which will use each mode of transportation. Thus, the objective of modal split analysis is the prediction of t_{ijm}, the number of trips from i to j by mode m, given a prediction of I_{ij}. Clearly, modal split has considerable implications for transportation policy, particularly in large metropolitan areas. For example, the decision as to whether to invest in a new subway system requires predictions obtained from the modal-split phase of the travel forecasting model. Similarly, the decision of whether to implement an exclusive high-occupancy-vehicle lane (for buses and/or car pools) should be informed by predictions of the number of people who will switch to the high-occupancy vehicles.

Two basic model types have been used to predict the distribution of trips by mode of travel: modal split and mode choice models. The former term generally refers to aggregate models and the latter to disaggregate model forms. In general terms, the distribution of trips by mode is expected to be dependent upon the transportation system characteristics (**T**) and the characteristics of the users (**U**). That is, a modal split model may be expressed as follows:

$$P_{ijm} = f(\mathbf{T}_{ij}, \mathbf{U}_i) \tag{8}$$

where P_{ijm} is the proportion of travelers from i to j using mode m; $\mathbf{T}_{ij}$ is a description of the relative performance of the alternative modes of travel from i to j; $\mathbf{U}_i$ is a description of the characteristics of the travelers in zone i; and f is some function. The performance characteristics of the alternative modes ($\mathbf{T}_{ij}$) include such variables as the relative speed of travel and the relative cost of travel between zones i and j. The users' characteristics ($\mathbf{U}_i$) include such variables as average household income.

Mode choice models are concerned with the travel behavior of individuals. They attempt to predict the probability that individual n will use mode m (for a specific trip) as a function of the individual's characteristics and the attributes of the alternative modes for making that trip. These models may be represented mathematically as follows:

$$p_{nm} = f(\mathbf{X}_K, \mathbf{S}_n) \tag{9}$$

where p_{nm} is the probability that individual n will choose mode m for a specific trip; $\mathbf{X}_K$ is a description of the attributes of all modes (K) available for the specific trip (including mode m); $\mathbf{S}_n$ is a description of the characteristics of individual n; and f is some function.

The attributes of the alternative modes typically included in such models are the travel time and travel cost by each mode for the specific trip. The user's characteristics typically incorporated in such models include the availability of an automobile and household income.

The most commonly used form for the mode choice model is termed a multinomial logit model. The general form of this model is given by

$$p_{nm} = \frac{e^{f(\mathbf{X}_m, \mathbf{S}_n)}}{\sum_{k=1}^{K} e^{f(\mathbf{X}_k, \mathbf{S}_n)}} \quad (10)$$

where $\mathbf{X}_m$ is a description of the attributes of mode m.

The function f in equation (10) represents the utility (or attractiveness) of mode m (with attributes $\mathbf{X}_m$) to individual n (with characteristics $\mathbf{S}_n$). Thus, the multinomial logit model essentially says that the probability that individual n chooses mode m for a particular trip is a function of the utility of mode m to individual n relative to the utilities of the other modes available to individual n for this trip. You will encounter further discussion of this model in chapters 8 and 9.

The development and application of individual choice models such as the multinomial logit model represents the major advance in travel demand modeling during the 1970s. Many of the applications of these choice models have been in the context of mode choice, primarily for the work trip. The use of disaggregate choice models in the context of mode choice modeling is accepted practice today. In addition to the multinomial logit model, a number of other random utility models have been developed and applied, including the multinomial probit model (Daganzo, 1979). A comprehensive text on the multinomial logit and related models (such as the nested logit model), has been published recently (Ben-Akiva & Lerman, 1985).

Trip Assignment (or Route Choice)

The final step in the conventional four-step model sequence is generally referred to as trip (or traffic) assignment (or route choice). The objective of this phase is to predict the number of the trips from each origin i to each destination j by each mode m that use each route from i to j by mode m. Thus, for example, if there are three different routes by automobile from origin i to destination j, the network assignment stage is concerned with predicting how many of the automobiles traveling from i to j will use each of the three routes. The assignment of traffic to the various routes between all origin–destination pairs results in an estimate of the quantity of traffic on each link of a specified transportation network, because each route between a given origin–destination pair comprises a specific set of links. The link volume estimates, in turn, are used to predict the impacts of the particular transportation system alternative being tested.

The proportion of vehicles using each route between a particular origin–destination pair depends upon a number of attributes of the alternative routes, including travel time, travel distance, number of stops or traffic signals, aesthetic appeal, safety, and road signs (Ben-Akiva, Bergman, Daly, & Ramaswamy, 1984). Travel time is the attribute more commonly considered in network assignment models. Now, the travel time on any route is the sum of the travel times on the links that comprise that route, while the travel time on any link is, in general, dependent upon the volume of traffic on that link. In particular, the travel time on any link of the highway network increases nonlinearly with the quantity of traffic using that link in some time period. Therefore, the assignment of traffic to the different routes between a given origin–destination pair depends upon the link travel times, which in turn depend on the link volumes. The latter, however, are a function of the routes chosen. Furthermore, the traffic using any link in the network consists, in general, of traffic traveling from various origins to various destinations. Thus, the trip assignment problem is an extremely complex equilibrium-type problem.

Early trip assignment techniques were based on extreme simplifications of the problem. For example, in some techniques link travel times were assumed to be independent of link volumes. In the past decade, considerable progress has been made in the development of sophisticated trip assignment techniques. A comprehensive text on trip assignment methods, including modern techniques employing mathematical programming formulations, was published recently (Sheffi, 1984).

Impact Prediction Models

The UTMS predicts the quantity and quality (in terms of travel time) of flow on the links of a specified transportation network, given the activity system as an input. Assessment of the consequences of alternative transportation system options, however, requires estimates of a broad range of impacts. These impacts include construction and operating costs, energy consumption, air quality, noise levels, and accident rates. In general, the models used to predict these impacts require as input the quantity and quality of traffic flow on the links of the transportation network. In addition, estimates are required of the energy, labor, land, and materials consumed in providing and operating a particular transportation system.

The energy impacts of alternative transportation system options became a concern only relatively recently, and the techniques used to predict energy impacts are still very much under development. The approach most commonly used to predict energy impacts is to estimate the change in the number of vehicle-miles of travel and multiply this change by a factor which reflects the average fuel consumption per vehicle-mile. The latter is often disaggregated by vehicle type and model year.

Air quality impact models include emission models (U.S. Environmental Protection Agency, 1981) and more complex dispersion models (Benson, 1979; U.S. Department of Transportation, 1976). Many of these models are computerized, but graphical solution procedures also exist for rough calculations. The noise impacts of transportation systems are difficult to predict, although a number of methods are available. Again, both manual techniques and computer models are available (Bowlby, 1980; Bowlby, Higgins, & Reagan, 1982; Kugler, Commins, & Galloway, 1976). Some of these environmental impact models are described in chapter 14.

POSTANALYSIS PHASE

The output of the technical analysis phase is predictions of the impacts of alternative plans and policies. The purpose of these predictions is to inform the decision-making process. The postanalysis phase of the urban transportation planning process includes evaluation of the impacts of the alternatives analyzed, selection of the alternative to be implemented, programming, budgeting and implementation of the chosen alternative, and monitoring of system performance. Each of these aspects is addressed below.

Evaluation of Alternatives

The output from the technical analysis phase that we have just described answers the question, "What will happen if the transportation system is changed/expanded in a particular manner?" In particular, the travel forecasting model is used to predict the flows (quantity) and travel times (quality) on the various links of a set of specified transportation systems. The flows and travel times, in turn, may be used to predict other impacts, including the monetary and other resources (such as land) consumed, energy consumption, noise levels, air quality levels, accident rates, and so on. Thus, for each alternative transportation system, consisting of physical elements and operating policies, the technical analysis phase produces a set of predicted impacts. During the evaluation stage of the urban transportation planning process the impacts of each alternative are summarized and compared.

Historically, evaluation included selection of the "best" alternative, which is a straightforward task only if all the impacts can be combined into a single numerical index for each alternative. Many early urban transportation studies reduced the positive and negative impacts (benefits and costs) to a common measurement scale (dollars) and thus derived a single numerical index which allowed for easy selection of the best alternative. This approach is commonly referred to as "economic evaluation." Two alternative approaches are "goal achievement" and "cost effectiveness." These three approaches are introduced briefly below.

Economic Evaluation

The overall benefit and cost of each alternative can be determined on the assumption that the benefits and costs of each alternative may be expressed in dollars. The time stream of costs and benefits are considered because of the time value of money. That is, because a dollar bill will be worth less next year than it is today, the costs and benefits in each year are discounted to obtain the present value of the costs and benefits of each alternative.

Two methods are commonly used to compare alternatives based on the present values of their costs and benefits: the net present value method and the benefit/cost ratio method. Wohl and Hendrickson (1984) note that if the latter method is applied properly, both methods lead to selection of the same alternative as best, in the economic sense.

Both the net present value and benefit/cost ratio methods require the analyst to select the discount rate for determining the present values of future costs and benefits. This is a very difficult task, and the choice of the discount rate has major implications for the results. The higher the discount rate, the less onerous are costs incurred well in the future. That is, a higher discount rate favors alternatives which have low capital costs and high operating costs relative to alternatives which incur substantial capital costs early in the analysis period and which have low operating costs. Thus, the choice of the discount rate can have a major influence on the conclusions drawn from an economic analysis.

The internal rate of return method does not require prior selection of the discount rate. Rather, an iterative procedure is used to determine the rate of return for each project for which the present value of benefits equals the present value of the costs. That is, the method determines the rate of return yielded by each alternative. The alternative with the highest internal rate of return is selected for implementation, assuming that this rate of return exceeds some minimum acceptable value.

A major problem associated with all economic-based evaluation methods is the requirement that all costs and benefits be expressed in monetary terms. Many of the costs and benefits associated with a transportation system change cannot be expressed satisfactorily in monetary terms. For example, what is the monetary value of one less fatality or a reduction in average noise levels of 10 dB? As a result of such difficulties, many early urban transportation studies ignored these impacts in evaluating the alternatives. Thus, the analyses generally included only the capital and operating costs of each alternative and the user benefits in the form of reduced travel time. Travel time reductions were converted to dollar terms using an assumed or inferred value of time.

The minimal attention given in many early studies to environmental and social impacts is an important consideration in understanding the so-called freeway revolt that began in San Francisco in the 1960s. Basically, citizens were telling transportation planners that factors other than user travel time and capital and operating costs were important community concerns. Below, we introduce alternative approaches to evaluation.

Noneconomic Evaluation

The common feature of non–economic-based evaluation procedures is that the impacts may be expressed in any units. Furthermore, the impact measures are explicitly related to the goals and objectives of the community. Thus, each alternative is described by a matrix like that depicted in Figure 3-5. The methods differ primarily in whether or not

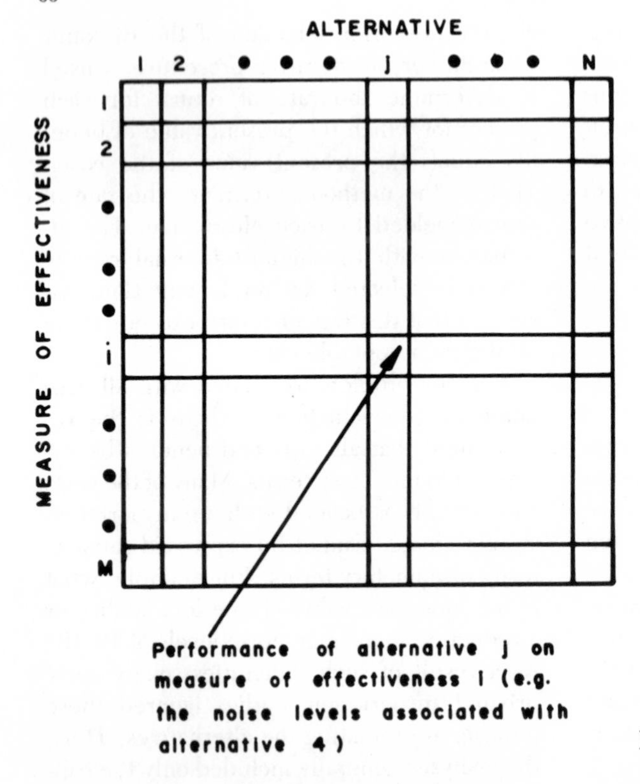

Figure 3-5. An impact matrix.

the impacts of each alternative are combined to produce a single numerical index.

The goal-achievement method (Steger & Stuart, 1976) uses the information contained in the impact matrix (Figure 3-5) to derive a single numerical index for each alternative, where the index is a measure of the degree to which an alternative achieves the goals formulated for the study area. Each objective is assigned a weight representing its perceived relative importance. Each alternative is scored according to how well it meets each objective relative to the performance of the best alternative on each objective. A single numerical index is obtained for each alternative by summation of the weighted scores on each objective. Selection of one alternative out of the set is a relatively simple task once the alternatives are ranked according to their scores on the single numerical index which describes the weighted degree of objective attainment. The obvious difficulty introduced by such a method is the subjective nature of the weighting scheme. Wachs and Schofer (1969) note that the assignment of weights to the various objectives is equivalent to the assignment of dollar values to the various nonmonetary impacts of a change in the transportation system.

Cost effectiveness (Thomas & Schofer, 1970) is an approach in which the analyst organizes information concerning the impacts of the alternatives and provides this information to the decision maker(s) and the community. The planner does not attempt to reduce all the information in the impact matrix to a single numerical index; rather, the decision makers must make the trade-offs necessary to reach a decision. The performance of each alternative on each measure of effectiveness may be compared with the cost of the alternative by means of a cost-effectiveness ratio. This is an index of the degree of objective attainment per dollar, for each measure of effectiveness. The cost-effectiveness approach recognizes that urban transportation planning is a political decision-making process, and it implies a particular viewpoint concerning the role of the technical team in urban transportation planning (see Figure 3-1).

The major advantage of the cost-effectiveness approach is that those individuals accountable to the public are responsible for making the trade-offs necessary in selecting a preferred alternative. Furthermore, if the performance of each alternative on each measure of effectiveness is specified according to socioeconomic groups, the decision makers can see who gains what and who loses what. Of course, it is difficult to integrate such information and make a decision, but many consider this to be the proper responsibility of elected public officials. The planner, however, is responsible for presenting the information in a manner that facilitates decision making, particularly because the predicted impacts are only a part of the information provided to decision makers.

Decision Making

As noted earlier, decision making historically was essentially integrated with evaluation, in the sense that if one scores each alternative on a single numerical index, such as a benefit/cost ratio, selection of the "best" alternative is a simple task. On the other hand, in the absence of a single numerical measure of the performance of each alternative, decision making is an extremely complex process appropriately performed by elected officials.

A number of models of decision making may be identified, ranging from the rational-comprehensive model to a political bargaining model (Meyer & Miller, 1984). It is unclear which model is actually employed in urban transportation decision making in the public sector, but given the social and economic implications of urban transportation decisions, the importance of political bargaining in urban transportation decision making should not be underestimated. In other words, the urban transportation planner must be aware of political realities if he or she is to have an impact on the decision-making process. (See Pill, 1978, for a good discussion of this subject.)

Implementation

Implementation of the selected transportation plan includes two considerations. First, implementation requires compliance with the regulations and requirements of the appropriate institutions, such as the preparation of an Environmental Impact Statement. Second, implementation requires consideration of the time when various actions will be taken and ways by which they will be financed. This is generally termed programming. As noted earlier, preparing for action (planning) and taking action (implementation, specifically programming) were integrated into the urban transportation planning process by the joint FHWA and UMTA regulations of 1975 that specified that the transportation plan should include a Transportation Improvement Program (TIP).

The major question to be addressed at the implementation stage is the programming of the transportation system changes. Programming has been defined as "the matching of available projects with available funds to accomplish the goals of a given period" (Transportation Research Board, 1978). One of the goals of every period is to use available resources as efficiently as possible; thus, the consideration of the time-staging of urban transportation changes is very important. However, this aspect has received little attention in the past, and quantitative techniques are not much used.

System Monitoring

The evaluation stage described above deals with preimplementation evaluation, that is, the assessment of the performance of alternative transportation systems based on predictions. In one sense, system monitoring may be considered to be postimplementation evaluation, that is, the assessment of the performance of the transportation system as the programmed changes are implemented.

System monitoring, therefore, is the ongoing collection of data to be used in tracking the performance of the urban transportation system over time. Such monitoring facilitates the identification of problems and opportunities, and thus this stage emphasizes the continuing nature of the urban transportation planning process.

System monitoring is important not only in tracking long-term changes and trends; it is also particularly useful in the case of Transportation System Management (TSM) options. As described earlier, these options are relatively low-cost, short-term alternatives that can often be modified relatively easily and cheaply. In these cases system monitoring can provide information useful in fine-tuning a TSM scheme—for example, modifying the number of passengers needed in order for a vehicle to qualify for using a high-occupancy-vehicle lane.

SUMMARY AND OUTLOOK

This chapter introduces the urban transportation planning process. In particular, the discussion focuses on the evolution of this process and on an introduction to the quantitative tools that support the planning process. The chapter attempts to show how the planning process and the modeling and analysis tools have been shaped by changes in societal concerns, particularly the environmental and civil rights movements of the 1960s. Although urban transportation planning has undergone many changes during the past 30 years, many of the basic notions developed during the 1950s have endured.

The role of technical analysis and model building in the urban transportation planning process has been highlighted here in order to place the chapters that follow in perspective. The reader should appreciate that technical analysis is only one component of the planning process and that analysis is not an end in itself. We have distinguished the analysis phase from the other elements of the planning process by pointing to the technical nature of the analysis phase. It is important, however, to note that even mathematical models of travel and related behavior implicitly employ subjective judgments and reflect particular perspectives on human behavior.

One might well ask whether urban transportation planning has been effective. Hassell (1980) addresses this question and concludes that there is little doubt about the effectiveness of urban transportation planning. However, the question cannot really be answered without knowing what would have happened in the absence of urban transportation planning studies. Would the same decisions have been made, or would the decisions have been better or worse? What is probably more important, however, is addressing the question "How can the urban transportation planning process be effective in the future?"

To say that we live in a world of change is a well-worn cliché. However, many changes that are currently underway, or that can be anticipated, will have an impact on urban travel needs and desires and thus influence urban transportation planning in the coming years. In addition, the sociopolitical environment within which urban transportation planning exists is changing and thus shaping the planning process.

A major influence on changes in urban travel behavior is the changes taking place in the American household, including decreasing household size, more single-adult households, and increasing proportions of women in the work force (Spielberg, Weiner, & Ernst, 1981). In addition, the increasing average age of the American population is an important influence on travel needs and desires, particularly when one considers that tomorrow's senior citizens grew up with the automobile. In addition to sociodemographic changes, future urban travel behavior will be affected by the increasing sophistication of telecommunications technology and the substitution of electronic communications for travel. The latter is encouraging employees to live even farther from work, beyond traditional suburbs.

The continued population movement from the Northeast to the South and Southwest necessitates an urban transportation planning process that is capable of dealing with very different issues. In the Northeast, the question to be addressed in the coming years may well be "which facilities/services should be repaired/reconstructed/maintained?" On the other hand, in the rapidly growing and increasingly congested cities in the South, the question is a more familiar one to transportation planners: "How should the transportation system be expanded/managed to deal with the increased travel generated by the increased population?"

The second term of the Reagan administration will no doubt ensure continued delegation of responsibility for urban transportation to state and local governments. In addition, there has been increasing participation by the private sector in the provision of urban public transportation (Orski, 1980). There is also renewed concern today with efficiency (rather than equity) as budget costs dominate development in many cases. At the same time, the trend toward openness in the planning process is being reversed, while less attention is being devoted to social, environmental, and ethical issues in transportation planning and policy (Flyvbjerg, 1984). The urban transportation planning process will need to adapt if it is to be effective in the changing environment.

References

Allen, J. G. (1985). Post-classical transportation studies. *Transportation Quarterly, 39,* 451–463.

Atherton, T. H., Ben-Akiva, M. E. (1976). Transferability and updating of disaggregate travel demand models. *Transportation Research Record, 610,* 12–18.

Ben-Akiva, M., Lerman, S. R. (1985). *Discrete choice analysis: Theory and application to predict travel demand.* Cambridge: MIT Press.

Ben-Akiva, M., Bergman, M. J., Daly, A. J., & Ramaswamy, R. (1984). Modelling inter urban route choice behavior. In J. Volmuller & R. Hamerslag (Eds.), *Proceedings of the Ninth International Symposium on Transportation and Traffic Theory,* Utrecht, The Netherlands: VNU Science Press.

Benson, P. E. (1979). *CALINE 3—A versatile dispersion model for predicting air pollutant levels near highways and arterial streets.* Sacramento: California Department of Transportation.

Birch, D., Atkinson, R., Sandstrom, S., & Stack, L. (1974). *The New Haven Laboratory: A testbed for planning.* Lexington, Mass.: Lexington Books.

Bolan, R. S. (1967). Emerging views of planning. *Journal of the American Institute of Planners, 33,* 233–246.

Bowlby, W. (1980). *SNAP 1.1: A revised program and user's manual for the FHWA level 1 highway traffic noise prediction computer program.* Washington, D.C.: Federal Highway Administration, U.S. Department of Transportation.

Bowlby, W., Higgins, J., & Reagan, J. (1982). *Noise barrier cost reduction procedure STAMINA2.0/OMTIMA: User's manual.* Washington, D.C.: Federal Highway Administration, U.S. Department of Transportation.

Colcord, F. (1979). Urban transportation and political ideology: Sweden and the United States. In A. Altshuler (Ed.), *Current issues in transportation policy.* Lexington, Mass.: Lexington Books.

Daganzo, C. F. (1979). *Multinomial probit.* New York: Academic Press.

deBettencourt, J. S., Mandell, M. B., Polzin, S. E., Sauter, V. L., & Schofer, J. L. (1982). Making planning more responsive to its users: The concept of metaplanning. *Environment and Planning A, 14,* 311–322.

Flyvbjerg, B. (1984). The open format and citizen participation in transportation planning. *Transportation Research Record, 991,* 15–22.

Gakenheimer, R. (1976). *Transportation planning as response to controversy: The Boston case.* Cambridge: MIT Press.

Hansen, W. G., & Lockwood, S. C. (1976). Metropolitan transportation planning: Process reform. *Transportation Research Record, 582,* 1–13.

Hassell, J. S. (1980). "How effective has urban transportation planning been?" *Traffic Quarterly, 34,* 5–20.

Herald, W. S. (1980). "Generating alternatives for alternatives analysis." *Transportation Research Record, 751,* pp. 17–27.

Hill, D. M., Brand, D., & Hansen, W. B. (1966). Prototype development of statistical land use prediction model for the Greater Boston Region. *Highway Research Record, 114,* 51–70.

Hillegass, T. (1969). Urban transportation planning—A question of emphasis. *Traffic Engineering,* June.

Ingram, G. K., Kain, J. F., & Ginn, J. R. (1972). *The Detroit Prototype of the NBER Simulation Model.* New York: National Bureau of Economic Research.

Koppelman, F. S., & Wilmot, C. G. (1982). Identification of conditions for effective transferability of disaggregate choice models, Report No. UMTA-429-I. Evanston, Ill.: Transportation Center, Northwestern University.

Kugler, B. A., Commins, D. E., & Galloway, W. J. (1976). Highway noise: A design for prediction and control. National Cooperative Highway Research Program, Report 174, Washington, D.C.: Transportation Research Board.

Lerman, S., & Manski, C. (1979). Sample design for discrete choice analysis of travel behavior: The state-of-the-art. *Transportation Research A, 13,* 29–44.

Lowry, I. S. (1968). Seven models of urban development: A structural comparison, in *Highway Research Board Special Report, 97,* 151–174.

Manheim, M. L. (1979). *Fundamentals of transportation system analysis.* Cambridge: MIT Press.

Meyer, M. D., & Miller, E. J. (1984). *Urban transportation planning: A decision-oriented approach.* New York: McGraw-Hill.

Oi, W. Y., & Shuldiner, P. W. (1962). *An analysis of urban travel demands.* Evanston, Ill.: Northwestern University Press.

Orski, C. K. (1980). Urban transportation: The role of major actors. *Traffic Quarterly, 34,* 33–44.

Pill, J. (1978). *Planning and politics: The Metro Toronto transportation plan review.* Cambridge: MIT Press.

Sheffi, Y. (1985). *Urban transportation networks: Equilibrium analysis with mathematical programming methods.* Englewood Cliffs, N.J.: Prentice-Hall.

Spielberg, F., Weiner, E., & Ernst, U. (1981). "The shape of the 1980s: Demographic, economic and travel characteristics." *Transportation Research Record, 807,* 27–34.

Steger, C., & Stuart, R. C. (1976). Goals achievement as an integrating device for technical and citizens input into urban freeway location and design. *Man-Environment Systems, 6.*

Stopher, P. R. (1983). Data needs and data collection—State of the practice. In *Travel Analysis Methods for the 1960s,* Transportation Research Board Special Report 201, Washington, D.C.

Stopher, P. R., & Meyburg, A. H. (1975). *Urban transportation modeling and planning.* Lexington, Mass.: Lexington Books.

Thomas, E. N., & Schofer, J. L. (1970). Strategies for the Evaluation of Alternative Transportation Plans. *National Cooperative Highway Research Program Report 96,* Highway Research Board. Washington, D.C.

Transportation Research Board. (1978). National Cooperative Highway Research Program Synthesis of Highway Practice 48. Washington, D.C.: Transportation Research Board.

U.S. Department of Transportation. (1975). Federal Highway Administration, Guidelines for Urban Transportation Planning, *Federal Register, 40,* No. 181, pp. 42976–42984.

U.S. Department of Transportation. (1976). Federal Highway Administration, *Fundamentals of Air Quality.* Washington, D.C.

U.S. Department of Transportation. (1977). Urban Mass Transportation Administration. *Urban Transportation Planning (UTPS) Reference Manual,* Washington, D.C.

U.S. Department of Transportation. (1983). Federal Highway Administration and Urban Mass Transportation Administration, Urban transportation planning, *Federal Register, 48,* No. 127, pp. 30332–30343.

U.S. Environmental Protection Agency. (1981). *User's Guide to Mobile 2,* Washington, D.C.

Wachs, M., & Schofer, J. L. (1969). "Abstract values and concrete highways." *Traffic Quarterly,* January, 133–145.

Weiner, E. (1983). *Urban transportation planning in the United States: An historical overview.* Washington, D.C.: U.S. Department of Transportation.

Weiner, E. (1984). Devolution of the federal role in urban transportation. *Journal of Advanced Transportation, 18,* No. 2, 113–124.

Wohl, M., & Hendrickson, C. (1984). *Transportation investment pricing principles: An introduction for engineers, planners and economists.* New York: John Wiley.

SECTION II

UNDERSTANDING MOVEMENT PATTERNS IN CITIES

4 AGGREGATE CHARACTERISTICS OF URBAN TRAVEL

GERALD BARBER
University of Victoria

Urban transportation consists of the movements of both people and goods between various origins and destinations. These movements can take place at various times of the day, use different modes of transportation, and serve many needs. At the level of the *individual traveler* or goods shipment, urban transportation can therefore be conceptualized as a *trip*, beginning at some origin and ending at some destination, where some activity is performed. At the level of the *metropolitan area*, urban transportation is expressed as the aggregate of thousands or even millions of these individual trips. The net effect of these individual trip decisions is the creation of vehicle or passenger trips over the transportation system of the city. These are the *travel flows* that constitute the subject of this chapter.

To understand the nature of transportation in cities, we must have a basic understanding of the characteristics of these urban travel flows. The analytical framework used in conventional urban transportation planning is based on the relationship between these flows and the transportation system. Generating solutions to the transportation problems of North American cities depends on our ability to understand this complex relationship thoroughly. This chapter describes the regularities that characterize the pattern of travel flows in North American cities. Before documenting these patterns, it is first necessary to emphasize that not all these patterns will be exhibited in every North American city for two major reasons. First, there is the tremendous diversity in the size, functions, and structure of these cities. Among the city characteristics with important implications for the pattern of urban travel flows are the absolute size of the city, its principal functions, the geographical setting, and particularly the proportion of the growth of the city that occurred in each of the major transportation technology eras (see chapter 2). For example, we would expect that there would be significant differences between the patterns of travel flow in an older city of northeastern United States and a rapidly growing Sunbelt city in southwestern United States. The second major factor leading to variations in these travel flow patterns is the strategy used by local planners in developing their transportation systems. Travel flows in cities whose planners have attempted to maximize the use of public transit and/or discourage the use of the automobile are likely to be much different from the flows in a city dominated by private automobile transportation.

There are five characteristics of urban travel flow that deserve detailed attention—trip purpose; the temporal distribution of travel flows; the selection of travel mode; trip lengths in terms of time, cost, or distance; and spatial patterns. Information concerning these five characteristics relies on three principal sources. The most useful sources are undoubtedly the reports generated by the large-scale urban transportation studies carried out in the 1955–1970 period. Data on household travel behavior were collected using roadside, home, and/or telephone interviews of a representative sample of city residents. Characteristics of respondents' trips made on a single day formed an important component of the questions used in these interviews. The origin, destination, purpose, duration, and mode of all trips made by household members was recorded. This information was used to complement physical inventories of both the transportation system (road capacities, traffic volumes, service characteristics, and so on) and the land use pattern (type, intensity, and location). Many of the procedures developed in the earliest studies, for example, Pittsburgh (Pittsburgh Area Transportation Study—PATS), Chicago (Chicago Area Transportation Study—CATS), and Toronto (Metropolitan Toronto and Region Transportation Study—MTARTS), were followed in virtually all subsequent studies undertaken in most major urban areas of North America. Because these studies are so time-consuming and expensive, they have not been duplicated in recent years. Instead, relevant information for specific problems or issues is now collected more or less on an ad hoc basis by various transportation planning agencies. Many of these comprehensive land use/transportation planning agencies emerged from these one-time urban transportation studies.

There are two other sources of information. Some material can be obtained from the U.S. Bureau of the Census and Statistics Canada. For example, the U.S. Census Annual Housing Survey (1976) and the U.S. Nationwide Personal Transportation Survey (1970) contain useful information. A third set of sources of information is the research reports generated by government agencies such as Transport Canada and the U.S. Department of Transportation. The Urban Mass Transit Administration and the Federal Highway Administration, in particular, have produced a number of important reports on travel behavior. Information drawn from each of these primary sources is used in this chapter.

TRIP PURPOSE

It is common to classify individual trips made in a city into one of several categories based on the *purpose* of the trip, characteristics of both the trip origin and destination, or just the trip destination. Though the actual categories may vary from study to study and thus city to city, most classifications include the following common trip types:

- *Work trips* are made to a person's place of employment. The place of employment may be a manufacturing plant, a retail store, an office, or a public or private institution such as a hospital or university.
- Shopping trips are trips to any retail outlet, regardless of the size of the store (or shopping center) and whether or not a purchase was actually made.
- Social trips are made to social activities such as parties or simply visiting friends.
- Recreation trips are trips to recreation, entertainment, or cultural facilities to attend such activities as concerts, sporting events, and civic meetings. Sometimes this trip purpose category is combined with social trips in an aggregate category.
- School trips includes trips made by students at any level to an institution of learning. Typically, students from elementary school through university are grouped in this category.

- Business trips usually include trips made from the place of employment to other destinations in the city.
- Home trips includes any trip ending at home. This category must be included as a trip purpose because trips are always defined as one-way movements.

When these categories are used in models to predict travel behavior at the aggregate level, it is quite common to distinguish the role that *home* plays in the trip. For example, some transportation studies group these purposes into three categories: home-based work trips, home-based other trips, and non/home-based trips. The level of aggregation used depends very much on the purpose of the model being built. Usually, finer disaggregations are possible in very large studies of large cities and coarser groupings in more modest studies of smaller cities.

A typical summary of travel flows by trip purpose for a metropolitan area is shown in Table 4-1. Using data from the Chicago Area Transporation Study (1976), this table classifies trips by seven purposes. The single most important destination is the home trip, accounting for approximately 40% of the daily trips in Chicago. Assuming that people begin and end the day at home, often termed the assumption of symmetry, the home is one end of about 80% of all trips made each day in the city. To appreciate the absolute numbers of trips involved, we might note that over 15 million trips are made in Chicago on an average day! The implication of this is that if we control the location of residential land use in the city, we are expressing some control over a large portion of urban travel flows.

Table 4-1. Trip purpose categorization, Chicago, 1976

Trip purpose	Total trips	Percentage
Work	2,388,378	15.0
Shopping	2,092,875	13.1
Social/recreation	2,134,461	13.4
School	392,813	2.4
Business (work)	503,558	3.2
Business (personal)	1,101,375	6.9
Home	6,772,736	42.8
Other	518,128	3.2
Total	15,904,324	100.0

Note. From *Transportation Improvement Program FY76–FY80* by Chicago Area Transportation Study, 1976, Chicago.

Three other trip purposes, work, shop, and social/recreational, account for about 12–15% of the travel flow on a typical day. The remaining trip purposes account for smaller, but significant, proportions of daily urban passenger travel. Another way of illustrating the importance of home-based trips and the distribution of trips by purpose is shown in Figure 4-1. The proportion of trips by purpose is roughly the same as in Chicago, though in this case trip purpose is defined in terms of the origin and the destination of the trip. As we shall see throughout this chapter, the stratification of aggregate travel flows by trip purpose often helps to identify the underlying patterns in mode choice, trip timing, and trip length.

Urban *goods* movements reflect the extreme diversity in the requirements of the many establishments of metropolitan areas. These movements can be at least partially understood by examining the movement of trucks. Table 4-2 indicates the percentage

Table 4-2. Percentage distribution of purposes for urban truck trips in 11 urban areas

Trip purpose at trip destination		Percentage of total daily trips
Home base		19.3
Personal use		9.1
All pick-up and delivery		41.1
Retail	17.3	
Wholesale	16.3	
Merchandise	7.5	
Mail and express		6.1
Construction		4.9
Maintenance and repair		8.0
Business use		7.2
Other		4.3
Total		100.0

Note. From *Transportation and Traffic Engineering Handbook* (1982), 174. Englewood Cliffs, N.J.: Prentice-Hall, Inc. Copyright © 1982 by the Institute of Transportation Engineers. Reprinted by permission.

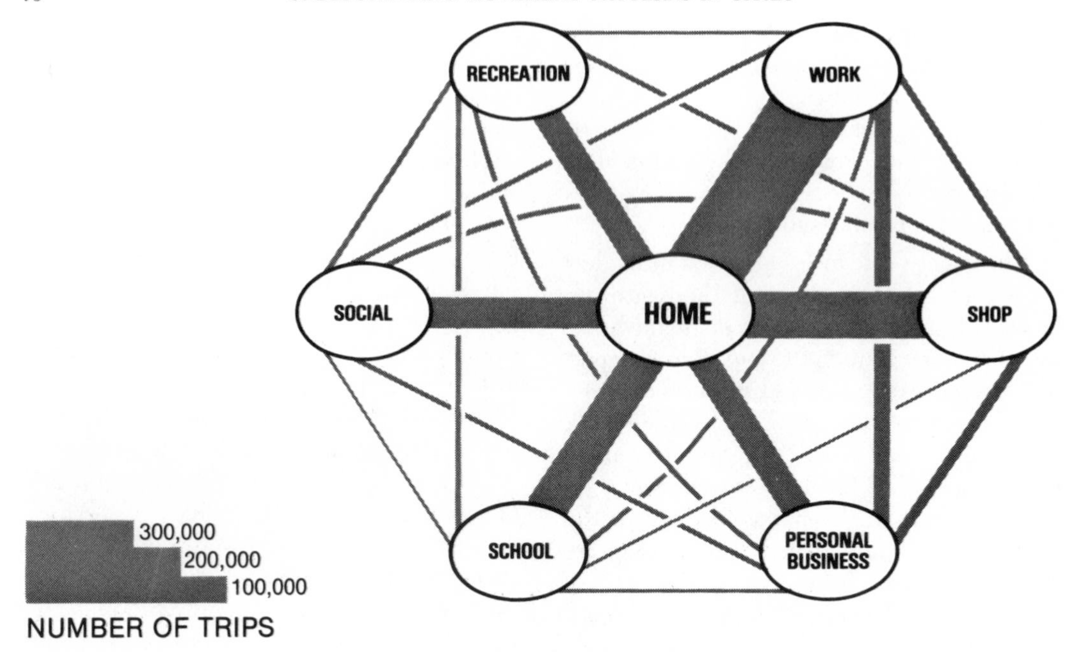

Figure 4-1. Trip purposes in a typical North American city.

distribution of trip purposes for urban truck movements in eleven U.S. cities. By far the most important purposes are pick-up and delivery operations. Together, these purposes account for over 40% of all truck trips. Other important purposes are mail and express services, construction, and maintenance and repair services. As we shall see, goods movements by truck have a markedly different temporal distribution than individual passenger trips.

THE TEMPORAL DISTRIBUTION OF TRIPS

If the travel flows in a city on a typical day were to be evenly distributed throughout the day (and night), the problems associated with urban traffic congestion would certainly be less severe than is characteristic of most North American urban areas. Unfortunately, this is not the case: the morning and evening rush hours are familiar to most North Americans. The principal, but by no means the only, reason for this double-peaked distribution of trips over the day is the morning journey to work and evening journey home. Most employers require their employees to begin work between 7 and 9 A.M. and, after an 8-hour shift, to return home between 4 and 6 P.M. This phenomenon is clearly illustrated in the generalized distribution of trips of Figure 4-2.

First, if we examine the distribution of *total* daily trips, we can easily see the double-peaking phenomenon. There tend to be two peak demands, on the average about twice the nonpeak demand. The morning peak is characteristically more pronounced; there is a much more gradual buildup to the evening peak, and a slower decline. If we stratify these trips into different trip purpose categories, it is possible to illustrate differential temporal peaking. For example, the journey to work and return trips are extremely peaked, with peak demand three to four times nonpeak trips. Since these work trips tend to be localized in space as well, particularly into and out of the central business districts and to be concentrated in certain modes of transportation, it is easy to see why we have traffic congestion. For shopping trips, there is a very different pattern. There are relatively few such trips until just after the morning

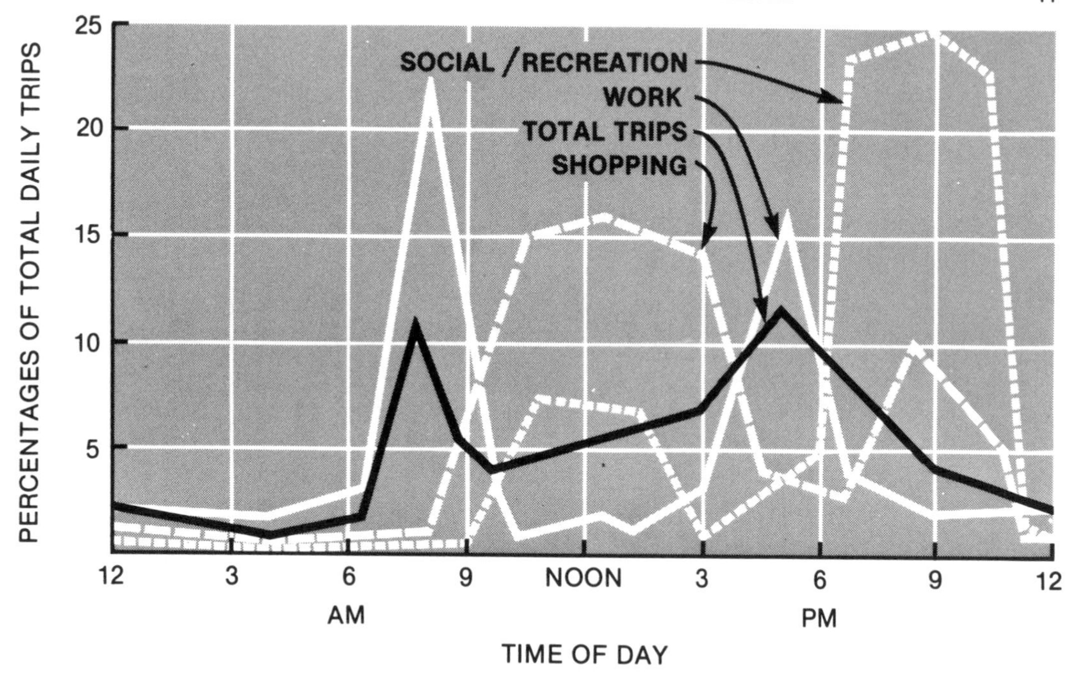

Figure 4-2. The temporal distribution of trips.

rush hour, a significant share between 10 A.M. and 4 P.M. with a slight noon-hour peak, and a smaller early evening peak after the dinner hour. Nevertheless, shopping trips represent an important part of rush hour trips, particularly in the afternoon. Social and recreational trips are peaked slightly over the lunch hour, but are most pronounced in the 7 P.M.–11 P.M. period. It should also be noted that there is a positive relationship between the degree of peaking in a city and the size of the city. Peak demands are less noticeable in smaller urban areas.

The temporal distribution of truck trips and thus goods movement does not follow this pattern. These trips are constrained by the operating hours of the businesses that receive the goods or services they provide. A typical example of the distribution of truck trips by time of day is reproduced in Figure 4-3. These data describe the temporal distribution in Pittsburgh. The peak begins just *after* the morning rush hour, has a significant drop over the lunch hour, and then picks up again until a rapid decline just *before* the evening rush hour. Since these trips account for about 15% of the total vehicular trips in a city, it is easy to see why the double peaking of total travel is not as pronounced as the work-trip pattern.

Different parts of the city may experience peaking at slightly different times, depending upon the mix of traffic. Many business/retail arterials experience significant congestion throughout the day. Non-rush-hour parking spaces usually reduce street capacities by one lane in each direction. Coupled with truck deliveries and pedestrian traffic, this often leads to virtual continuous traffic congestion.

The peaking of demand for travel at specific times of the day leads to congestion, which is probably still the most important problem in urban transportation. Congestion of a transport facility is said to occur when the users of the facility experience delays. Very low traffic volumes do not produce congestion; only when the traffic begins to approach the capacity of a facility (or even exceed it) do traffic delays and congestion occur. The implications of the peaking of travel flows are extremely serious. Because these flows make twice the demand of flows at other times of day, we must make a deci-

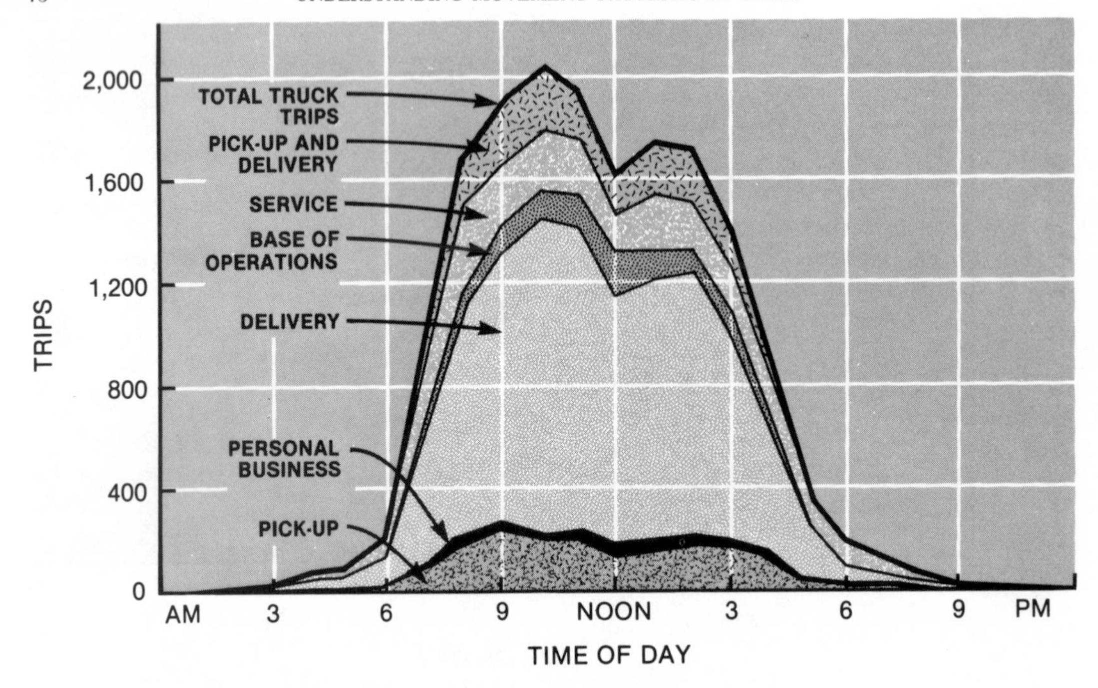

Figure 4-3. The temporal distribution of truck trips.

sion about the proper capacity to provide in the transportation system. If we provide sufficient capacity to handle only the *average* hourly traffic, there would be tremendous time delays during rush-hour periods. However, if we provide sufficient capacity to handle peak travel flows, then we will have *excess* capacity for most of the day. This is a particularly pressing problem for public transit companies, faced with peak requirements three to four times the average requirement.

In the 1950s and 1960s the relief of traffic congestion was seen as the prime goal of urban transportation planning. Most transportation planners and engineers advocated massive investments in the highway and transit networks to increase the capacity of the system and thereby meet the existing vehicular demands. Unfortunately, these investments often led to even greater demands for transportation and more congestion. Urban protests, often by inner-city neighborhood residents, gradually led to the abandonment of this policy. The adverse social and environmental impacts in disrupted neighborhoods along the path of these facilities seemed to be much more severe than anticipated. Today, these concerns are playing a much more important role in transportation planning decision making. A variety of actions are now being used to ease traffic congestion without promoting massive investments in transportation system capacity. These include (1) staggering work hours to spread demand more evenly over the day, (2) ridesharing programs and preferential treatment for vehicles occupied at or near capacity, (3) one-way street systems and other traffic engineering improvements, and (4) increased enforcement of parking violations. Policies such as these are discussed in some detail in chapter 16.

MODAL SPLIT

A third important characteristic of urban travel is the distribution of travel flows among the various travel modes available in the city. One of the key statistics is the so-called mode split between the private automobile and various public transit alternatives. Only about 10% of all passenger trips in North American cities are made by public transit, and the share of total passenger trips made by transit has been declining since the end of World

War II (see Figure 4-4). It now appears that this long-run decline is over, and the proportion of trips by transit has stabilized. However, such gross statistics can be somewhat misleading. Declines in smaller cities have been greater than those of metropolitan areas. In very large metropolitan areas the situation can be dramatically different; during peak hours, the proportion of trips into and out of the central business district can be very high. In cities with highly developed, integrated bus and rail rapid transit systems such as Chicago, Toronto, and New York, the percentage of all trips by transit into and out of the central business district can be in the 70–90% range.

The split of total trips among the various travel modes varies from city to city depending upon the size of the city, its growth history and spatial structure, and the transportation planning strategies employed by local authorities. Modal-split statistics from the transportation planning studies of several urban areas are indicated in Table 4-3. Smaller cities tend to have very limited public transportation systems with a transit share usually less than 5%. While larger cities *tend* to have more developed public transit systems with possible transit shares in the 15–30% range, newer, fast growing western cities such as Los Angeles and Denver have dominantly auto trips.

Modal split also shows significant variation by trip purpose. Very few social/recreation trips and only modest proportions of shopping and personal business trips are made by public transit. Table 4-4 examines a more detailed breakdown of trips for the journey to work in eight U.S. metropolitan areas. Work trips generate a greater proportion of transit trips than *all* other purposes. Note that almost one-half of the work trips in New York use public transit. Also, walking trips can account for significant numbers of work trips. Unfortunately, data on all aspects of pedestrian travel are very limited. To focus on the gross transit/auto modal split causes one to lose sight of the range of travel modes and transportation technologies used in North American cities. Trips which *combine* two or

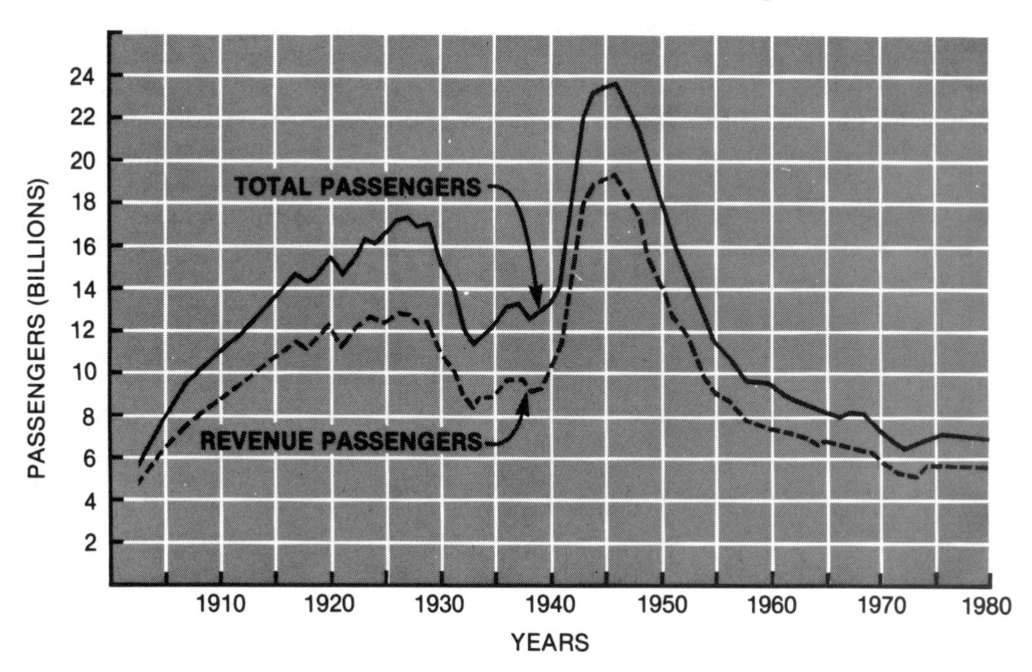

Figure 4-4. Trends in transit patronage in the United States, 1902–1976. From "The development of public transportation and the city" by George M. Smerk, 1979, in G. E. Gray and Lester A. Hoel (eds.), *Transportation: Planning Operations and Management*, Englewood Cliffs, N.J.: Prentice-Hall.

Table 4-3. Gross modal split in selected cities

City	Year	Study area population	Thousands of daily person trips	Percentage by Auto	Percentage by Transit
New York	1963	16,300,000	29,570	67	33
Los Angeles	1967	9,000,000	20,578	96	4
Chicago	1970	7,600,000	18,616	86	14
Toronto	1964	1,800,000	3,877	81	19
Denver	1971	1,100,000	3,097	98	2
Fresno	1971	295,000	857	96	4
Stockton	1967	170,000	401	96	4

Note. Adapted from *Transportation and Traffic Engineering Handbook* (Tables 10-10 and 10-11) by the Institute of Traffic Engineers, 1982, Englewood Cliffs, N.J.: Prentice-Hall, Inc. Copyright 1982 by Institute of Transportation Engineers. Reprinted by permission.

even more modes are also possible. To understand why a particular city has a certain pattern of modal use, it is extremely useful to examine the available *transportation technologies* in the city and their *service characteristics.*

The diversity of transportation technologies typical of North American cities is indicated in Table 4-5. There are many variations on the five basic technologies. Any particular city may have a more or less unique combination of these modes, given its number and the relative emphasis placed on them in that city. Of more importance are the service characteristics of these modes. Each has a particular service "window" to offer urban travelers. This window consists of a vector of different characteristics of the modes including accessibility, speed, comfort, and reliability. Two of the more important components of this vector are the ability of the mode to offer accessibility to all locations in the city and the average speed of the mode. Some modes offer door-to-door service and therefore serve all locations in an urban area; other modes are limited in their accessibility. Most forms of public transit offer only what might be termed intermediate access, since they do not cover all parts of the city all the time. Also, most forms of public transit require a certain amount of walking time for access/egress. For example, local transit buses are often routed on arterials and generally require a 5- to 10-minute walk at either end of a passenger trip. Modes such as rail rapid transit offer extremely limited access, except perhaps in certain inner-city areas of large cities.

Table 4-4. Mode of travel (%) of commuting workers in selected U.S. standard metropolitan statistical areas, 1976

City	Private auto or truck	Public transport	Walk	Other
New York	47	44	8	1
Baltimore	83	12	5	1
Houston	93	4	3	1
Raleigh	95	2	3	1
Oklahoma City	95	1	3	1
Seattle	87	8	3	1
Honolulu	83	11	4	2
Denver	87	5	5	2

Note. Figures may not add to 100% due to rounding.

Note. From *Current Population Reports, Selected Characteristics of Travel to Work in 20 Metropolitan Areas* by U.S. Bureau of the Census, 1976, Washington, D.C.: Government Printing Office.

Table 4-5. Transportation technology and travel modes for North American cities

Guideway characteristics	Human powered	Highway driver steered	Rubber tired guided, partially guided	Rail	Other
Surface street with mixed traffic	Bicycle	Automobile Truck Motorcycle Moped Paratransit Regular bus Express bus	Trolley bus	Streetcar Cable car	
Physical separation of traffic types, traffic crossings permitted	Walking (on sidewalks) Bicycle (on lanes)	Bus, car, or van pool on preferential lanes on arterial street		Light rail transit	Ferry Hydrofoil Helicopter
Complete physical separation of traffic types	Walking (pedestrian bridges) Bicycle (paths)	Bus, car, or van pool on busways	Rubber-tired rapid transit Automated guided transit Group rapid transit Personal rapid transit	Commuter rail Rapid transit	Tramway

Note. As presented in *Urban Transportation Planning: A Decision-Oriented Approach* (p. 30), by M. D. Meyer & E. J. Miller, 1984, New York: McGraw-Hill; adapted from *Urban Public Transportation: Systems and Technology* by V. Vuchic, 1981, Englewood Cliffs, N.J.: Prentice-Hall.

The second major dimension of service is the average speed offered by the mode. This is an especially important factor in individual decisions of travel mode choice. The *relative* speeds of the modes are often the deciding factor for many individuals. Though auto travel offers door-to-door service, the average speed of auto travel depends on the nature of the traffic stream. Under free-flow conditions where there is no congestion, average automobile speeds approach (or exceed) speed limits. As the level of traffic approaches network capacity, the average speed of auto travel falls to the point where it is comparable to walking and cycling. A comparison of different modes of urban transportation with these two dimensions is shown in Table 4-6.

A great deal of public policy in urban transportation has been directed toward improving the service characteristics of public transit vis-à-vis the automobile. Many of the innovations in public transport in the last two decades are specifically oriented toward reducing the modal share of the automobile. Bus lanes, for example, can be interpreted as a means of improving the average speed of buses to increase their competitiveness with the auto. Metropolitan Toronto is one of the more successful cities in North America in public transit development and planning. Ridership on the Toronto transit system has risen dramatically over the last decade (see Figure 4-5). Part of this increased ridership can be attributed to changes in the public transit system, including the introduction of monthly bus passes, service extensions, and marketing efforts. The other portion of this increase is most likely due to changes in the attractiveness of the automobile brought on by increases in the cost of gasoline and downtown parking rates. Together, these two sets of factors have clearly improved the competitive position of public transit, particularly for rush-hour trips into and out of the central business district.

It is impossible to analyze modal choice in

Table 4-6. Service characteristics of transportation modes of North American cities

Average speed	Accessibility		
	Limited	Intermediate	Door to door
Low	Ferry	City bus Trolley bus	Walking Bicycle Auto in very congested conditions
Medium	Helicopter Hydrofoil	Streetcar	Dial-a-bus Auto in mildly congested conditions
High	Rail rapid transit	Express bus Light rail	Auto under free-flow conditions

cities independent of the other major characteristics of urban travel such as trip purpose and timing. For example, Figure 4-6 illustrates the percentage of hourly trip volumes by mode of travel over the day. The impact of peaking is clearly most significant for the three public transit modes and least significant for auto passengers and drivers. Variability of peaking by trip purpose and location can also be observed. If we were to limit ourselves to peak period, public transit work trips into and out of the central business

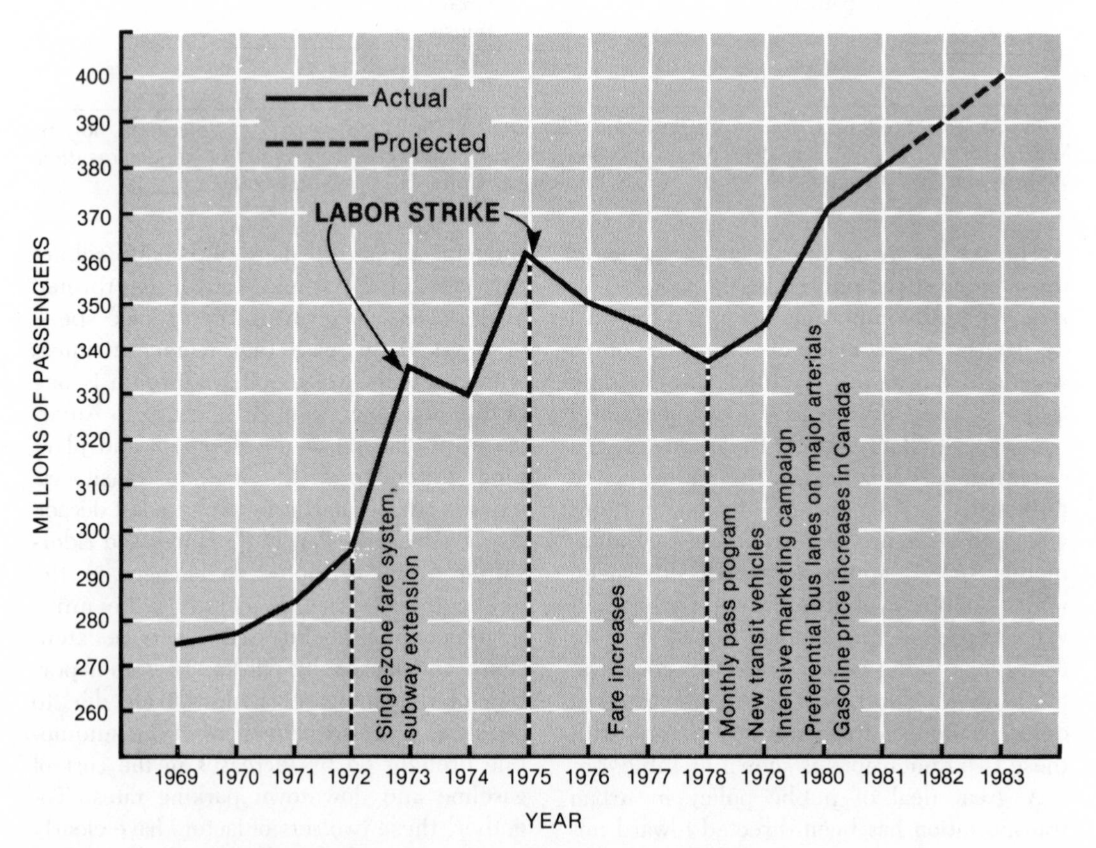

Figure 4-5. Transit ridership in Toronto, 1969–1982. Drawn for the author from data reported by the Toronto Transit Commission in *Transit in Toronto,* Toronto: Toronto Transit Commission, 1982

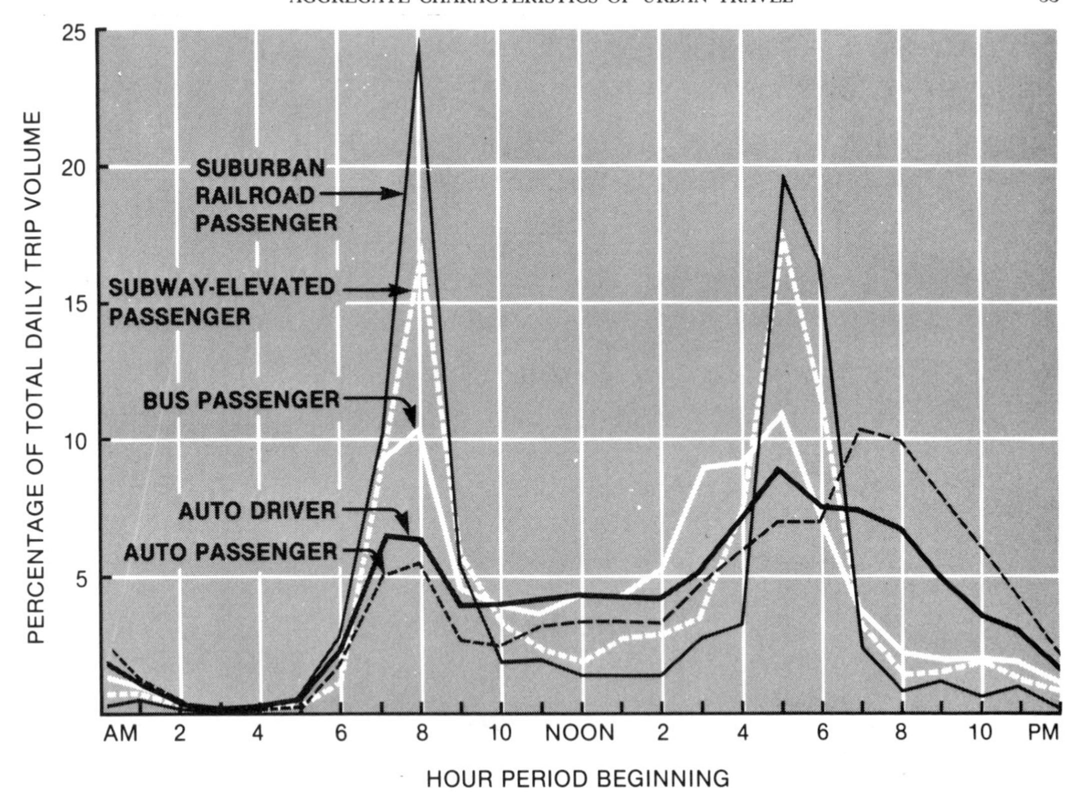

Figure 4-6. Hourly percentages of total daily trip volume by mode of travel in a large North American metropolitan area.

district of a large metropolitan area, we would notice particularly pronounced peaks.

Finally, in this section on mode use, although comprehensive information on the nature and characteristics of pedestrian travel in cities is not readily available, we should consider what is known about trips made on foot. We do know that pedestrian flows are far more localized than either automobile or transit flows. By far the largest share of these trips is made in the CBD. Even within the CBD, trips are extremely concentrated in the retail-office core areas and decline rapidly in a matter of three to four blocks. Pedestrian destinations and trips reflect two basic patterns of linkage: (1) the flows between principal transit stops, parking garages, and other transport terminals to and from places of work, and (2) the flows between stores, restaurants, and offices within the commercial core area. Trips of the first type tend to dominate flows in the morning and afternoon rush hours, whereas trips of the second type tend to be characteristic of the midday period. These two patterns are clearly illustrated in the results of a study of Boston shown in Figure 4-7.

Pedestrian walking distances reflect people's desire to minimize the travel time and general inconvenience of their trips. They vary according to city size and are highly conditioned by the locations of public transit terminals, parking garages, and major retail-office concentrations. Median distances seem to vary between 400 and 1,200 feet. The spacing of rapid transit stations controls the length of a large percentage of pedestrian flows in the largest North American cities. Very little is known about the nature of pedestrian travel in smaller cities and towns, as well as in suburban areas of the larger metropolitan areas.

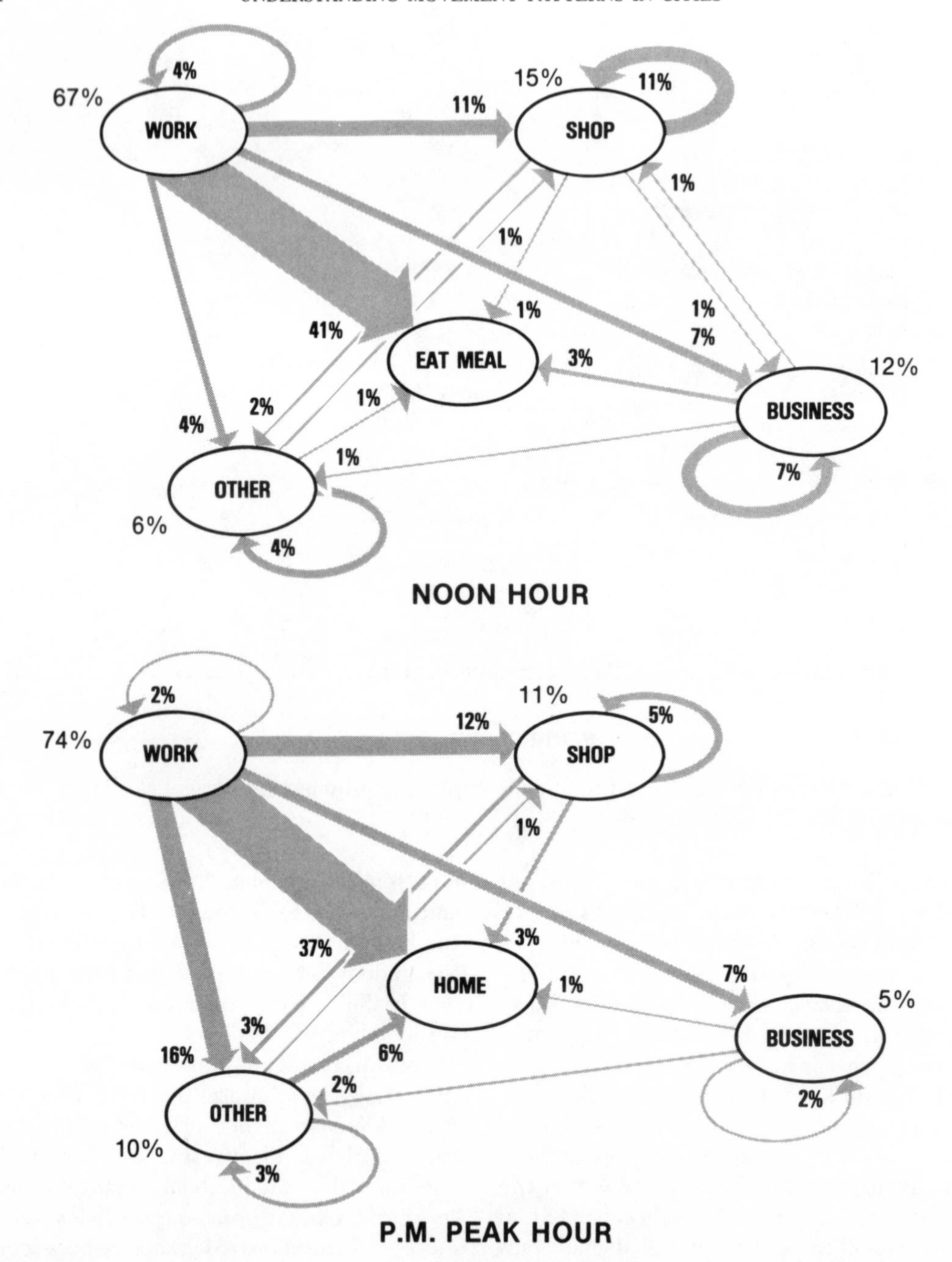

Figure 4-7. Pedestrian trip purposes within the Boston CBD. From *Traffic Circulation and Parking Plan, Central Business District Urban Renewal Area Boston, Massachusetts*, 1963, Boston: Barton-Aschman Associates.

DISTRIBUTION OF TRIP LENGTHS

Associated with any trip is a measure of desirability that we may call its *utility*. The utility of any trip depends on a number of factors including the purpose of the trip or the activity to be performed at the destination, personal characteristics of the person undertaking the trip, and especially the time, cost, or distance of the trip. Because increases in time, cost, or distance of a trip reduce its utility to a tripmaker, we say they add *disutility* to a trip. Put another way, the demand for mobility or travel in a city is a derived demand. Travel itself is only an unnecessary by-product of the desire to accomplish some activity, brought about by the spatial separation of land uses in the city. All other things being equal, shorter trips are to be preferred over longer trips. This phenomenon should be apparent in a summary of aggregate travel flows in a city. Figure 4-8 presents a summary of the reported trip lengths in Metropolitan Toronto as found in the Metropolitan Toronto and Region Transportation Study. Note the significant decline in the reported trip lengths in minutes for trips of all purposes. The average trip time is about 25 minutes, and very few trips are more than an hour in duration. These long trips may be cross-town transit trips taken at times of infrequent bus service, or simply very long automobile trips during a peak period. Again, when total trips are disaggregated by trip purpose, we can see some variance in this pattern. For example, from Figure 4-8 it can be seen that nonwork trips are more sensitive to trip time than are work trips. Work trips tend to be longer than trips for most other purposes. If the average duration of work trips is 27.8 minutes, or about one-half hour, then the average resident in a city spends about 1 hour in the journey to work and return trip home, at least in Metropolitan Toronto in the late 1960s.

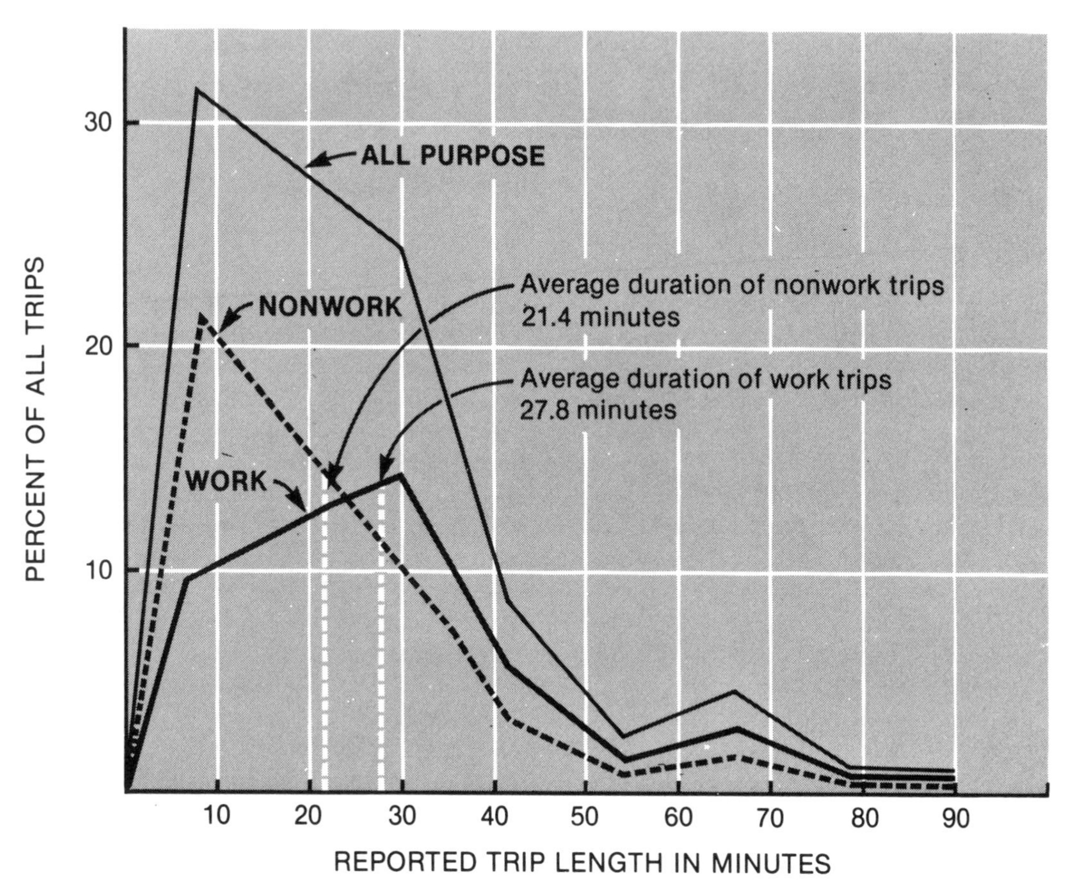

Figure 4-8. Frequency distributions of all-purpose, nonwork, and work-trip lengths, Toronto. From *Metropolitan Toronto and Region Transportation Study*, 1966, Toronto: The Queen's Printer.

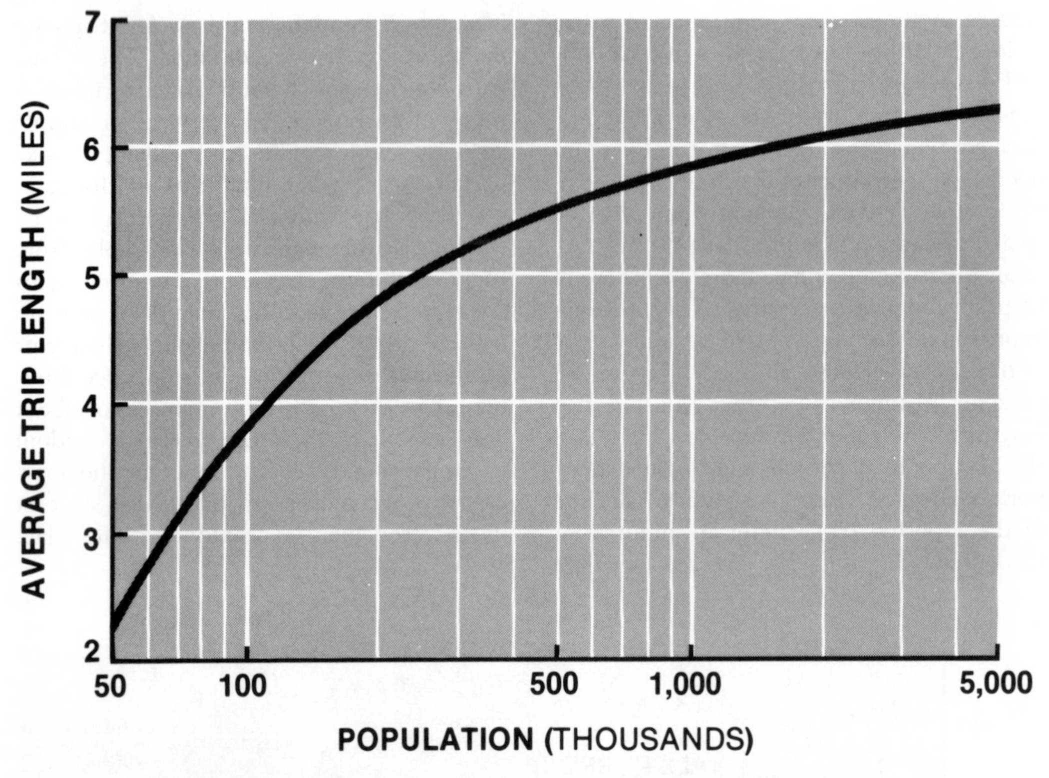

Figure 4-9. The average trip length as a function of size of city.

The length of trips in time, distance, or cost tends to increase as the size of the city increases, though not in a linear fashion (see Figure 4-9). This relation is as expected. The land areas of larger cities are obviously greater than those in smaller cities. Offsetting the effect of increases in land area with city size are increases in densities. Both residential and employment densities in larger cities exceed those found in smaller urban areas. Together, these two factors imply a curvilinear relationship between city size and average length of trip. More importantly, even in cities of about the same size there can be a tremendous variability in average trip lengths. The spatial structure of the city is very important. For example, a study of average trip lengths on the journey to work by Voorhees et al. (1961) found the average distance of work trips in New Orleans to be 2.5 miles and the average duration to be 7.4 minutes. These are much lower than the reported distance of 8.1 miles and 15.7 miles in Forth Worth, a much smaller city.

When trip times and distances are classified by *mode* of travel, the superiority of auto travel becomes readily apparent. Table 4-7 compares travel times and distances in selected standard metropolitan statistical areas of the United States in 1976. The figures in this table probably underestimate the advantage of the automobile, since they are restricted to work trips. Most public transit systems are at their best comparative advantage for work trips, since auto congestion is highest and transit headways lowest during the rush hour. The pattern is the same for all cities. Public transit patrons take one and a half to two times longer than auto travelers, even though their trips may be less the half the length!

It is also well known that workers traveling by public transit have significantly lower incomes than those commuting by car. Trans-

Table 4-7. Trip time and distance for selected U.S. standard metropolitan statistical areas, 1976 (work trips)

City	Private auto or truck	Public transport	Walk	Other means
Baltimore				
Median time (min.)	23.1	39.7	9.1	15.8
Median distance (miles)	10.0	6.5	0.6	3.3
Denver				
Median time (min.)	19.5	28.4	9.4	16.1
Median distance (miles)	18.2	7.8	0.6	4.0
New York				
Median time (min.)	21.9	42.0	10.4	17.2
Median distance (miles)	8.6	9.4	0.6	3.8
Houston				
Median time (min.)	21.7	35.8	7.5	14.1
Median distance (miles)	9.6	9.5	0.6	3.5
Sacramento				
Median time (min.)	18.0	30.3	8.1	13.8
Median distance (miles)	7.8	10.3	0.5	2.7

Note. From *Current Population Reports: Selected Characteristics of Travel to Work in 20 Metropolitan Areas* by the U.S. Bureau of the Census, 1976, Washington, D.C.: U.S. Government Printing Office.

portation systems in urban areas apparently serve higher-income residents much better than lower-income residents. In fact, some even argue that the structures of public transit systems also serve higher-income residents better than lower-income residents, since they offer excellent rush-hour accessibility between high-income residential areas and white-collar employment concentrations such as the CBD.

Interest in the time/cost/distance of urban trips is twofold. First, the analysis of travel demand incorporated into the conventional urban transportation planning process uses this type of regularity in various forecasting mechanisms. The choice of destination (trip distribution), choice of mode (modal split), and choice of route (route assignment) all include the relative costs of different trips in their formulations. The distance decay exponent included in gravity models of trip distribution, for example, reflects the preference for trip lengths in the city being modeled. The second reason why the time/distance/cost of urban trips is important has to do with the principal motivation for undertaking transportation investments in cities. The major component of the benefits of most of these projects is the savings in travel times afforded by these projects. By associating a value with these travel-time savings, it is possible to measure the benefits of any potential transport improvement project. By minimizing the disutility of travel time, we are therefore maximizing the travel-time saving and hence the benefit of the project.

SPATIAL PATTERNS OF TRIP MAKING

The distribution of trip lengths cannot be interpreted independently of the distribution of these trips in space. Each trip made in a city has a specific origin and destination located at particular points within the city. Just as we can identify significant patterns in travel flows when we disaggregate by aspatial categories such as trip purpose or time, so too can we identify typical patterns of the spatial distribution of trip making. The pattern to be observed in any city depends on two interrelated factors: (1) the *spatial structure* of the urban land use pattern, and (2) the *spatial configuration* and service characteristics of the transportation system. As noted in chapter 2, these are two mutually supporting systems. The distribution of land use places demand on the transportation system in the form of travel flows, and in turn, the structure of the transport network facilitates certain types of trips and encourages certain locational patterns of land use.

Because of the extreme variability in spatial structure and transport technologies in North American cities it is impossible to make broad generalizations about the spatial structure of flows that are valid for all cities. A simple example of the relationship between travel flows, the transportation system, and spatial structure is illustrated in Figure 4-10. In (a) the distribution of automobile traffic volumes in Edmonton, Alberta, in 1975 mirrors the underlying urban structure. The concentration of trips into and out of the central business district along radial lines is clearly illustrated. But, there are also significant nonradial components to the pattern of travel flows, reflecting the multicentric nature of the urban structure and the form of the road network. Note that virtually all places in the city appear to be well served by the automobile. In Figure 4-10 (b) the daily transit volumes on the public transit system illustrate a markedly different pattern. The transit network does not serve or connect all areas of the city. It is primarily oriented toward providing capacity for the journey to work in the Edmonton central business district. Transit flows have a strong radial component reflecting this characteristic. There are few cross-town or nonradial trips. This pattern can be seen more clearly in Figure 4-11, which decomposes the transit trips in Winnipeg, Manitoba, into central business district (CBD) and non-CBD trips. Transit travel is obviously more spatially restrictive than automobile travel.

Significant changes in the spatial structure of cities since the turn of the century have dramatically altered the trip-making patterns in cities. The dominance of radial trips to and from the central business district has

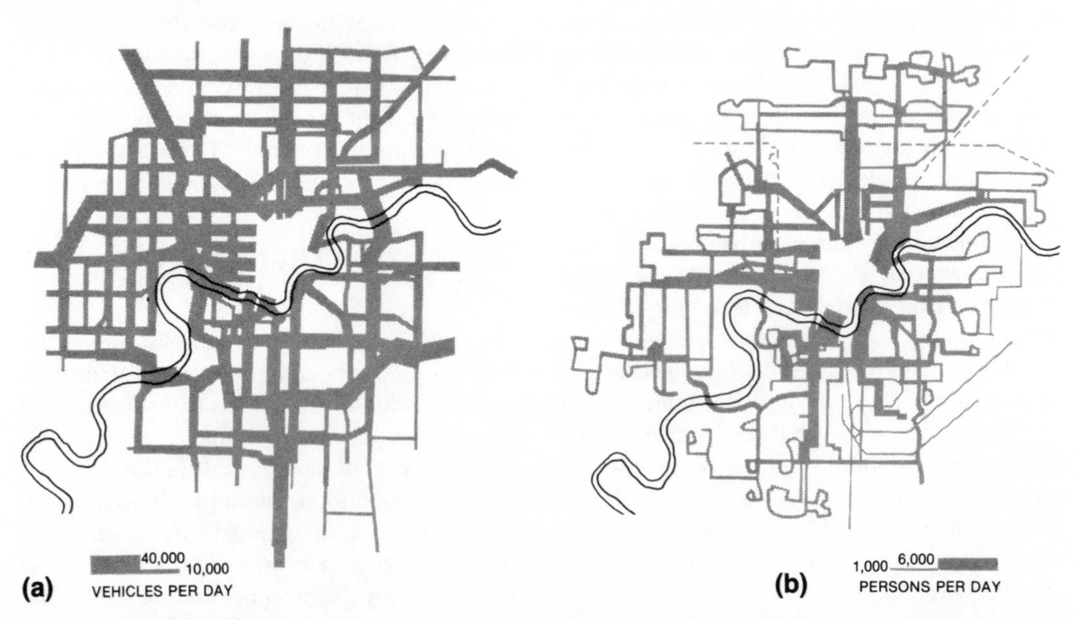

Figure 4-10. (a) Daily automobile volumes for Edmonton, Albérta, 1975, and (b) Daily transit volumes for Edmonton, Alberta, 1975. From *Canadian Transit Handbook* by R. M. Soberman and H. Hazard, 1980, Toronto: University of Toronto/York University Joint Program in Transportation. Copyright 1980 by University of Toronto/York University Joint Program in Transportation. Reprinted by permission.

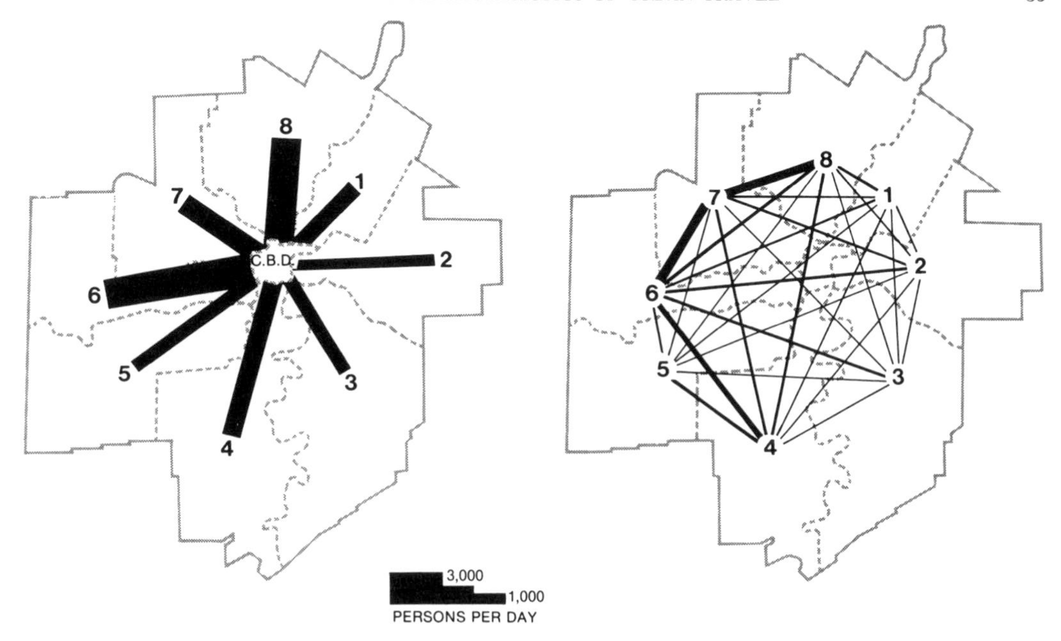

Figure 4-11. Transit patterns for CBD and non-CBD trips, Winnipeg, Manitoba, 1965. From *Winnipeg Area Transportation Study*, 1965, Winnipeg, Manitoba: City of Winnipeg.

been declining steadily. Cross-town trips are best served by auto travel and are extremely difficult using conventional public transit.

The spatial distribution of urban travel flows is extremely complex, but understanding it is an obvious prerequisite to effective urban transportation planning. Transportation problems themselves have a spatial component. Improving the performance of the transportation system means specifying where action must be taken to improve urban mobility.

SUMMARY

A basic knowledge of the characteristics of aggregate travel flows within cities is essential to understanding the transportation problems of contemporary North American cities and the strategies that planners employ to solve them. Five characteristics of these aggregate travel flows are especially important, though they can never be completely separated. Unraveling the tremendous complexity of aggregate urban travel is facilitated by disaggregating the total flow pattern along these five dimensions. Trip *purpose* is important because it defines the origin and destination of trips together with the land use pattern. The *temporal distribution* of both passenger and truck trips is directly related to the nature of urban activity systems and is the principal cause of urban traffic congestion. The *modal* distribution of trips is related to the size and spatial structure of the city, the available technologies, and the goals of local planners. The *length* of trips in the city also varies by mode, city size, and trip purpose. Fifth, the *spatial distribution* of aggregate travel reflects both characteristics of urban spatial structure and hence the transportation system. Since transportation planning is by its very nature a spatial problem, a comprehensive land use–transportation planning process is required if we are to solve the transportation problems of our cities.

Any attempt to analyze the aggregate flow pattern in a large metropolitan area must somehow reduce the inherent complexity of these patterns. The processes that underpin these patterns differ significantly by mode of travel, purpose, and time of day, and may be specific to urban areas of a given size. In order to come to a more complete understanding of these travel flows, the problems they create, and their potential solutions, it is necessary to use some sort of simplifying device. Transportation planners have found that a number of mathematical models can be used for this purpose. In fact, the conventional urban transportation planning process is decomposed into a set of sequential tasks, each based on a model exploiting the empirical regularities in travel flows. These models will continue to play an important role in policy analysis. Several of these models are presented in subsequent chapters.

The characteristics of urban travel described in this chapter are typical of those found in North American cities, but it is worth repeating that these characteristics can vary substantially from one city to the next. They are least evident when analyzed at the gross level and most evident when travel flows are stratified along a number of lines simultaneously. That is, there is more similarity between the peak-hour work trips into and out of the central business districts of cities than there is between work trips in general in these same cities. Finally, it is to be emphasized that all the relationships described in this chapter are purely cross-sectional. They therefore ignore all dynamic aspects of cities and their transportation systems. Nevertheless, the persistence of these patterns among North American cities attests to the stability of the underlying forces shaping aggregate urban travel patterns.

References

Chicago Area Transportation Study (CATS). (1976). *Transportation Improvement Program FY76–FY80*, Chicago, Ill.

Gray, G. E., & Hoel, L. A. (Eds.). (1979). *Public transportation: Planning operations and management*. Englewood Cliffs, N.J.: Prentice-Hall.

Institute of Traffic Engineers. (1982). *Transportation and traffic engineering handbook*. Englewood Cliffs, N.J.: Prentice-Hall.

Meyer, M. D., & Miller, E. J. (1984). *Urban transportation planning: A decision oriented approach*. New York: McGraw-Hill.

Smerk, George M. (1979). The development of public transportation and the city. In G. E. Gray & L. A. Hoel (Eds.), *Public transportation: Planning operations and management*. Englewood Cliffs, N.J.: Prentice-Hall.

Soberman, R. M., & Hazard, H. (1980). *Canadian transit handbook*. Toronto: University of Toronto/York University Joint Program in Transportation.

Toronto Transit Commission. (1982). *Transit in Toronto*. Toronto: Toronto Transit Commission.

U.S. Bureau of the Census. (1982). *Current Population Reports, Selected Characteristics of Travel to Work in 20 Metropolitan Areas*. Washington, D.C.: Government Printing Office.

Voorhees, A. M., S. J. Bellomo, Schofer, J. L., and Cleveland, D. E. (1961). Factors in work trip lengths. *Highway Research Record No. 141*, pp. 24–46. Washington, D.C.: Highway Research Board.

Winnipeg Area Transportation Study. (1966). Winnipeg: City of Winnipeg.

Zupan, J., & Pushkarev, B. (1975). *Urban space for pedestrians: A report of the regional plan association*. Cambridge: MIT Press.

5 MODELING AND PREDICTING AGGREGATE FLOWS

ERIC SHEPPARD
University of Minnesota

INTRODUCTION

Imagine watching a movie of a metropolitan area, taken from a satellite at about the height at which a commercial jet airliner flies, some 7 miles up, located over the central business district of an American metropolis. If this movie could be taken on film that is sensitive primarily to the type of heat generated by the internal combustion engine, it would be dominated by lines representing traffic flowing through the city. These types of flows, at the scale of detail captured in such a movie, are the subject of this chapter.

Any observer of such a movie would immediately begin to pick out regularities in the spatial distribution of flows, suggesting the possibility of constructing an aggregate model that could predict those regularities. What might some of these regularities be? If the movie was made using time-lapse photography, so that 24 hours in the life of a city were captured in a 20-minute movie, the lines of transportation would seem to pulsate as the day advanced. During the morning rush hour, the strongest lines would focus on places of employment and education, fanning out across the city to connect these to residential areas. Between 10 A.M. and 4 P.M. there would be a far more complex pattern representing such things as trips from home to shop, service calls to residences, trips between businesses, and social trips. In the evening rush hour a mirror image of the morning's job-focused pattern would reemerge, to be replaced by a pattern converging on places of eating and entertainment as the evening advances. Finally, this too dies out leaving a chaotic pattern of night trips including "journeys to crime," police patrols, bar hoppers, and the ever-present throughflow of interurban traffic on the interstate highways.

How can we make sense of this complex sequence and make practical use of our knowledge? For transportation planning the most immediate question is always whether transportation facilities are being appropriately used. Congestion on some roads, and underusage of mass transit would be examples of such problems. In this case the task is to be able to describe and predict the number of trips (including vehicle type) using each street, or link, of the urban transportation network. This number will depend on three factors: the number of trips of each type, the places where these trips start or end, and the route followed by each trip. All of these aspects depend on the spatial structure of the city. Thus a first, social-engineering-oriented, use of aggregate models would be *to determine how traffic flows are related to the internal geographical structure of a city.*

A second issue that should concern any planner is equity. In any city inequalities exist in the accessibility of people to both the amenities of a city and its negative features. This inequity is compounded by income and class differences among city residents and by the structure of society. These differences not only dictate the type of location within a city where certain people can afford to live, but also their ability to afford certain types of transportation. The equitability of any transportation plan turns on the degree to which people's observed travel behavior is a result of freely chosen acts as opposed to behavior conditioned by external constraints. As will be discussed in the next section, the answer to this question has crucial implications about how to alter the system to meet the welfare of urban residents. A second, welfare-oriented, use of aggregate modeling is *to determine the inequities associated with any pattern of transportation.*

SOME BASIC CONCEPTS

As noted above, *trips taken* are the basic unit of analysis for aggregate models. In this chapter we shall see that similar methodologies are used to model the components of trip making no matter what the purpose of the trip might be. However, since the factors influencing trip making presumably vary depending on trip purpose, it is useful to make some distinctions. Conventionally the following four trip purposes are identified:

1. Home-based work trips (the journey to and from work),
2. Home-based nonwork trips (trips to and from home for other purposes),
3. Non-home-based trips (passenger trips between two locations neither of which is the home), and
4. Goods movements (the movement of freight or the use of service vehicles other than private cars or mass transit), which may be conveniently divided into:
   a. Goods trips terminating at a commercial location, and
   b. Goods trips terminating at a residence (such as UPS or mail deliveries)

The number of trips taken in the city is usually referred to in the trade as *travel demand.* However the word "demand" should be used with some care. Conventionally it implies the notion that consumers of travel have freely chosen one possibility over all others, which in turn suggests that the observed pattern of trips represents the best possible set of actions that individuals could have taken given their preferences and the spatial structure of the city. However, observed behavior will not only include this case. To see why, consider a particular example. Suppose it is observed that the proportion of all blue-collar workers living in a (nearby) suburban residential area who journey to work at a particular suburban factory exceeds the proportion of all blue-collar workers living in a (distant) inner-city residential area who journey to work at that plant (see Figure 5-1). A reasonable interpretation of this "distance-decay" phenomenon would be that the inner-city residents have made the choice not to work there, because of the inconvenience of a long commute. From this would perhaps follow the recommendation that if road improvements are necessary, they should reflect the wishes of the largest group and thus be made within the suburbs. There is, however, at least one other possibility. It could as easily be the case that the greater distance faced by inner-city workers precluded many of them from considering the suburban plant at all as a possible site of employment. Perhaps they cannot afford the cost of transportation, or perhaps that plant does not advertise heavily in the central city, leaving its residents unaware of vacancies.

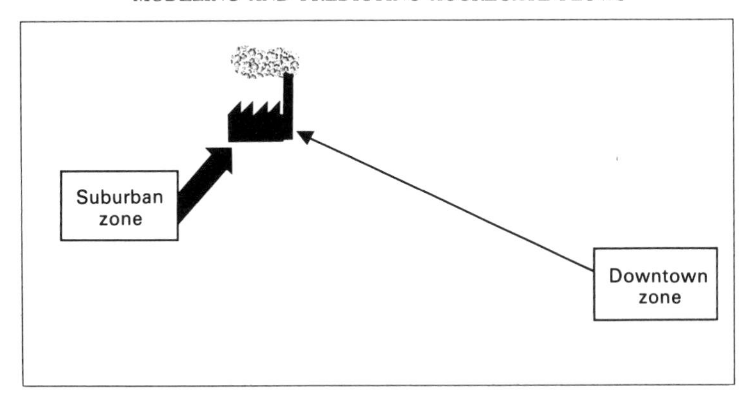

Figure 5-1. Distance decay and the journey to work.

Whatever the explanation, the implication is that the different behavior is due to some constraint rather than rational choice. If the above recommendation of improving suburban roads were followed, this would reinforce the inequality that inner-city workers face compared to suburban workers because the recommendation is based on the assumption that observed behavior reflects free choice when in fact it does not. (For another example of this phenomenon see Olsson, 1975.)

What can we learn from this mental game? Two things stand out. First, it is possible that an observed aggregate pattern of behavior can be the result of contrasting situations faced by an individual. Technically, we refer to aggregate models as being *overidentified;* more than one different causal explanation can lead to the same aggregate pattern. Second, it is necessary to identify the degree to which observed action represents free choice or constrained behavior. As the example shows, to simply assume that free choice dominates observed behavior can lead to policy recommendations that reinforce rather than alleviate urban inequalities—clearly something we would like to avoid (Sheppard, 1980). The only way to avoid these problems and to identify which explanation of behavior is correct is to treat each individual separately and to identify the limits on the choices available to him or her (the choice set). By definition, however, aggregate models cannot achieve this. We should therefore treat with considerable suspicion any attempt to infer the causes influencing individual behavior from aggregate data.

What exactly is an aggregate model, and to what use can it be put? An aggregate model of travel is a model that uses as its basic unit of analysis *zones* rather than individuals—zones that represent geographic subareas of a city such as census tracts or city blocks. These models cannot provide an adequate explanation of travel behavior, but they may be useful for predictive purposes. In other words they may provide an accurate prediction of what travel patterns will be like after a highway is built, even if they cannot tell us why that change occurs. We should, then, evaluate our models by their ability to give empirically accurate forecasts, but we should not be fooled into thinking that they necessarily explain what happened. Indeed, even the predictive ability of a model is a subject of some controversy, as there exist a number of different measures of the empirical accuracy of a model, none of which is useful for all purposes (Table 5.1).

One reason we use aggregate models is that disaggregate models are often not a pragmatic option. Disaggregate models have two disadvantages for applied research: Such models are often rather complicated when they attempt to specify all the causal variables and relationships that might explain observed behavior; and furthermore their empirical application requires, by definition, detailed individual-level data. These two characteristics together imply a requirement

Table 5-1. Measures of goodness to fit

Goodness to fit measure	Description (example of coefficient)
Correlation coefficient	The square root of the proportion of the variance in trip-making behavior explained by independent variables (0.975)[a]
Average proportional error	The size of the error in the prediction of the number of trips made, divided by the number of trips made added up over all observations (0.54)[a]
Root mean square error	The standard deviation of the size of the error of prediction, divided by the average number of trips made (0.23)[a]
Proportion of predictions misallocated	The percentage of the trips made that were incorrectly allocated to the wrong place, or wrong route (0.079)[a]

[a]These values are taken from a study by Ayeni (1979) which compares the performance of the different measures empirically. The actual numbers do not refer to trip frequencies but to the spatial distribution of housing. But they do show how the conventional correlation coefficient can give misleadingly high values. The best possible values are 1.0 for a correlation coefficient and 0.0 for all other measures.

Note. Adapted from Ayeni (1979) and Webber (1984).

for data that can be met only rarely. While it is true that use of disaggregate models can avoid the problem of incorrect inferences about the causes of individual behavior, even they do not necessarily avoid this. For example, if I simply assume that behavior is governed entirely by conscious choice and I choose to use a disaggregate model based on this assumption, then I can also end up making the same kind of mistaken inferences as those described above if that assumption turns out to be incorrect. Finally, even if an appropriate disaggregate model is chosen, and even if I have the data available for calibration, it is not always easy to add together predictions of individual actions in order to make predictions of aggregate urban travel patterns. For example, the effect of a change in, say, bus fares will have a different effect on a person's propensity to use the bus depending on that individual's preferences and income, and also depending on the nature of the choices available to him or her. Thus each predicted change must be modeled separately for each different type of individual and for each different choice situation, making the aggregate prediction a complicated affair (Hensher, 1979, pp. 82–83). Over all, it still remains to be seen whether the use of disaggregate models leads to an improvement in predictive performance that justifies the extra effort involved. Until we can be sure of this, aggregate models will continue to be extensively applied.

In order to understand the variables that are used in aggregate models, it is necessary to have some idea of the types of data that are available to the modeler. Table 5.2 lists the types of information that are typically at hand.

Having established the reasons why aggregate models are used and the types of data that are input, we can now discuss the way in which aggregate models incorporate the four components of travel (trip generation, trip distribution, modal split, and route assignment) outlined in chapter 4. There is an unending discussion of this issue. It is without question easier to model these components successively. Typically trip generation is modeled first. Then as a second step these trips might be allocated to destinations, followed by selecting travel modes for these flows, and concluding by assigning these flows to routes. This is, of course, only one possible sequence. A number of alternate sequences have been tried by different researchers, and some of the more common ones are summarized in Figure 5-2. The reason that no single sequence has been found to be best is that individual travelers will regard these elements differently in different situations. For example, in the journey-to-work decision the workplace (or destination) is typically chosen before the mode or the route. But for the delivery of commodities, the nature of the goods dictates that the mode must be chosen first, which then constrains the choice of route and even destination. As a third example, even the decision of how many trips to make (say, for shopping) may depend on the type of transport available to the shopper and on how accessible shops are to the home. If the sequence varies from one situation to another, or if sometimes more than one component may be decided on simultaneously, then the logical alternative is to have an aggregate approach that models all four components *simultaneously*. This has often been proposed but only rarely operationalized, perhaps because many researchers believe that it is important to identify how changes in the internal structure of the city affect each component separately.

Table 5-2. Typical transportation study data

Household information
Number of families per dwelling unit
Type of dwelling
Number of residents under, and over, 5 years old
Age/sex of household members
Number of years household head has lived there
Annual take-home family income
Number of residents employed full or part time
Employment status of household members
Occupation of each household member (type of job; industrial sector)
Number of hours worked on previous day for each member
Places of employment
Place of residence
Owner or renter of housing
Places of schooling
Number of privately owned cars
Number of company- or government-provided cars
Number of persons 5 years and older taking trips
Number of car-driver trips by purpose
Number of car-passenger trips by purpose
Main mode of travel by trip purpose
Trip-making propensity of each household member on day of survey
Number of public transport trips
Type of parking at work
Employer information
Type of company (S.I.C. classification)[a]
Number of employees, by occupation
Location of company
Trip information
Place of origin and destination
Trip purpose
Trip mode
Vehicle type
Number of persons/size of commercial load in vehicle
Time of day of trip
Trip duration

[a]S.I.C. is "Standard Industrial Classification"—the system used by the census to classify industries in its economic census.

Note. Adapted from Hutchinson (1974) and Black (1981).

However, when sequential models are used, we should be aware of their limitations. If each of the four components is modeled sequentially, mistakes that occur in predictions of those components that come early in the sequence can be very critical because these errors are compounded in each of the subsequent stages. Thus, for example, errors in predicting trip generation can lead to significantly more serious errors in overall predictions than might errors in route assignment. There are also some implications for policy. Thus if we were interested, for policy purposes, in how proposed changes in the city would affect one particular component of travel, then the sequence should ideally be arranged to allow for the dependencies of this component on the others.

For purposes of exposition, the sequence outlined on the left of Figure 5-2 will be followed in the next two sections of this chapter. This sequence has been widely used in urban travel models in the United States.

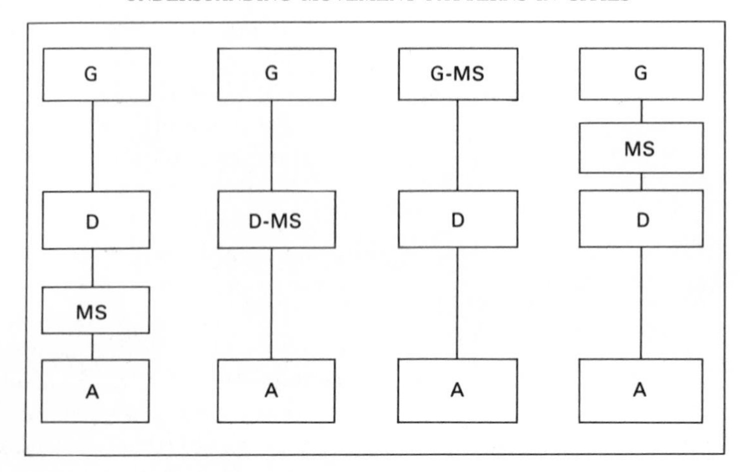

Figure 5-2. Alternative sequences for aggregate trip modeling. G is trip generation; D is trip distribution; MS is modal split; A is route assignment.

TRIP GENERATION MODELS

Trip generation models attempt to predict the amount of movement leaving a zone, based on attributes of that zone. We refer to the variable to be predicted as the *dependent variable,* measured as the number of trips of a certain kind leaving a zone during a fixed time period. We refer to the attributes of a zone that are used to predict this as the *explanatory variables.* The explanatory variables used will differ depending on the type of trip being modeled. However, in every case the variables used attempt to represent one or more of three different features of a zone. First, the potential number of trip makers in a zone should be identified by considering the land use mix or the number of residents. Secondly, the degree to which a potential trip maker's characteristics affect his or her propensity to make a trip should be considered. This may depend on whether or not he or she owns an automobile, on family size, on income, and so on. Third, the geographical accessibility of the zone to places where the trip purpose will be satisfied can also affect the number of trips made. If a trip may be generally more easily carried out because there are places nearby that fulfill its purpose, then that activity is relatively cheap and an individual would be inclined to perform that activity relatively more often, other factors being equal.

Table 5-3 summarizes some variables that are used as explanatory variables, classified into the three groups defined above, for each of the five different trip types outlined earlier in this chapter. Two things should be noted in examining this table. First, the basic unit of trip making for trips originating from the home is usually the household, whereas for trips originating from commercial establishments it is the individual firm. Second, it should be noted that variables measuring geographical accessibility have often been excluded from trip-generation studies. Only recently has the potential importance of this factor become widely recognized (see, for example, Vickerman, 1974; Wachs & Koenig, 1979). For any individual study, the precise variables used of course depend on data availability; often the information available is at best a poor substitute of what one ideally would like. This introduces an extra margin of error and bias into predictions, beyond that due to their aggregate nature. This is most important and should always be taken into consideration in evaluating the potential reliability of the resulting forecast.

There are two methodologies used for modeling the generation of trips based on such explanatory variables: *linear regression*

Table 5-3. Factors influencing trip generation

	Type of variable		
Trip type	Number of trip makers	Trip-making propensity	Accessibility
Home-based work	Residential density Size of population	Income Labor force participation Car ownership Transit availability	Access to jobs
Home-based nonwork	Residential density	Household size Income Car ownership Transit availability Occupation % nonworkers	Access to shops Access to leisure Access to friends
Non-home-based	Land use mix of origin zone Employment in zone by occupation	Income Car ownership	Accessibility within the city
Goods: Industry-oriented	Land use mix Employment in zone by occupation	Size of factory	Access to commercial activities
Goods: Consumer-oriented	Land use mix Employment in zone by occupation Number of residences	Size of factory	Access to consumers

Note. Adapted from Black (1981), Hensher (1977), and Hutchinson (1974).

analysis, and *category analysis*. The difference between these approaches is in the assumptions underlying each approach, and in the type of data that can be used in each approach. Let us examine each in turn.

Linear Regression Analysis

This approach attempts to predict trip generation by a best fit combination of explanatory variables. This can be represented formally as:

$$T_i = \beta_0 + \beta_1 x_{1i} + \beta_2 x_{2i} + \beta_3 x_{3i} + \beta_4 x_{4i} + \ldots + u_i \quad (1)$$

where T_i represents the number of trips generated from zone i; x_{ki} represents the value of the kth explantory variable (as taken, for example, from Table 5-3) observed for zone i; β_0 represents a constant; and β_k is a coefficient representing the numerical effect on trip generation that would result from a unit increase in the variable x_{ki}. For example, if the kth variable is the number of households and if β_k had a value of $+2.9$, then an increase by one in the number of households in zone i would lead to an increase by 2.9 in the average number of trips leaving zone i. Finally, u_i, known as the residual, represents the difference between the actual level of trip generation (T_i), and the level that would be predicted on the basis of the values of the explanatory variables and the coefficients (the β's).

It can be seen from equation (1) that this approach assumes a *linear* (additive) relation between the explanatory variables and the dependent variable; that the effect of any one explanatory variable on trip generation can be added to the effect of each other explanatory variable.

This approach is applied in the following way. Data are collected for all zones of the city, for explanatory and dependent variables, and for a particular point in time. This is taken to be a representative sample at one point in time of trip-generation patterns and their correlates. Based on these data, the methodology of linear regression produces *estimates* of the coefficients (β_k's) and an estimate of the residual (u_i) for this sample, estimates that lead to a predictive equation that minimizes the sum of the squares of the estimated residuals. The result of this estimation process is a set of estimates that are the *best linear unbiased* estimates possible, providing that a set of statistical assumptions are satisfied by the data collected. These assumptions are listed in Table 5-4.

As an example, consider the data shown in Figure 5-3. These data show levels of population density, income, and household trip generation for a hypothetical city with ten zones. The first two of these variables are independent variables, which are hypothesized to influence trip generation.

The results of a regression analysis of these two on trip generation are:

$$t_i = 8.74 + 0.11HI_i - 0.12PD_i + u_i \qquad (2)$$

where HI_i is the median household income in zone i, and PD_i is the population density in zone i. These results can be interpreted to tell us that an increase of $1,000 in median household income leads to an average of 0.11 extra trips being made per household in this city. Similarly, an increase in population density of one dwelling per acre leads to an average of 0.12 *fewer* trips per household. The overall proportion of the geographical variance in trip-generation behavior explained by the joint linear influence of these two variables (given by the R^2 of the regression) is 64.9%, a fairly reasonable level of explanation.

If assumptions 1–4 of Table 5-4 hold, then we have obtained an accurate and reliable predictor of trip generation, which could be applied in any other situation where the same causal relationships are believed to hold. If, in addition, the estimated residuals turn out to have an approximately normal (bell-shaped) distribution (assumption 5), some statistical inference is possible. We can determine whether each of the explanatory variables contributes significantly to our prediction of trip-generation rates (and thus whether we need to consider all these variables in our prediction). We can then also determine the statistical accuracy of our prediction by calculating *confidence intervals,* defined as a range of error, around our prediction of trip generation, within which we can expect the

Table 5-4. Assumptions of linear regression and the implications

Assumption	Effect on statistical estimates when assumption violated
1. Explanatory variables are measured without error	The estimates of the regression coefficients are biased
2. The true relationship is linear	The estimates of the regression coefficients are biased
3. Explanatory variables are statistically independent of one another (no multi-collinearity)	The estimates of the regression coefficients are not 'best': they may fluctuate wildly from one sample to the next, even though they are not biased
4. Residuals are independently randomly distributed:	
(a) Mean of residuals in population is zero	The estimates of the regression coefficients are biased
(b) Variance of residuals about the true values is constant (homoscedasticity)	The estimates of the regression coefficients are not 'best': they may fluctuate wildly from one sample to the next, even though they are not biased
(c) No spatial or temporal patterns to the residuals (no autocorrelation)	The estimates of the regression coefficients are not 'best': they may fluctuate wildly from one sample to the next, even though they are not biased
5. Residuals normally distributed	Statistical significance tests may be unreliable

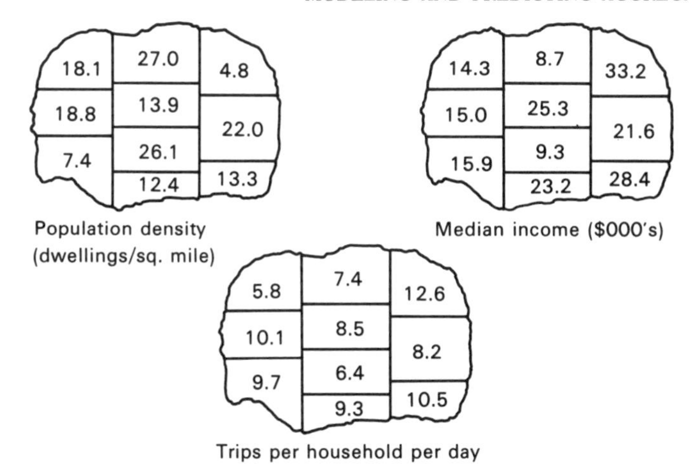

Figure 5-3. Trip generation rates and associated predictors: A hypothetical city.

actual level of trip generation to occur with a certain degree of confidence.

Turning again to our example of Figure 5-3, if we can assume that the relationships will not change significantly. If we project that the population density and median income in the downtown zone increase due to downtown redevelopment, then we can use our regression results to make a prediction of the effect of these changes on trip-generation rates in that zone. Thus, if we project the new population density to be 16 and the median income to increase to $35,000, then we would predict that the number of trips made per household would change to:

$$8.74 + (0.11).35 - (0.12).16 = 7.69$$

This provides us with a single "point" estimate. Given the uncertainty associated with any sample, we should also state how accurate this value is likely to be. The results of the regression also provide us with a confidence interval for our prediction. In this case we can say that there is a 0.95 chance that the new trip generation rate will be between 4.31 and 11.07.

This approach has been widely applied, and it is possible to make some evaluation of its effectiveness as a predictor of trip generation. It cannot be said that the results have been spectacularly successful. In many cases, a good fit of the relationship to the sampled data set has been possible. However, attempts to use the coefficients estimated from one sample of data in order to predict trip generation by applying them to data on the same explanatory variables collected in another sample have not led to particularly accurate predictions. The model has thus been a fairly good describer but a poor predictor, and it is accurate predictions that we want. This is true both when we are attempting to predict how trip generation will change at a future time in the same city and when we are attempting to predict trip generation in a different city.

There are two explanations for this poor performance. First, the assumptions underlying application of the regression model may not hold. Indeed it has been found on many occasions that there is multi-collinearity between different explanatory variables (Meyer & Straszheim, 1971). In this situation, where two or more explanatory variables may be duplicating one another by measuring roughly the same thing, reference to Table 5-4 indicates that this results in estimates of the β's that are unreliable; they can vary dramatically from one sample to the next. Secondly, the aggregate nature of the model can lead to unreliability, since causal factors influencing individual behavior are not being estimated. For example Kanafani (1983) reports a study where such a regression approach works adequately to describe aggregate data but provides a disturbingly poor fit when the data were disaggregated to predict trip generation by individual households. It seems that errors of prediction at the individual level cancel one another out somewhat at the aggregate level. Thus not only do errors creep in because aggregate models for zones try to make a single estimate for a somewhat heterogeneous group of individuals, but the inaccuracy of the model for predicting individual level behavior may also be hidden by its apparent good performance at the aggregate scale.

Table 5-5. Cross-classified rates of trip generation

Family size	Cars owned[a] 0	1	2+
1	1.0	2.9	5.6
2	1.9	4.5	5.9
3	2.9	6.2	7.7
4	4.1	8.5	10.7
5	5.8	10.2	13.7

[a]The numbers in the table represent the average number of home-based trips generated per household per day, Miami, 1973.

Note. From *Transportation Demand Analysis* by A. Kanafani, 1983, New York: McGraw-Hill. Copyright 1983 by McGraw-Hill. Reprinted by permission.

Category Analysis

The methodology of prediction followed by this approach is somewhat different. Suppose that data have been collected based on a survey of individuals and that a table is drawn up showing the average number of trips generated by households of different types (see, for example, Table 5-5). Suppose, then, that we wish to predict the number of trips generated in a zone at some future time. If we know how many people with each level of car ownership and household size will be residing in that zone, we simply apply the generation rates of Table 5-5 to these numbers to calculate a predicted number of trips (see Table 5-6). This approach can also of course be extended to more than just the two independent variables shown in Table 5-5.

What are the advantages and disadvantages of this compared to regression analysis? The major advantage is that no particular mathematical relation is assumed between the explanatory variables and the dependent variable (whereas regression analysis assumes a linear relation; see equation 1). The disadvantages are several. First, individual-level survey data are necessary in order to draw up the basic relation as in Table 5-5. Second, all relevant explanatory variables must be identified before the survey; it is not possible simply to add another variable later as can be done in regression, because the cross-tabulated relationships of each explanatory variable with each other must be known. Third, category analysis works best for independent variables that can only take on discrete values, such as those of Table 5-5. But many relevant variables show continuous

Table 5-6. Calculating trip generation through category analysis

Predicted population[a]	Trip generation rate[b]	Predicted number of trips generated
145 HH's of type (1.0)	1.0	145
73 HH's of type (1.1)	2.9	211.7
204 HH's of type (2.1)	4.5	918
5 HH's of type (2.2+)	5.9	29.5
14 HH's of type (3.0)	2.9	40.6
324 HH's of type (4.1)	8.5	2,754
23 HH's of type (5.0)	5.8	133.4
Total number of trips		4,232

[a]Classified by: family size, number of cars.

[b]From Table 5-5.

variation (for example income and accessibility), and to include these in a category analysis requires first breaking them into discrete class intervals. This is not a desirable step, since it throws away information.

As with regression analysis, it is possible to determine statistically whether all of the explanatory variables are contributing to the prediction. This can be done by applying analysis of variance to the original survey data (Kanafani, 1983). We can also estimate the statistical accuracy of the prediction by calculating confidence intervals (Hutchinson, 1974, shows how to do this).

The Problem of Hidden Demand

There has been a special problem that has recurred when urban transportation planners have applied trip generation models to predict the effects of proposed highway improvements on the travel patterns in cities. This is the way in which these models have persistently underpredicted the number of people who would travel on the newly built roads. Again and again urban travel models suggested that if a new highway were built in the city, this would reduce problems of congestion. However, once the highway was built, the traffic using it was so much more than predicted that the new road itself would become rapidly congested. This then led to a push to build yet another highway (Bennett, 1978). This has not only been a planning problem but also a social one, given that these highways have often been located in such a way as to displace working-class and black urban residents (for an example in Nashville see Seley, 1983).

What explains this inability of trip generation models to predict the hidden demand that is released by transportation improvements? Two reasons can be cited. First, the absence of accessibility variables in many trip-generation models will lead to underprediction. If accessibility is positively related to trip generation and accessibility increases when new roads are built, then this alone would lead to an increase in trips not accounted for by such models. Second, there is an interrelation between the transportation system and land use patterns in a city that is not captured in most aggregate urban travel models. In these models, land use is an independent, explanatory variable. In fact, however, new highways lead to the construction of new shopping centers, housing developments, and office buildings at the highway intersections, and these new land uses attract more traffic than could be predicted if we use current land use patterns as a variable in our trip generation model (Putman, 1983).

TRIP DISTRIBUTION MODELS

A number of models have been developed for the purpose of determining how many of the trips leaving a given zone will terminate at each possible destination. The most popular are gravity models, intervening opportunity models, and entropy models. However different these approaches are, they all apply the same basic logic: that there are three basic elements that must be understood in order to model the distribution of trips: the number of trips generated by a place, the degree to which the *in situ* attributes of a particular destination attracts trip makers, and the inhibiting effect of distance. The first of these three is, at least in principle, taken care of in transportation models by the trip-generation model. The second refers to what it is about a place that pulls urban travelers to the opportunities offered there. This will vary depending on the type of trip, and Table 5-7 gives some examples of such destination-specific pull factors. The third element refers to how the remoteness of a destination may inhibit its ability to attract travelers, even when its *in situ* attributes might be favorable. In geographical terms, the second element tends to refer to *site* characteristics of a destination, whereas the third element refers to attributes of its *situation*. Gravity and intervening opportunity models differ primarily in their definition of situation. Entropy models differ from

Table 5-7. Destination-specific pull factors in a trip distribution model

	Trip type				
	Home-based work	Home-based nonwork	Non-home-based	Goods: industry-oriented	Goods: consumer-oriented
In situ destination pull factors	Employment	Shops Population Entertainment	Shops Employment (by type) Population	Employment (by type)	Number of households Income

the other two in that they are appropriate in situations where incomplete data mean that gravity or intervening opportunity models cannot be empirically calibrated.

Gravity Models

The gravity model states that the number of trips made between an origin i and a destination j is positively related to the number of trips leaving i (which we will call O_i) and to the pull attributes of a destination (A_j), but is inversely related to the distance between i and j (d_{ij}). This can be formally summarized as:

$$I_{ij} = \frac{O_i \cdot A_j}{f(d_{ij})} \tag{3}$$

where I_{ij} stands for the interaction, as measured by the number of trips made from i to j, and $f(d_{ij})$ is a mathematical function that increases with distance. It can be seen that in Equation 3 an increase in O_i or A_j does indeed lead to an increase in I_{ij}, whereas an increase in d_{ij} results in a fall in the number of trips I_{ij}. Many different functions have been used for $f(d_{ij})$, of which perhaps the best known is $f(d_{ij}) = d_{ij}^{\beta}$. This is the traditional way in which the effect of distance has been represented by researchers. The coefficient β is an empirical constant representing the severity of the inhibiting effects of distance on trip making, otherwise known as the *distance decay effect*. Other things being equal, the higher the value of β, the faster the fall in the number of trips with distance.

As an example, consider Figure 5-4. This describes the effect of distance on the trip-making behavior in a situation where the attractiveness of destinations is the same everywhere. Each curve in the figure represents the situation for different values of β. The vertical axis measures the size of the distance decay effect, $d_{ij}^{-\beta}$. The number of trips to a destination 3 miles away is, by equation 3 equal to $I_{ij} = O_i \cdot A \cdot 3^{-\beta}$, where $3^{-\beta} = 0.333$ if $\beta = 1$, and 0.111 if $\beta = 2$. Thus for every 100 trips made that are 1 mile in length, 33 are made of 3 miles in length if $\beta = 1$, whereas only 11 are made of 3 miles in length if $\beta = 2$. The figure then shows that for larger values of β, the number of trips decays more rapidly as distance increases. Thus, when $\beta = 0.5$, for every 100 trips of a distance of 1 mile 58 trips will be made to destinations 3 miles away, and 33 trips will be to destinations 9 miles away.

In practice if we draw a graph showing the number of trips made in a city on the vertical axis and the length of those trips on the horizontal axis, this graph need not show a regular decline with distance, even when the distance decay effect is significant. Rather it could look like Figure 5.5. This would be true, for example, for journey-to-work trips in a city when the internal structure of that city is such that a plurality of the residential zones happen to be located approximately 5 miles from places of work. Figure 5-6, then, incorporates the joint effects of site (the number of jobs) and situation (their distance from residences). However, if we are to develop an accurate aggregate model of the geographical distribution of urban trips, we would

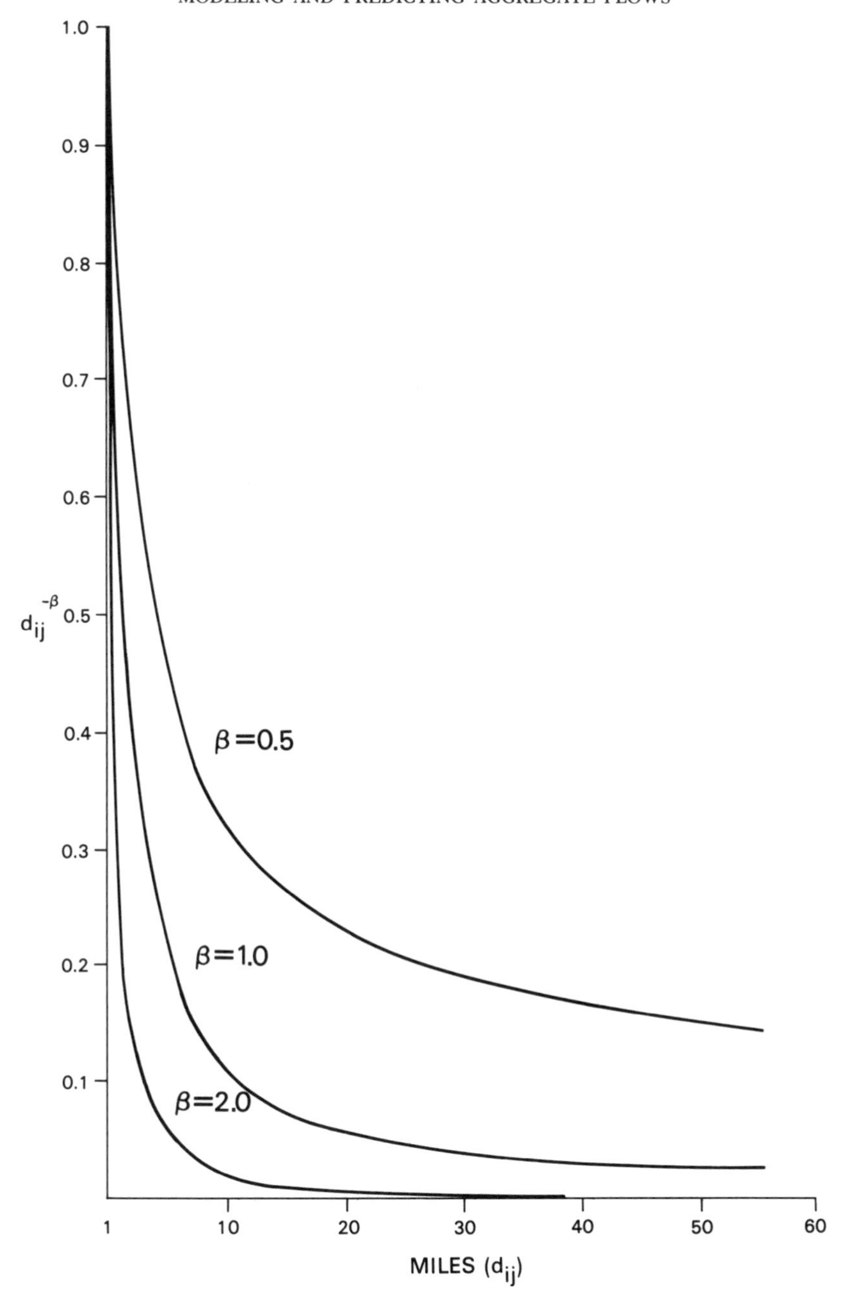

Figure 5-4. Traditional distance decay functions: Different values of β. From "The Distance Decay Gravity Model Debate" by E. Sheppard, 1984, in G. Gaile & C. Willmot (eds.), *Spatial Statistics and Models,* Dordrecht: D. Reidel. Copyright 1984 by D. Reidel. Reprinted by permission.

have to be able to separate the effects of these two different types of factors.

The way we do this is similar to that used in trip-generation models. If we have available at one point in time detailed data on trip-making patterns in a city and on various explanatory variables which influence trip distribution, we can develop a best-fit model

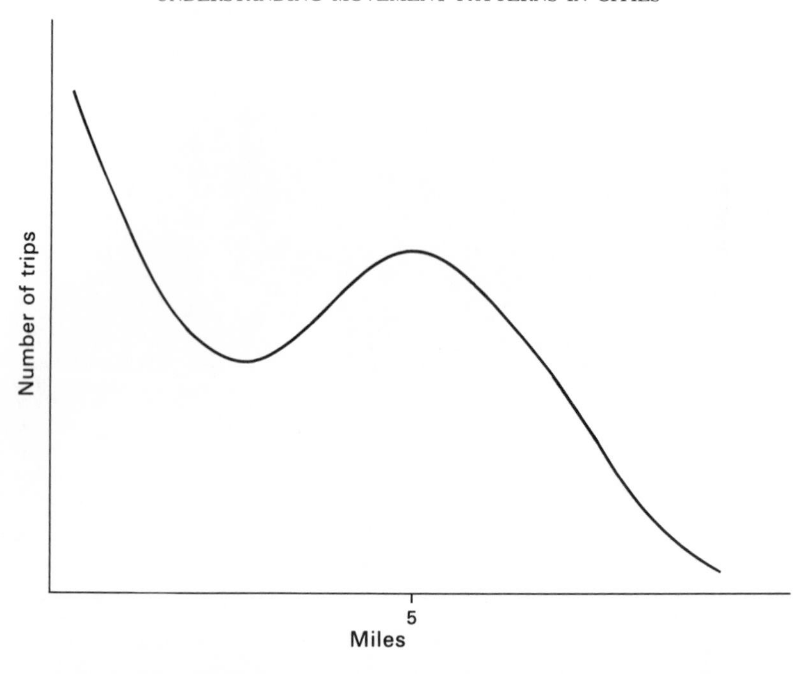

Figure 5-5. The effect of site and situation on distance decay.

predicting trip distribution using these variables. We then hope that this relationship will allow us to predict how trip patterns might change as a result of changes in these explanatory variables. What variables should be used? Recall that we have three types of variables: the number of trips generated, *in situ* attributes of the destinations, and distance. The first of these three is an input from our trip-generation model. The second can be drawn from variables such as those in Table 5-7. The distance variable involves two choices: a choice of how distance should be measured, where typically a choice is made between physical distance, travel time, and travel cost; and a choice of the type of mathematical function used. One disadvantage of the power function as given in equation 3, for example, is that this function equals zero when distance equals zero; and dividing by zero in equation 3 would lead to a prediction of an infinite number of trips! Many different distance decay functions have been tried, each of which represents the effect of distance on travel in a different way (see Taylor, 1975; Hutchinson, 1974). A graphical comparison of some commonly used functions is given in Figure 5-6. This graph is much like Figure 5-4. The vertical axis measures the attenuating effect of distance on trip-making behavior. For curve two, then, c_{ij} on the vertical axis equals $d_{ij}^{-\beta}$, and the figure shows the general shape of the distance decay function when the effect of distance is measured as a negative power (as in equation 3). If instead of a power function, distance is measured by a negative exponential function, then $c_{ij} = e^{-\beta d_{ij}}$, producing a distance decay that is less rapid as distance increases (line 1 in Figure 5-6). Finally use of the distance decay function $d_{ij}e^{-\beta d_{ij}}$ can allow for a situation where the number of trips actually increases over short distances, but then decreases thereafter (line 3 in Figure 5-6). This last case can represent a situation of auto trips where people only take the car if the journey-to-work trip is long enough to warrant it (more than a mile, say), and thus there are few car trips of very short lengths.

Suppose that we are modeling home-based nonwork trips and choose as *in situ* destination pull variables shopping floor space (S_j)

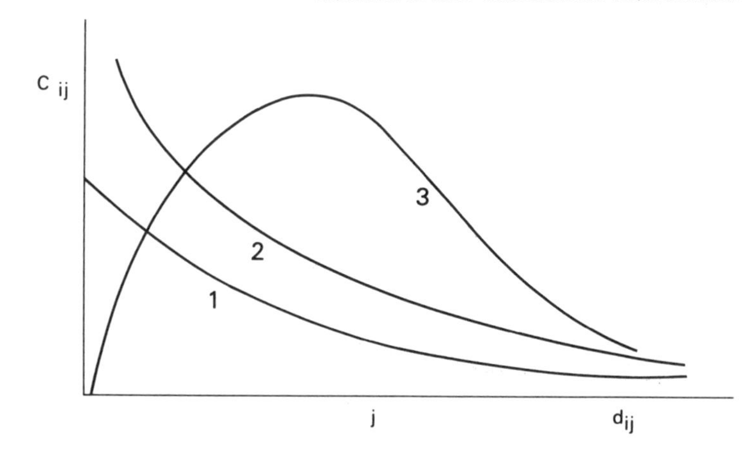

Figure 5-6. Some different distance decay functions. From *Transportation Demand Analysis* by A. Kanafani, 1983, New York: McGraw-Hill. Copyright 1983 by McGraw-Hill. Reprinted by permission.

and population (P_j), and as the distance variable an exponential function of travel time (t_{ij}). Then our gravity model might look like:

$$I_{ij} = \frac{K \cdot O_i(S_j)^a(P_j)^c}{e^{\beta(t_{ij})}} \cdot u_{ij} \qquad (4)$$

In this equation K, a, c, and β are empirical parameters to be estimated when the model is calibrated to give a best fit to observed travel patterns, I_{ij}. u_{ij} is the residual representing the difference between the predicted level of travel as estimated by the model and the observed number of trips I_{ij}. The negative exponential form of the distance decay function, a form that we already came across in Figure 5-5, is used here because the gravity model is a nonlinear equation, and in order to obtain estimates of the relationship using linear regression, a way must be found to transform it mathematically into a linear equation. Transportation modelers use the exponential form in equation 4 because if logarithms are taken of that equation, then it takes on the linear form:

$$\log(I_{ij}) = \log K + \log(O_i) + a \log(S_j) + c \log(P_j) - \beta\, t_{ij} + u_{ij} \qquad (5)$$

Note that one term on the righthand side of equation 5 has no parameter attached to it. This is because by definition the number of trips for i to any destination j, I_{ij}, is strictly proportional to the number of trips leaving i, O_i. For this reason, it is useful to move this term to the lefthand side by subtracting it from both sides. Since $\log(x) - \log(y)$ equals $\log(x/y)$:

$$\log(I_{ij}/O_i) = \log K + a \log(S_j) + c \log(P_j) - \beta \cdot t_{ij} + u_{ij} \qquad (6)$$

The model now attempts to predict the proportion of trips leaving i that terminate at $j(I_{ij}/O_i)$ using information on floor space, population, and distance. The parameters $\log K$, a, c, and β in equation 6 can now be estimated with linear regression methods. Suppose we collect data for a number of zones on shopping floor space and population in the zones, on travel times between the zones, and on trip-generation rates (using a trip-generation model). This will give us data for our explanatory variables. If we also collect information on the number of shopping trips made from each zone to each other zone, then we can use linear regression to estimate the four empirical parameters of equation 6, applying the methodology outlined in the section on linear regression analysis. Each origin–destination pair would be an observation for this model. If we have reason to believe that our results are reliable, they can then be used to predict future trip patterns, given known changes in the explanatory variables.

As an example of this, consider again the example of Figure 5-3. For simplicity, let us aggregate the ten zones of that hypothetical city into three zones: one representing the three western zones of Figure 5-3 (zone 1), one representing the four central zones (zone 2), and one representing the three eastern zones (zone 3). We hypothesize that shopping behavior can be predicted by a linear function of the logarithm of shopping space (S_j) and travel time (t_{ij}). This differs from equations 4 to 6, since population is excluded as a variable. The observed data on spatial interaction for shopping purposes, on the travel times between the three zones, and on the floor space of available shopping are given in Table 5-8. The result of a regression analysis of these data is:

$$\log (I_{ij}/O_i) = -2.29 + 0.305 \cdot [\log (S_j)] - 1.07 \cdot (t_{ij}) \quad (7)$$

The amount of the variance in shopping behavior explained by travel time and shopping space in this case is 88.9%. Once again predictions of shopping behavior can be made by entering into this equation new values for shopping space and travel time, although in this case the prediction is a prediction of the proportion of all shopping trips from i that terminate in zone j (I_{ij}/O_i), not of the actual number of trips. The latter can be deduced by multiplying the predicted value by O_i.

Empirical experience with this approach has produced similar findings as that for trip-generation models: The regression approach for trip-distribution models will often provide a very good fit to observed data but is a significantly poorer predictor of future trip patterns. There are three logical problems in using the linear regression approach, which help account for this poor performance. First, the spatial distribution of opportunities is typically such that the *in situ* variables may be correlated with the travel time or distance variable. This can cause problems of multicollinearity, or mis-specification (treating the variables as linear and independent from one another when in fact they are not), leading to unreliable and/or biased parameter estimates (Cliff, Martin, & Ord, 1974; Fotheringham, 1981; Sheppard, 1984).

One way to avoid these problems with calibrating the distance function is not to use a general function at all, but rather to determine separately, for each origin–destination pair, a number representing the distance between them. This is only possible if we are confident that we have accurately specified the elements measuring *in situ* attractiveness of destinations. Many applications of the gravity model for urban travel forecasting make this assumption and then estimate a set of numbers for the friction factor between all places that best fit the observed data. This is done using an iterative procedure as follows:

1. Divide the measure of separation, say travel time, into a number of discrete distance classes.
2. Determine the proportion of observed trips in each class. This is called the observed trip frequency distribution.

Table 5-8. Data for the example of shopping behavior

Origin	Destination	Travel time (minutes)	Shopping space at destination (square feet)	$\log(S_j)$	I_{ij}/O_i
1	1	5	4,000,000	6.62	0.5
1	2	15	1,000,000	6.00	0.225
1	3	30	4,000,000	6.62	0.275
2	1	15	4,000,000	6.62	0.33
2	2	5	1,000,000	6.00	0.34
2	3	15	4,000,000	6.62	0.33
3	1	30	4,000,000	6.62	0.275
3	2	15	1,000,000	6.00	0.225
3	3	5	4,000,000	6.62	0.5

3. Choose an initial friction factor (D_t) for each travel time class t.
4. Use the gravity model to compute the number of trips (N_{ij}) between each origin and destination, using the initial friction factors.
5. Compare the observed and computed trip-length frequency distributions. If the two distributions are not sufficiently close, modify the friction factors as follows:

$$D_{t(\text{new})} = D_{t(\text{old})} \cdot \left[\frac{o_t}{P_t} \right] \qquad (8)$$

where $D_{t(\text{new})}$ is the revised friction factor for distance class t, $D_{t(\text{old})}$ is the friction factor for class t in the previous iteration, o_t is the observed number of trips in class t, and P_t is the number predicted in the previous iteration.

Steps four and five are repeated until a closure criterion is satisfied. The typical closure criterion is:

$$\frac{\sum_{c_{ij} \in t} (N_{ij} - I_{ij})}{\sum_{c_{ij} \in t} (N_{ij})} \leq \epsilon \qquad \text{for each } t \qquad (9)$$

where ϵ is the closure standard, such as 0.05. Note that this procedure achieves a fit by class of distance, but not along individual links. This means that a further adjustment factor, sometimes termed a socioeconomic adjustment factor, must be added. Even this, however, is not explicitly related to the characteristics of origin–destination pairs; thus it must be assumed to be constant over time.

The second logical problem is that gravity models like that of equation 4 produce results that are logically inconsistent with prior information about the trips. We know that the number of trips leaving any origin equals O_i. Yet when we count up how many trips from i are predicted by the model to terminate at all destinations and then we compare it to O_i, the two numbers are generally not equal to one another, although they should be. Mathematically we would like the following constraint to hold:

$$\sum_j I_{ij} = O_i \qquad \text{for all } i \qquad (10)$$

Figure 5-7 compares the observed trips made between the origins and destinations from the data of table 5-8 with regression estimates of the trip distribution, as predicted using equation 7. Even in this simple example, while 4,000 trips are known to leave zone 1, the model estimates this number as 3,256. Similarly, the number of trips leaving

Observed trip patterns from Table 5-8

from \ to	1	2	3	
1	2000	1600	550	
2	900	1800	450	
3	1100	1600	1000	
O_i	4000	5000	2000	

Estimated trip patterns using regression equation 7

	1	2	3	
1	1895	1854	808	
2	957	1534	479	
3	404	1854	948	
O_i	3256	5242	2235	

Figure 5-7. A comparison of observed and predicted trips.

zone 2 is estimated at 5,242 instead of the known figure of 5,000, and for zone 3 the estimated number is 2,235 instead of 2,000. These are significant discrepancies, especially given the high overall level of explanation achieved by the regression model.

This inconsistency can be easily eliminated if the gravity model of equation 4 is changed as follows:

$$I_{ij} = O_i \cdot \frac{(S_j)^a (P_j)^c e^{-\beta(t_{ij})}}{\sum_k (S_k)^a (P_k)^c e^{-\beta(t_{ik})}} \cdot u_{ij} \qquad (11)$$

The difference between this equation and equation 4 is the addition of the expression $\Sigma(S_k)^a(P_k)^c e^{-\beta(t_{ik})}$ as a denominator. This summation, over all destinations k, essentially adds up the attractiveness (taking into account *in situ* characteristics and distance) of all possible destinations. The righthand side of equation 11 then equals the number of trips leaving i, multiplied by the attractiveness of j expressed as a proportion of the attractiveness of all destinations, multiplied by the possible residual error. Now if equation 11 is added up over all j, that number will always equal the number of trips generated, O_i, whereas for equation 4 that is not generally the case. This logically consistent version of the gravity model is referred to as an *origin-constrained* gravity model, since it is adjusted to ensure that the total number of trips originating in i that arrive at all destinations is constrained to equal the predefined number (O_i). If it is also known exactly how many trips terminate at each destination, a second constraint would have to be built in to the model to take this into account:

$$\sum_i I_{ij} = E_j \qquad (12)$$

where E_i is, say, the number of jobs filled by workers in zone j. In this event the model, constrained at both the origins and the destinations, is known as a *doubly constrained* interaction model.

The third problem, which is particularly relevant in the light of the previous two, is as follows. The transformation used to change equation 4 to equation 5 is a convenient way of creating a linear relationship, so that standard package computer programs can be used to estimate a *linear* regression. However, that is just a trick and is not really the most rigorous way of making a statistical estimate. Ideally one should find the best estimate for the nonlinear equation directly, rather than trying to convert it into a linear relation. There are methods of nonlinear regression analysis that can be used for this purpose. In fact these methods must be used if we are going to adjust our model to remove the first two types of inconsistency: if we are to allow for the (generally nonlinear) correlation between *in situ* attributes and distance, and if we are going to use the equation form 5, then it is no longer possible to translate our gravity model into a linear form for statistical estimation purposes.

We do not know by how much the predictive abilities of the gravity trip-distribution model will be improved once these three inconsistencies are eliminated, because no one has ever done that. However, we cannot expect that these corrections will eliminate all errors because of the limitations that any aggregate descriptive model possesses as we have seen earlier in this chapter.

Intervening Opportunities Models

There are really just two differences between an intervening opportunities model and a gravity model. The first is that the *in situ* characteristics of a destination are measured as simply the number of opportunities available there. The second is in the way by which the effect of geographical separation between places on the number of trips made between them is conceptualized. Whereas in the gravity model distance per se is supposed to influence the chance of a trip being made, in the intervening opportunities model the critical factor is the number of other opportunities closer to an origin than any particular destination being considered by a traveler. It was argued by Stouffer (1940) that when there are a large number of opportunities closer to the origin of a trip than this destination, then it is more likely that a trip maker will stop off at one of these instead of going on to that more distant opportunity. Thus, Stouffer suggested that instead of using distance in the denominator of an interaction model, that variable should simply be replaced by a count of the number of intervening opportunities.

Stouffer's proposal was taken up by Schneider (1960), who used it for estimating trip distributions for the Chicago Area Transportation Study. He modified it somewhat, however, to make it more consistent with rational choice behavior. He argued that a trip maker leaving a particular origin i will consider each possible opportunity sequentially, starting with the closest. The probability of stopping at the closest possibility is said to be proportional to the number of opportunities there. The probability of stopping at the next nearest place depends on the number of opportunities there and on the probability that the trip maker did *not* stop at the previous destination. This logic is then carried through for each destination in turn (see Figure 5-8 for an example). The resulting formula for the intervening opportunities model is:

$$P_{ij} = Pr(\text{trip maker does not stop at the } j-1\text{th closest place}) - Pr(\text{trip maker does not stop at } j\text{th closest place}) = e^{-L \cdot V(j-1)} - e^{-L \cdot V(j)} \quad (13)$$

where P_{ij} represents the probability of traveling from i to j; Pr refers to "the probability that"; L is an empirical constant, and $V(j)$ is the total number of opportunities to be found at all destinations up to and including the jth closest place. The exponential form is used here not for convenience of estimation (as for the gravity model), but because the probabilistic theory of trip-making behavior used by Schneider requires it.

If equation 13 describes the probability that a trip will be made from i to j, then the total number of trips made from i to j is simply equal to this probability multiplied by the number of trips leaving i, O_i:

$$I_{ij} = O_i \cdot [e^{-L \cdot V(j-1)} - e^{-L \cdot V(j)}] \quad (14)$$

Clearly the intervening opportunities model is somewhat similar to the gravity model. In both cases the number of trips from i to j will increase as the number of opportunities at j increases and will generally decrease as distance between i and j increases. This is true of the intervening opportunities model as well, since we would generally expect that as distance increases, so would the number of intervening opportunities. Thus, it is perhaps not surprising that both the gravity and the intervening opportunities models have performed about equally well as descriptors of trip-distribution patterns. Indeed, researchers typically choose one rather than the other not because it performs better, but

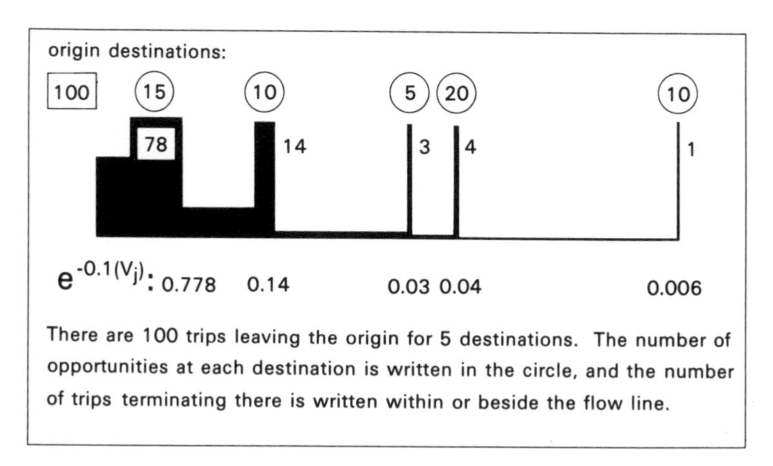

Figure 5-8. An example of an intervening opportunities model.

because they prefer it on conceptual grounds.

The intervening opportunities model shares with the classical gravity model the disadvantage that it is not internally consistent. Ideally if one were to add up equation 13 over all destinations the total should be equal to O_i. Again this is only true once a correction factor is added. The consistent, *origin-constrained*, intervening opportunities model that should be used is:

$$I_{ij} = O_i \cdot \frac{[e^{-L \cdot V(j-1)} - e^{-L \cdot V(j)}]}{[1 - e^{-L \cdot V(J)}]} \qquad (15)$$

where in the denominator $V(J)$ is the total number of opportunities in the city.

How would we use the intervening opportunity model to predict trip distribution patterns? As with the gravity model, if the appropriate data are collected, a regression model can be used to determine how the explanatory variables influence trip distribution. In this case there are just three explanatory variables for which we need information: the number of trips leaving an origin (given from our trip-generation model), the number of opportunities at each possible destination, and a rank ordering of our destinations with respect to how far they are away from the origin—the closest destination would have rank one, the next closest rank two, and so on. Our dependent variable is again the number of trips leaving the origin and terminating at each destination. If all this information is collected for each origin, we would have a complete data set and could calculate $V(j)$ for each j and for each origin. A nonlinear regression could then be run to find the value of the regression coefficient L, which gives the most accurate prediction of observed trip distribution patterns. Again, if we can be sure that our regression estimate is statistically reliable and if we have reason to believe that the relationship will stay the same, we can use this estimate to predict trip distribution patterns for some new situation.

The equation used for estimating L empirically would be:

$$P(j) = \frac{[1 - e^{-L \cdot V(j)}]}{[1 - e^{-L \cdot V(J)}]} \qquad (16)$$

where $P(j)$ is the proportion of all trips leaving j that stop at the jth closest destination.

Entropy Models

As can be seen from the discussion of the previous two sections, the gravity and intervening opportunities models can only be applied in situations where data have been collected on the observed distribution of trips at some point in time and on attributes of the destinations. The collection of trip distribution data, however, is a time-consuming and expensive affair, typically requiring an extensive survey. What can be done when these data are not available? The so-called entropy-maximizing approach has been developed for just this purpose. The aim of the approach is to provide the least biased estimate of some missing data (such as trip distribution patterns), given just partial information on these data.

Suppose, for example, we know (or have an estimate of) trips generated from each origin (O_i) for a particular purpose, and we also know that the average length of a trip taken for this purpose in the city is C miles. The least biased estimate of the trip pattern is calculated by finding those values I_{ij} which maximize $H = \sum_i \sum_j I_{ij} \cdot \ln(I_{ij})$, where $\ln(I_{ij})$ is the natural logarithm of I_{ij}, and the expression $I_{ij} \cdot ln(I_{ij})$ is added up for all possible destination pairs i and j. This maximization is carried out while subjecting the final outcome to the constraints that the number of trips leaving each origin does indeed equal O_i, and that the average trip cost is in fact equal to C. This process is detailed in Figure 5-9. H is known as the entropy of the trip distribution pattern and can be interpreted as the probability that this particular pattern of trips could occur by chance in a system where the above mentioned constraints hold. Thus by maximizing H, we are choosing that trip distribution that is most likely if these constraints are true and if no other con-

Equation (5.10) is deduced by maximizing:

$$H = \sum_i \sum_j I_{ij} \ln(I_{ij})$$ (which is known as the entropy of the trip distribution pattern)

subject to the constraints:

$$\sum_i \sum_j I_{ij} t_{ij} / \sum_i \sum_j I_{ij} = C$$ (Average trip length equals C)

and

$$\sum_j I_{ij} = O_i \quad \text{for each } i$$ (Number of trips generated equals O_i)

If there are N origins, this implies N + 1 constraints. The maximization process also estimates a coefficient (Lagrangian multiplier) for each constraint. For the cost constraint this coefficient is β (see Eqn.5.10).

Figure 5-9. Calculating a maximum entropy trip distribution.

straints on the trip-making pattern exist. The least biased estimate of trip distribution patterns given information only on O_i and on average trip length is:

$$I_{ij} = O_i \cdot \frac{A_j e^{-\beta(d_{ij})}}{\sum_k A_k e^{-\beta(d_{ij})}} \quad (17)$$

Equation 17 says that the number of trips made from i to j equals the number of trips leaving i, multiplied by the attractiveness of j (discounted by an exponential distance function) expressed as a proportion of the attractiveness of all other destinations. Now compare equation 17 with equation 11. Equation 17 looks basically like an origin-constrained gravity model. The one difference in this case is that the use of an exponential distance function is not a choice of convenience, but is dictated by the mathematics of the entropy-maximization process.

How can we call this a least biased estimate of the missing trip distribution data? It is least biased in the same way that our least biased estimate of the probability of a coin toss coming up as heads or tails is to give each possibility a probability of one-half. We assume that the coin is fair unless we are given definite information otherwise; similarly, we assume that the trip-distribution pattern is as evenly scattered as is possible given the information we have on trip patterns. For a further discussion see Figure 5-10.

The β parameter in equation 10 is calculated from the information on the average length of a trip (see Figure 5-9). The smaller the average length of a trip, the larger the value of β, since with generally shorter trips the distance friction effect is obviously smaller.

If in addition to the above information, we also know the number of trips terminating at each destination, D_j, then the least biased estimate of trip distribution patterns is given by the following entropy maximizing model:

$$I_{ij} = E_i B_j O_i D_j e^{-\beta(d_{ij})} \quad (18)$$

where

$$B_j = \frac{1}{\left[\sum_m O_m e^{-\beta(d_{mj})}\right]}$$

$$E_i = \frac{1}{\left[\sum_k D_k e^{-\beta(d_{ij})}\right]}$$

B_j and E_i are constants that can be estimated if β is calculated; as before, β can be calculated as part of the process of developing the least biased estimate of trip distribution patterns. Equation 18 is known as the *doubly constrained* model, since the ratio E_i ensures that the number of trips leaving i always equals O_i as before, whereas the ratio B_j ensures that the number of trips terminating at j always equals D_j. These models are discussed in detail by Wilson (1970), and their use is reviewed by Webber (1984). It should always be remembered that such

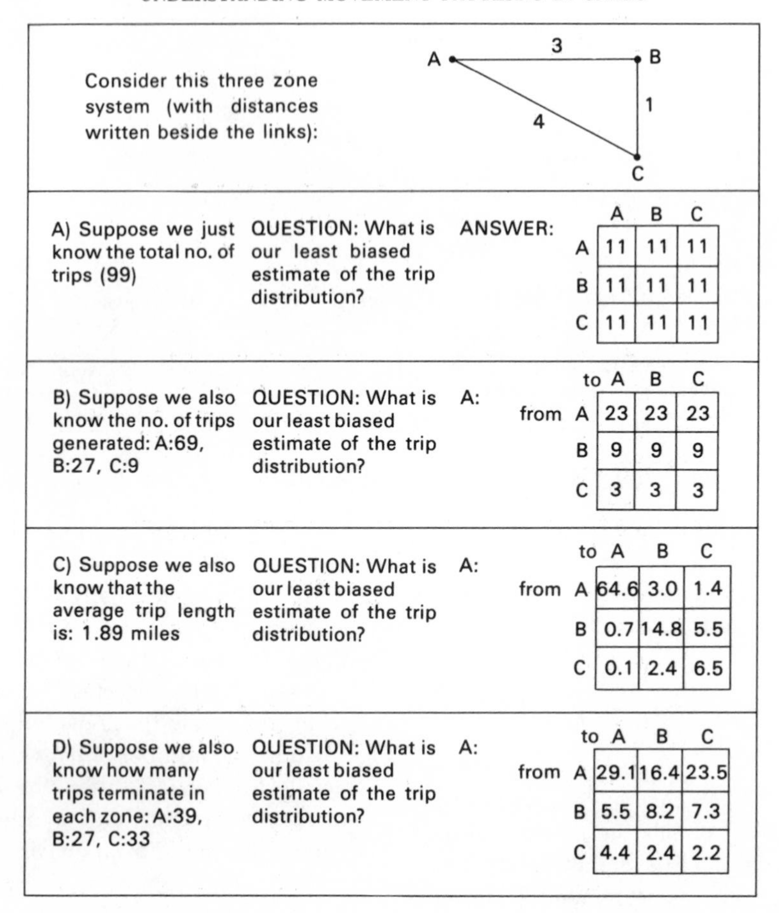

Figure 5-10. Least biased (entropy-maximizing) estimates of trip distributions with varying degrees of information.

models are not a substitute for the gravity and intervening opportunities models. Entropy models are for use only when limitations on the available data do not allow these other approaches to be applied.

MODAL SPLIT AND ROUTE ALLOCATION

Once it is estimated how many trips go from each origin and to each destination for a particular purpose, the final steps in an aggregate model are to estimate how many trips use different modes of transport and which routes are followed by that traffic. Mode use is important because it converts trips into the number of vehicles of each type in use, whereas knowledge of the routes tells us which links in the urban highway system must cope with these vehicles. Thus the two together help us predict how much traffic load will be carried by each street.

Modal Split

Mode selection is usually seen as selection between just two broad possibilities; private cars and trucks on the one hand, and public mass transit of the other. Certain groups of urban travelers can be virtually eliminated before considering mode selection, simply because they have little choice. Passenger travel undertaken by those who cannot afford, or cannot drive, an auto must generally take mass transit. Goods movements, on the other hand, are usually not possible, or not efficient, by public transit, so in this case private trucks and vans will be used. Thus the first step in mode selection is to identify these subpopulations, allocating them entirely to the one mode that they use (see Hutchinson, 1974, for examples). These fixed subpopulations must be identified as fractions of the population of each zone. For the remaining fractions, modal split models can then be applied.

It is argued that if the transit-captive fraction can be identified and eliminated, then the remainder will indeed be making a conscious choice between the two modes. The factors that have generally been found to be most important in influencing this choice are measures of the difference in travel time and cost associated with using the two modes between each origin and destination. Two methods have been widely used to predict modal split from this information: *diversion curves* and *choice models.*

The diversion curve is constructed by drawing a graph plotting the percentage of people using public transit against some measure of the difference in the attractiveness of the two modes. For example, if we measure this difference as equal to ratio of the door-to-door travel time by mass transit $(T(m))$, to that for the automobile $(T(a))$, then a diversion curve will look like that shown in Figure 5-11. These curves are typically not just drawn

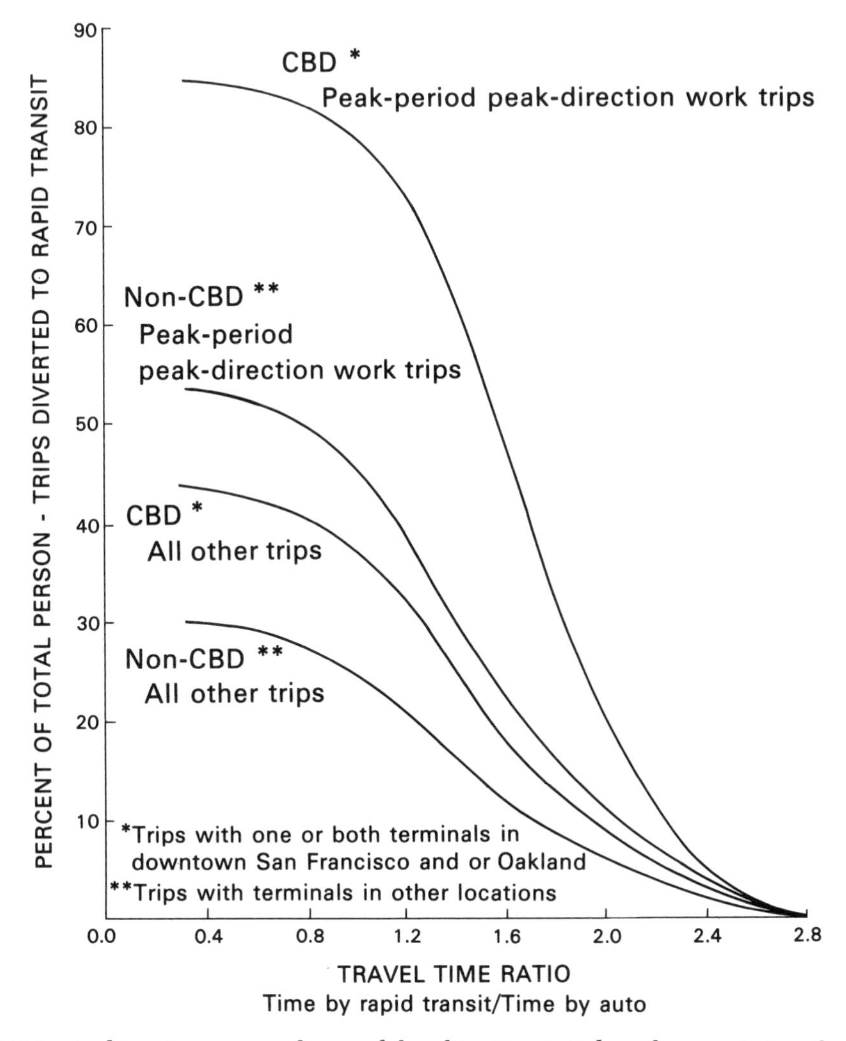

Figure 5-11. A diversion curve for modal split. Reprinted with permission from Traffic Engineering, Vol. 31, No. 5, p. 27, published by the Institute of Transportation Engineers.

for one measure of this difference, but are recalculated for different combinations of service and monetary cost for the two modes. Such diagrams are plotted based on data collected at one point in time and are then used to predict future modal splits. For example, if one effect of some new transport plan is that the ratio of travel times for the two modes between two particular zones is predicted to be 0.8, then from Figure 5-11, the predicted modal split along that route for CBD peak-period peak-direction work trips will be 81% by mass transit (and thus 19% by private car).

If 20% of the population in the zone of origin cannot afford to own an auto and thus travel by mass transit, then the total number traveling by mass transit will equal this 20% plus 81% of the remaining 80%, or 84.8% of the zonal population. Thus, 15.2% of the zonal population drive to work.

Choice models use a regression-type format to estimate the probability that a randomly selected traveler between two zones will use one of the two modes. A commonly used version of this hypothesis is:

$$P^m = \frac{e^{[a + b\{T(m) - T(a)\} + c\{C(m) - C(a)\}]}}{1 + e^{[a + b\{T(m) - T(a)\} + c\{C(m) - C(a)\}]}} \qquad (19)$$

This is known as a *logit model*, a concept that will be discussed in detail in chapter 8. P^m is defined as the probability that a traveler on a route will use mass transit; $T(a)$ and $T(m)$ are defined above; and $C(a)$ and $C(m)$ are the monetary costs of travel for the auto and mass transit modes, respectively. a, b, and c are empirical parameters to be estimated by regression techniques. If we have data on the proportion of auto and mass transit riders between a number of destinations, and on the time and monetary costs of those trips, these data can be entered into equation 19, and the empirical parameters can be estimated. These values can then be used to predict modal splits between other origin destination pairs, or at some future point in time.

This approach generates a curve, showing how modal split responds to the differences between travel convenience on the two modes, of the reverse S-shape shown in Figure 5-12. Obviously, with only two alternatives, if we estimate the proportion of people using mass transit, the proportion using automobiles (P^a) is easily calculated as:

$$P^a = 1 - P^m \qquad (20)$$

Route Assignment

Route assignment is the method used for deciding exactly which routes will be taken by the automobiles, vans, and trucks traveling between each origin and destination. Typically mass transit trips are not treated by these methods, since the mass transit system is rarely so congested and so complex that passengers feel the need to, or have the chance to, choose other than the shortest route. For auto travel, however, the geometry of urban road systems is such that if all travelers were to take simply the shortest

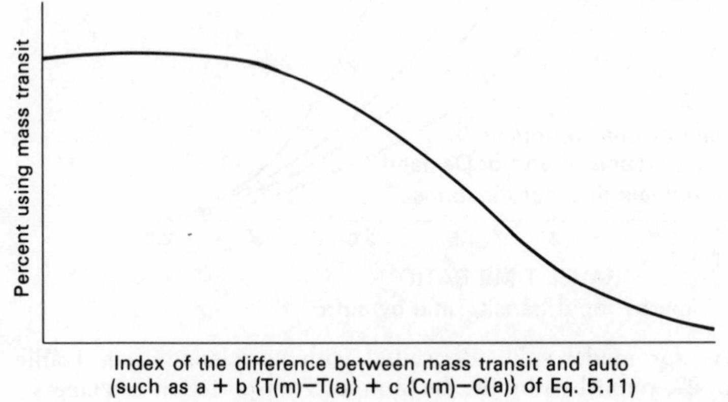

Figure 5-12. Response curve for a logit modal split model.

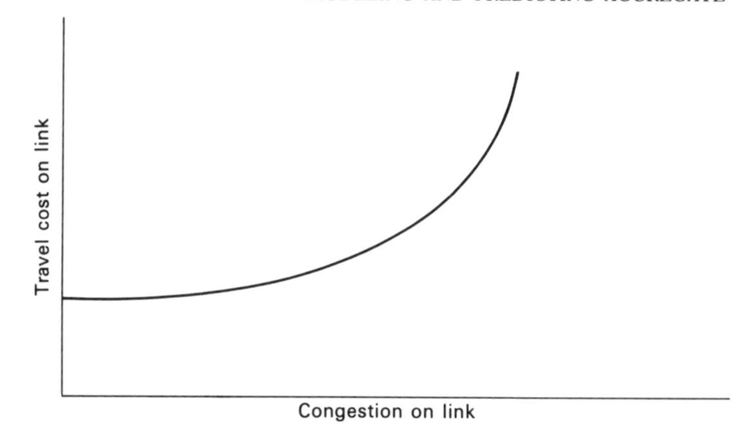

Figure 5-13. Relation between congestion and travel cost.

route from home to work, say, they would all converge on a few bottlenecks. These would then become so congested that travelers would be compelled to seek an alternative route. We wish to model how these alternatives are selected. Obviously, we are going to be particularly interested in trips made during the rush hour, since it is at this time that roads become most congested.

The basis of this approach is to recall that a route from any origin to any destination is made up of a sequence of links. Any different sequence of links that connects the origin and destination is a possible route. The amount of traffic on a link is equal to the number of trips made by people traveling between all origins and destinations in the city who use routes that include that link. Finally, there is a relationship between link useage (or congestion) and the cost of travel on that link (see Figure 5-13).

Other than the strategy of just assigning all travelers along what (in the absence of congestion) would be the shortest route between each origin and each destination, thus neglecting the problems of congestion altogether, there are two basic approaches to the assignment problem. Each is an incremental approach whose goal is to discover that pattern of traffic assignment that leads to a situation where the (perceived) costs of travel between each origin and each destination are the same no matter which route travelers use between those places; there is no cheaper unused alternative. This situation of network equilibrium is known as *Wardrop's first principle of network equilibrium* (Wardrop, 1952).

The *incremental assignment* approach allocates a limited number of trips at a time along the cheapest available route, updating after each allocation the new travel costs on each link as a result of these allocations. Thus one might start with the first origin and allocate all of its trips to all destinations along the shortest routes. Having done this, graphs such as that of Figure 5-13 would be used to calculate the travel costs on each route as a result of these allocations. Then, based on these new costs, all trips from the second origin would be placed along the cheapest routes available to them. Then the costs would be updated again, origin three's trips would be allocated, and so on, proceeding in this manner until all trips have been allocated. Such an approach would obviously result in a biased pattern, since the first origin's trips would conform more closely with the theoretical congestion-free shortest routes than would the trips from any other origin. Thus a more flexible variant of this approach is not to allocate all of the trips from one origin at once. For example (see Putman, 1983), we could go through the entire sequence of origins once, only allocating 40% of the trips from each origin. Then the next 40% could be allocated on a second round, with the final 20% being allocated on a third round.

The *multipath assignment* approach is a fairly simple extension of the incremental approach, based on the proposition that at any stage in the allocation process all trips from an origin will not be assigned along the least-cost route to each destination. Thus instead of proceeding as above (allocating all trips at each step along the cheapest route available at that time) different routes between the origin and the destination will be examined. The proportion of trips allocated along each route available will then depend on the relative length of that route compared to others, taking into account the costs of travel on each link at that stage in the allocation procedure. The justification of this approach is that travelers' perceptions of the cost of a particular route may vary from our measurement of its cost, and therefore we cannot expect all of them to use what we calculate to be the cheapest route.

Putman (1983), in a comparison of these alternatives, finds that the latter approach yields the more satisfactory results, but that it is more time consuming to operationalize. There is indeed an ongoing debate among transportation planners as to whether the increment in knowledge that is obtained from trip-assignment procedures is worth the effort involved (Hutchinson, 1974).

TRAVELERS' WELFARE

As pointed out in the introduction, there are two somewhat different policy-related questions that emerge from aggregate transportation modeling. The first set of questions has to do with imbalances in the use of transport facilities in the city. The aggregate approach outlined in this chapter provides basic data that can be used to evaluate transport use, allowing planners to make proposals about increasing the efficiency of that use. If it were true that all travelers in the city were equally able to, and did, exercise free choice in traveling about the city, then we could interpret observed behavior as being people's preferred behavior as well, and any increase in efficiency of the transport system could generally be seen as being in the interest of the city's inhabitants.

Indeed, under these assumptions economics has developed a methodology for measuring the welfare that consumers of any good obtain from changing the way that the good happens to be provided. This methodology of *welfare economics* is based on two concepts: that a welfare-maximizing pattern of provision is one that maximizes the sum of *consumer and producer surplus* (defined as the benefits that some consumers or producers have when compared to the consumer in the worst position in the city, and the least efficient producer in the city, respectively), and that such a welfare-maximizing pattern of provision will be *Pareto optimal* (defined as a situation where consumer and producer surplus are at a maximum, subject to the condition that no one is worse off after changes in the pattern of provision than he or she was beforehand). If the good we are considering is the provision of transport facilities in a city, then researchers have shown that in making recommendations to improve the transportation system, the best possible improvement from the point of view of welfare economics is the one that maximizes the overall accessibility of all places to all other places in the system, taking into account people's choice-making behavior (Coelho & Williams, 1978; Neuburger, 1971).

These conclusions, however, are based on the assumption that all participants in the urban transportation system are equal participants in some sense; this is a necessary condition in order for the invisible hand of the market to work. However, market-based, capitalist societies do not meet these ideals. There are differences in income, differences in political power, and differences in economic position between, say, a blue-collar worker who needs employment in order to pay the bills and an entrepreneur who may or may not choose to hire particular workers depending on the profit he or she can make. Such questions may seem remote from considerations of transportation in the city, but they are not.

Consider, for example, the use of Pareto optimality as a criterion by which planners decide whether a transportation system is welfare maximizing or not. By definition this criterion does not allow anyone, rich or poor, to be worse off than he or she was beforehand. This may seem to treat everyone equally, but it does so assuming that the status quo should indeed be maintained. Thus for all its apparent equity this is fundamentally a conservative strategy. If you choose to use this approach for transport planning as outlined you are only doing the right thing if you believe that society is optimally organized as it is. If you do not believe that it is, then your equity criterion must allow for a more radical redistribution. Indeed, any more radical plan is likely to be ineffective unless it is combined with political action on a broader scale.

Such issues are not only important in planning for the future, but they also require us to look historically at how the current urban transportation system has evolved. For example, it is often noted how mass transit is a rather redundant mode of transportation in many American cities, since large parts of these cities are developed at too low a density for mass transit to be able to compete economically with the private car. This reason is then used to put money into highway construction rather than mass transit in planning for future transport needs. But consider for a minute how this situation came to be. As Yago (1983, pp. 49–76) has shown, the decline of mass transit in American cities was in part the result of actions by General Motors, who bought up and then ran down mass transit facilities, reducing their competitiveness with the automobile, during the 1930s and 1940s (see also Snell, 1974; Whitt, 1982). This made autos artificially cheaper (for those who could afford them) than mass transit, auto ownership boomed, and soon thereafter low-density suburbanization also exploded (further subsidized for middle-income families by tax breaks introduced on home mortgages), giving rise to this low-density phenomenon. In the light of this, and considering that this has made American cities into much higher consumers of energy for both transportation and heating than cities elsewhere, is it really such a good idea to plan for this status quo to continue? Only by considering such questions from a point of view that allows for the possibility of criticizing the status quo can we come to a balanced evaluation of how an equitable urban transportation system should be constructed.

When this side of transportation planning is considered, aggregate models of the type outlined in this chapter are often of limited use, and indeed they are no substitute for analyzing these questions. They sketch the empirical surface of actual use patterns but do not probe deeper to ask about the processes in society that bring this about. They can still be useful, however, in evaluating the likely use of any transit system that is planned with these deeper issues in mind.

References

Ayeni, B. (1979). *Concepts and techniques in urban analysis*. London: Croom Helm.

Bennett, R. J. (1978). Forecasting in urban and regional planning closed loops: The examples of road and air traffic forecasts. *Environment and Planning A, 10*, 145–162.

Black, J. (1981). *Urban transport planning*. Baltimore: Johns Hopkins University Press.

Coelho, J. D., & Williams, H.C.W.L. (1978). On the design of land use plans through locational surplus maximization. *Papers of the Regional Science Association, 40*, 71–85.

Cliif, A. D., Martin, R. L., & Ord, J. K. (1974). Evaluating the friction of distance parameter in gravity models. *Regional Studies, 8*, 281–286.

Fotheringham, A. S. (1981). Spatial structure and distance decay parameters. *Annals of the Association of American Geographers, 71*, 425–436.

Hensher, D. A. (1979). Demand for urban passenger travel. In D. A. Hensher (Ed.), *Urban transport economics* (pp. 72–99). Cambridge: Cambridge University Press.

Hutchinson, B. G. (1974). *Principles of urban transport systems planning*. New York: McGraw-Hill.

Kanafani, A. (1983). *Transportation demand analysis*. New York: McGraw-Hill.

Meyer, J. R., & Straszheim, M. R. (1971). *Pricing and project evaluation*. Washington, D.C.: Brookings Institution.

Neuberger, H.L.I. (1971). User benefit in the evaluation of transport and land use plans. *Journal of Transport Economics and Policy, 5*, 52–75.

Olsson, G. (1975). Servitude and inequality in spatial planning. *Antipode, 8.*

Putman, S. H. (1983). *Integrated urban models.* London: Pion Press.

Schneider, M. (1960). Appendix to panel discussion on inter-area travel formulas. *Highway Research Board Bulletin, 253,* 136–138.

Seley, J. E. (1983). *The politics of public-facility planning.* Lexington, Mass.: Lexington Books.

Sheppard, E. (1980). The ideology of spatial choice. *Papers of the Regional Science Association, 45,* 197–213.

Sheppard, E. (1984). The distance decay gravity model debate. In G. Gaile & C. Willmot (Eds.), *Spatial statistics and models* (pp. 367–388). Dordrecht: D. Reidel.

Snell, B. C. (1974). *American ground transport.* Washington, D.C.: Subcommittee on Antitrust and Monopoly of the Judiciary Committee, U.S. Senate.

Stouffer S. (1940). Intervening opportunities: A theory relating mobility and distance. *American Sociological Review, 5,* 845–867.

Taylor, P. J. (1975). *Distance decay models in spatial interaction.* Norwich, England: Geo-Abstracts.

Vickermann, R. W. (1974). Accessibility, attraction, and potential: A review of some concepts and their use in determining mobility. *Environment and Planning A, 6,* 675–691.

Wachs, M., & Koenig, J. G. (1979). Behavioral modelling, accessibility, mobility and travel need. In D. A. Hensher & P. R. Stopher (Eds.), *Behavioral travel modelling* (pp. 698–709). London: Pion Press.

Wardrop, J. G. (1952). Some theoretical aspects of road traffic research. *Proceedings, Institute of Civil Engineers, London, Part II,* 1.

Webber, M. J. (1984). *Explanation prediction and planning: The Lowry model.* London: Pion Press.

Whitt, J. A. (1982). *Urban elites and mass transportation.* Princeton, N.J.: Princeton University Press.

Wilson, A. G. (1970). *Entropy in urban and regional analysis.* London: Pion Press.

Yago, G. (1984). *The decline of transit.* Cambridge: Cambridge University Press.

6 ANALYSIS OF AGGREGATE FLOWS: *THE ATLANTA CASE*[1]

RONALD L. MITCHELSON
JAMES O. WHEELER
University of Georgia

Atlanta in the 1980s is a low-density, rapidly growing metropolitan area serving as the transportation hub of the southeastern United States. The low densities of population and the dispersed economic activities have evolved as a result of heavy reliance on the automobile and its associated street, highway, and freeway patterns. The city spread in all directions beyond the boundaries of the political city of Atlanta to become a metropolitan area of over 2.4 million people in the mid-1980s. In addition to the Metropolitan Area Regional Transportation Authority (MARTA) bus service, MARTA opened a rail rapid transit system in 1979, which will eventually extend a total of 53 miles.

The rapid growth of Atlanta into a major metropolitan center is related to the city's early role as a transportation hub. The first settlement in 1843 in what is now downtown Atlanta occurred at the terminus of a rail line stretching eastward from Charleston, South Carolina. Soon other rail lines entered the city, and Atlanta gained the position as a rail distribution center for the southeastern United States. With the development of the paved road system during the 1920s and 1930s, Atlanta developed better road connections with other places than did any other city in the region. As early as 1923 paved roads spread out from Atlanta in six directions, and in subsequent years the accessibility advantages of Atlanta via paved roads ensured the city's continued growth and expansion (Figure 6-1). During the 1930s Atlanta emerged as a major center for air passenger service. Flying from anywhere in the Southeast to another part of the country entails first flying to Atlanta to connect with flights to destinations in Los Angeles, Chicago, New York, London, or wherever. Today Hartsfield International Airport is the second busiest airport in the United States in number of passengers enplaned.

The development of the Interstate Highway System gave Atlanta yet another major growth impetus. With the availability of federal funds for interstate construction in the late 1950s and 1960s, Atlanta gained two ad-

[1]The material in this chapter is based upon information provided in the Atlanta Regional Commission publications listed at the end of this chapter.

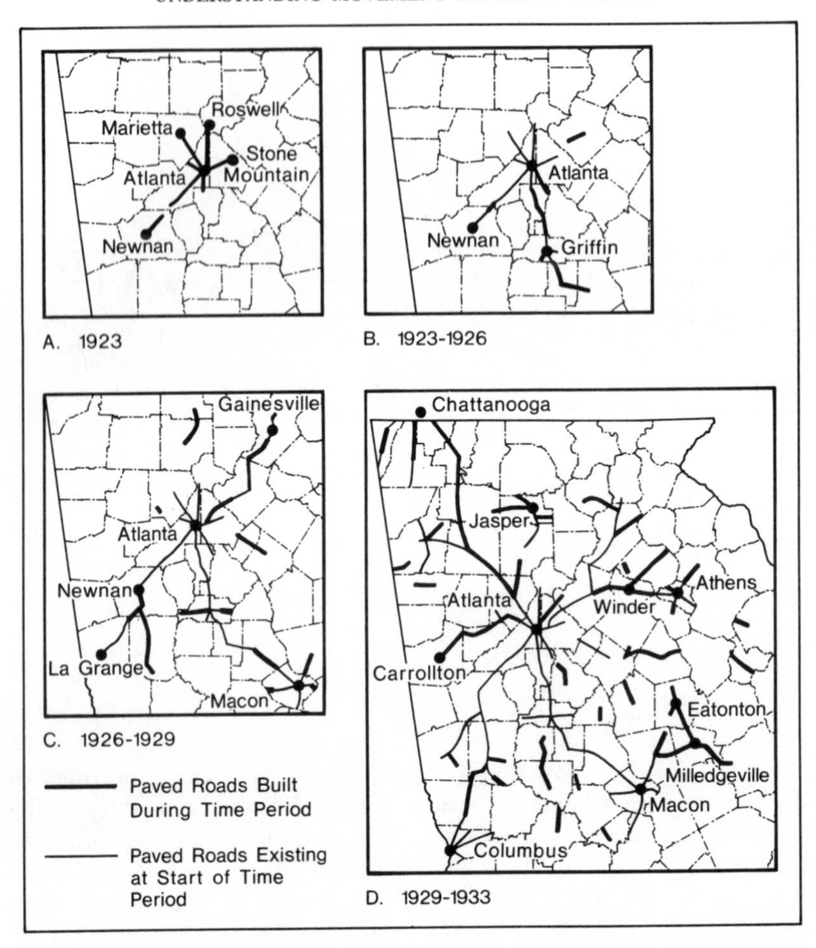

Figure 6-1. The diffusion of paved roads out of Atlanta and the growing connectivity of the highway system, 1923–1933. From: James A. Liesindahl, "Expansion of the Paved Road in the State Highway System of Georgia, 1923 to 1962," M.A. thesis, University of Georgia, 1964.

vantages. First, the routing of the interstate highways gave Atlanta superior connections with other metropolitan areas in the region. I-75, I-85, and I-20 converge on the Atlanta metropolitan area from six directions. Especially significant is I-75, one of the major north–south arteries connecting the Middle West with Florida. Second, the construction of the interstates and other freeways within the metropolitan area provided excellent *internal* accessibility. The six-spoke radial interstate system is connected by a 64-mile perimeter highway, I-285, to enhance lateral and circumferential movement (Figure 6-2). The Lakewood Freeway in the southwest, the Stone Mountain Freeway to the east, and Georgia 400 to the north round out the limited-access highway system of the metropolitan area. Because Atlanta got an early start in the construction of interstates, the metropolitan area was able to provide needed automobile transport service earlier than many cities, before lawsuits brought by neighborhood groups delayed or halted construction as happened in some cities. Not only did Atlanta get an early start in its interstate program, but the metropolitan area has continued to spend federal funds to add lanes to the existing system and to modernize several interchanges. For example, the new interchange at I-285 and I-85 will cost $500 million.

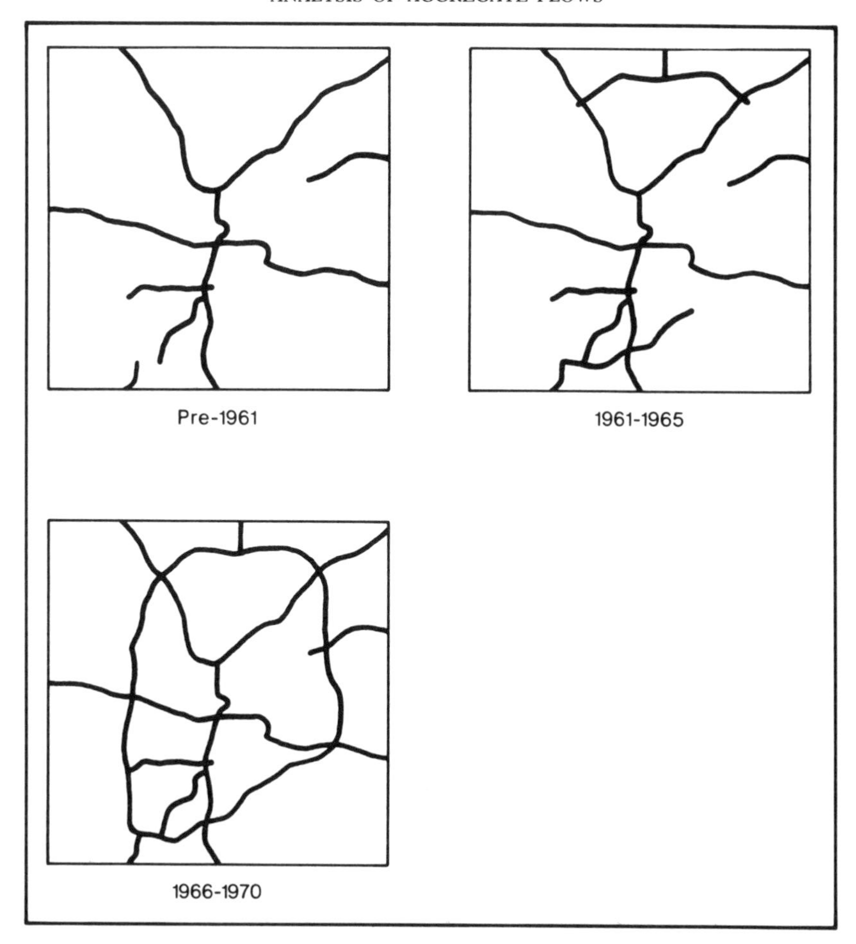

Figure 6-2. The evolution of Atlanta's freeway system. The system was completed relatively early compared to most metropolitan areas. During the 1970s and 1980s the interstates have been widened and improved, though no new routes have been constructed.

Although the interstate system and other highway construction have encouraged dispersal of population and economic activities throughout the metropolitan area, growth has by no means been uniform. First, the city of Atlanta is 66% black, and most blacks live to the south and west of the downtown (Figure 6-3). Whites have been moving to and establishing suburban areas in north Atlanta; blacks have spread toward the southwestern part of the metropolitan area. Second, the growth of retail shopping and employment, especially office employment, has been greatest in the northern portions of the metropolitan area. With the decline of the central business district (CBD) as a retail center, several huge integrated shopping malls were opened in the late 1960s and 1970s. The largest and most spectacular of these malls are located in the northern area of the metropolitan area (Figure 6-3). It is no accident that these malls are located along the interstates and major interchanges. Though there are of course shopping malls in southern Atlanta, they are not of the size or quality of the malls in northern Atlanta. Thus, Atlanta may be roughly characterized as an affluent, white, economically vibrant north and a largely black, usually less affluent, and economically more stagnant southern part. Nonetheless, Atlanta has a significantly large, upper-income black population.

One of the most significant developments in the Atlanta metropolitan area since the late

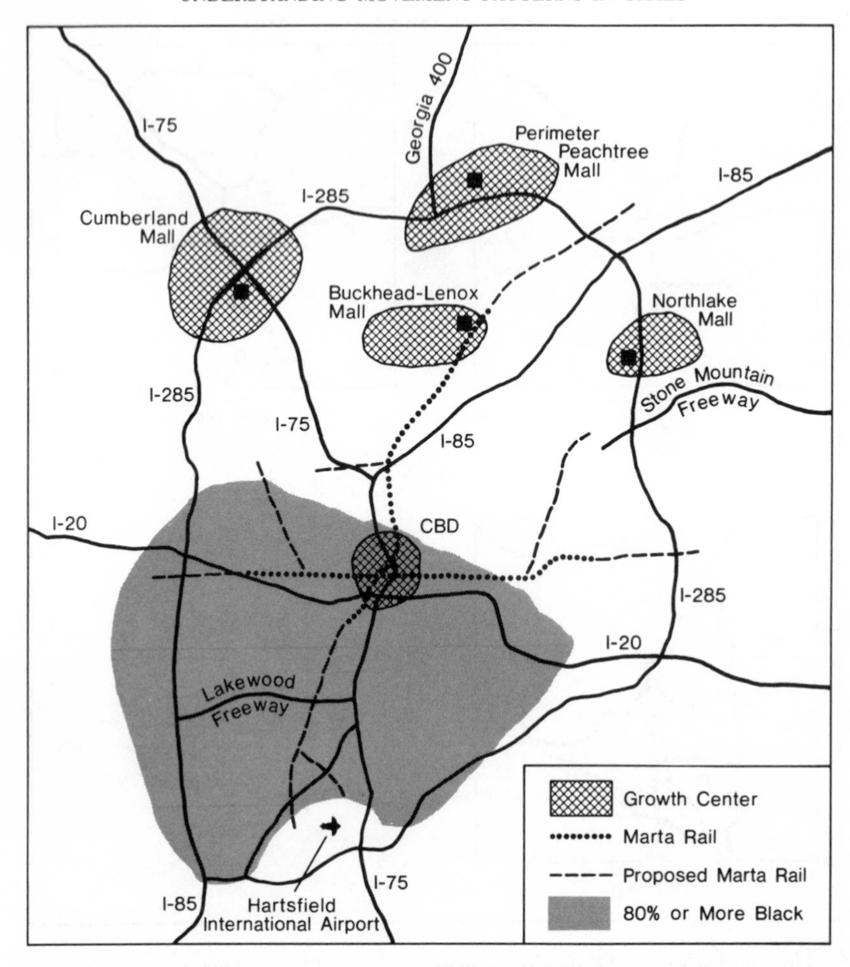

Figure 6-3. The freeway system in Atlanta and the location of major growth centers, especially suburban downtowns. The black residential area is to the west and south. Also shown here are the current (1985) and proposed routing of MARTA's rapid rail system.

1970s has been the development of suburban downtowns. Centered around suburban shopping malls, these suburban downtowns boast high-rise office buildings, hotels, banks, condominiums, restaurants, and apartments. Incredible traffic volume has been the result. Two-thirds of the new office space being constructed or planned in Atlanta in the next few years will be located within the Golden Crescent from Cumberland Mall to Perimeter to Northlake. Single-family houses built within the past 15 years are being razed to make room for office complexes and hotels. Atlanta, with the third largest convention trade of any U.S. city, is seeing more of the conventioners in the suburban downtowns. Except during rush hours, it is only about a 20-minute taxi ride from the airport to the hotels along I-285 to the north. The new and still expanding suburban downtowns have added a fundamental change to the spatial structure of the metropolitan area, a trend already observable in other centers such as Houston. Atlanta has become a truly multinodal metropolitan area. One result is that Atlanta has a significantly higher proportion of suburb-to-city and suburb-to-suburb commuting than does the average American city (see Table 6-1). The peripheral journey to work, that is, from suburb to suburb, is especially pronounced in the Atlanta case. In fact, the 64-mile outer belt (I-285), which was

Table 6-1. Generalized commuting flows, Atlanta, 1980

	Area (%)	
Commuting flow	United States	Atlanta
City to city	33	12
City to suburb	7	5
Suburb to city	20	25
Suburb to suburb	40	58

Note. From *Regional Review* (vol. 2, p. 1), by U.S. Bureau of the Census (Atlanta Regional Office), 1984.

completed in 1969, had certain segments (especially those in the rapidly growing Northeast sector) that serviced 120,000 vehicles per day in 1980. Average flows for the Atlanta beltway were higher than those in any other city examined in a recent U.S. Department of Transportation (1980) report (including such highly decentralized cities as Columbus, Ohio).

Downtown Atlanta, of course, continues to be a major node with many high-rise hotels, bank buildings, office buildings, and headquarters of major corporations. But the CBD is now of only minor significance as a retail node. The governmental function remains important downtown. Areas and buildings in the downtown are being redeveloped for offices, and in some cases for condominiums. Some gentrification has been occurring in inner-city neighborhoods of the near north side primarily by young professionals who work in downtown offices. Fringing the CBD are educational institutions, medical facilities, and cultural attractions.

The MARTA rapid rail transit system currently operates approximately 25 miles of track (Figure 6-3). The trains operate at 6-minute intervals on weekdays (10 minutes on weekends) for 25 stops or stations. The system generates approximately 12 million vehicle miles per year, or an average of 185,000 riders for a weekday. The system has a flat fare of 60 cents. Associated with most of the stations are a total of nearly 12,500 parking spaces. The stations interconnect with the MARTA bus lines, allowing the passengers to travel by bus to destinations beyond the rapid rail lines. In addition to rapid rail, MARTA has increased its bus fleet from less than 500 in 1970 to roughly 850 buses by 1980. At the present time, only about 5% of all trips made in Atlanta are taken on public transportation although 35 percent of peak-hour trips to and from the central business district are on public transit.

Where do the routes go and why? The rail system obviously focuses upon downtown, although there are several new suburban downtowns emerging. The system stretches west almost to the I-285 perimeter along I-20 and likewise to the east through Decatur and Avondale, Georgia. The north line now extends through the Buckhead–Lenox Mall growth area, though access to the mall via MARTA involves crossing a busy street. The south line is presently only a stub, but eventually will extend to Atlanta's busy airport.

THE ATLANTA CASE STUDY

In this section of the chapter, we describe portions of the most commonly used tools and methods in the urban transportation planning process to project the demand for personal travel from some base period (the early 1970s in the case of Atlanta) to some future period (the year 2000). In the empirical work that follows, Atlanta is defined to include the counties that comprise the Atlanta Regional Commission (ARC) planning area (see Figure 6-4). Because this seven-county region is changing rapidly in terms of population, employment, and provision of transportation services, it provides a great challenge to successful planning.

The following subsections essentially conform to the major steps in the urban transportation planning process:

1. A brief inventory of Atlanta in the base years—the early 1970s;
2. A forecast of the essential features of the Atlanta landscape (number of households, household size, income, auto ownership, land uses and their respective intensities),

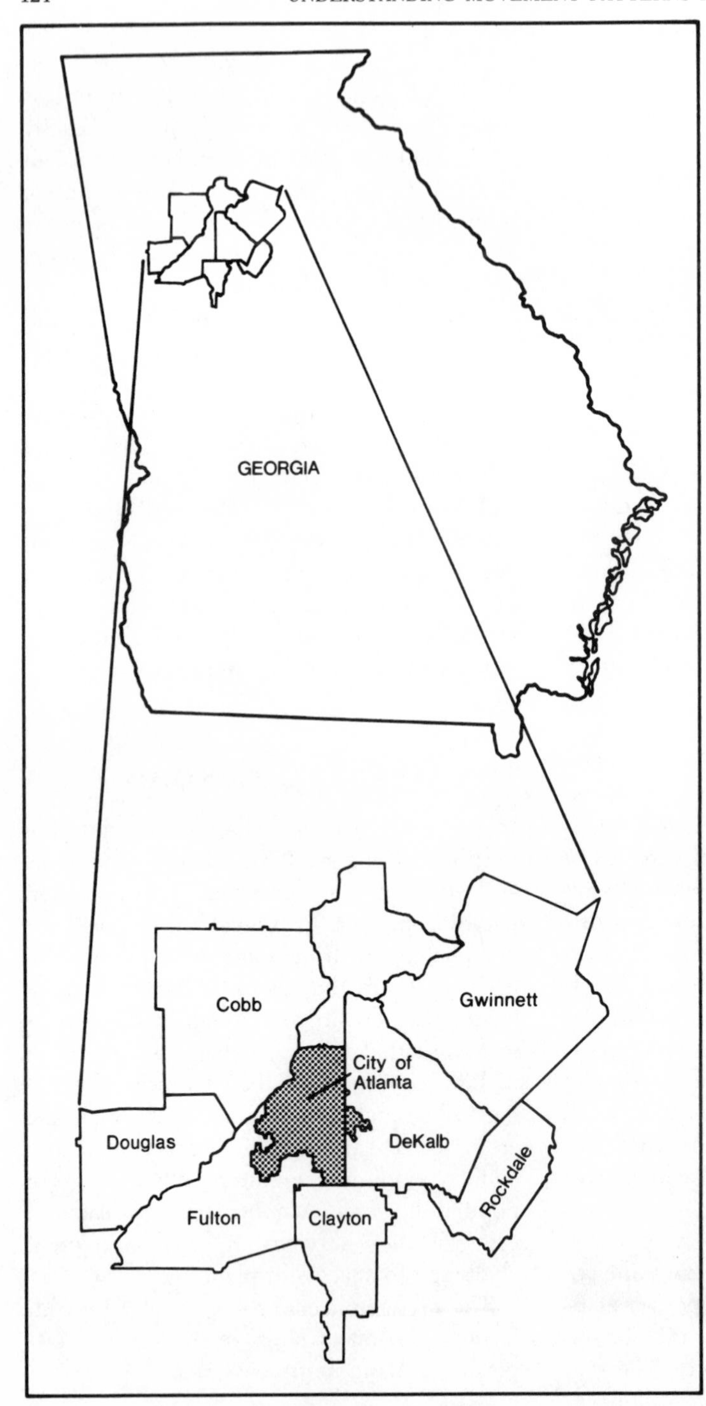

Figure 6-4. The Atlanta Regional Commission's planning area within the state of Georgia. The seven-county region is one of the most dynamic areas in the U.S.

needed in order to project travel demand in the year 2000;

3. Description of the fundamental trip-generation relationships in the base period (the early 1970s) required to estimate person-trip productions and attractions by relatively small geographic areas (comprising the study area) for the year 2000; and
4. Calibration and application of the most

commonly used model to estimate the spatial distribution of person trips for the year 2000, that is, the gravity model.

Atlanta Inventory, the Early 1970s

The first step in the urban transportation planning process is to project to the year 2000 those urban features that will shape travel patterns. This subsection provides a brief inventory of the urban features necessary to forecast travel needs in the year 2000 for the city of Atlanta. Included is a discussion of population, employment, household characteristics, land use, and the highway and transit systems as manifest in the study area during the base planning period, the early 1970s. Subsequently, the values of these basic planning features are extrapolated to the year 2000.

Figure 6-3 shows a detailed map of the immediate Atlanta environment. Limited-access highways, major state roads, the (then) proposed rail rapid transit system, major retail centers, and industrial centers are all indicated. Figure 6-5 is a companion map of residential development in Atlanta with the urbanized area outlined and recent housing developments highlighted with respect to location of Atlanta's outer belt (I-285) and other limited-access routes (I-20, 75, and 85). Note the residential void on the northwest side of the city, which corresponds to the Chattahoochie River floodplain.

The greater Atlanta area has enjoyed rapid population and employment growth since 1960. The growth history of the seven-county

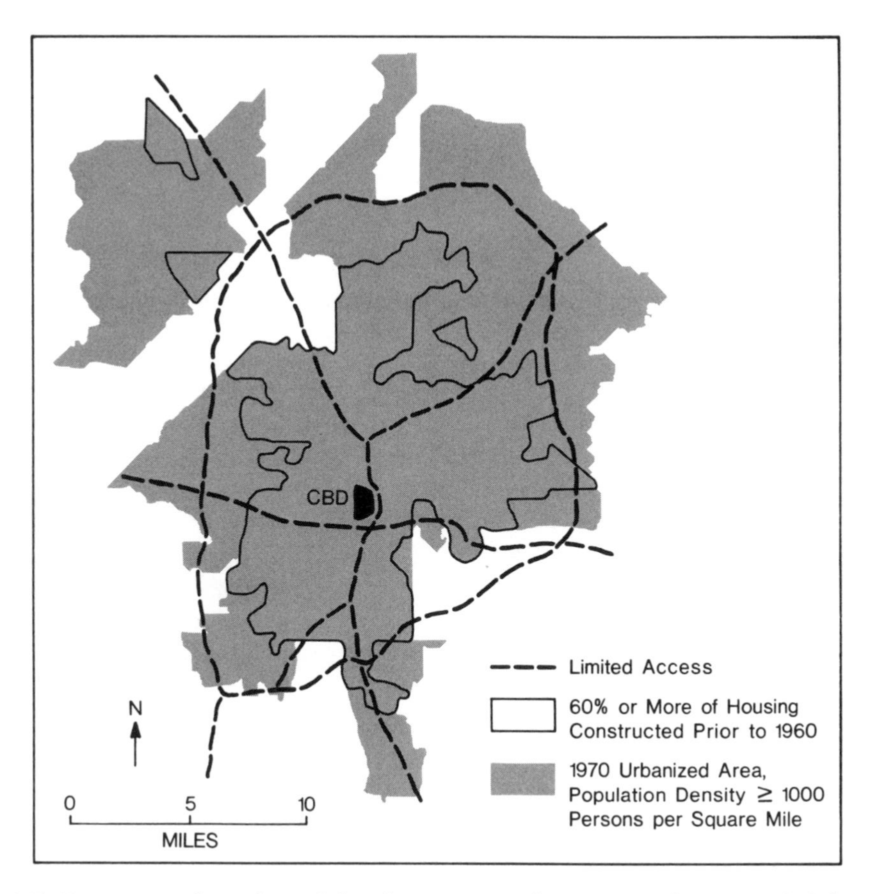

Figure 6-5. Freeway and residential development in Atlanta. Note that post-1960 development is peripherally located. From *Urban Atlas* by U.S. Bureau of the Census, Washington, D.C.: U.S. Government Publishing Office, 1974.

Table 6-2. Atlanta's growth and change, 1960–1980

Census year		Atlanta city	ARC[a] planning area	Atlanta as percentage of ARC
1960	Population	487,495	1,044,501	46.7
	Area (sq. mi.)	126	2,070	6.1
	Jobs	232,023	421,860	55.0
1970	Population	496,973	1,436,975	34.6
	Area (sq. mi.)	132	2,070	6.4
	Jobs	336,626	634,101	53.0
1980	Population	425,022	1,779,226	23.9
	Area (sq. mi.)	132	2,070	6.4
	Jobs	335,026	884,100	37.9

[a]ARC = Atlanta Regional Commission.

Note. From *Census of Population and Housing,* by U.S. Bureau of the Census, 1960, 1970, 1980, Washington, D.C.: U.S. Government Printing Office.

ARC planning area since 1960 is contained in Table 6-2. The City of Atlanta's relative decline in population and jobs is evident throughout the period, and an absolute decline in population and jobs took place during the 1970s. This ongoing decentralization of residences and jobs has resulted in a great many peripherally focused work trips.

The ratio of population to jobs for the counties of the ARC planning area is roughly 2.3, but considerable spatial variation exists about this aggregate. Table 6-3 contains the population/jobs ratio for each of the seven counties. Fulton County, with 1.5 people per job, serves as the major employment center for the study area (nearly 63% of all jobs). In contrast, counties such as Gwinnett (4.8 people per job) and Douglas (7.1 people per job) serve primarily as residential areas with significant out-commuting to jobs located in Fulton and/or DeKalb counties. In 1970, 57% of the labor force residing in Gwinnett County commuted to jobs in Fulton and DeKalb (the percentage had fallen somewhat to 53% by 1980). Again, the high level of decentralization that characterizes the ARC planning area is highlighted by such statistics.

Table 6-3. Population and jobs by county, Atlanta, 1970

	1970 values		
Area	P[a]	J[a]	P/J
Fulton County	605,582	396,649	1.5
DeKalb County	415,387	125,423	3.3
Cobb County	196,793	62,805	3.1
Clayton County	98,043	25,116	3.9
Gwinnett County	72,349	14,907	4.8
Rockdale County	18,152	5,165	3.5
Douglas County	28,659	4,045	7.1
Region (ARC)[b]	1,436,975	634,101	2.3

[a]P, population; J, jobs.
[b]ARC = Atlanta Regional Commission.

Note. From *Regional Development Plan, Final Small Area Forecast of Population, Jobs, and Housing Units,* by ARC 1975, Atlanta, GA: ARC.

County-level land use characteristics are shown in Table 6-4 along with regional aggregates. The level of vacant lands (60% of total) is, once again, testimony to the highly decentralized urban form of Atlanta. The central counties of Fulton and DeKalb have the greatest intensity of land use, that is, the largest percentage of combined residential and commercial–industrial acreages, in the region.

In 1970 the existing transportation system of the Atlanta region included bus service and private forms of highway transportation. Construction of Metropolitan Atlanta Rapid Transit Authority's (MARTA's) rail system did not begin until 1975, and the first portions of the system did not open until 1979. In 1970 of the 4.2 million person trips taking

Table 6-4. 1970 land use in acres[a]

Category	Fulton	DeKalb	Cobb	Clayton	Gwinnett	Rockdale	Douglas	Total
Single family	47,648 (14)	36,305 (21)	25,689 (12)	9,745 (10)	12,664 (4)	3,019 (4)	4,672 (4)	139,742 (11)
Multifamily	6,583 (2)	2,766 (2)	920 (1)	487 (1)	51 (0^+)	33 (0^+)	24 (0^+)	10,864 (1)
Commercial industrial	11,745 (3)	5,873 (3)	3,699 (2)	1,719 (2)	2,252 (1)	418 (1)	463 (0^+)	26,169 (2)
Vacant	189,332 (56)	79,427 (46)	130,825 (59)	60,881 (64)	177,437 (63)	64,947 (79)	89,777 (69)	792,626 (60)
Unsuitable soils, Vac.	45,381 (13)	21,882 (13)	36,537 (16)	12,371 (13)	72,520 (26)	9,393 (11)	29,588 (23)	227,672 (17)
Other	39,305 (12)	25,884 (15)	10,237 (11)	14,754 (11)	4,110 (5)	4,854 (4)	4,854 (4)	124,232 (9)
Total	339,994	172,137	222,758	95,440	279,678	81,920	129,378	1,321,305

[a] Figures in parentheses represent percentages.

Note. From *Regional Development Plan Final Small Area Forecasts of Land Use*, by ARC December 1975, Atlanta, GA: ARC.

place during the average workday, only 3% were transit trips. The percentage of work trips made by transit was significantly larger at roughly 9% (in 1970). Vehicle occupancy rates reflect the dominance of the private auto in servicing person trips. There were only 1.13 people per vehicle in the aggregate for the average Atlanta work trip in 1970 although the nonwork average (1.65 people per vehicle) was somewhat greater as is typically the case.

Table 6-5 and Figures 6-6 and 6-7 summarize basic traffic flows (vehicle miles of travel [VMT]) in the Atlanta region in 1970. This information is based on an expanded home interview trip file completed in 1970 by the ARC. Daily per capita vehicle miles traveled was 16.2 for Atlanta in 1970. Approximately 53% of that travel was on limited-access highways. Traffic on much of the highway system was well under design capacity, that is, the density of traffic beyond which traffic engineers expect unreasonable delay, hazard, or restriction of movement. Planners often use measures of traffic congestion to reflect actual or forecasted use as compared with design capacity. One such measure is the volume/capacity or V/C, ratio. Actual VMT is the numerator while design capacity (also measured in VMT units) is placed in the denominator. A V/C ratio greater than 1.0 generally implies bumper-to-bumper driving conditions during peak-hour traffic. During 1970 in the Atlanta case, only 17% of total vehicular movement (VMT) experienced V/C ratios greater than 1.0, while 11% of VMT on the limited-access portion of the region's highway system experienced V/C ratios above 1.0.

Table 6-5. Daily vehicle miles of travel and congestion levels, 1970

Type of road	VMT[a]	Percentage VMT at V/C[a]>1
Freeways	12,463,766	11.4
Arterials	8,005,817	25.7
Collectors	2,883,117	14.9
Total	23,350,048	16.8
Daily per capita	16.2	

[a]VMT, vehicle miles of travel; V/C, volume/capacity ratio.
Note. From *Regional Transportation Plan*, by ARC, 1977, Atlanta, GA: ARC.

Figure 6-7 reveals the essential features of the spatial distribution of the traffic volumes for the typical weekday during 1970. Again, the importance of the freeway system to efficient circulation in the Atlanta region is highlighted. Interstates 75 and 85, especially to the north, along with east–west-oriented I-20, provide access to jobs in Atlanta proper from outlying areas. Interstate 285 (completed in 1969) provides a feeder to Interstates 20, 75, and 85 while most importantly facilitating peripheral journeys to work (or suburb-to-suburb flows). V/C ratios are mapped in Figure 6-7. Note the extreme capacity problem that existed between the suburban communities of Smyrna, Marietta, and Kennesaw and Atlanta proper along I-75. Interstate 20, approaching Atlanta from the west, provided another set of capacity problems at the interchanges with I-75, 85, and the outer belt (I-285)—only one year after its opening.

Forecast of Population and Employment in the Atlanta Region, 2000

The first step in arriving at the set of values necessary to provide a satisfactory picture of the transportation demands of Atlanta in the year 2000 is to estimate aggregate regional characteristics. These aggregate characteristics of the region are the basis for determining the likely economic activities to be found in Atlanta in the year 2000 and the travel to be derived from those activities. Such critical planning variables as jobs, households, and household characteristics must be included. These regional aggregates are then spatially disaggregated, and appropriate portions are allocated to small areas that comprise the seven-county ARC planning area.

Aggregate employment projections were prepared by using a simple step-down method. This method attempts to determine the level of future economic activity in the study area

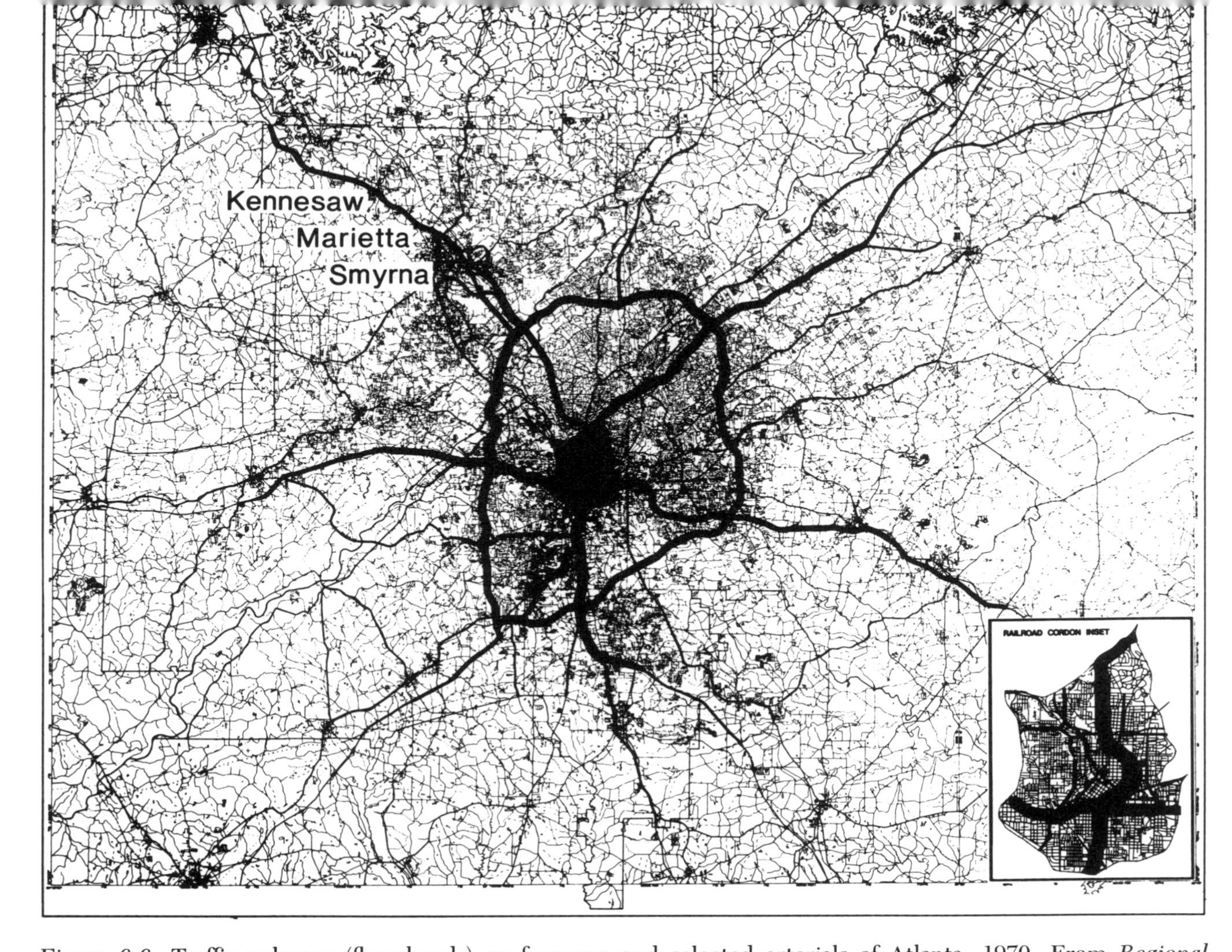

Figure 6-6. Traffic volumes (flow bands) on freeways and selected arterials of Atlanta, 1970. From *Regional Transportation Plan*, by ARC, 1977, Atlanta, GA: ARC.

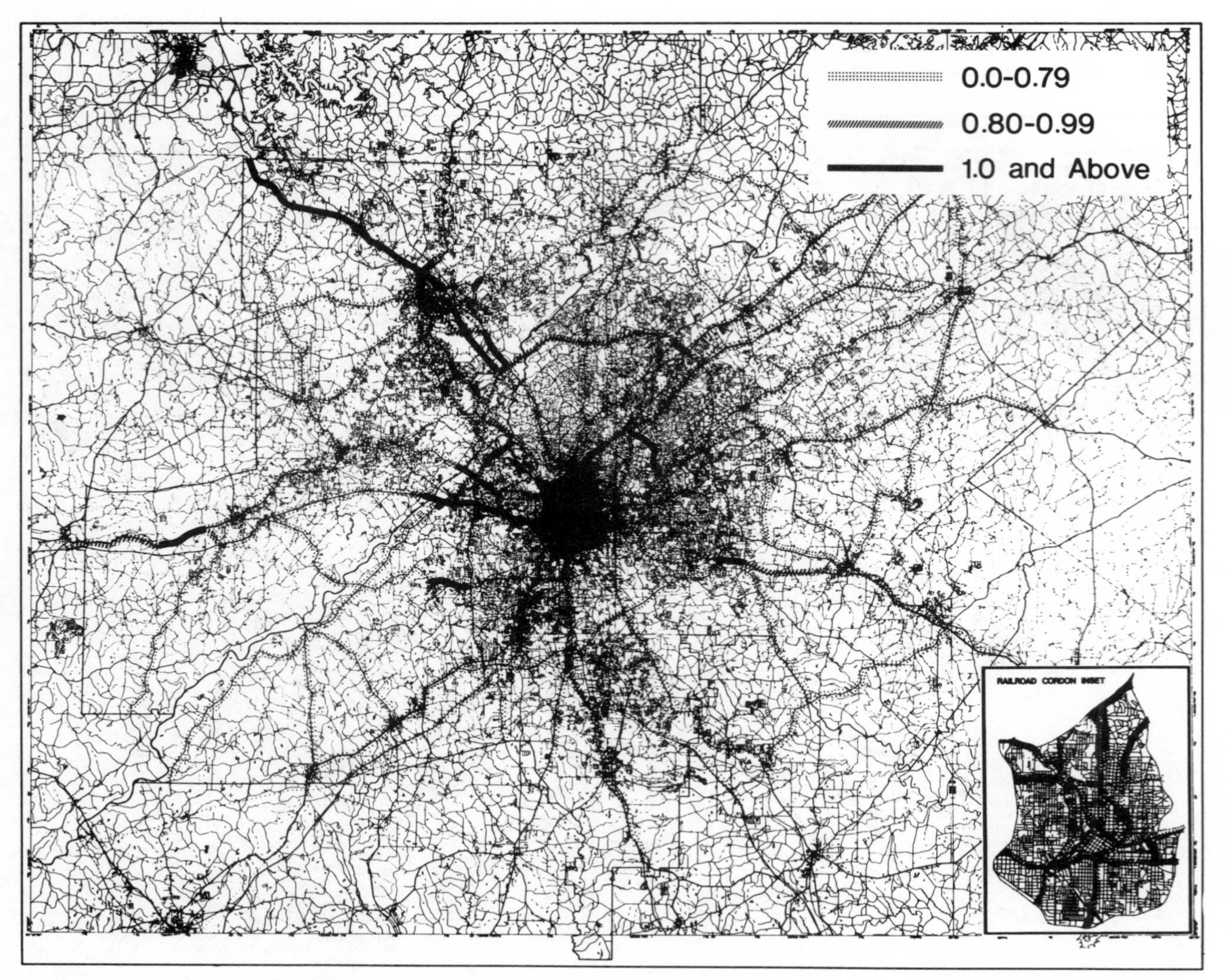

Figure 6-7. Volume/Capacity ratios on freeways and selected arterials of Atlanta, 1970. From *Regional Transportation Plan*, by ARC, 1977, Atlanta, GA: ARC.

by examining the current ratio of such activity to that in larger areas of which the study area is part. The National Planning Association provides yearly projections of U.S. employment. Growth in the Southeast's, Georgia's, and the Atlanta region's share of the U.S. employment growth is based on historic trends.

Employment statistics for the Southeast, Georgia, and Atlanta regions come from the U.S. Bureau of Labor Statistics and its affiliate, the Georgia Employment Security Agency. This time series extends back to 1939 for Georgia and 1949 for the Atlanta region. Table 6-6 is a summary of these projections, formulated by the ARC in March 1975, based on 1970 employment values. The projected 1980 total for the Atlanta region (860,000 jobs) is quite close to the actual 1980 value (880,000 jobs). Manufacturing is the only category of employment that provides a significant discrepancy between 1980 projected employment (153,000) and 1980 actual employment (130,000). ARC planners appeared to be a bit optimistic in their estimates of future growth in this case. In general, however, the estimates have thus far performed well, providing some confidence in extrapolation of their values to the year 2000. By the year 2000 the number of jobs in the seven-county area will increase to roughly 1.6 million from the 1970 base of little more than 0.6 million—an increase exceeding 154%.

With a healthy growth in employment opportunity within the ARC's planning area, it is not surprising that substantial population growth is anticipated through the year 2000 as well. By assuming modest but steady increase in labor force participation rates and assuming moderate unemployment rates throughout the 1970–2000 period, population growth is directly tied to employment growth.

Based on the Atlanta region's anticipated share of U.S., Southeastern, and Georgian growth, the population of the seven-county area will grow by 2.5 times between 1970 and 2000. This involves an increase from 1.4 to 3.5 million (see Table 6-7). By the year 2000 Atlanta's population could well exceed one-half of Georgia's total.

Two of the most important variables in the entire planning procedure are estimates of the number of households in the Atlanta region along with their size and income characteristics for the year 2000. The number of trips that a household undertakes in a 24-hour period is normally estimated on the basis of household size and automobile ownership. However, auto ownership is in large part a function of income level. Therefore, in order to forecast the production of trips, we must forecast income, so that auto ownership levels can then be used in the trip-forecasting procedure. Good forecasts of household size are crucial because, other things being equal, large household units make more trips than do smaller units.

Since the early 1960s the number of persons per household has been declining sharply in the United States. This trend is expected to continue, although a deceleration is expected by the latter half of the 1980s. The Atlanta region will share in these general events, so that the number of households is expected to grow at a significantly faster rate than population, per se. Table 6-8 provides a summary of anticipated regional household growth between 1970 and 2000. The number of households will grow by roughly 300%, but household size will shrink from 3.2 persons to 2.8 persons.

The Atlanta region's share of U.S. and Georgia income is expected to continue the historic growth of the 1960s and 1970s. Table 6-9 shows the income distribution of households for the Atlanta region in 1970 and that expected for the year 2000 (in constant 1970 dollars). The median household income is expected to climb from $10,600 to $21,800. A general upward creep of households into higher income categories is evident. Assumptions involving a greater share of households with two wage earners, a sectoral shift toward more high-paying technical jobs, and a greater share of population in the 25-to-64-year age group (which has fewer nonearners) are fundamental to this projected income growth.

Table 6-6. Civilian nonfarm wage and salary employment, United States, Southeast,[a] Atlanta Region, 1970–2000

	United States (1000s)				Southeast (1000s)				Atlanta region (1000s)			
	1970	1980	1990	2000	1970	1980	1990	2000	1970	1980	1990	2000
Manufacturing	19,369.4	22,210.4	25,261.6	28,514.6	2,815.6	3,682.4	4,575.0	5,571.3	123.8	152.9	214.7	291.9
Government	12,535.0	15,455.0	18,565.0	21,855.0	1,675.9	2,163.7	2,710.5	3,322.0	95.8	129.0	174.9	230.6
Services	11,630.0	15,180.0	18,970.0	22,950.0	1,278.3	1,776.1	2,371.2	3,075.3	94.3	143.8	205.3	284.0
Retail trade	11,098.0	13,348.0	15,728.0	18,208.0	1,431.3	1,855.4	2,343.5	2,895.1	103.4	148.0	198.5	257.3
Wholesale trade	3,824.0	4,608.0	5,413.0	6,238.0	490.0	654.3	849.8	1,072.9	66.3	93.9	128.6	170.1
T.C.U.[b]	4,504.0	4,944.0	5,419.0	5,929.0	543.2	697.1	867.0	1,061.3	59.4	82.1	109.2	142.6
F.I.R.E.[c]	3,690.0	4,640.0	5,660.0	6,750.0	429.7	580.0	758.4	965.2	45.0	64.3	90.1	122.8
Construction	3,345.0	3,965.0	4,610.0	5,280.0	542.4	733.5	935.8	1,161.6	32.6	45.1	60.9	79.5
Mining	622.0	592.0	612.0	652.0	43.1	41.4	49.0	58.7	0.8	0.8	0.9	1.0
Total	70,617.4	84,942.4	100,238.6	116,376.6	9,249.5	12,183.9	15,460.2	19,183.4	621.4	859.9	1,183.1	1,579.8

[a]Alabama, Florida, Georgia, Mississippi, North Carolina, South Carolina, and Tennessee.
[b]Transportation, communications, and utilities.
[c]Finance, insurance, and real estate.
Note. Economic Base Study of Atlanta, by ARC, 1975, Atlanta, GA: ARC.

Table 6-7. Population projections for the Atlanta region

Year	Population	Percentage change
1960	1,044,321	—
1970	1,436,975	37.6
1980*	1,779,326	23.8
1990*	2,678,200	50.5
2000*	3,478,500	29.9

*Forecasted values.

Note. From *Economic Base Study of Atlanta,* by ARC, 1975, Atlanta, GA: ARC.

Table 6-8. Households and household size, Atlanta region forecasts

Year	Number of households	Persons per household
1960	298,518	3.42
1970	442,813	3.18
1980*	670,625	2.92
1990*	927,689	2.86
2000*	1,214,717	2.83

*Forecasts

Note. From *Economic Base Study of Atlanta,* by ARC, 1975, Atlanta, GA: ARC.

With regional year 2000 forecasts of employment by category and households by income category and size category in hand, the next task is to allocate this total activity to small areas comprising the Atlanta region. In this particular case, each employment category and each household income/household size category is regarded as an activity. The EMPIRIC activity allocation model was used to generate year 2000 forecasts for each of 525 traffic zones that make up the seven-county area under investigation.

The EMPIRIC model was originally developed by the Traffic Research Corporation and has been used in a large number of urban transportation (and related) projects involving many metropolitan areas including Atlanta. The conceptual foundation upon which EMPIRIC is based is quite simple. Basically, it reorganizes that the locations of different activities are interrelated and are affected by public policy (for example, provision of sewer, water, and transportation facilities).

From a more technical perspective, EMPIRIC is based on a simultaneous linear regression model in which the change in any small area's share of a total regional activity (where activity is a particular commercial activity or household income/size category) is statistically related to the change in other selected activities found there and the change in sewage and water facilities along with accessibility to other small areas and their respective activities. The final form of the model is calibrated (that is, coefficients of the simultaneous linear regression equation are

Table 6-9. Distribution of households by income category, Atlanta region

Income category	Percentage of households in category			
	1970	1980*	1990*	2000*
Less than $3,000	8.2	5.0	2.2	0.0
$3,000–4,999	8.1	5.6	3.4	0.7
5,000–6,999	10.5	7.9	5.4	2.8
7,000–9,999	19.2	14.8	10.0	5.1
10,000–11,999	12.9	11.9	10.4	8.6
12,000–14,999	15.5	14.6	13.3	12.1
15,000–24,999	19.8	23.4	26.9	30.3
25,000–49,999	4.8	13.4	22.2	31.3
50,000 or more	1.0	3.4	6.2	9.3
Median HH income	10,620	12,953	16,953	21,843

*Forecasts

Note. From *Economic Base Study of Atlanta,* by ARC, 1975, Atlanta, GA: ARC.

estimated) using historical data for two points in time.

In EMPIRIC application to the Atlanta case, the model is calibrated on the basis of historic trends between 1961 and 1970. This procedure explicitly makes the assumption that the same processes (and manifest trends) at work during the calibration time period (1961–1970) will also be at work during the planning period (1970–2000). The forecasted year 2000 activity levels for each subregion are derived from solution of the set of simultaneous equations adopted from the 1961–1970 calibrations in concert with anticipated broad-gauged decisions on year 2000 water, sewage, and transport services. The equations do not directly yield activity levels, but rather computed changes in the shares of the activity level for each subregion between 1970 and 2000. The changes in shares are then added to the base year (1970) share for each subregion and activity. The new shares are multiplied by the exogenously forecasted regional totals for each activity. This produces an absolute subregional activity level, for example, the number of households in each income/size group and the number of employees in each employment category.

Table 6-10 summarizes the land use forecast for the seven-county ARC planning area. These county-level estimates have simply been aggregated from EMPIRIC output. In general, these statistics suggest the anticipated and continued dominance of Fulton and DeKalb counties as seats of commercial and industrial development. Significantly, population/jobs (P/J) ratios for DeKalb, Gwinnett, and Douglas counties, however, indicate more rapid anticipated growth of employment opportunities there and are testimony to the high probability of accelerated decentralization of employment opportunity in the Atlanta region. Figures 6-8 and 6-9 show the directions of relative population and employment growth for the Atlanta region between 1970 and 2000, respectively. Again, this summary is based on EMPIRIC results for small areas that have been aggregated to directional sectors. Note the disproportionately strong employment growth in the north-northeast and the east-northeast directions corresponding to current and anticipated growth along Georgia Route 400 and I-85. Population growth, in similar fashion, is biased toward northern directions but not nearly as strongly as employment growth because of current and anticipated residential growth along I-20 East and West and I-85 South. This significant mismatch in growth directions is at least in part responsible for current and anticipated capacity problems for I-285 as it serves peripheral (suburb-to-suburb) work trips.

Table 6-10. Forecasted features (year 2000) of Atlanta, county level

County	Land Use Acreage[a]			Population	Jobs	Population/Jobs (P/J) ratio
	Single-family	Multifamily	Commercial–Industrial			
Fulton	64,166 (35)	19,135 (191)	19,937 (70)	1,081,620 (79)	770,613 (94)	1.4 (−7)
DeKalb	46,344 (28)	12,786 (362)	13,316 (127)	788,965 (90)	347,032 (177)	2.3 (−3)
Cobb	42,314 (65)	9,552 (938)	8,493 (130)	611,998 (211)	184,410 (194)	3.3 (+6)
Clayton	18,765 (93)	5,454 (1,020)	4,383 (155)	332,757 (239)	97,555 (288)	3.4 (−13)
Gwinnett	31,761 (151)	4,619 (8,957)	7,426 (230)	384,206 (431)	119,839 (704)	3.2 (−33)
Rockdale	9,520 (215)	1,412 (4,180)	1,338 (220)	116,135 (540)	26,687 (417)	4.4 (+26)
Douglas	12,632 (170)	1,733 (7,121)	1,938 (319)	153,150 (434)	34,859 (762)	4.4 (−38)
Regional total	225,502	54,691	56,831	3,505,051	1,593,456	2.2 (−4)

[a]Figures in parentheses are percentage change, 1970–2000.

Note. From *Regional Transportation Plan*, by ARC, 1977, Atlanta, GA: ARC.

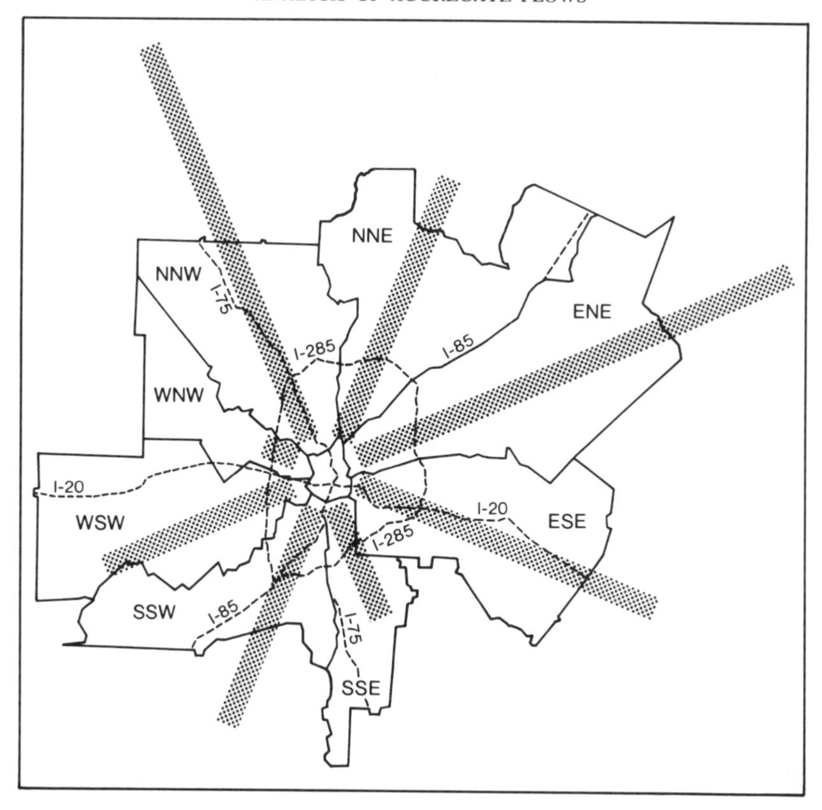

Figure 6-8. Directions of population growth in Atlanta between 1970 and 2000. Growth is anticipated to be most rapid in the north and east. From *Atlanta Region Population and Employment Trends*, by ARC, 1980, Atlanta, GA: ARC.

THE DEMAND FOR TRAVEL IN ATLANTA

Trip generation is the first phase in the actual travel-forecasting process. It generally involves estimation of the total number of trips entering (attractions) or leaving (productions) a parcel of land. In the Atlanta case, person trips are classified into one of five categories: work, shopping, school, and social-recreational trips that either begin or end at home; and one category of trips of any purpose that neither begin nor end at home. The trip-generation process takes as input the predicted distributions of zonal population, employment, activities, and land uses for the design year (year 2000 here) and by extrapolating relationships between these distributions and zonal trip making for the base year (1970 here), generates estimates of design-year productions and attractions. Note that it is explicitly assumed that the relationships between trip rates and zonal characteristics remain stable over time. Finally, these trip-generation relationships are constructed for each of the five trip categories.

Residential Trip Generation—Productions

Conventional modeling of residential trip generation assumes that the propensity of a household to make trips is related to characteristics of the household. It is assumed that the propensity to make trips is not related to the transportation system. In the Atlanta case, the two key variables thought to dictate person trips produced from resi-

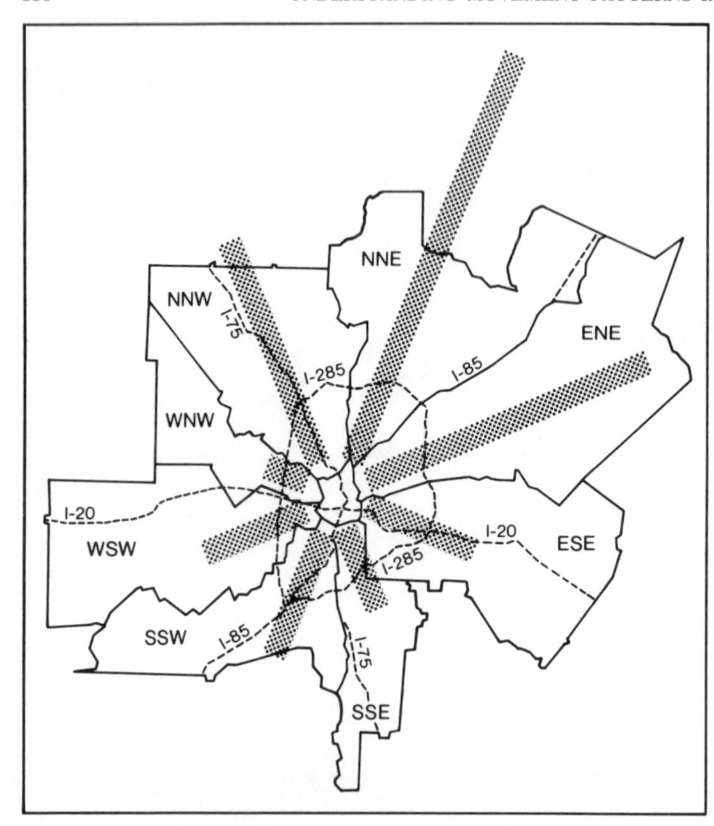

Figure 6-9. Directions of employment growth in Atlanta between 1970 and 2000. The northeast is the dominant direction of anticipated growth. From *Atlanta Region Population and Employment Trends*, by ARC, 1980, Atlanta, GA: ARC.

dential areas are household size and number of automobiles owned. Five levels of household size are adopted: 1, 2, 3, 4, and 5 or more members. Three levels of automobile ownership are adopted: 0, 1, and 2 or more autos.

Through simple cross-classification procedures, and based on 1970 home interview trip file, the number of households in each size-by-auto ownership category is tabulated along with the number of trips (disaggregated by work, school, shop, and social) made by those households. In this manner, a household trip rate is established for the "typical" 24-hour period. For example, in the Atlanta case, households of one member owning no cars average 1.05 trips while households with two or more autos and five or more persons average 13.27 trips per 24-hour period. Table 6-11 summarizes the total person trip rates for Atlanta using this cross-classification procedure.

In order to use the trip rates to predict residential productions in the year 2000, the number of households in each cell of the cross-classification table is predicted for the year 2000. The appropriate trip rate is then applied to each group of households. Prediction of the total number of households and their size and income distributions at the zonal level was accomplished through use of

Table 6-11. Trip rates for Atlanta, 1970

	Auto ownership		
Household size	0	1	2+
1	1.05	2.67	4.13
2	1.45	5.08	6.68
3	1.72	7.01	8.49
4	2.87	7.97	12.15
5+	5.05	9.44	13.27

Note. From *Interdepartmental Technical Report, Trip Rates*, by ARC, 1976, Atlanta, GA: ARC.

the EMPIRIC model described in the previous section. Auto ownership distributions are obtained by relating auto ownership in the base year to income in the base year. In this way, when household income distributions are predicted for Atlanta's traffic zones through EMPIRIC, so are auto ownership distributions. Application of trip rates to the numbers of households in categories for the year 2000 is the most important part of trip generation analysis because between 80 and 90% of total person trips are home based.

Nonresidential Trip Generation—Attractions

The previous section of this chapter described the residential trip-generation procedure for estimating Atlanta's trip productions in the year 2000 by purpose. In this section, the approach to trip attraction estimation is described. Trip attraction is normally related to nonresidential land use for most trip purposes. On the other hand, trips for social and recreational purposes, by their nature, include considerable travel to residential land.

In the Atlanta case, multiple regression analysis is used to relate the number of trips attracted to various categories of land use (such as industrial, retail, residential). The intensity of each land use category in an area can be measured by either acreage or employment in that category. Regression is performed for each of the five trip types (that is, home-based work, shop, social, school, and non-home-based). The observation unit is the traffic zone, of which there are 525 in the Atlanta case. Regression coefficients indicate the number of person trips per employee attracted or the number of person trips per acre attracted for relevant land uses included in each trip type model. In the particular case of social and recreational trips, residential land use is measured by the number of households in an area. As an example of primary interest here, the following equation is used to relate the attraction of home-based work trips to land use characteristics:

$$\begin{aligned} HBWT_i = 2650 &+ 1.62\,(NIND)_i \\ &+ 1.68\,(NSHOP)_i \\ &+ 1.74\,(NOFFICE)_i \\ &+ 0.36\,(NHH)_i + e_i \\ R^2 = .88 & \end{aligned} \tag{1}$$

where $HBWT_i$ is the number of home-based work trips attracted to zone i; $NIND_i$ is the number of industrial employees working in zone i; $NSHOP_i$ is the number of retail employees working in zone i; $NOFFICE_i$ is the number of office employees working in zone i; NHH_i is the number of households located in zone i; and R^2 is the proportion of explained variation in *HBWT*-attracted to zones. As was the case in prediction of residential person trip productions for the year 2000, the number of employees in each of the land use categories must be forecast before the regression coefficients can be applied to obtain year 2000 attraction values for each zone. These coefficients essentially are trip attraction rates (for example, 1.62 home-based work trips are attracted per industrial employee). Again, that sort of forecast was accomplished using the EMPIRIC model, reviewed in the previous section.

Finally, forecasted (year 2000) total person trip productions and attractions are shown in Table 6-12. These attractions and productions, although estimated at the traffic zone level (525 zones), are aggregated here to the district level (34 districts) for expository purposes. Figures 6-10 and 6-11 are a general cartographic illustration of the spatial distributions of these most fundamental planning variables. Note that total regional productions (more than 9 million per 24-hour period) are exactly equal to total regional attractions. For purposes of trip distribution, these estimates should have an areawide balance for each trip purpose and in total. As was explained in the previous chapter, in order to attain the desired balance, trip productions are used as the control and zonal attractions are generally deflated in a proportion dictated by the ratio of regional productions to attractions.

Table 6-12. Person trip productions and attractions at the district level, Atlanta, 2000

	Productions		Attractions	
District	Work	Total	Work	Total
CBD-Atlanta	7,930	110,669	249,331	441,967
NE-Atlanta	49,722	243,030	182,567	451,002
NW-Atlanta	79,754	329,372	127,300	407,338
SE-Atlanta	64,407	267,576	100,742	359,299
SW-Atlanta	109,524	423,760	103,300	433,674
Tri-Cities	63,518	269,538	112,419	366,240
S. Fulton	160,136	609,690	83,329	461,774
Buckhead	47,490	211,093	99,285	371,710
S. Springs	72,266	279,708	59,440	239,784
N. Fulton	91,943	355,597	45,769	258,201
A.-DeKalb	23,942	88,731	16,185	72,229
Decatur	16,288	66,842	26,725	78,125
Chamblee	100,077	418,641	140,793	520,955
NE-DeKalb	102,019	408,403	94,563	433,747
NW-DeKalb	54,501	327,340	83,155	291,122
SE-DeKalb	76,291	301,023	54,516	254,376
SW-DeKalb	88,006	331,367	59,217	307,067
S-DeKalb	78,447	302,726	49,040	274,724
Marietta	38,350	157,601	41,338	179,570
S. Cobb	176,493	674,327	125,982	590,885
N. Cobb	118,904	466,265	73,817	395,163
NW-Cobb	76,561	192,042	32,237	205,010
Airport	15,180	68,674	37,806	104,219
NE-Clayton	94,990	372,114	67,747	328,122
Riverdale	57,002	205,138	15,201	116,818
S-Clayton	60,574	222,881	26,276	150,258
Buford	43,713	172,109	30,211	143,391
SW-Gwinnett	90,966	362,492	72,021	319,148
Lawrenceville	113,994	444,988	67,566	334,893
N. Rockdale	19,286	71,079	5,657	38,948
S. Rockdale	64,726	247,453	35,507	187,986
N. Henry	26,931	108,076	18,895	92,578
E. Douglas	67,657	260,820	37,209	210,892
W. Douglas	40,550	153,704	16,822	103,584
Region total	2,392,072	9,524,437	2,392,072	9,524,437

Note. From *Regional Transportation Plan*, by ARC, 1977, Atlanta, GA: ARC.

DISTRIBUTING TRIPS IN ATLANTA

The next stage in the travel-modeling process attempts to link trips produced in each zone with trips attracted by each zone. The most widely used trip distribution model, and the one adopted for the Atlanta region, is the gravity model, in this case the traditional gravity formulation. There are, however, several subtleties of its use that must be made clear from the start, so that we shall provide the explicit gravity approach used here (and commonly used) and speak to these subtleties before moving on to a description of modeling results. The gravity formulation employed is:

$$T_{ij} = O_i \cdot \left(\frac{A_j \cdot D_{ij}}{\sum_j A_j \cdot D_{ij}} \right) \quad (2)$$

where T_{ij} is the number of trips from zone i to j; O_i is the number of trips originating in zone i; A_j is the number of trips attracted to

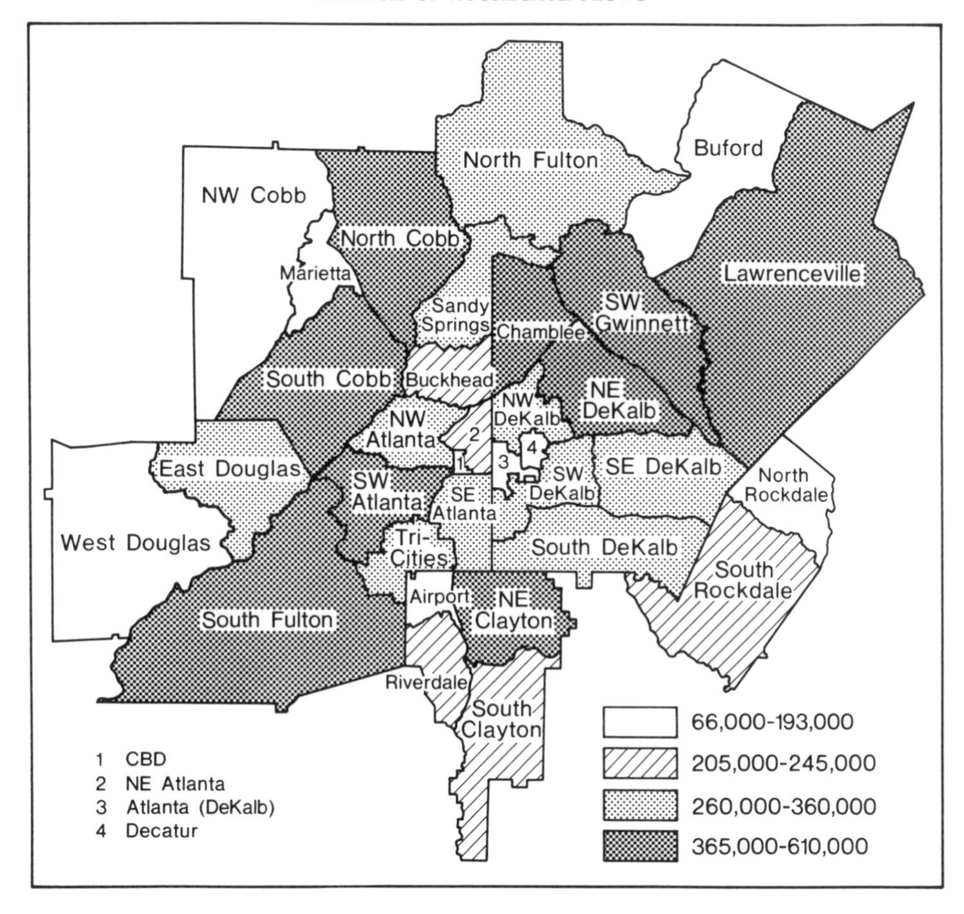

Figure 6-10. The spatial distribution of total trip productions for the year 2000.

zone j; and D_{ij} is a measure of the effect of separation (travel time) on trips made between zone i and zone j. In turn, let us be more explicit about the nature of the D_{ij} values contained in the above expression. Algebraically,

$$D_{ij} = \frac{1}{t_{ij}^{b}} = t_{ij}^{-b} \qquad (3)$$

where D_{ij} is the separation factor between zones i and j; t_{ij} is the total (driving + terminal) time separating zone i from zone j; and b is the travel time exponent. Note that the b values provide distinctive measures of D for each level of travel time. Once the set of b values is known, so is the set of D values.

Within these two formulas, values for O, A, and t are known from the 1970 (base year) inventory of the Atlanta region (in this particular case). What is desired is a set of values for the unknown, that is, the set of b values, which ultimately yield D values. Once a set of values for D is obtained that satisfactorily represents the importance of separation (travel time) on trip making in Atlanta for 1970, these D values can and will be used to distribute trips for year 2000 projected values of O, A, and t. Clearly the critical assumption is that the D values, representing the importance of travel time in urban trip making, are constant between base year estimate and forecast year use, in other words, no change in the importance of travel time to trip makers takes place between 1970 and 2000.

To continue our discussion of this critical set of unknowns, the D values, it is critical to understand that urban planners permit D

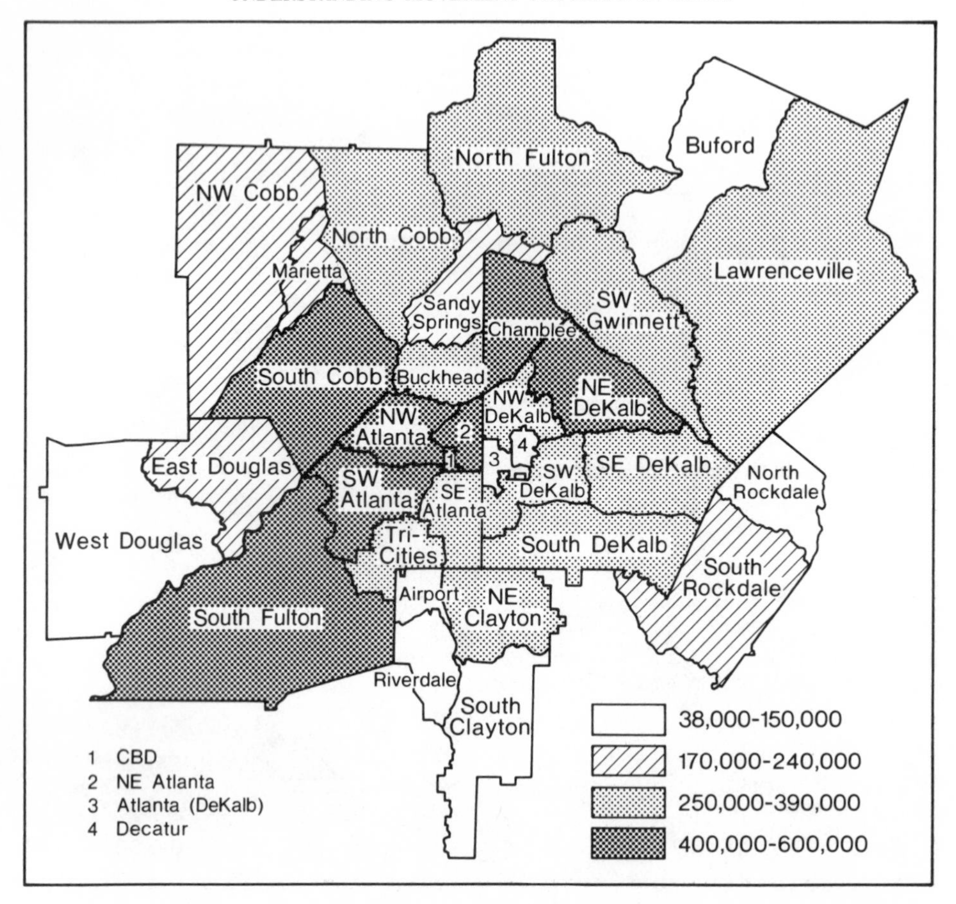

Figure 6-11. The spatial distribution of total trip attractions for the year 2000.

values to vary in two distinctive ways. First, these friction-of-distance (time) values vary by trip purpose (there are five trip types used in the Atlanta case—see the previous section). For instance, all else being equal, *D* values are larger for work trips than they are for shopping trips (reflecting the dominance of convenience-type movements in the shopping category). This feature is a clear reflection of the greater sensitivity of trip makers to additional units of travel time when going to work as opposed to shopping.

The second manner in which the *D* values vary is with respect to travel time itself. Travel time exponents, *b* values, are assumed to be a smooth and negative function of *t*. Therefore, friction factors (*D* values) are numerically larger than expected for longer trips (largest) reflecting the empirical reality that a uniform increment of travel time has less of an effect on impeding travel between zones that are closer together than it does for zones that are farther apart, but the effect of travel time increases at a decreasing rate! The implication is a different, that is, larger, value of *D* for each additional unit of *t*. Thus, an entire set of *D* values must be estimated on the basis of 1970 data before year 2000 trips can be distributed. In the Atlanta case, total travel times vary between 1 and 90 minutes, so that 90 separate *D* values must be found. Note that two zones that are separated by equal travel time share identical *b* values and hence identical *D* values.

Once year 2000 estimates of general trip productions, attractions, and interzonal travel times are established, it remains to have a good set of *D* values to complete the requisite information for the gravity model used to distribute these trips across the Atlanta landscape. Trip productions and attractions from the 1970 home interview trip file are required to estimate this set of good D values (we shall give added meaning to the phrase "good D values" below). Not only are O and A values taken out of the 1970 trip file, but the actual number of trips made between each zonal pair (interzonal) and within each zone (intrazonal) are also known from this file. These actual flow volumes, as we shall see, play a crucial role in assessing the quality of the D values chosen. One last item of data is required before actually calibrating the gravity model, that is, finding good D values. These are the actual travel times, and they are obviously needed if we are to assess the friction of temporal separation on urban trip making. Aggregate travel time for interzonal flows is composed of ride time and terminal time (at origin and destination). Intraregional travel time is a bit more difficult to assess and is calculated in the Atlanta case to be one-half the average travel time to adjacent zones. Terminal times (in Atlanta) are assessed to the production and attraction ends of each zone-to-zone impedence as shown in Table 6-13.

In summary, base year (1970) values of trip productions, attractions, interzonal travel times, and intra- and interzonal flow volumes are known quantities. We wish to assess the aggregate and areawide effect of temporal separation on trip interchanges between zones. This effect is embodied in a set of D values, which vary not only by trip purpose but also

Table 6-13. Terminal travel times, Atlanta, 1970

Area (see Figure 6-12)	Terminal time (minutes)
CBD	6
CBD to inner cordon	4
Inner cordon to boundary of urbanized area	2
Boundary of urbanized area to seven-county ARC boundary	1

Note. From *Regional Transportation Plan*, by ARC, 1977, Atlanta, GA: ARC.

by travel time between zones. Once the D values are known, they are applied to forecasted (year 2000) values of productions, attractions, and travel times in order to estimate intrazonal and interzonal flow volumes. The manner in which the D values are estimated from the 1970 (base year) data is briefly described in the next few paragraphs.

Recall that the friction factor, as embodied by the time exponent, is not a constant, but is assumed to vary with t. This is in contrast with some gravity applications where the gravity model is calibrated for a single "distance exponent," which would then be the same for all values of t. Additionally, the benchmark against which performance of the calibrated gravity model is assessed is the actual trip frequency distribution as graphed over travel time, that is, percentage of trips versus travel time (see Figure 6-13). If our derived D values are good ones, then the actual trip frequency curve should resemble the (gravity model) predicted trip-frequency curve. Again, Figure 6-13 exhibits the type of comparison being referred to. Please note that this method of assessing "goodness" of the D values is somewhat different from the common approach taken in geography where specific actual interzonal flows are compared to specific predicted interzonal flows when assessing the model's performance.

The important reasons for (1) permitting D values to vary with time, and (2) assessing goodness of the D values by comparing actual to predicted trip frequencies versus time (as opposed to comparing actual versus predicted specific interzonal flows), are quite pragmatic. With regard to (1), better fits are obtained simply because more parameters, that is, D values, are being used to model the flows. With regard to (2), there is a tremendous saving in the number of arithmetic operations required by the comparison being used here, that is, actual versus gravity-model predicted trip frequencies over time. Instead of comparing expected versus actual flow volumes for thousands of possible origin–destination pairs, only 60 to 90 such comparisons are required, in other words, one comparison between percentage of ac-

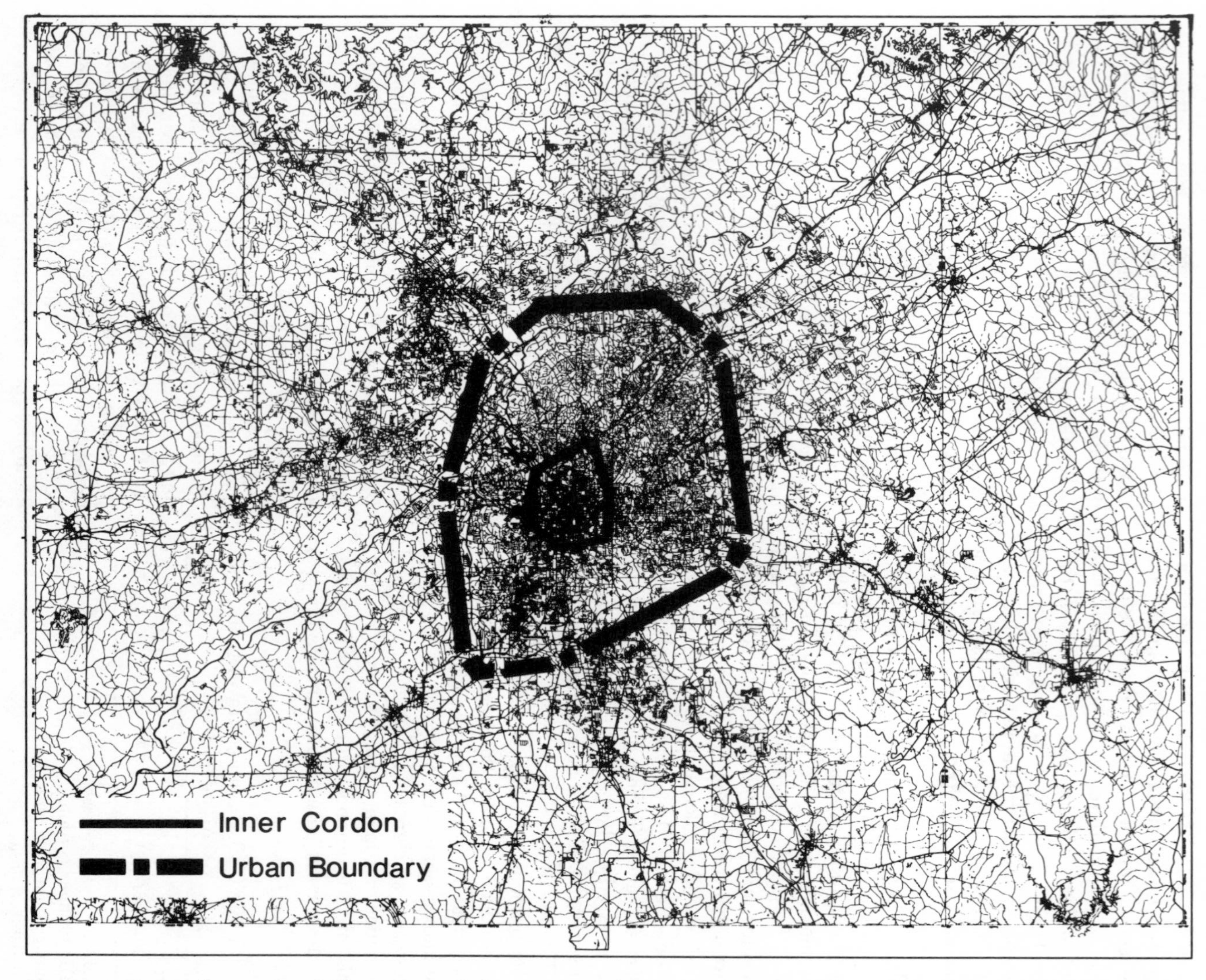

Figure 6-12. Cordons selected by ARC to delimit terminal time regions of Atlanta. From *Regional Transportation Plan*, by ARC, 1977, Atlanta, GA: ARC.

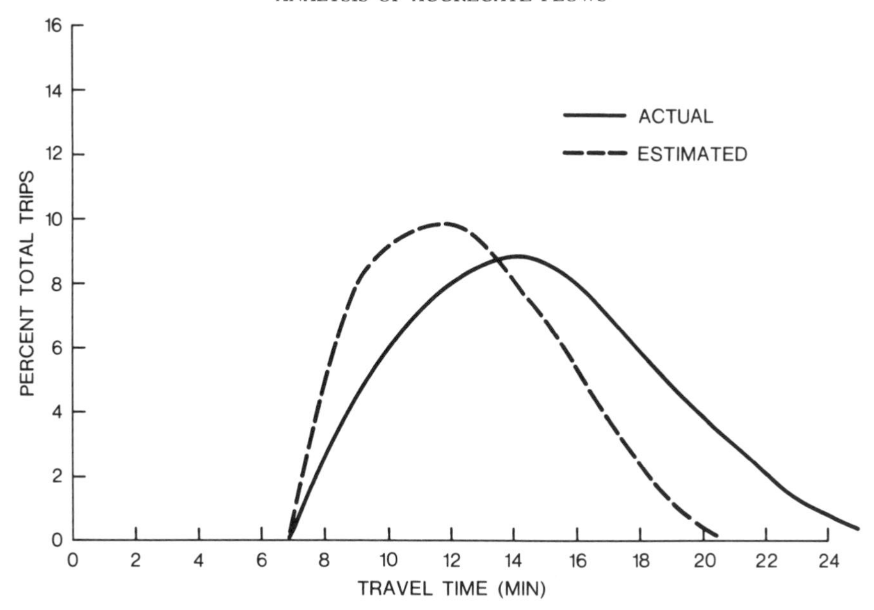

Figure 6-13. Actual and estimated trip length frequency distributions. This is a hypothetical example of the type of comparison made when assessing estimated values.

tual trips and percentage of gravity-predicted trips for each additional minute of travel time between 0 and 90 minutes.

So, there exists an entire set of *b* values, and hence *D* values, to be estimated for each of five trip types as we attempt to assess the friction of temporal separation on travel in Atlanta. The benchmark against which we will evaluate the quality of our selected *D* values is direct comparison with actual trip making in 1970 (as reflected by the distribution of trips versus time) and the predicted trip making we obtain when plugging *O*, *A*, and estimated *D* values into the gravity model (see the earlier portion of this section). We can then use this set of good 1970 *D* values in our attempt to distribute forecasted values of *O*, *A*, and *t* spatially for the year 2000.

More precisely, in order to estimate the values of *D* for a particular trip type, an iterative calibration process is used as follows. First, initial *D* values are selected for each increment in travel time of one minute (values range in the Atlanta case from 1 to 90 minutes). Initial values of *D* in the Atlanta case were arbitrarily set at a value of 1.0 for all values of *t*. Other starting solutions would have been permissible. Next, the actual 1970 frequency distribution of person trips versus travel time is constructed. A graph such as that contained in Figure 6-13 summarizes this information. Remember that these actual values come directly from the 1970 base-year trip file.

The third step in this iterative procedure for finding good *D* values involves construction of the predicted (gravity-model) frequency distribution of person trips versus travel time. Given zonal trip productions, trip attractions, inter- and intrazonal travel times, and the initial *D* values (arbitrarily set at 1.0) in conjunction with the gravity formula, a predicted set of inter- and intrazonal trips is attained. Then it is an easy matter to construct the predicted frequency distribution of trips versus travel time. Note the type of discrepancy that can occur between the predicted and actual frequency distributions when improper *D* values are used (see Figure 6-13). Typically, the initial (arbitrary) *D* values provide just this type of mismatch, which is obviously unacceptable. So, the objective now is to correct the values of *D*, so that the predicted trip distribution is more similar to the actual; that is, we want a better fit between the two curves of Figure 6-13.

Movement toward convergence of the two (actual versus gravity-model estimate) distributions is accomplished by altering the distance decay values, as was described in the previous chapter. Convergence is obtained in the following manner:

$$D_{t(\text{new})} = D_{t(\text{old})} \cdot \left[\frac{\%\text{ observed}}{\%\text{ gravity}}\right] \qquad (4)$$

where $D_{t(\text{new})}$ is the new travel time factor for a particular travel time, t; $D_{t(\text{old})}$ is the previous travel time factor for a particular travel time, t; % observed is the percentage of observed (actual) person trips occurring with length t; and % gravity is the percentage of predicted person trips occurring with length t. When the percentage of actual trips taking place at a particular value of t surpasses the gravity expectation, the old D value is multiplied by a value greater than 1.0 and the new D value is enlarged. When D is larger, the gravity model predicts more trips, so that the new value of D should bring actual and predicted percentages closer together at that particular value of t. The D value associated with each t value is adjusted in the same manner.

Once the new D values are obtained, new predicted trip profiles (percentage of trips versus travel time) are constructed. Of course, the actual trip frequency (versus travel time) remains as the benchmark against which the quality of chosen (new) D values is assessed. The predicted trip-frequency curve is altered in this iterative fashion until essential characteristics of the predicted distribution are sufficiently "close" to the actual benchmark distribution. The quality of fit can be assessed in a variety of ways including such summary statistics as correlation coefficients or differences between means, and so on. Remember, too, that the calibration procedure is carried out separately for each trip type (there are five in the Atlanta case).

When a suitable set of friction factors is obtained for each trip type using the base year data, these friction factors are applied to trip productions and attractions for the design year (Atlanta in the year 2000). In such a manner, the forecasted spatial distribution of trips is made explicit.

Results of Trip Distribution in Atlanta

The reported characteristics of the 1970 base trip file, the benchmark against which all calibration is measured, are shown in Table 6-14. Note the number and duration of home-based work trips. The longer trip lengths reflect longer-distance trips made during peak travel periods. Keep in mind that these are the types of characteristics that the accepted gravity model predictions must replicate within tolerable limits.

Tolerable limits, at least in the Atlanta case, were well defined. First, average predicted trip length from the final gravity model should be within 3% of average actual trip length. Second, the number of intrazonal trips should be within 5% of the actual number of intrazonals. Third, the models (one for each trip type) should have satisfactory statistical measures; in other words, the correlation between predicted and observed trip interchanges should be high.

Fine tuning the gravity models required approximately 20 iterations each (one for each trip type). Friction factors were incrementally adjusted especially at the lower range of time values in order to achieve the proper quantity of intrazonal trips while preserving the trip-length frequency distribution and the associated statistical qualities of each model. Table 6-15 shows statistical results of the final set of selected distribution models. You can see by explicit comparison of entries in Tables 6-14 and 6-15 that the models do satisfy the predetermined criteria with the single exception of intrazonal shopping trips, which are underestimated by a significant margin.

Figure 6-14 illustrates the close fit of the actual work-trip duration frequency distribution and the predicted frequency distribution derived from gravity expectations. The two distributions are remarkably similar. Such evidence provides added confidence in use of the derived travel time factors when distributing year 2000 work-trip productions and

Table 6-14. Characteristics of 1970 Atlanta trip file, 24-hour period

	Characteristic		
Trip purpose	Total trips[a]	Average length (minutes)	Number of intrazonals
Home-based work	1,037,206 (25)	23.05	53,085
Home-based shop	1,048,984 (25)	15.36	247,182
Home-based social	779,802 (18)	18.52	95,214
Home-based school	552,711 (13)	13.29	160,135
Non-home-based	785,672 (19)	16.81	120,154
Total	4,204,375	17.84	675,770

[a] Figures in parentheses represent percentages.
Note. From *Interdepartmental Technical Report, Trip Distribution Models*, by ARC, 1976, Atlanta, GA: ARC.

attractions. Finally, Figure 6-15 shows the manner in which the Atlanta work-trip travel-time factors (that is, $D=t^{-b}$) vary with respect to travel time. In Figure 6-15 the values of D^{-1}, which is t^b, are plotted versus travel time. When t is equal to 10 minutes, $D^{-1}\cdot(100)$ is equal to 10,500 so that D^{-1} is 105.5, and b (at 10 minutes) must then be 10.21. In contrast, when t is 30, b is equal to 0.41. The dramatic peak of the function at roughly 10 minutes may be suggestive of the tremendous sensitivity of Atlanta's labor force to marginal adjustments in travel time at this intermediate range in travel-time values.

Given person-trip productions and attractions for the year 2000 in concert with travel-time factors, the gravity model provides the general spatial distribution of trips (total or by trip type). When these interzonal volumes are assigned to the existing transportation network, flow bands can be cartographically displayed as on the map of traffic volumes of Atlanta in the year 2000 (Figure 6-16). Note the remarkable growth in the intensity of flows over those in 1970 (compare with Figure 6-6). As we might anticipate, the 1970 highway stock is wholly inadequate to meet the demands of Atlanta's trip makers in the year 2000. This is evident when we compare the projected flows with the highway capacity of 1970. Figure 6-17 illustrates the extent of capacity problems to be expected by the year 2000, given the 1970 highway network. Every radial extending from the heart of the seven-county ARC region is characterized by lengthy stretches of V/C (flow volume/design capacity) ratios in excess of 1.0—indicating bumper-to-bumper conditions during peak-hour traffic. Of particular importance are the capacity problems evident at the interchanges between the I-285 outer belt and radials extending in northerly directions.

Table 6-15. Attributes of selected distribution models

	Predicted characteristics		
Trip purpose	Average length	Number of intrazonals	r
Home-based work	22.40	51,523	.961
Home-based shop	15.08	207,219	.992
Home-based social	18.14	96,103	.956
Home-based school	13.23	157,464	.982
Non-home-based	17.12	121,312	.991

Note. From *Interdepartmental Technical Report, Trip Distribution Models*, by ARC, 1976, Atlanta, GA: ARC.

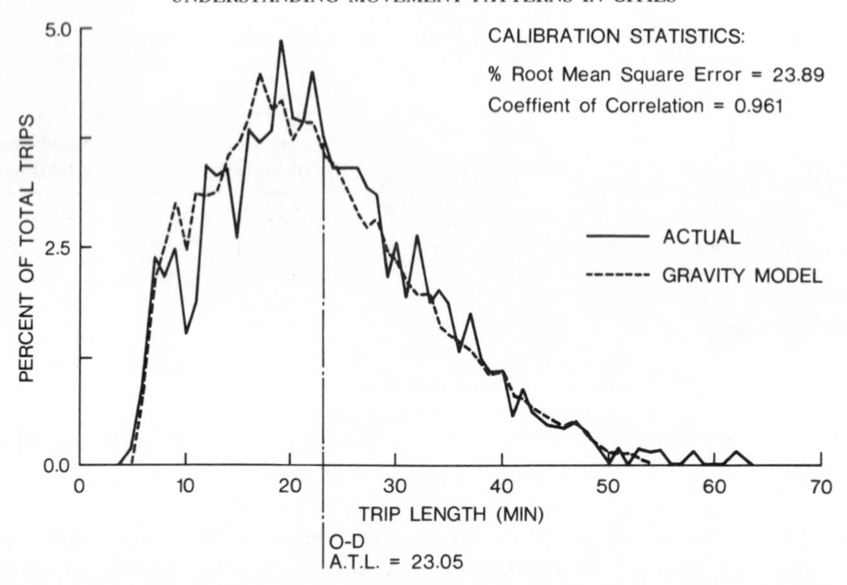

Figure 6-14. Actual (1970) and estimated (2000) work-trip-length frequency distributions for Atlanta. Note the good match between the two curves. From *Interdepartmental Technical Report, Trip Distribution Models*, by ARC, 1976, Atlanta, GA: ARC.

Future Plans

Clearly, there exists a large assortment of ways to try to avoid the sorts of user costs depicted in Figure 6-17. Highway construction projects and construction of rapid rail facilities are large-scale, capital-intensive responses that have been stressed in the At-

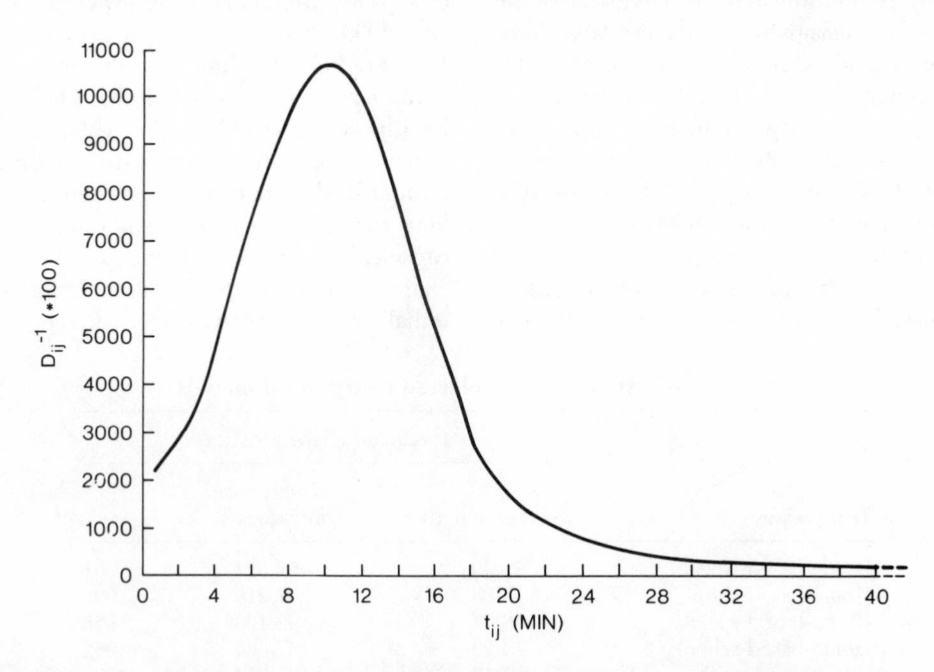

Figure 6-15. The distribution of travel time factors versus trip duration for work trips of Atlanta in 1970. From *Interdepartmental Technical Report, Trip Distribution Models*, by ARC, 1976, Atlanta, GA: ARC.

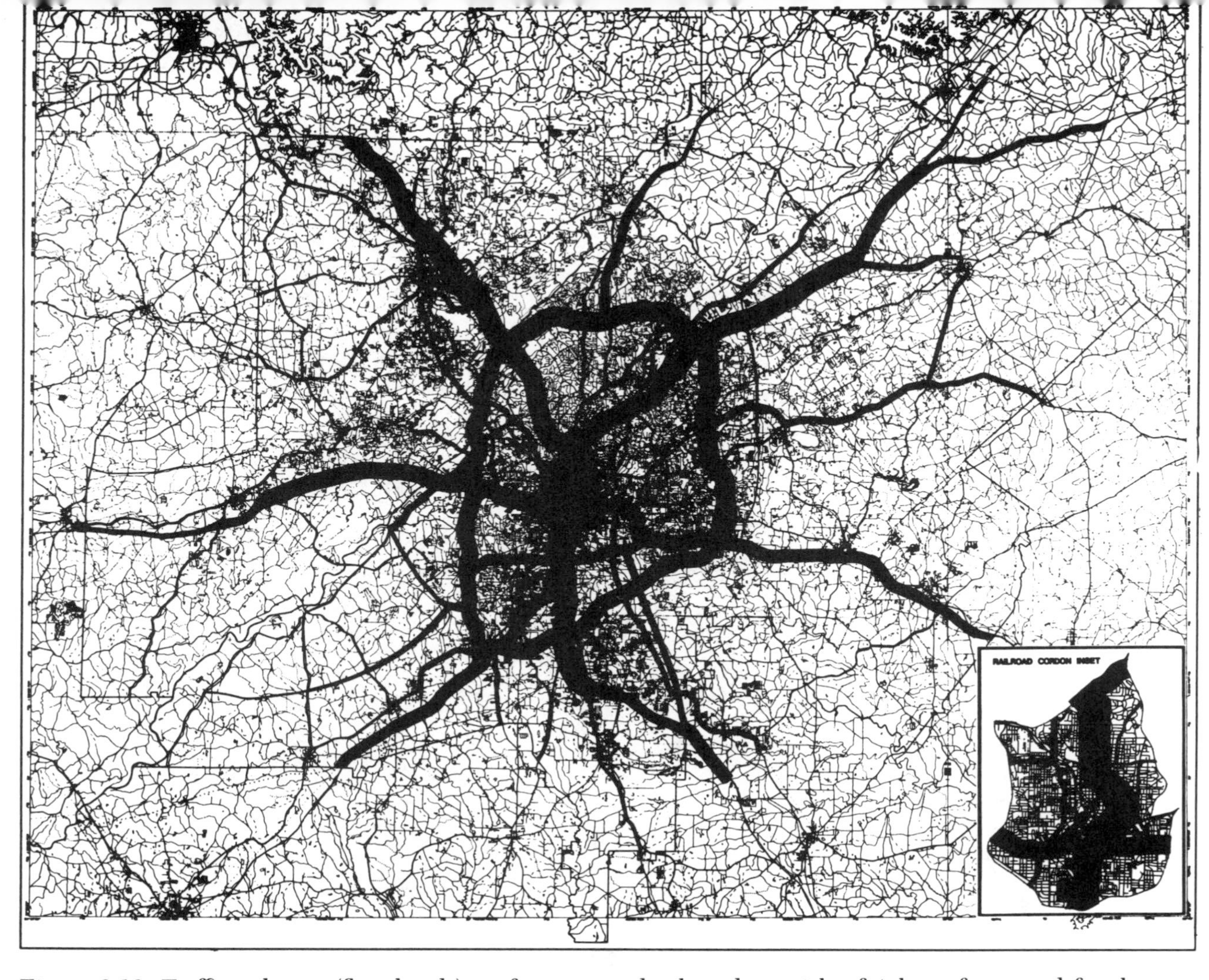

Figure 6-16. Traffic volumes (flow bands) on freeways and selected arterials of Atlanta forecasted for the year 2000. Compare to Figure 6-6. From *Regional Transportation Plan*, by ARC, 1977, Atlanta, GA: ARC.

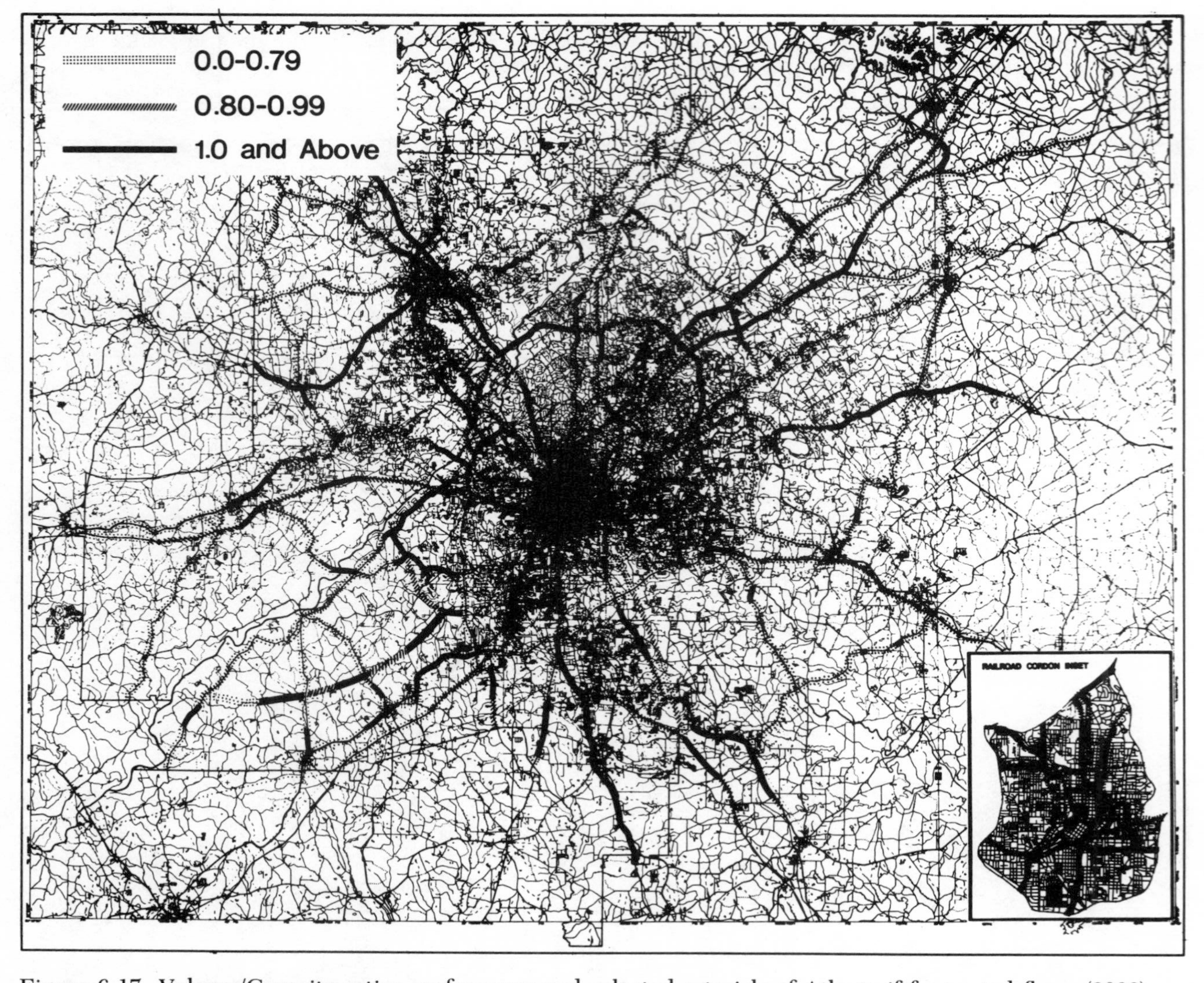

Figure 6-17. Volume/Capacity ratios on freeways and selected arterials of Atlanta if forecasted flows (2000) are serviced by existent (1970) highway stock. From *Regional Transportation Plan*, by ARC, 1977, Atlanta, GA: ARC.

lanta case. Transit Systems Management (TSM) approaches are designed to maximize patronage of a given transit stock through operational improvements such as bus lanes, synchronized traffic-light systems, radio communications, fine-tuning of headways, and ramp metering. The aim would be to increase vehicle occupancy rates through increased ridership of public forms of transportation and enhanced participation in private ridesharing schemes. If programs like these were fully implemented and positively received, then vehicle occupancy rates would climb from 1.13 to 1.50 persons per vehicle between 1970 and 2000.

As mentioned earlier, however, the primary response of the Atlanta region to its impending transportation problem has been to add tremendously to the capacity of its highway stock. Figure 6-18 shows freeway improvements that have been made since 1978 or will be accomplished prior to the year 2000. The primary highway alterations involve completion of largely an eight-lane (four in each direction) system with ten lanes necessary for the central-city portions of the Atlanta downtown connector. The obvious intent is to service the needs of a population continuing to decentralize out of Atlanta proper and engaging more and more in the peripheral journey to work (suburb-to-suburb) or the more traditional trek to the CBD. Suburb-to-suburb work trips now constitute 58% of all Atlanta work trips. More specifically, Atlanta has planned for 32 miles of new freeways, 122 miles of widened existing freeways, 163 miles of new arterials and collectors, and 609 miles of widened existing arterials and collectors. The interstate portion of the highway improvement project (started in 1978) will total $1.4 billion and should be completed by the end of 1988. Interchange renovations are costing as much as $100 million each. The other capital-intensive response to Atlanta's imminent transportation problem is construction of the MARTA rapid transit system, which is now partially operational. Given completion of a proposed 105-mile system being fed by MARTA's bus system, it is expected that patronage will involve 1.1 million daily trips, or 12% of the total, by the year 2000. Furthermore, 56% of all CBD work-trip attractions should be made on public transit by the year 2000.

It would appear that even public transportation is being designed and implemented to provide ease of access by suburban population to downtown Atlanta jobs. In fact, because of these limited services, the Private Industry Council of Atlanta (PIC) maintains that lack of adequate public transportation for central-city residents to expanding job opportunities in the suburbs is the single key barrier to reducing unemployment in the Atlanta area. Commuting times for transit work trips to suburban activity centers currently average anywhere from 50 to 70 minutes. These travel times are between 2 and 2.5 times the duration of the average work trip by car. Quite clearly, the transit-dependent population is more than ever the transportation-disadvantaged population.

Before leaving this section, we should take at least a cursory look at the MARTA rapid rail system and related prospects for its role in Atlanta's transportation future. What is MARTA rapid rail's current and projected role in reducing congestion? First, each rider boarding the MARTA train costs the system about one dollar. Thus, the system is heavily subsidized, primarily from federal funds. To an extent, the MARTA routes follow the freeway system, running east-west and north-south of the central business district. To a limited degree, this rail system has reduced automobile traffic congestion, but the number of route miles is far too small to have any metropolitan-wide effect. The average freeway commuter has noticed no real change in congestion, except that it is getting worse.

Where do the rapid rail routes go and why? In general, the MARTA rapid rail lines are not going to connect with those parts of the metropolitan area that are growing most rapidly, the suburban downtowns. The present and proposed routes were selected before it was realized that those suburban downtowns would emerge on such a gigantic scale. For example, the north line, instead of going to the Perimeter Mall complex, avoids

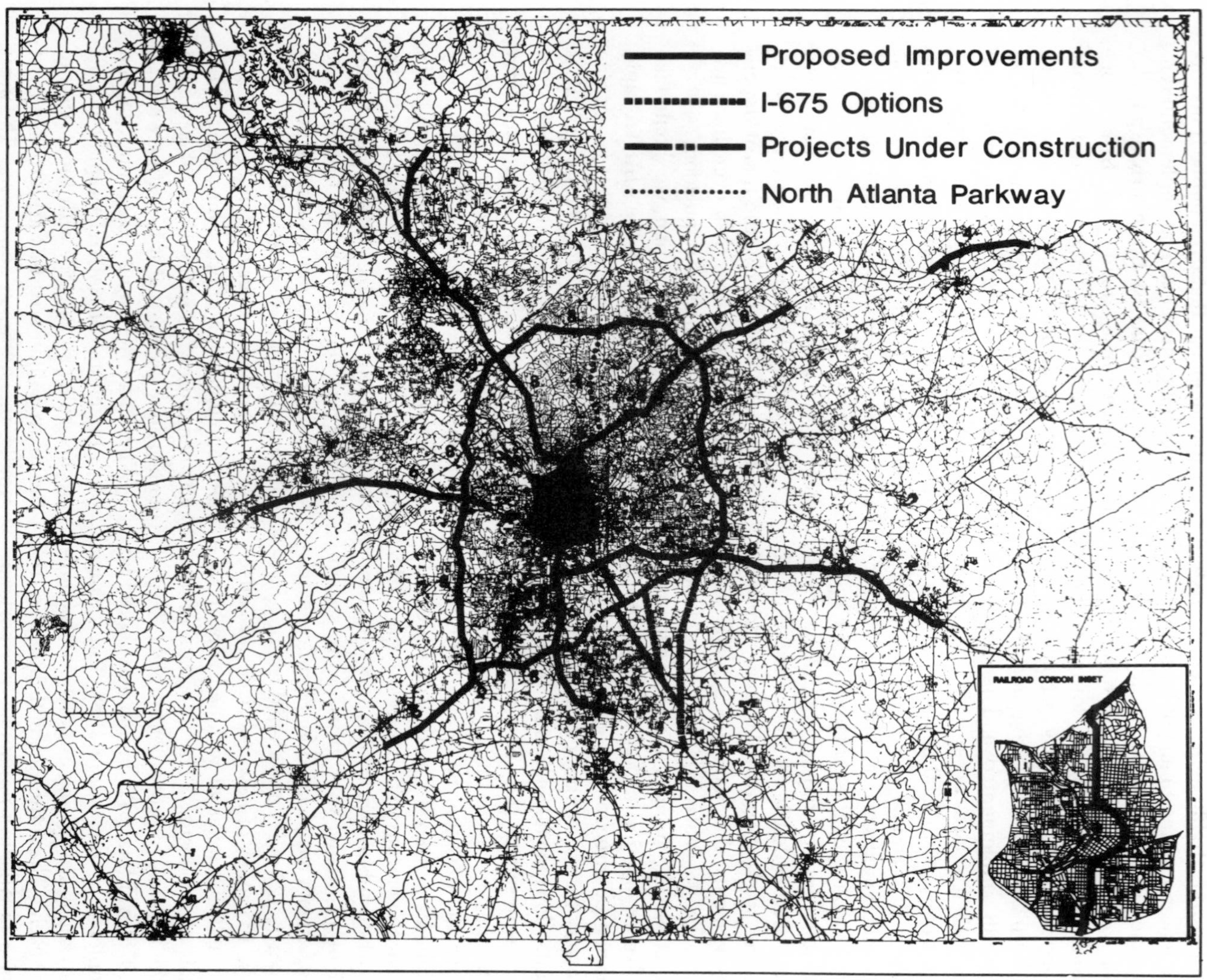

Proposed Improvements
I-675 Options
Projects Under Construction
North Atlanta Parkway
RAILROAD CORDON INSET

Figure 6-18. Plans for improvements to Atlanta's freeway system between 1978 and 2000. From *Regional Transportation Plan*, by ARC, 1977, Atlanta, GA: ARC.

it by swinging eastward to the edge of DeKalb County (see Figure 6-2). A major reason it is proposed to go eastward is to give additional route miles to DeKalb County, as most of the current route miles are in Fulton County. Perimeter Mall is in Fulton County. MARTA rapid rail is sponsored by these two counties along with the City of Atlanta (which occupies a portion of each county).

MARTA rapid rail lines also are not designed to provide job access for the many blacks living in western and southern Atlanta. The proposed south line, although it will pass through black neighborhoods, is in fact going to connect the Atlanta airport with downtown hotels and associated convention sites. The west line carries predominantly black riders and the north line mainly white riders. It is too early to know whether the MARTA rapid rail experiment will bring centralization to Atlanta or contribute to ongoing centrifugal spread.

In addition to the apparent inappropriateness of Atlanta's rapid rail system to meet the job-related needs of Atlanta's black and poor population, other MARTA decisions also reveal an insensitivity to making jobs accessible to the transport disadvantaged. For instance, MARTA recently revealed a plan to move its headquarters from the current central-city location to a plush northside suburban location. Several citizens' groups have voiced their dissatisfaction with removal of hundreds of jobs to an even more distant and inaccessible location to Atlanta's southside black concentration. Despite this opposition MARTA officials continue to plan the move.

SUMMARY AND CONCLUSION

The modern metropolis is a complex spatial system with many specialized parts necessitating tremendous levels of spatial interaction. In the preceding sections, we have illustrated the predominant methodology that has evolved to understand, predict, and manipulate those parts and ultimately those interactions. The Atlanta case study serves as a good example of the strengths and weaknesses evident in the dominant approach to studying urban transportation. Criticisms of the approach can be leveled at many aspects of the procedure highlighted on the previous pages.

First, what are the goals of analyzing aggregate flows for Atlanta? The immediate concern has been to stimulate various aspects of the demand for travel in the Atlanta region in the year 2000. But why? Certainly the anticipation of required transportation services is the primary motive. But this begs the question, what are required services? When the concern for required services is disaggregated, what are the more specific objectives in supplying these services for the year 2000 demand? For example, more specific goals of the transportation system might include:

1. Provision of a "good" level of accessibility to all of Atlanta's resources (housing, jobs, shopping, education, and so on) for all of Atlanta's population;
2. Energy conservation;
3. Minimization of undesirable environmental and social impacts of ground transport facilities; or
4. Provision of effective and efficient transportation services to the region's handicapped and elderly.

The lack of specific goals that would influence the nature of the estimation methods employed is often a problem in such aggregate analyses. The policy sensitivity of the procedures highlighted by the Atlanta case is woefully inadequate to suggest the possible impacts of a specific urban transportation design on attainment of stated goals. In most cases, there actually may be direct trade-offs wherein one goal is attained but only at the expense of another.

For example, in the Atlanta case the emphasis given to circumferential, limited-access highway construction and improvements enhances the mobility of residents in suburban areas while simultaneously reducing the accessibility of central-city residents to the economic opportunity that has been fostered by the outer belt. Not surprisingly, the Georgia Department of Transportation is currently planning to construct an additional circumferential highway around Atlanta. The current outer belt is roughly 64 miles in circumference while the planned outer belt will be nearly 280 miles in length. Although not guaranteed, continued decentralization of economic opportunity and housing is expected as a result of the completed project. Decreased accessibility of Atlanta's transport-disadvantaged and central-city residents can result.

Aside from an apparent lack of strong connection between goals and method, there are several other aspects of the typical aggregate procedure for analyzing travel demand that are evident in the Atlanta case. In analysis of trips generated, for example, the number of trips leaving an origin zone is modeled as a function of automobile ownership and household size. Transportation network variables such as accessibility are not considered, so that changes in trip frequency are assumed to be independent of changes in the transportation system. The same criticism can be leveled at the estimation of trip attractions.

The trip distribution model employed, that is, the gravity model, is also bothersome in certain respects. For example the impedance factors are based on trip duration as measured by automobile travel, so that changes by MARTA in provision of improved transit service and associated impacts cannot be accurately assessed. An additional consideration is that the impedance mechanism is based on a fixed distribution of trip duration measured in minutes. That distribution is simply the one that prevailed at the equilibrium between supply and demand in Atlanta during 1970. Thus the impedance factors serve as distributive factors that are descriptive rather than causal. If Atlanta's travel times are altered through transportation system changes or if the spatial distributions of trip productions and attractions significantly change, planners are not in a position to predict accurately the spatial distribution of travel demand under the new equilibrium conditions.

So, the conventional urban transportation demand model as highlighted by the Atlanta case study must be viewed as having a number of faults. The individual models basically replicate Atlanta's 1970 travel demand with the assumption that the relationships, that is, trip generation, production, and distributional friction, will remain unchanged through the year 2000. The models are not causal. They provide little guidance when attempting to simulate the effect on travel demand should the transportation environment be significantly altered. The models are anything but policy-sensitive. Variables over which policy makers have more direct control are excluded from trip generation and attraction modeling efforts. Atlanta planners have no control over household size and automobile ownership, but they can manipulate parking costs and space, land use densities and types, highway capacities, and transit service levels. The models are also based on zonal aggregates, which smooth variability and obscure much of the information that actually exists. Aggregation biases are currently being studied by geographers among others, but as yet are not well understood.

In summary, the beauty of the aggregate approach as exemplified by the Atlanta case study lies in its simplicity. The volume of information alone required to plan for a rapidly developing metropolis like Atlanta is excessive. The conventional model that has evolved provides for a reductionist simplification to examine a complex base of data. The tools embodied by the conventional approach are descriptive as opposed to explanatory. Evaluation of the Atlanta case study, which is a typical example of the aggregate approach to modeling and understanding future urban transportation demand, suggests several basic inadequacies. First, better goal definition is required. What is the urban transportation system of the year 2000 supposed to accomplish? In turn, there needs to be a better link established between forecasting methods and policy sensitivity. Planners need to be able to assess the effects of transportation alternatives upon stated goals. This requires a set of forecasting methods that incorporate a causal approach and include policy sensitive variables. These appear to be lacking in the current case.

References

Atlanta Regional Commission (1975). *Economic Base Study of Atlanta*. Atlanta, Georgia: Atlanta Regional Commission.

Atlanta Regional Commission (1975). *Regional Development Plan, Final Small Area Forecast of Population, Jobs, and Housing Units*. Atlanta, Georgia: Atlanta Regional Commission.

Atlanta Regional Commission (1976). *Interdepartmental Technical Report, Trip Rates*. Atlanta, Georgia: Atlanta Regional Commission.

Atlanta Regional Commission (1976). *Interdepartmental Technical Report, Trip Distribution Models*. Atlanta, Georgia: Atlanta Regional Commission.

Atlanta Regional Commission (July, 1977). *Regional Transportation Plan*. Atlanta, Georgia: Atlanta Regional Commission.

Atlanta Regional Commission (1980). *Atlanta Region Population and Employment Trends*, Atlanta, Georgia: AKC.

James A. Liesindahl (1964). *Expansion of the Paved Road in the State Highway System of Georgia, 1923–1962*. M.A. thesis, University of Georgia.

U.S. Bureau of the Census (1974). *Urban Atlas: Atlanta*. Washington, D.C.: U.S. Government Printing Office.

U.S. Bureau of the Census (1960, 1970, 1980). *Census of Population and Housing*. Washington, D.C.: U.S. Government Printing Office.

U.S. Bureau of the Census—Atlanta Regional Office (1984). *Regional Review*, vol. 2. Atlanta, Georgia: U.S. Government Printing Office.

U.S. Department of Transportation (1980). *The Land Use and Urban Development Impacts of Beltways*. Washington, D.C.: U.S. Government Printing Office.

7 DESCRIBING DISAGGREGATE FLOWS: *INDIVIDUAL AND HOUSEHOLD ACTIVITY PATTERNS*

SUSAN HANSON
MARGO SCHWAB
Clark University

In the aggregate approach outlined in the previous three chapters, the focus has been on *zones* as generators of travel and as destinations for travel. Such a focus is appropriate for the sort of large-scale, long-range transportation planning that dominated planning in the past and that is described for the case of Atlanta in chapter 6. It is less suitable, however, for doing the kind of transportation planning that is now so important—the finer-scaled, shorter-term planning epitomized by TSM (Transportation System Management) and described in chapter 3. As demonstrated in chapter 6, the purpose of aggregate analyses is to predict overall flow patterns within a metropolitan area in order to help planners know where to construct new transportation facilities or where to add capacity to existing facilities. While an aggregate approach serves the important purpose of providing insights into the workings of the urban transportation system as a whole, it cannot help us answer a whole range of questions that pertain to *people* rather than to *zones*. For example, zonal data cannot tell us whether certain groups of people such as the elderly are unable to obtain basic goods and services (such as food and medical care) because they are unable to travel.

Thanks to John Pipkin for his comments on this chapter.

WHAT ARE DISAGGREGATE FLOWS?

A disaggregate perspective takes individuals or households rather than zones as the unit of analysis. Instead of examining flows between zones, the disaggregate approach is concerned with the movements of individuals as they participate in different activities at different locations over the course of a day, or sometimes a longer time period such as a week or a month. As you can imagine, descriptions of the daily travel–activity patterns of different types of people can provide considerable insight into the nature of daily life in the city and into the quality of life experienced by different groups of urban residents. Only through such individual-level studies can we begin to understand whether a travel pattern we observe is an expression of choice or of constraints; only through such detailed studies can we assess the impact of policies on the lives of urban residents. Chapin (1971), for example, has suggested studying the temporal and spatial aspects of urban activity patterns as a way of evaluating the success or failure of urban planning policies and projects. A theoretical perspective that focuses on individuals requires data that are collected for individuals and households rather than for zones.

Disaggregate data, then, are used to describe and analyze the travel of individuals or households within an urban environment. Such data contain detailed information on travel patterns, on the individuals making trips, and on the urban environment within which travel decisions are made. *Travel* characteristics include the timing, duration, sequencing, and purpose of trips. The focus on individuals' trips means that the travel data-collection procedure can recognize that often people make multistop trips, visiting several different places in the course of a home-to-home journey. The travel mode used on each leg of such a journey might not be the same, and the disaggregate approach recognizes this. Trip origins and destinations are often coded to street addresses rather than to zones. Disaggregate travel data, therefore, provide a more detailed, finer-scaled picture of movement patterns in the city. They enable us to examine movement patterns *within* zones (such as travel between shops) that are of particular interest to geographers seeking to understand relationships between travel and land use configurations.

As we have seen with zonal data, travel patterns are not random, but display certain regularities, which enable them to be predicted (to some extent). Trip characteristics are known to vary with the characteristics of the traveler (and the traveler's household) and with the nature of the urban environment and transportation system. There are many variables describing the individual's personal or household characteristics which are of interest because they have been found to affect travel: income, sex, employment status (employed or not), age, occupation, household size, and auto availability. These variables are used as surrogates for more complex factors, such as gender roles or social status, which affect people's needs for or ability to obtain various goods and services, and hence their demand for travel. Among the aspects of the urban environment that help to shape travel patterns are (1) the location and quality of different activity sites or potential destinations (such as employment places, retail outlets, parks, banks, and schools), and (2) the location and quality of the transportation facilities available (the highway network and the public transit system). These variables are used to measure the supply of transportation facilities and destinations within which the demand for travel is (or is not) satisfied. Though not every disaggregate study will include measures of all of these factors, the data requirements needed to undertake even a simple disaggregate transportation analysis can be quite formidable. Disaggregate data-collection procedures are outlined in greater detail later in this chapter.

WHY STUDY TRAVEL PATTERNS AT THE INDIVIDUAL LEVEL?

If greater simplicity and ease of analysis are not to be gained by adopting the disaggregate approach, what, then, are the advantages of moving to a finer scale of analysis? There are two main reasons for shifting the focus away from zones and onto individuals or households. The first is related to theory building and stems from a desire to understand how and why flow patterns emerge. The second is related to (and certainly depends on) the first but has a more practical bent—the desire to be able to predict the impacts of proposed transportation policies on travel. Each of these reasons is elaborated below.

Theory-Related Reasons for Disaggregate Studies

Recall the aggregate trip generation models described in chapters 5 and 6. In this kind of model the movements studied are flow

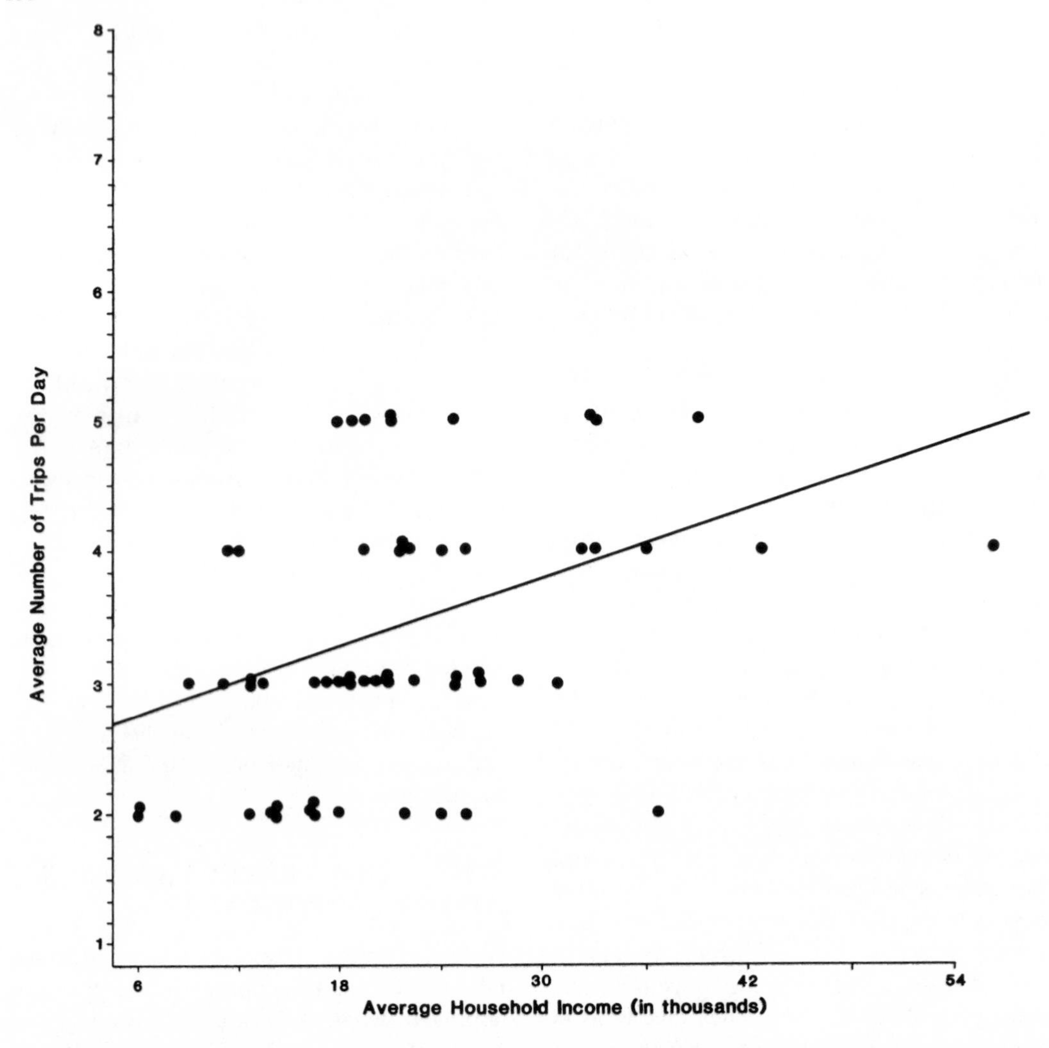

Figure 7-1a. Relationship between household income and number of trips per person per day in Baltimore, Maryland, for 70 planning districts.

levels generated by zones (for example, zone *i* generates an average of three trips per person per day), and the size of these flows is predicted on the basis of the characteristics of the households living in the zone, but, like the trip rate, these characteristics are averaged over all households in the zone (for example, the average household income for zone *i* is $15,000). Using zonal data such as these, we can find certain relationships between, for example, daily trip rates and household income: as zonal income rises, so too do the number of trips per household per day. But the relatively smooth relationship depicted in Figure 7-1a masks considerable within-zone variability in trip rates. Figure 7-1b shows the relationship between trip generation and income at the individual level. Notice how large the range of trip rates is for any given income level; this variability stems in part from the fact that there are other factors (such as household size, sex, employment status, and age) aside from income that affect an individual's propensity to make trips.

With zonal data it is difficult, in fact impossible, to examine the relative importance of these different variables, because there is

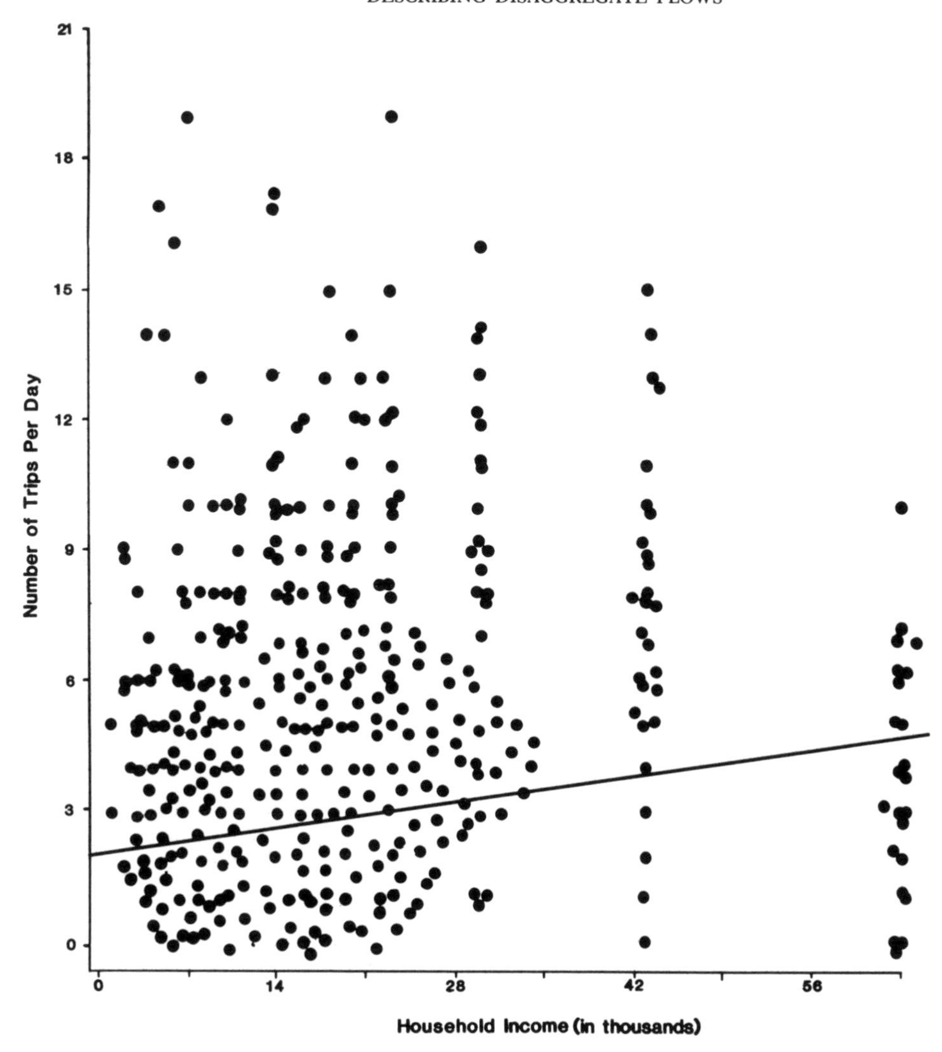

Figure 7-1b. Relationship between household income and number of trips per person per day in Baltimore, Maryland, for a random sample of 2,000 adults.

no way of associating a particular trip rate with a particular *set* of characteristics (for example, employed female without a car, living in a two-person household near the city center). If, for example, we find a strong negative relationship at the zonal level between percentage of elderly and number of trips per day (that is, trip rates are lower in zones with a high percentage of elderly people), we have no way of knowing if, within a zone, the elderly are the ones making fewer trips. Perhaps such zones also have a high percentage of unemployed people, and *they* are the ones who are contributing to the low zonal trip rate. If we assume that relationships known to exist at the aggregate level also exist at the individual level, we are making an unjustified inferential leap—a leap often referred to as the *ecological fallacy* ("ecological" because areal or zonal data are also called ecological data).

Thus, in order to understand how both personal/household characteristics and transportation/urban system characteristics affect an individual's travel, individual-level studies are needed. After all, people make decisions about when, where, and how to travel; zones don't. The shift to disaggregate transportation studies was motivated in large part by a desire to get a better grasp on how travel decisions are made. As you will see in the next chapter, models of individuals' travel decision making see people as choosing among alternatives (such as alternative modes or alternative destinations), but all such choices are made within certain constraints (a certain mode might not be available, one's income is limited, and one might have only a few minutes a day open for discretionary travel). Disaggregate studies that attempt to uncover the ways in which people make travel decisions are called *behavioral* because they focus explicitly on the attitudes and preferences that lie behind decision making.

One of the motivations for developing disaggregate models was dissatisfaction with the absence of any explicit behavioral theory that might help explain the patterns, such as the distance decay of trips, that aggregate models described. Moreover, the behavioral assumptions implicit in some aggregate models seemed intuitively unreasonable. For example, the entropy-maximizing principle postulates equiprobable microstates: within the set of given constraints, all patterns of trips are equally likely to occur. This randomness appears inconsistent with the idea that people's travel choices are at least partly rational.

There is now mounting evidence that incorporating more realistic behavioral assumptions in aggregate models can help explain certain aspects of aggregate behavior that were previously poorly understood. Although in principle the distance decay parameters in gravity models should not depend on the spatial configurations of origins and destinations, in fact there are systematic relationships between spatial structure and the magnitude of the distance-decay exponent. In attempting to explain this spatial structure effect (or spatial structure bias), Fotheringham (1983) proposed a new competing destination model based on the assumption that people make trip decisions in two stages. First they choose a general target region, and then they select a specific destination within that region. This seems quite plausible as an account of how many types of travel decisions are made, including how shopping destinations are selected or how residential locations may be chosen within a particular city, affecting the future configuration of work trips. This simple and strictly behavioral assumption helps us to understand the source of spatial structure bias in aggregate models, although issues of misspecification and spatial effects in gravity models are complex and multifaceted (see, for example, Sheppard, 1984).

Despite these perfectly logical reasons for wanting to study individual movements and individual decision making, at both the theoretical and the practical level we ultimately want to be able to make statements about aggregates—about groups—not about isolated individuals. But here again disaggregate data can actually be quite useful, for they permit a great deal of flexibility in grouping schemes (that is, methods of aggregation). Zonal data come to us already aggregated, but on the basis of one variable only—spatial proximity. Individuals in a disaggregate data base can be aggregated on any one of the variables describing that person; location is but one of many characteristics, then, that could be the grounds for defining groups of similar individuals. This flexibility is particularly important for policy analysts who need to be able to identify very specific groups (such as people whose only means of travel is public transit or people who commute to the CBD by car) in order to design and evaluate transportation policy options. This brings us to the second reason for pursuing the disaggregate approach in transportation studies.

Policy-Related Reasons

Some transportation policies are aimed at places, and some are aimed at people (and to the extent that certain kinds of people are spatially clustered, some policies are aimed at both places and people). An example of a place-oriented policy might be the location of Atlanta's MARTA described in the previous chapter; an example of a people-oriented policy would be the provision of special vans to transport the elderly to stores and clinics. While aggregate studies with zonal data are appropriate for designing and evaluating place-oriented policies, disaggregate studies are needed to assess people-oriented policy options, such as those designed to relieve traffic congestion, to reduce energy consumption, to lower pollution levels, or to increase the mobility of a particular group.

A disaggregate approach can be very useful in designing effective policies. Such policy crafting depends on an understanding of the relationship between travel behavior and its sociodemographic and spatial determinants; this understanding should reveal at which groups policies should be aimed and which policies are likely to be most effective. The elderly, for instance, have often been identified as a group that has low mobility and that makes fewer and shorter trips than younger people do. Among the reasons suggested for this lower mobility are the elderly's lower income, inability to drive, poor health, and absence from the labor force. If it is the elderly's retirement status that is the primary reason for their having low trip rates, then policy makers need not worry about how to "improve" the mobility of the elderly because their reduced trip making simply reflects the absence of the work trip and does not imply reduced access to shopping, socializing, or recreation. If, however, poor health and the lack of a driver's license are the major reasons, then some form of paratransit (such as door-to-door van service) is probably needed to ensure the elderly's access to essential goods and services.

Disaggregate data can also help policy makers to assess the likely impacts of transportation policies. If bus headways are increased on a certain route, how will ridership be affected? If a park-and-ride lot is built, who is likely to use it and by how much will auto traffic to certain parts of the city be reduced? If a kiss-and-ride service is implemented instead, will traffic be reduced at all or will another adult in the household now use the car that was formerly parked at a workplace all day? If a "diamond lane" (a high-speed lane for buses and carpools only) is added to a freeway, will commuters switch from single-occupancy auto to bus or carpool? Which kinds of people living in which kinds of places are most likely to switch? If a firm is considering introducing some form of flex-time in order to reduce its workers' commute times, what kind of policy would be most effective for the particular types of workers involved? If the nation experiences energy shortages, what are the implications of rationing fuel? How will the travel patterns of different types of people be affected?

These are but a few examples of the kinds of policies that can be designed and evaluated with disaggregate analysis. Because most policies like these are expected to generate only incremental changes in people's travel behavior, they require a level of analysis that is sensitive enough to detect small-scale changes.

A CONCEPTUAL FRAMEWORK FOR DESCRIBING DISAGGREGATE FLOWS

In order to achieve the goals of (1) understanding the determinants of travel, and (2) designing effective transportation policies, we need to study not only trip making but also the context within which travel decisions are made. In this section we outline first how we conceptualize travel and then how we conceptualize the decision-making context.

Characteristics of Disaggregate Flows

Disaggregate flows are made up of the daily travel-activity patterns of individuals, which are continuous in time and space and can therefore be represented as a time-space path. One such daily path, along with the map from which it is derived, is shown in Figure 7-2. Many of the components of a daily travel pattern are illustrated in this figure. Over the course of the day this person made two journeys (a journey is a home-to-home circuit); one of these was a single-stop journey (at the bakery), and the other was a multiple-stop journey (consisting of stops at work, a

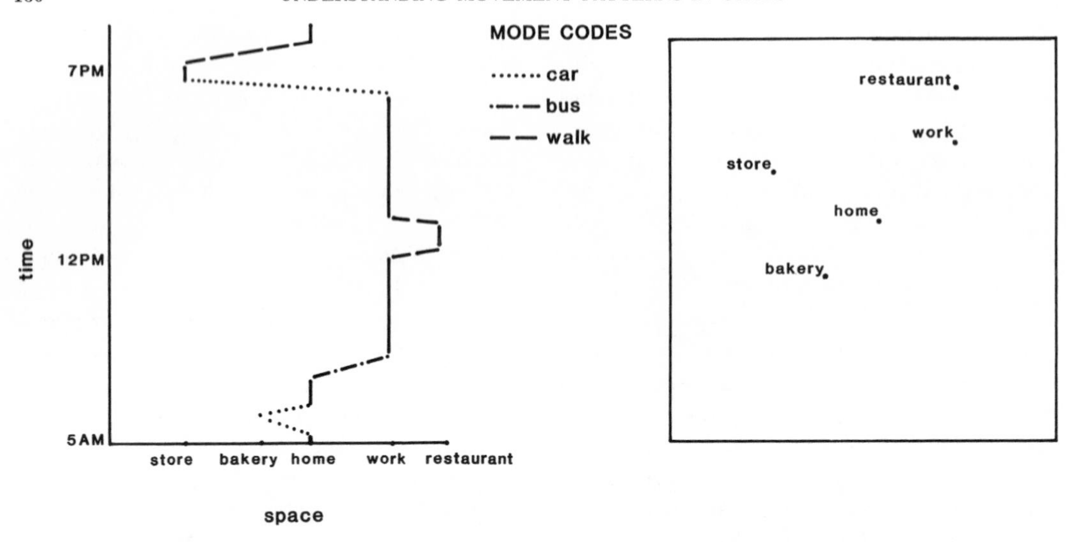

Figure 7-2. The travel activity pattern of a hypothetical individual shown as a path in space and time.

restaurant, back to work, and then at a food store before returning home). Each *stop* on each journey has a number of attributes: the *time* of arrival at and departure from the destination, the *mode* used to travel to the stop (note that the slopes of the different travel lines in Figure 7-2 denote speed of travel), the *distance* traveled from the origin to the destination, the *purpose* of the stop (activity undertaken at the destination), and the *location* of the stop. (Because two-dimensional space has been compressed to only one dimension in Figure 7-2, stop location is shown in the space–time path only in terms of distance.) The sequencing and the duration of stops are clearly evident from the space–time path. These are the components of the complex daily travel–activity pattern that comprise the disaggregate flows we describe below.

Although for purposes of this discussion these components have been separated, it is important to remember that they are interrelated in a person's travel decision making and that they do not all carry equal weight in the decision-making process. For example, the timing of a trip for a particular purpose (for example, to buy milk before breakfast) might dictate the mode and destination (walk to corner store rather than drive to supermarket). Also, the amount of time one wants to spend at a destination might dictate the mode chosen: a faster mode might be used so that time at the destination can be increased.

One question that cannot be answered with data like these for only a single day's path is how typical this pattern is for this person. Is pretty much the same daily path etched out day after day with only minor variations from day to day; does the individual generate several "typical" daily paths, or is every day quite different from every other day? Recent analysis of multiday travel data (Huff & Hanson, 1986) suggests that the second of these alternatives is the one that best describes most people's behavior; certain aspects of the daily path are repeated, but there is no single space–time path that adequately captures the essence of the individual's habitual behavior. It is important to keep in mind the fact that day-to-day variability does exist in travel–activity patterns because most of the disaggregate models described in the next chapter assume that habitual, highly repetitive behavior prevails, and it is habitual behavior that is modeled. In other words, although a mode-choice model might assume that one mode is always used on the journey to work and a destination-choice model might assume that one food store is always selected, people might use different modes or visit different food stores from time to time. Because this kind of variability is a regular feature of urban travel, it should be incorporated in disaggregate models, but this has not yet been accomplished.

Nature of the Travel-Decision Context

Although the terms "mode choice" and "destination choice" are often used, it is important to remember that these and other travel decisions are rarely truly free choices, but choices made within constraints. Sometimes (and for some people perhaps always) the constraints are such that there is no "choice" at all. As mentioned above, the two main sets of constraints within which travel decisions are made are personal/household characteristics and the nature of the spatial/transportation environment within which transportation is to take place.

Personal and household characteristics such as gender, stage in the life cycle, income, employment status, and even religion can affect people's ability, desire, and need to engage in certain activities at certain times, in certain places, and by certain modes. The amount of time available for travel and the effects of travel costs vary with these sociodemographic characteristics. For example, numerous studies have documented very real differences in the travel patterns of women and men. Women make shorter trips, work closer to home, make a higher proportion of their stops for the purposes of shopping and personal business, and conduct more of their travel on foot or by public transit. The impact of traditional gender roles has been revealed in all studies that have compared the travel patterns of women and men. As another example, consider the ways in which income can influence an individual's travel. Lower incomes mean that there is less money available to spend on gas, vehicles, parking fees, or on purchasing goods and services at trip destinations. In addition, income and employment status can jointly affect the time and money available for travel (because many low-income people work long hours). In fact, income affects mode "choice," trip frequency, and travel distances, with lower-income people relying more on public transportation, making fewer trips, and traveling shorter distances than higher-income people (Davies, 1969; Nader, 1969; Potter, 1977).

The nature of the transportation system and of the spatial distribution of activity sites is also an important constraint on observed travel patterns. Consider the three hypothetical distributions of shops depicted in Figure 7-3. Visits to the individual shops within the cluster on the right could easily be made on foot, whereas the distributions shown in the middle (typical of strip development in U.S. cities) and on the left (typical of destinations in rural areas) make pedestrian travel more difficult and auto travel more likely. The spatial distribution of activity sites around the home (and, for employed people, around the workplace as well) is likely to affect not only the modes used but also the distances traveled. Consider, for instance, how changes in the retail landscape go hand in hand with altered travel patterns.

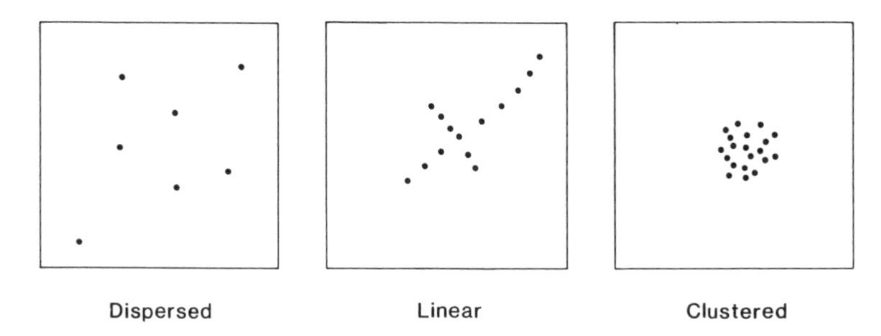

Figure 7-3. Hypothetical distributions of shops (a) dispersed, (b) linear, and (c) clustered.

Several decades ago numerous small food stores dotted the urban landscape; these stores, which were easily accessible on foot, have now given way to fewer, larger, more widely spaced stores surrounded by parking lot moats—clearly not designed for pedestrian access.

The ease of reaching potential destinations is a function not only of the location of facilities but also the nature of the transportation network and the modes available to travelers. Is there a public transportation system available? If so, what parts of the city are covered by public transit routes, and how frequent is the service on each route? Forer and Kivell (1981) have shown how the accessibility levels in various parts of Christchurch, New Zealand, are different for bus-system users versus private auto users. Figure 7-4 illustrates those parts of the city that are more accessible by bus than they are by car (the striped areas) and those parts of the city that can be reached more easily by car than by the public transport network (the hatched areas). It is also easy to imagine how the quality or the very existence of roads might affect travel decisions.

Another way in which travel decisions are constrained is the fact that certain activities are fixed in time and space. Most workers have to be at work for certain set hours; arrival at and departure from work cannot take place at will. Most stores and offices are open for business only part of every 24-hour day. Cullen and Godson (1975) investigated the impact of time–space fixity on the individual's travel–activity patterns and found that activities fixed in time were more important (than those fixed in space) in structuring the daily schedule (that is, the organization and sequences of activities), but a larger proportion of the individual's activities were fixed in space than in time.

DISAGGREGATE DATA

Both aggregate and disaggregate urban transportation models require staggering amounts

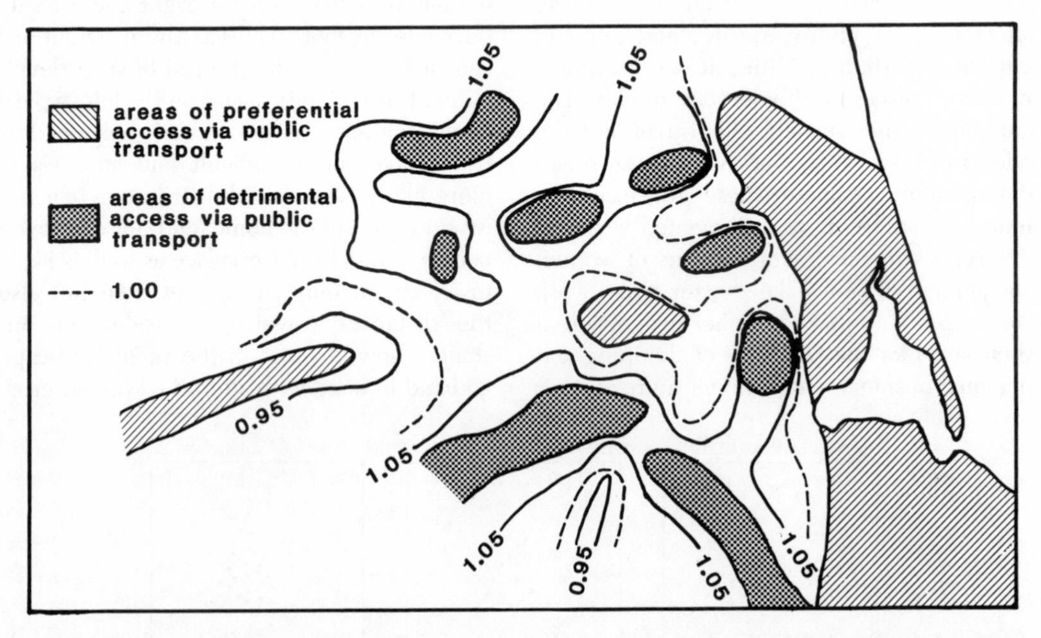

Figure 7-4. Accessibility levels within Christchurch, New Zealand. Hatched areas are more easily reached by car than by public transit; striped areas are more easily reached by public transport than by car. From "Space-Time Budgets, Public Transport, and Spatial Choice," by P. L. Forer and H. Kivell, 1981, *Environment and Planning A*, 13, p. 506. Copyright 1981 by *Environment and Planning*. Reprinted by permission.

of data—on travel patterns, on the traveler's characteristics, and on the existing transportation system and land use pattern. Although the two types of studies often collect very similar information, they store and analyze the resulting data quite differently.

Information on travel and travelers can be collected in a number of ways. Several methods involve getting information from people who are actually traveling (on-board surveys or cordon surveys—where travelers are stopped and questioned at certain places on the road network), but the disadvantage of this approach is that it does not provide information on immobile people, precisely those who are most in need of ameliorative transportation policies. A preferable method of collecting data for overall transportation planning is the household travel survey, in which people are surveyed at their places of residence.

Because it would be prohibitively expensive to collect the needed information for every person living in the city under study, it is necessary to draw a sample. In order to be able to make inferences from the sample to the metropolitan population as a whole, the sample households must be carefully chosen so that they are representative of the general population. A representative sample should include, for example, people from all parts of the metropolitan area, from all different socioeconomic backgrounds, and from all stages in the life cycle. A simple random sample drawn from a list of all households in the study area (such as the list of electricity company patrons) should yield such a representative sample. Sampled households are then contacted by phone, by mail, or in person, but the survey method chosen will depend on the kind of travel data to be collected.

Collecting Travel Data

Although travel data collected via on-board or cordon surveys frequently ask the traveler only about the trip he or she is making at the moment, household travel surveys typically attempt to gather information on all out-of-home travel. These data are usually collected in the form of a self-administered travel diary, an example of which is shown in Figure 7-5. Most often the travel diary covers only one day's travel, but sometimes diaries last for longer time periods. The advantage of collecting travel data for more than one day is of course that some of the variability in day-to-day patterns, mentioned earlier, can be observed. One type of variability that may be particularly valuable to observe is the occurrence of the "no-trip" day interspersed among days on which trips are made. Of course, the costs of data collection rise as the number of days is extended, because respondents must be monitored to ensure that they continue to record their trips faithfully and accurately. For this reason, travel diaries rarely last longer than one week. In one household travel survey conducted in Sweden, however, the sample respondents kept travel diaries for five consecutive weeks (Marble, Hanson, & Hanson, 1972); some of the results of that study are described below.

Notice that it is possible to derive from the information recorded in the diary form (Figure 7-5) all the travel-pattern dimensions identified earlier as being of interest—trips, stops, travel distances, purposes, destination locations, modes, time of travel, and so on. Notice also that the diary is designed to elicit a response from the person who made no trips on the travel-logging day.

Another important issue that is evident from a glance at the diary form concerns how various activities, times, locations, destinations and modes are to be coded. What level of detail should be used? What level of detail can we realistically expect of respondents? Take travel times as one example. We could ask people to write down their arrival and departure times to the nearest second, the nearest minute, the nearest five minutes, the nearest quarter hour, and so on, but people are not really aware of time to the nearest second or minute (digital watches may be changing this). Thus, even if we ask for the nearest minute, most people will round to the nearest five minutes. The study that used the diary form in Figure 7-5 gave respon-

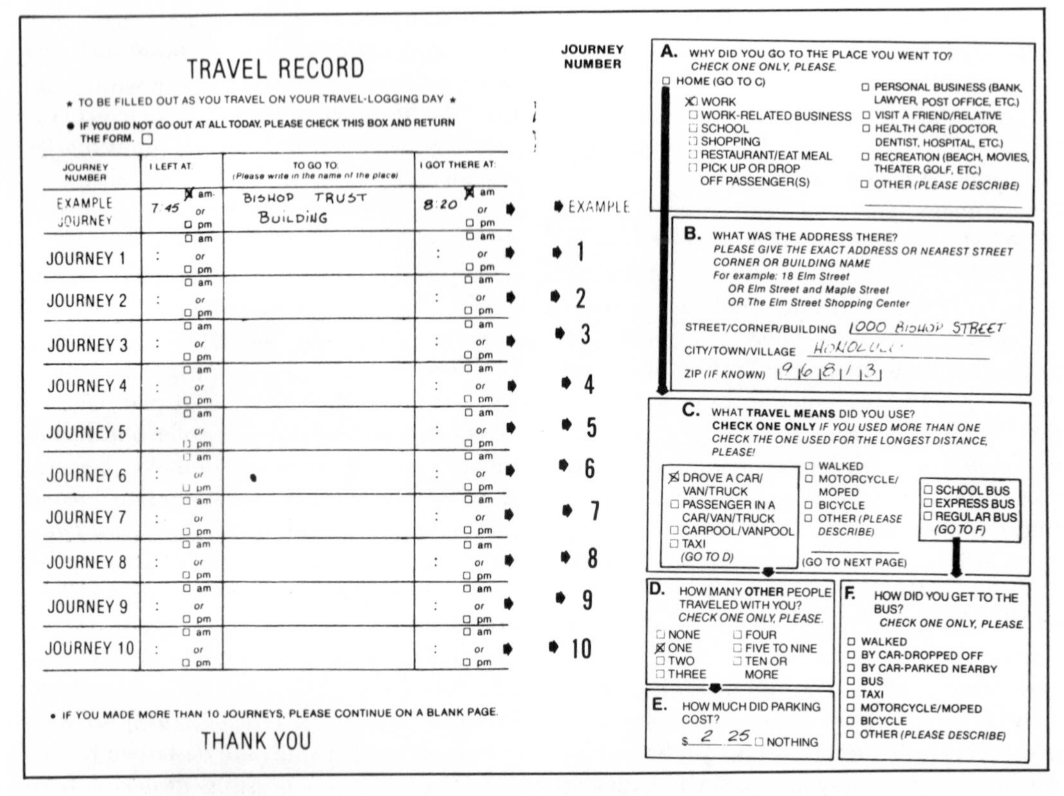

TRAVEL RECORD

★ TO BE FILLED OUT AS YOU TRAVEL ON YOUR TRAVEL-LOGGING DAY ★

● IF YOU DID NOT GO OUT AT ALL TODAY, PLEASE CHECK THIS BOX AND RETURN THE FORM. ☐

JOURNEY NUMBER	I LEFT AT	TO GO TO (Please write in the name of the place)	I GOT THERE AT
EXAMPLE JOURNEY	7:45 ☒ am or ☐ pm	BISHOP TRUST BUILDING	8:20 ☒ am or ☐ pm
JOURNEY 1	: ☐ am or ☐ pm		: ☐ am or ☐ pm
JOURNEY 2	: ☐ am or ☐ pm		: ☐ am or ☐ pm
JOURNEY 3	: ☐ am or ☐ pm		: ☐ am or ☐ pm
JOURNEY 4	: ☐ am or ☐ pm		: ☐ am or ☐ pm
JOURNEY 5	: ☐ am or ☐ pm		: ☐ am or ☐ pm
JOURNEY 6	: ☐ am or ☐ pm		: ☐ am or ☐ pm
JOURNEY 7	: ☐ am or ☐ pm		: ☐ am or ☐ pm
JOURNEY 8	: ☐ am or ☐ pm		: ☐ am or ☐ pm
JOURNEY 9	: ☐ am or ☐ pm		: ☐ am or ☐ pm
JOURNEY 10	: ☐ am or ☐ pm		: ☐ am or ☐ pm

• IF YOU MADE MORE THAN 10 JOURNEYS, PLEASE CONTINUE ON A BLANK PAGE.

THANK YOU

JOURNEY NUMBER: ➧ EXAMPLE, ➧ 1, ➧ 2, ➧ 3, ➧ 4, ➧ 5, ➧ 6, ➧ 7, ➧ 8, ➧ 9, ➧ 10

A. WHY DID YOU GO TO THE PLACE YOU WENT TO? CHECK ONE ONLY, PLEASE.

☐ HOME (GO TO C)
☒ WORK
☐ WORK-RELATED BUSINESS
☐ SCHOOL
☐ SHOPPING
☐ RESTAURANT/EAT MEAL
☐ PICK UP OR DROP OFF PASSENGER(S)
☐ PERSONAL BUSINESS (BANK, LAWYER, POST OFFICE, ETC.)
☐ VISIT A FRIEND/RELATIVE
☐ HEALTH CARE (DOCTOR, DENTIST, HOSPITAL, ETC.)
☐ RECREATION (BEACH, MOVIES, THEATER, GOLF, ETC.)
☐ OTHER (PLEASE DESCRIBE)

B. WHAT WAS THE ADDRESS THERE? PLEASE GIVE THE EXACT ADDRESS OR NEAREST STREET CORNER OR BUILDING NAME
For example: 18 Elm Street
OR Elm Street and Maple Street
OR The Elm Street Shopping Center

STREET/CORNER/BUILDING 1000 BISHOP STREET
CITY/TOWN/VILLAGE HONOLULU
ZIP (IF KNOWN) 96813

C. WHAT TRAVEL MEANS DID YOU USE? CHECK ONE ONLY IF YOU USED MORE THAN ONE CHECK THE ONE USED FOR THE LONGEST DISTANCE, PLEASE!

☒ DROVE A CAR/VAN/TRUCK
☐ PASSENGER IN A CAR/VAN/TRUCK
☐ CARPOOL/VANPOOL
☐ TAXI
(GO TO D)
☐ WALKED
☐ MOTORCYCLE/MOPED
☐ BICYCLE
☐ OTHER (PLEASE DESCRIBE)
(GO TO NEXT PAGE)
☐ SCHOOL BUS
☐ EXPRESS BUS
☐ REGULAR BUS
(GO TO F)

D. HOW MANY OTHER PEOPLE TRAVELED WITH YOU? CHECK ONE ONLY, PLEASE.

☐ NONE
☒ ONE
☐ TWO
☐ THREE
☐ FOUR
☐ FIVE TO NINE
☐ TEN OR MORE

E. HOW MUCH DID PARKING COST?
$ 2.25 ☐ NOTHING

F. HOW DID YOU GET TO THE BUS? CHECK ONE ONLY, PLEASE.

☐ WALKED
☐ BY CAR-DROPPED OFF
☐ BY CAR-PARKED NEARBY
☐ BUS
☐ TAXI
☐ MOTORCYCLE/MOPED
☐ BICYCLE
☐ OTHER (PLEASE DESCRIBE)

Figure 7-5. Travel diary form. Developed for the Oahu Metropolitan Planning Organization (OMPO) and reprinted by permission of the Oahu OMPO.

dents no specific directions as to how to code times, but note that the example times are rounded to five-minute intervals.

Activity at destination (or trip purpose) is another example. We can ask people to check off one of several general activity categories as is done in the diary shown here, or we can request a detailed accounting of what was done at the destination (such as purchased food, withdrew money, watched children play). These "free-response" trip purposes would then later be coded into detailed activity codes by the survey team. A final example is the coding of stop origin and destination locations. Should respondents identify locations by street addresses, by naming the nearest street intersection, or by neighborhood name? In the diary shown here respondents were asked to be as specific as possible, so that their movements could be mapped and studied from point to point rather than from zone to zone.

Generally, in all these cases, a finer rather than a more general categorization scheme is preferred for disaggregate studies. It is important to remember that fine categories can always be aggregated to more general ones, but the reverse cannot be done. Thus, if an activity-coding scheme discriminates between convenience shopping and shopping for durables (clothing, skis, bikes), these categories can easily be lumped together in a study in which the type of shopping is not important, but the two types of shopping behavior can also be studied separately.

Data on Personal and Household Characteristics

Most of the individual/household attributes on which data are gathered as part of a household travel survey are quite familiar; a few are particular to the goals of transportation studies (see Table 7-1). The standard measures of socioeconomic status are there (education, occupation, income), as are standard

Table 7-1. Personal and household characteristics often included in disaggregate travel studies

Occupation (for example, an occupation prestige score)
Education (years of formal education)
Income (household income before taxes)
Automobile availability (number of automobiles per household divided by number of drivers' licenses per household)
Employment status (number of hours worked per week)
Stage in life cycle (age, marital status, presence/absence of children)
Sex (male/female)
Household size (number of permanent residents)
Possession of drivers' licenses/monthly bus pass
Distance from home to nearest bus stop
Distance from home to nearest limited-access highway
Distance from home to nearest major arterial
Availability of parking at work place/at home

demographic variables (age, household size, number and ages of children, employment status, sex). Items peculiar to travel studies include questions about the availability of parking at the work place, whether the employer provides free parking, and the possession of a driver's license or a monthly bus pass. In addition to these relatively easy-to-measure factors, there are other, less straightforward measures, such as personality dispositions (status orientation) that some researchers (for example, Taylor, 1979) have argued are important in understanding spatial choice, especially destination choice for shopping.

In this part of the survey, information is also sometimes collected on the respondent's perceptions of and attitudes toward different destinations, routes, and modes. The next chapter describes examples of studies that required people to evaluate the characteristics of food stores (checkout service, variety, parking facilities, freshness of produce, layout, and so on) in an attempt to understand how people go about selecting destinations. There have also been numerous studies of mode choice that have posed seemingly endless queries about people's perceptions of the comfort, convenience, and safety of different modes (especially transit versus auto). The behavioral models mentioned earlier require this sort of data to analyze how people make decisions to visit one destination instead of another or to use one mode instead of another.

Collecting Data on Space–Time Constraints

Gathering data on space–time constraints is different from collecting data on travel and sociodemographic characteristics because it involves amassing information on the nature of the urban environment—the transportation and land use systems—rather than surveying people. Data on space–time constraints can usually be obtained from sources available in planning or transportation agencies, from maps, aerial photographs, field survey results, or public transportation operator records.

A transportation system inventory includes information on the highway system (the number of lanes, flow volumes, and travel speeds on each link in the network) and on the public transit system (routes, schedules, ridership levels, cost and revenue data, and system operation data). Land use data are gathered via an inventory of different land use types (residential, commercial, industrial) throughout the metropolitan region. These can be collected for zones of varying size or by street address. In disaggregate studies it is usually desirable for the land use data to be spatially quite detailed (at the level of the establishment) rather than generalized to zones. If data on potential destinations are available at the establishment level, then such data can be used to ascertain how travelers select certain destinations (such as department stores) from among the set of possible alternatives. Sometimes data on the quality or attractiveness of each establishment are also gathered, again for use in decision-making studies. As the operating hours of an establishment play no minor role in people's decision whether or when to visit it, these are also sometimes included as part of the inventory of the opportunity set (for example, Lenntorp, 1976). In such cases, however, data are usually collected on only one type of establishment, say, food stores.

DESCRIBING TRAVEL–ACTIVITY PATTERNS

With disaggregate data like those described here, one can carry out almost endless descriptive analyses. In this part of the chapter we shall touch upon but a few of the ways in which such data can be used to describe the travel–activity patterns of samples representative of either (1) the whole population of a city, or (2) population subgroups defined in terms of either location or sociodemographic attributes. The wealth of information contained in disaggregate data and the complexity of the patterns such data reveal point to the need for modeling. Without the simplification and the logical structure that models impose, one could easily end up drowning in computer printouts describing disaggregate flows in a million and one ways.

Many of the examples that follow draw upon the 35-day travel diary data mentioned earlier in this chapter. We shall use these data to describe selected aspects of the travel activity patterns, first for a sample representative of the general population of the study area—Uppsala, Sweden—and then of certain population subgroups. The discussion of the general population's travel–activity patterns is focused on travel distances.

Travel Distances of the General Population

The distance decay curve shown in Figure 7-6 illustrates the expected way in which movement frequency declines or "decays" with the increasing distance; the curve represents all movements made by the sample individuals over the five-week diary period. While the overall shape of this curve would be found in many settings, the precise way in which travel decays with distance reflects the spatial structure of the particular city in which the data were collected. At the time of the travel study, Uppsala was a city of about 120,000 people and the bulk of the city's commercial activity was concentrated in the CBD; very little suburbanization of retail activity had occurred.

This distance decay curve for *all* movements masks a number of patterns that shed light on the deterrence effect that distance has on trip making. People are willing to go longer distances when they travel by certain modes and for certain purposes. In Figure 7-7 the movement distances for each of several modes have been plotted as cumulative distributions, which is another way of looking at the decay of movements over distance. The figure shows how much more sensitive to distance walking trips are than are, say, trips made by bus or automobile. Fully 98% of all stops made on foot cover less than 2 kilometers (about 1.2 miles), whereas 85% of bicycle stops, 45% of car driver stops, and only 35% of all stops made by bus are less than 2 kilometers in length.

In interpreting these curves, it is important to know something not only about what sort of a city Uppsala was then but also how these data were collected. In particular, it is necessary to realize that respondents were asked to record every stop they made on every journey over the 35 days and that a new stop was defined as occurring when someone went to a new place or entered a new establishment. A new stop would occur, then, every time the traveler went into a different shop, which might happen several times on a single shopping expedition to the CBD. With all movements thus included in the travel–activity pattern, walking trips turn out to be a major contributor to intraurban mobility—nearly half (47%) of all out-of-home movements in the Uppsala study were made on foot. But these walking stops can be further distinguished by travel purpose, for not all purposes are equally likely to be the object of a pedestrian stop. More than 60% of all walk stops were to shop, whereas less than 30% of walk stops were made to work.

This differential use of travel modes for different trip purposes (not fully explored here) suggests that trip distances should also differ systematically by purpose. Figure 7-8

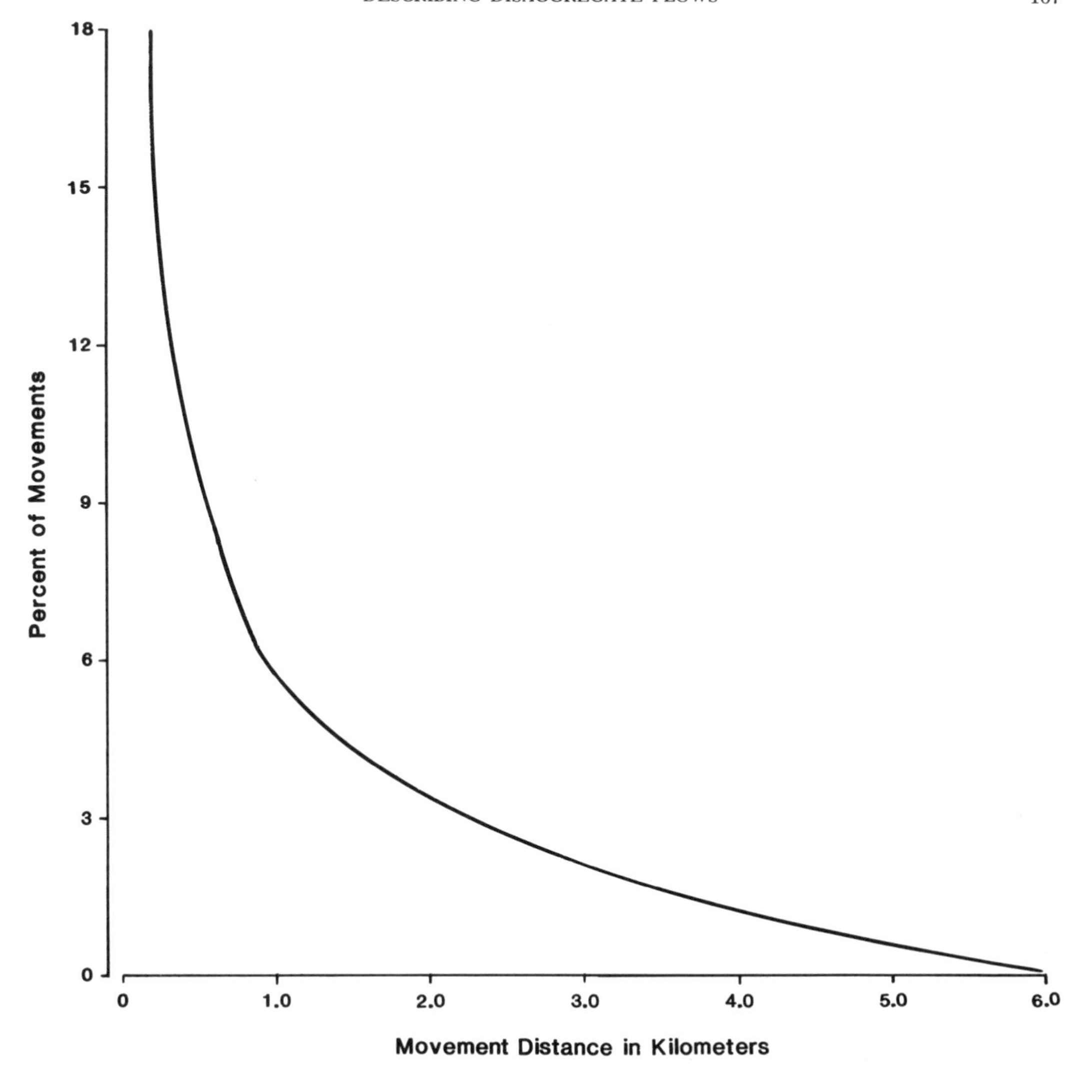

Figure 7-6. Decline of movement frequency over distance, sample of 149 people, Uppsala, Sweden, 1971.

shows the cumulative travel distances for each of several different purposes, and again it is evident that the smooth curve in Figure 7-6 masks some important patterns. For instance, social and recreation trips are far less sensitive to distance than are shopping and personal business trips (examples of personal business stops include visits to the bank, the post office, the laundromat, or the dentist). Whereas 84% of all shopping trips are less than 2 kilometers in length, only 54% of recreation trips are this short. These distinctly different distance decay curves for different travel modes and purposes indicate that travelers evaluate distance differently in different circumstances, and therefore suggest that disaggregate models of trip distribution should recognize this by not lumping all types of travel together.

The distances people travel reflect more than just mode and purpose; we also know that travel distances vary systematically with the nature of the urban environment (density of potential destinations) and with the personal/household attributes of the traveler. Because disaggregate data allow us to asso-

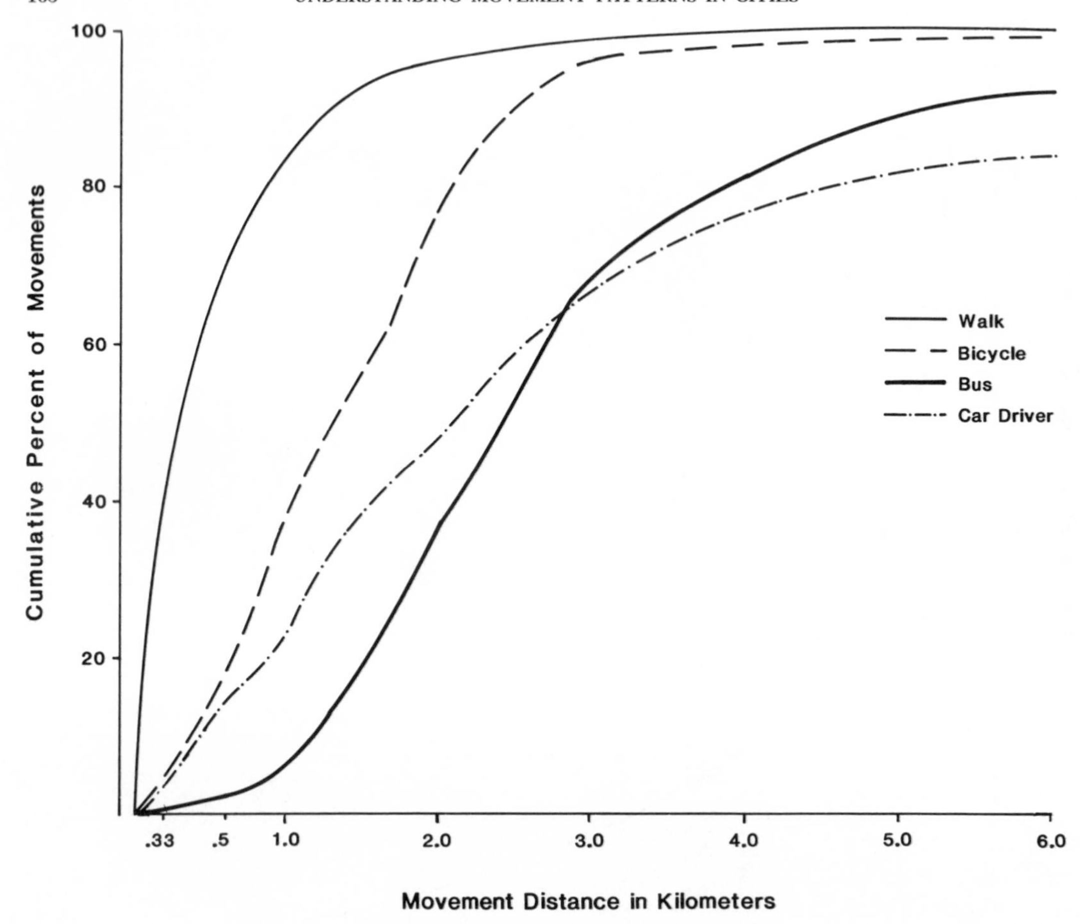

Figure 7-7. Cumulative distributions of distances traveled between stops by different modes. Representative sample of 149 people, Uppsala, Sweden, 1971.

ciate a particular trip characteristic with a specific individual's personal attributes and exact location within the city, such data enable one to separate the influences of personal attributes versus spatial structure on travel. This is not possible with zonal data, because there is always some degree of heterogeneity (with respect to types of people) within zones.

How Space–Time Constraints Affect Travel

The density of potential destinations (especially for discretionary travel like shopping and personal business) and their opening and closing hours have been shown to influence travel and therefore should be included in models of trip decision making. In the Uppsala study, the density of opportunities around the home was found to affect discretionary travel distances and travel modes used. People living in high-density settings traveled shorter distances for shopping and personal business than did those living in lower-density environments (Figure 7-9). Moreover, people living in higher densities also made a higher proportion of their stops by nonmotorized modes (on foot or by bicycle) and made fewer of their trips by automobile than did people living in lower-density arrangements (Figure 7-10). The density of opportunities also affects the size of the individual's activity space (the area within which activi-

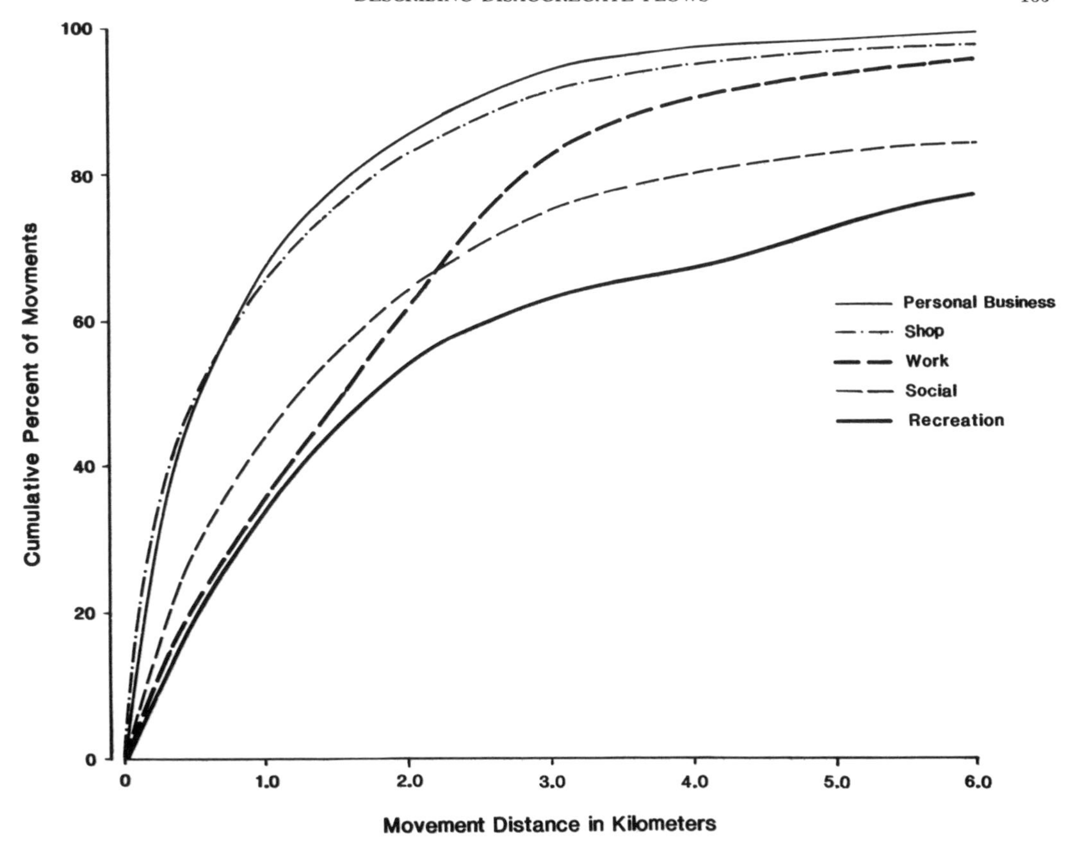

Figure 7-8. Cumulative distributions of distances traveled between stops for different purposes. Representative sample of 149 people, Uppsala, Sweden, 1971.

ties are carried out): people living in high-density environments have smaller activity spaces. These findings suggest that high-density arrangements, where people can accomplish much of their travel on foot or by bicycle relatively close to home, should promote lower levels of fossil fuel consumption.

In this same study the density of opportunities around home (measured as the number of establishments at a given distance from home, discounted by that distance) was found to be associated with the number of social trips made, but only for unemployed people without access to cars. This implies that only for this group of people does density affect the propensity to make social trips and illustrates the important point that spatial structure does not have the same impact on the travel patterns of all kinds of people.

In addition to influencing discretionary travel distances, mode use, and social trip frequency, the location of the traveler with respect to potential destinations has also been shown to affect overall trip frequency for nonworking people (Koenig, 1980); as Figure 7-11 shows, travel frequency increases as the density of opportunities around the home increases. Why might trip frequency vary as a function of density for unemployed people, but not for employed people?

A final example of how space–time constraints can affect travel comes from Lenntorp's (1976) extremely fine-scaled study of how the mode(s) and time available to the traveler and the location and operating hours of stores narrow down the number of opportunities (in this case food stores) that the traveler can realistically visit. Although the

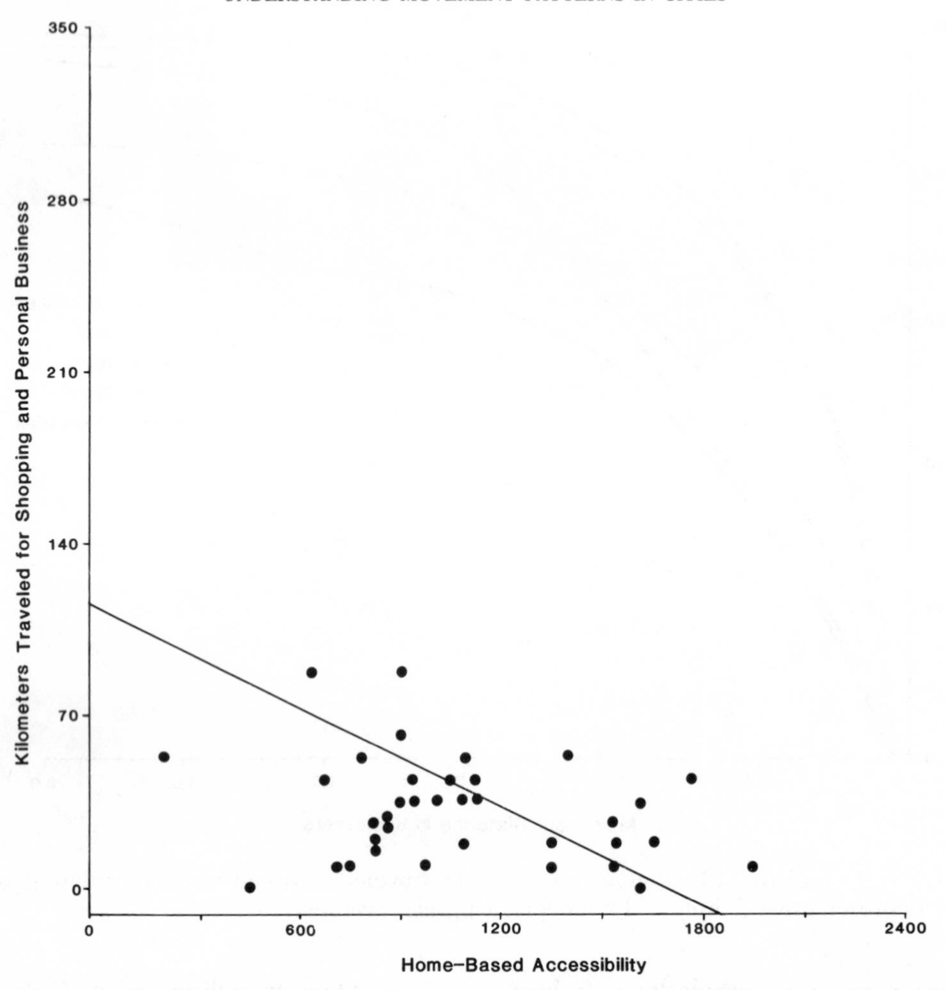

Figure 7-9. The relationship between accessibility to retail and service establishments and distance traveled for shopping and personal business over 35 days. Uppsala, Sweden, 1971.

number of food stores in the opportunity set may be large, Lenntorp shows that when the space–time constraints outlined above are considered, there are actually very few stores from which the traveler can select. As you will see in the next chapter, disaggregate models attempt to specify how an individual goes about selecting one item (be it a mode, a destination, or a route) from the set of available options, sometimes called "the choice set." What Lenntorp's work shows is the fact that the opportunity set and the choice set are usually quite different; many items in the opportunity set are not in the choice set at all.

Although, as we have seen, the location of the individual relative to opportunities does have an impact on travel, it is not the only—or even the primary—determinant of travel. Figure 7-12 shows the movements over 35 days of two individuals in the Uppsala sample; notice how different the two movement patterns are despite the fact that these people live in essentially the same location. If location alone were a major impact on travel, we would expect the maps to be far more similar than in fact they are. The characteristics of travelers also play a major role in molding travel patterns, and we next turn to a consideration of these.

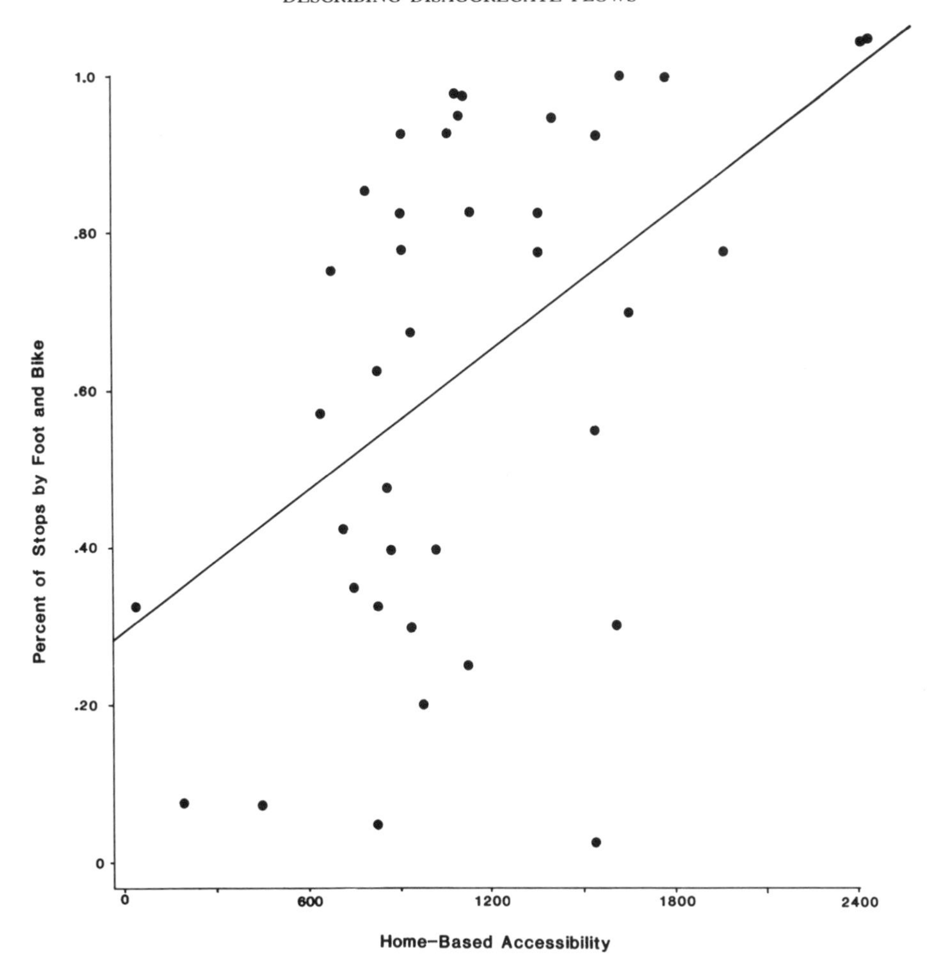

Figure 7-10. The relationship between acessibility to retail and service establishments and the proportion of individuals' stops made by nonmotorized modes (foot and bike) over 35 days. Uppsala, Sweden, 1971.

How Personal Characteristics Affect Travel

The relatively long list of personal attributes that are known to affect travel appears in Table 7-1. Our purpose here is not to provide an exhaustive litany of how each of these affects every aspect of travel, but only to sketch out a few examples of the links between travel and sociodemographic characteristics, focusing on the impacts of age, sex, race, and automobile availability.

Age

Numerous studies have examined the impact of the aging process on travel behavior and have found that as age advances, individuals' travel patterns undergo a number of changes. The focus of most of these studies has been on the travel patterns of the elderly (usually defined as those 65 years of age or older), compared with those of younger persons. Among the travel changes that these studies have documented as occurring with advancing age are reduced levels of trip making, a shift to different modes of travel, altered trip purposes, fewer multistop trips, and shifts in the timing of trips.

One of these studies was carried out with the Uppsala travel data (P. Hanson, 1977). To compare the intraurban travel patterns of the elderly and nonelderly, many trip char-

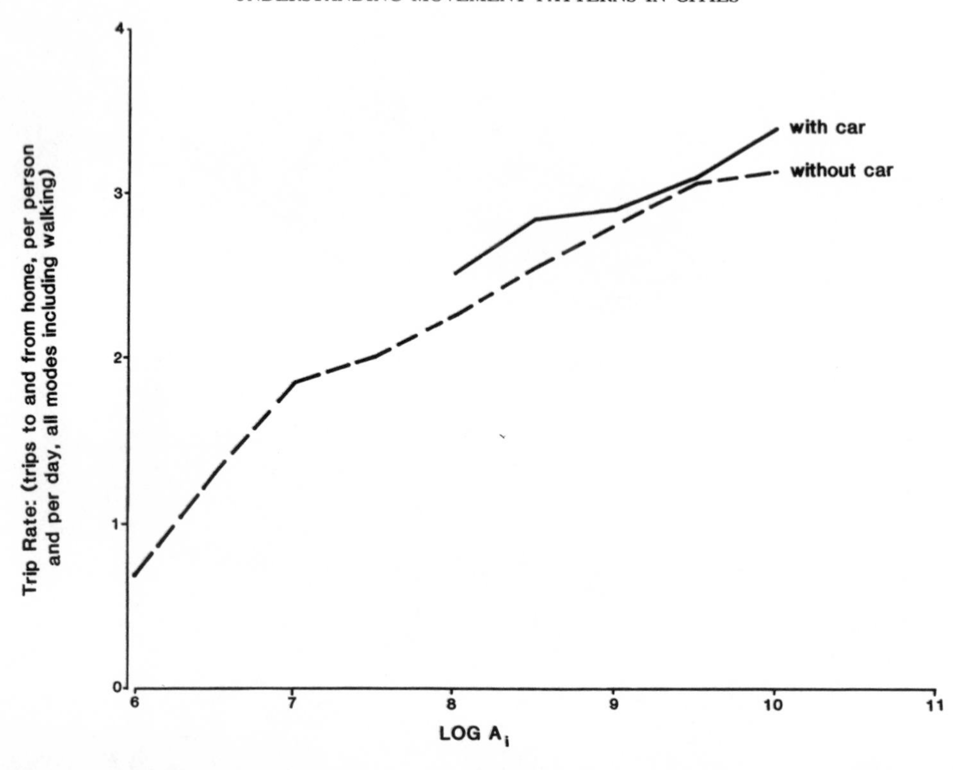

Figure 7-11. The relationship between accessibility (A_i) and trip rate for people with and without a car (Marseilles, 1966). Trip rate increases as access to opportunities increases. The A_i measure used here is similar to the one defined in equation 1 in Chapter 1. From "Indicators of urban accessibility: Theory and application," by J. G. Koenig, 1980, *Transportation*, 9, p. 167. Copyright 1980 by *Transportation*. Reprinted by permission.

acteristics were examined, including measures of the frequency, timing, and purpose of travel as well as the distances traveled to reach destinations, the amount of time spent in different activities, and the means of travel used. The analysis revealed that when all travel, including that done on foot, is examined, the lower trip frequency among the elderly is attributable entirely to the diminished importance of the work trip; the elderly participate in all other out-of-home activities as frequently as do younger persons. There are, however, some interesting differences between the travel patterns of the two groups, but these differences indicate that the elderly have made relatively minor adaptations by making trips with fewer stops, by using different modes of travel (see Table 7-2), and by traveling at different times of day than

Table 7-2. Mode use[a] by elderly and nonelderly people, Uppsala, Sweden, 1971

	Elderly ($N=64$)	Nonelderly ($N=119$)
Walk	56.7	40.9
Bike	6.2	11.5
Bus	17.2	9.8
Car driver	11.4	26.2
Car passenger	6.7	10.1
Totals	98.2[b]	98.5[b]

[a] Percentage of all movements; figures are group averages.

[b] Column does not add to 100 because of rounding.

Note. From "The Activity Patterns of Elderly Households," by P. Hanson, 1977, *Geografiska Annaler*, 59, p. 117. Copyright 1977 by *Geografiska Annaler*. Reprinted by permission.

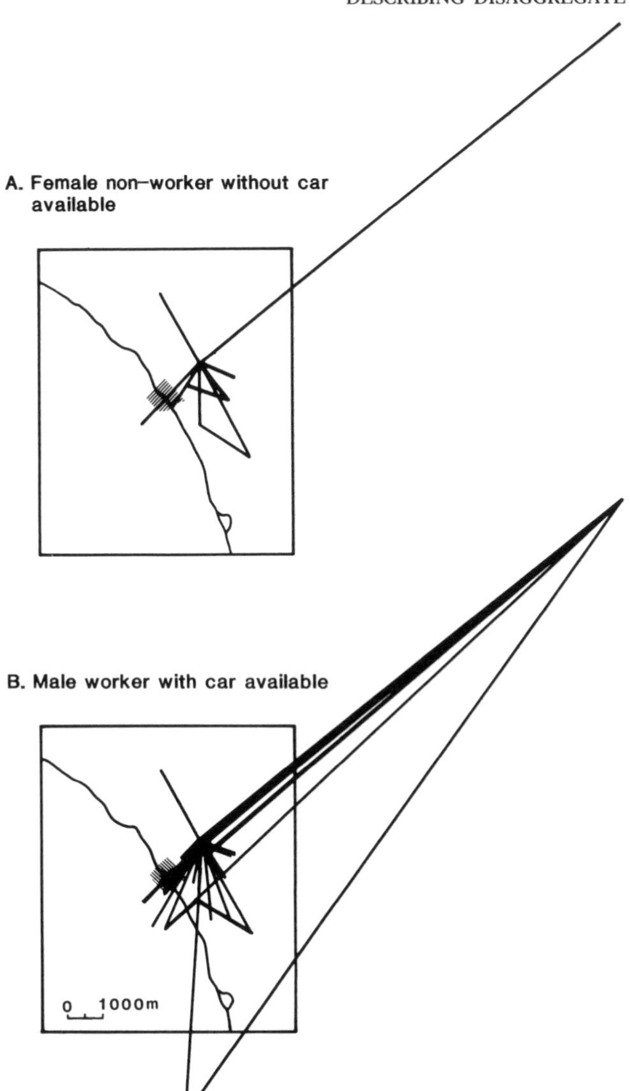

Figure 7-12. Movement patterns over 35 days for two individuals with the same residential locations, Uppsala, Sweden.

younger people do (mainly during off-peak hours).

Similar results have been obtained from studies in the United States (Zabinski, 1977) and Australia (Morris, 1979). In Britain Hopkin, Robson, and Town (1978) found that much of the reduction in travel by the elderly was attributable to the lack of car availability or license holding; however, a study of the elderly in New York (Hartgen, Pasko, & Howe, 1976) found no significant difference in car availability between older people and the rest of the population. In most Western nations, the evidence clearly points to a continuing increase in car availability and license holding among persons over age 65 (Wachs & Blanchard, 1977). Future studies might not, then, find the mode use differences illustrated in Table 7-2.

Sex

As one might expect—given the tenacity of traditional gender roles in our society and given that consequently the daily lives of women and men are quite different—the travel activity patterns of women are not the same as those of men. Moreover, studies of male/female travel patterns in several countries have turned up similar findings. One of these is a remarkable difference in access to the various travel modes. As Tables 7-3 and 7-4 show, women's mobility is largely based on public transportation and their own feet while men make greater use of the private automobile (see also Guiliano, 1979). Table 7-3 shows the mode use breakdown for the trips of all purposes over 35 days in the Swedish study, and Table 7-4 shows modes used on the journey to work in Baltimore, Maryland. The patterns are similar. You might reflect on why this pattern emerges again and again in different places and at different times.

Another consistent finding concerns travel distances—especially on the work trip. Several studies have found that women travel shorter distances than men overall (Fox, 1983; Hanson & Hanson, 1980, 1981) and, most particularly, that if they work, their workplaces are closer to home than are men's workplaces (see, for example, Black & Conroy, 1977; Hanson & Johnston, 1985; Howe & O'Connor, 1982; Madden, 1981). Table 7-5 demonstrates this point with data collected in Baltimore; women's work trips are significantly shorter than men's in both travel distance and travel time. Just why this is so is not clearly understood; most investigators hypothesize that the explanation lies in employed women's dual roles (as home maker/child rearer plus worker), which translates into an unwillingness or an inability to devote much time to a long commute.

The purposes for which trips are made also differ for men and women and reflect the impact of traditional gender roles. When employment status is held constant and the travel-activity patterns of only full-time employed women and men are examined, male-female differences in travel purposes are still evident (see Table 7-6), with women under-

Table 7-3. Mode use[a] by full-time employed women and men, Uppsala, Sweden, 1971

	Women	Men
Walk	43	22
Bike	13	15
Bus	14	6
Car driver	12	46
Car passenger	16	8
Totals	98[b]	97[b]

[a]Percentage of all movements; figures are group averages.

[b]Column does not add to 100 because of rounding.

Note. From "Gender and Urban Activity Patterns in Uppsala, Sweden," by S. Hanson and P. Hanson, 1980, *Geographical Review, 70*, p. 295. Copyright 1980 by the American Geographical Society. Reprinted by permission.

Table 7-4. Modes used[a] on work trip, Baltimore, Maryland, 1977

	Women ($N=303$)	Men ($N=484$)
Walk	6.0	4.0
Bus	19.5	13.0
Car driver	56.0	71.0
Car passenger	18.5	12.0
Totals	100.0	100.0

[a]Figures indicate percent of sample using each mode.

Note. From "Gender Differences in Work-trip Lengths: Explanations and Implications," by S. Hanson and I. Johnston, 1985, *Urban Geography, 6*, p. 208. Copyright 1985 by V. H. Winston. Reprinted by permission.

Table 7-5. Travel distances and times on the journey to work, Baltimore, Maryland, 1977

	Men ($N=484$)	Women ($N=303$)	t values
Average distance in miles	7.8 (6.4)	5.5 (4.9)	5.73*
Average time in minutes	28.8 (19.7)	25.7 (18.1)	4.61*

*Significant at $p = .01$

Note. Figures in parentheses are standard deviations. From "Gender Differences in Work-trip Lengths: Explanations and Implications," by S. Hanson and I. Johnson, 1985, *Urban Geography, 6*, p. 200.

Table 7-6. Mean number of stops per person for different activities, Uppsala, Sweden, 1971

Purpose of stop	Women	Men	*t* values
Social	9.8	10.2	−.25
Personal business	8.3	8.4	−.03
Shopping	27.8	19.2	3.80*
Work	23.3	29.2	−3.46*
Recreation	5.5	8.1	−3.67*

*Significant at the $p < .01$ level; two-tailed test.

Note. From "Gender and Urban Activity Patterns in Uppsala, Sweden," by S. Hanson and P. Hanson, 1980, *Geographical Review, 70,* p. 297. Copyright 1980 by the American Geographical Society. Reprinted by permission.

taking more travel for shopping and personal business and men emphasizing work and recreation.

Race

Any personal characteristic (like age, gender, or race) that means unequal access to resources will portend distinctive travel patterns. Like the difference observed between men and women, blacks and whites in Baltimore make different uses of the transportation modes on the journey to work (see Table 7-7). While only 9% of white commuters rode to work on the bus, 28.6% of black workers did so; and while almost three-quarters of the white sample drove to work, only half of the black sample did. The implications of these figures for time–space use are reflected in Table 7-8. Although bus riders in Baltimore travel significantly shorter distances to work than do automobile users, their work trips take almost twice the time the automobile drivers' work trips take. So it should not be a surprise that blacks travel shorter distances to work, but have significantly longer travel times to work than whites do (Table 7-9).

Table 7-7. Mode use[a] by race on journey to work, Baltimore, Maryland, 1977

	Race	
	White ($N=511$)	Black ($N=255$)
Pedestrian	4.1	6.3
Bus rider	9.0	28.6
Automobile:		
Driver	73.2	50.2
Passenger	13.7	14.9

[a]Percentage of sample using each mode.

Table 7-8. Travel distance and time by mode on journey to work, Baltimore, Maryland, 1977

	Distance (miles)	Time (minutes)
Pedestrians ($n=37$)	1.1 (1.3)[a]	12.8 (10.8)
Bus riders ($n=120$)	4.8 (3.3)	44.6 (21.9)
Drivers ($n=505$)	7.6 (6.0)	24.5 (14.6)
Passengers ($n=112$)	7.9 (7.1)	28.6 (24.6)
F ratios	22.1[b]	52.0[b]

[a]Figures in parentheses are standard deviations.
[b]Significant at $p = .01$.

This brief discussion of racial differences in journey-to-work patterns demonstrates the importance of access to the automobile. Our final look at how personal/household characteristics affect travel will focus on the impact of auto availability.

Automobile Availability

A person's choice of travel mode is a function of time (the time available for travel and the

Table 7-9. Racial differences in travel distance and time on journey to work, Baltimore, Maryland, 1977

	White[a] ($N=511$)	Black[a] ($N=255$)
Travel distance in miles	$\bar{x}=7.1$ $s=6.1$	$\bar{x}=6.3$ $s=5.4$
Travel time in minutes	$\bar{x}=24.4$ $s=17.1$	$\bar{x}=33.7$ $s=21.2$

[a]$\bar{x}$ = mean
s = standard deviation

time different modes require), cost (long-term costs such as auto purchases and short-term costs such as gas, tolls, or bus fare), comfort, safety, and environmental/social factors (such as concern about air pollution or ability to converse while traveling). As many studies have revealed, however, a person's choice of modes is often limited because either public transit service or an automobile is unavailable.

Automobile availability is related to both auto ownership and possession of a driver's license. The first of these is a function of income (ability to purchase a car and pay for its upkeep), the relative price of the automobile, and one's mobility needs. As an illustration one can look to Great Britain, where in 1973 "non car-owning households included nearly three quarters of income group, which comprised one-third of all households in Britain and nine-tenths of the elderly households" (Hillman, Henderson, & Whalley, 1973, p. 41). Possession of a license is a function of eyesight (often related to age), driving ability and record, societal norms about roles, and again mobility needs. In Great Britain, for example, women and older people are less likely to hold a license than are men and younger people (Hillman, Henderson, & Whalley, 1973, p. 40).

The availability of an automobile is a major determinant of travel behavior. As the distance decay curve in Figure 7-7 shows, people travel farther by automobile than by other modes; more activity sites are accessible, therefore, to a person who has an automobile. Also, as Figure 7-8 shows, the distance decay of movements is different for different trip purposes. Again this implies that those with autos will have a greater choice of social and recreation places and workplaces.

Not surprisingly, disaggregate travel surveys have borne out some of these predicted impacts of auto availability on activity patterns. In a study of Great Britain, Doubleday (1977) found that trip rates increased most between 1962 and 1971 for housewives with cars, whereas the trip rate of those without cars remained constant. Paaswell, Recker, and Milione (n.d.) found trip frequencies distinctly different for the car-owning versus the non-car-owning in Buffalo, New York. The carless went grocery shopping twice as often as did those with a car, probably because of the difficulty of carrying packages on foot. Neighborhood social visits were also made with a higher frequency by carless individuals, but social and recreational activities that involved an out-of-pocket expenditure were engaged in less often. When questioned, carless individuals said they would make more recreational trips if better transportation were available. Finally, these authors found that for the carless, more walking was done for grocery shopping, but buses were used more often for clothes shopping. Hillman et al. (1973, p. 57) also present evidence that most people who have a car use it to go to work. Thus, where there is only one car available in a household, the worker takes it, leaving the rest of the household to rely on other modes for most of the day.

CONCLUSION

Although the above discussion of travel patterns has been simplified by focusing on the impact of only one personal/household characteristic at a time, the examples given hint at how complex travel–activity patterns are when viewed at the level of the individual or household. There are many variables (both personal characteristics and attributes of the travel environment) that impinge on travel, and often the analyst's task is to determine which of a set of variables is/are most important in shaping a particular dimension of travel. Several studies have tried to assess the relative importance of sociodemographic versus location factors molding travel–activity patterns (see, for example, Hanson, 1982; Pas, 1984; Recker & Schuler, 1982; Wermuth, 1982). Despite the fact that these studies have used data collected at different times and in different places, the findings are remarkably consistent: variables, such as sex and employment status, that describe a person's social *role* have a greater impact on individuals' travel than do measures of social *status* (such as occupation or income). Moreover, the characteristics of the individual traveler have been found to be more important than attributes of the spatial environment (the nature of the transportation system or the location of potential destinations) in shaping travel patterns. This latter finding no doubt reflects the fact that the studies cited above have all been focused at the *intra*urban scale; there is the possibility that at that scale the variability in the density of activity sites or in the access to highways is not large enough to generate significant differences in travel.

Describing disaggregate flows and trying to untangle the many related factors that affect these flows is a useful preliminary to modeling such flows. The descriptive phase helps identify the independent variables that should be included in models. Description also helps to point out which modeling approaches are appropriate for which groups of people; for example, choice models are clearly most suitable for people who operate under relatively few constraints. In this regard, perhaps the most important role that description can play is to identify groups of individuals who can be expected to respond similarly to changes in the transportation system. Disaggregate models, such as those introduced in the next chapter, assume that all individuals sharing certain characteristics (for example, car-owning people living in suburban locations) will respond in essentially the same way to a policy-induced change in the transportation system (such as the opening of a diamond lane) because they all have essentially the same set of preferences and therefore evaluate alternatives in the same way. Descriptive studies can help to identify the bases for defining appropriate aggregates of individuals—appropriate in that a link has been found to exist between some personal or locational characteristic(s) and some aspects(s) of travel. A search for explanations of travel patterns led to the need for disaggregate data in the first place; the next chapter demonstrates how models are formulated at the level of the decision maker as the search for explanation continues.

References

Black, J., & Conroy, M. (1977). Accessibility measures and the social evaluation of urban structure. *Environment and Planning A, 9,* 1013–1031.

Chapin, F. S., Jr. (1971). Free-time activities and the quality of urban life. *Journal of the American Institute of Planners, 37,* 411–417.

Cullen, I., & Godson, V. (1975). Urban networks: The structure of activity patterns. *Progress in Planning, 4,* 1–96.

Davies, R. L. (1969). Effects of consumer income differences on shopping movement behavior. *Tijdschrift voor Economische en Sociale Geographie, 60,* 111–121.

Doubleday, C. (1977). Some studies of the temporal stability of person trip generation models. *Transportation Research, 11,* 255–263.

Fagnani, J. (1983). Women's commuting patterns in the Paris region. *Tijdschrift voor Economische en Sociale Geographie, 74,* 12–24.

Forer, P. L., & Kivell, H. (1981). Space-time budgets, public transport, and spatial choice. *Environment and Planning A, 13,* 497–509.

Fotheringham, S. (1983). A new set of spatial interaction models: The theory of competing destinations. *Environment and Planning A, 15,* 15–36.

Fox, M. (1983). Working women and travel: The access of women to work and community facilities. *Journal of the American Planning Association, 49,* 156–170.

Giuliano, G. (1979). Public transportation and the travel needs of women. *Traffic Quarterly, 33*(4), 607–616.

Hanson, P. (1977). The activity patterns of elderly households. *Geografiska Annaler, 59,* 109–124.

Hanson, S. (1982). The determinants of daily travel activity patterns: Relative location and sociodemographic factors. *Urban Geography, 3,* 179–202.

Hanson, S., & Hanson, P. (1980). Gender and urban activity patterns in Uppsala, Sweden. *Geographical Review, 70,* 291–299.

Hanson, S., & Hanson, P. (1981). The impact of married women's employment on household travel patterns: A Swedish example. *Transportation, 10,* 165–183.

Hanson, S., & Johnston, I. (1985). Gender differences in work-trip lengths: Explanations and implications. *Urban Geography, 6,* 193–219.

Hartgen, D., Pasko, M., & Howe, S. M. (1976). *Forecasting non-work public transit demand by the elderly and handicapped.* Research Report 108. Albany: New York State Department of Transportation.

Hillman, M., Henderson, I., & Whalley, A. (1973). *Personal mobility and public transport policy.* London: PEP Broadsheet No. 542.

Hopkin, J. M., Robson, P., & Town, S. W. (1978). *The mobility of old people: A study of Guildford.* Transport and Road Research Laboratory Report 850.

Howe, A., & O'Connor, K. (1982). Travel to work and labor force participation of men and women in an Australian metropolitan area. *Professional Geographer, 34,* 50–64.

Huff, J. O., & Hanson, S. (1986). Repetition and variability in urban travel. *Geographical Analysis,* 18, 97–114.

Koenig, J. G. (1980). Indicators of urban accessibility: Theory and application. *Transportation,* 9, 145–72.

Lenntorp, B. (1976). *Paths in space–time environments.* Lund: CWK Gleerup.

Madden, J. (1981). Why women work closer to home. *Urban Studies, 18,* 181–194.

Manning, I. (1978). *The journey to work.* Boston: George Allen and Unwin.

Marble, D., Hanson, P., & Hanson, S. (1972). *Household travel behavior study: Field operations and questionnaires.* Evanston, Ill.: The Transportation Center at Northwestern University.

Morris, J. (1979). *Personal mobility and transport requirements of the elderly.* Internal Report AIR 344-1. Canberra: Australian Road Research Board.

Nader, G. (1969). Socio-economic status and consumer behavior. *Urban Studies, 6,* 235–245.

Paaswell, R. E., Recker, W., & Milione, V. (n.d.). *A profile of a carless population.* Buffalo: SUNY/Buffalo, Department of Civil Engineering.

Pas, E. (1984). The effect of selected sociodemographic characteristics on daily travel activity behavior. *Environment and Planning A, 16,* 571–581.

Potter, R. (1977). The nature of consumer usage fields in an urban environment: Theoretical and empirical perspectives. *Tijdschrift voor Economische en Sociale Geographie, 65,* 168–76.

Recker, W., & Schuler, H. (1982). An integrated analysis of complex travel behavior and urban form indicators. *Urban Geography, 3,* 110–210.

Sheppard, E. (1984). The distance decay gravity model debate. In G. L. Gaile & C. J. Willmott (Eds.), *Spatial Statistics and Models* (pp. 367–388). Boston: Reidel.

Taylor, S. M. (1979). Personal dispositions and human spatial behavior. *Economic Geography, 55,* 184–195.

Wachs, M., & Blanchard, R. (1977). Lifestyles and transportation needs of the elderly in the future. *Transportation Research Record, 618,* 19–22.

Wermuth, M. J. (1982). Hierarchical affects of personal, and residential location characteristics on individual activity demand. *Environment and Planning A, 14,* 1251–1264.

Zabinski, R. J. (1977). Travel by the elderly and handicapped. *Research Report 128.* Albany: New York State Department of Transportation.

8 DISAGGREGATE TRAVEL MODELS

JOHN S. PIPKIN
State University of New York at Albany

INTRODUCTION

In the mid-1960s geography and the transportation planning sciences began study travel behavior at a disaggregate level. By the early 1970s transportation researchers had produced disaggregate, behavioral travel demand models (see, for example, Domencich & McFadden, 1975). The earliest and still the preeminent concern in transport planning was the study of mode choice, often in conjunction with the work trip. In geography cognitive–behavioral and decision-making approaches became popular at about the same time (for example, Cox & Golledge, 1981; Golledge & Rushton, 1976). For a review of cognitive–behavioral applications to travel see Pipkin (1981b). Geographers have been most interested in destination choice, usually in the context of discretionary travel such as shopping.

Many of the advantages of disaggregate models were described in the previous chapter. Aggregate principles such as entropy maximization often appear to be behaviorally questionable (Anas, 1981; Wilson, Coelho, McGill, & Williams, 1981). Ecological fallacies and aggregation biases arise in area-based flow models (for example, Okabe, 1977). Distance-decay parameter estimates decrease as zones of aggregation become larger and fewer (Openshaw, 1977). In a theoretical analysis of spatial aggregation problems, Webber calls emphatically for behavioral theory:

> A prerequisite of good spatial interaction theory is the identification of behaving objects. . . . Zones, counties, local authorities and regions do not behave: they are merely spatial aggregations of acting objects. (1980, p. 130)

There is now a consensus that the decision-making unit—whether an individual or a household—is the appropriate level at which to build travel theory.

Disaggregate models have different—and more exacting—data requirements than aggregate formulations. Usually detailed survey analysis is required to record people's travel and to elicit their preferences and attitudes toward the various attributes of routes, modes, and destinations. The latest activity models require extremely detailed records of household activities over time, often collected in lengthy interviews or travel diaries (see chapter 7). A good part of contemporary disaggregate research is concerned with development and testing of survey instruments of various kinds.

This chapter is divided into three parts, corresponding very broadly to historical stages in the development of disaggregate travel models, though research in each area continues today. The first presents some theoretical perspectives on the effects of distance on individual interactions. The second deals with mathematical models of the various choices underlying travel. The third section discusses more recent models that study travel in the whole context of household activity patterns.

DISTANCE AND INTERACTION

The decline of interaction in a concave-upward curve against distance has been verified for many kinds of movement and for various distance measures including miles, cost, and travel time. It is evident from previous chapters that the relationship is usually reported for aggregate data. Nevertheless, it holds up quite well when data are broken down, for example by trip type. Yuill (1967) showed that the regularity held for several kinds of retail travel in Ann Arbor. He fitted a power function of the form

$$T = cD^{-b} \quad (1)$$

where T represents trips per capita, D is distance, and c and b are parameters (see Figure 8-1).

This, and similar distance-decay functions, have figured in most gravity models of destination choice. As these models proved their worth in making predictions, the lack of a firm theoretical base became a nagging problem. Isard, an important early proponent who brought gravity models to the attention of economists and regional scientists, summarized the problems caused by the lack of micro-level theory:

> I received sharp criticism from my fellow economists when I was not able to provide any rationale for the model. . . . I have always remained unhappy with the rationale presented for explaining travel behavior. The current development of entropy maximization . . . leaves me still unhappy. (1975, p. 25)

The objectives of theorizing and practical modeling are not always the same. Models are judged by their predictive success; theorists lay more emphasis on simplicity and generality in the underlying structure of explanation. Practical models often contain many fitted parameters (such as distance-decay and weighting parameters discussed in previous chapters) that have no clear behavioral interpretation. Theorists are less tolerant of uninterpreted parameters. Theory building typically starts from idealized and simplifying assumptions, which usually include some axiom of rationality. Here we will discuss representatives of three theoretical styles.

Distance-Minimizing and Breakpoint Principles

Why does interaction decline over distance? What accounts for the distance-decay regularity? Perhaps the most straightforward way to idealize the effects of distance on travel is to assert as an axiom of rationality that travelers minimize some measure of distance or effort. Hubbard (1978) reviews some of the normative principles that have been proposed, often not in the context of travel *per se* but in attempts to explain urban spatial structure (such as central place theory or market area analysis).

Zipf (1949) proposed a principle of least effort as a general explanation of many kinds of behavior and social organizations. In a travel context the principle maintains that people minimize distance, cost, travel time, or some other disutility, either objectively or—which is not the same thing—as subjectively perceived. With the extra premise that disutility is proportional to distance, simple distance minimization is the result. For shop-

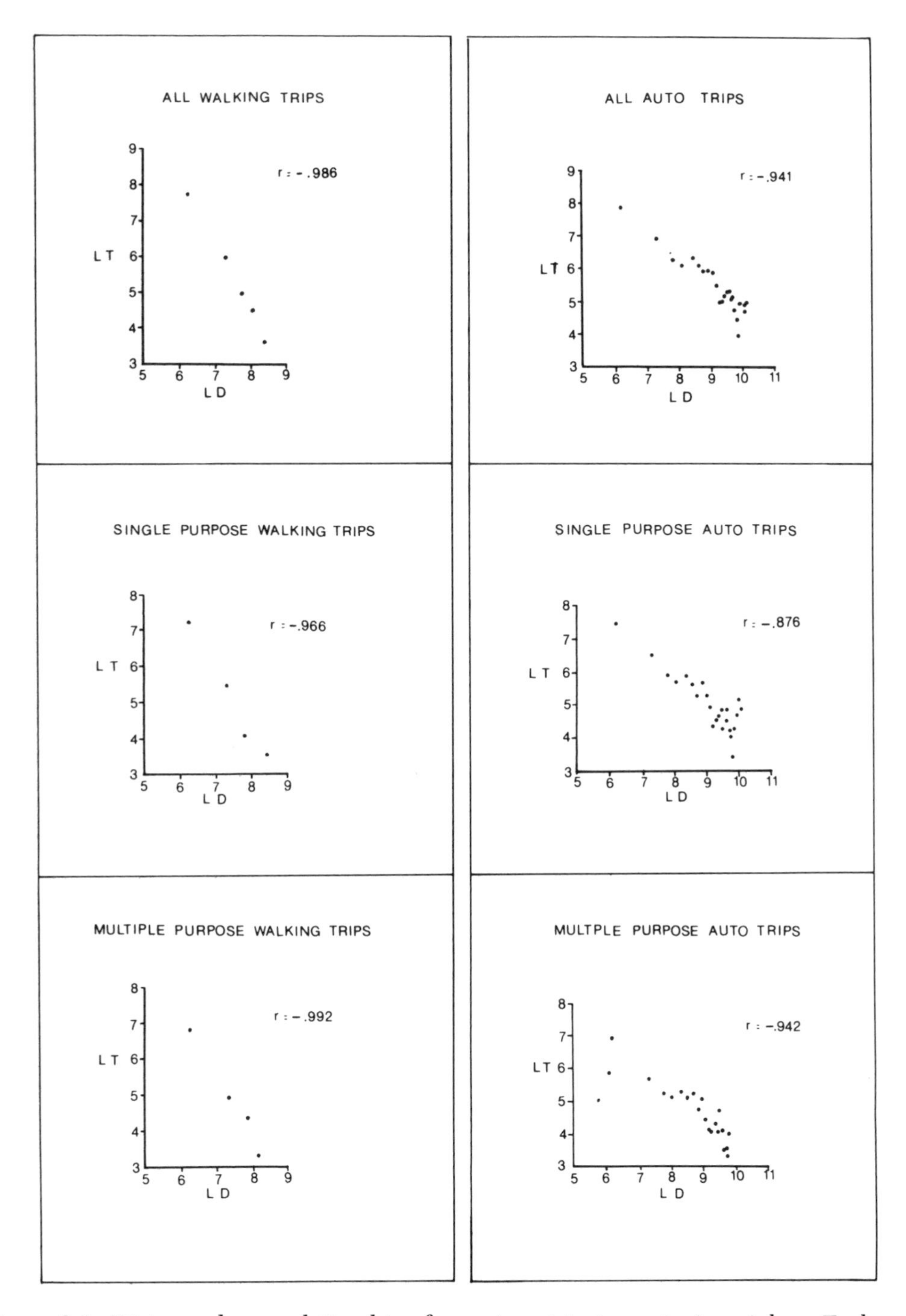

Figure 8-1. Distance-decay relationships for various trip types in Ann Arbor. Each vertical axis represents the logarithm of trips per capita. Each horizontal axis represents the logarithm of distance. From "Spatial Behavior of Retail Consumers: Some Empirical Measurements" by R. S. Yuill, 1967, *Geografiska Annaler B*, *49*, p. 112. Copyright 1967 by Almqvist & Wiksell, Stockholm. Reprinted by permission.

ping behavior this is equivalent to the nearest center postulate of central place theory (discussed and criticized by Clark & Rushton, 1970, and Hubbard, 1978). If people always go to the nearest among a set of comparable alternatives, predicting destination choices is a matter of simple geometry (see Figure 8-2). Unfortunately all evidence shows that at least in modern, mobile societies, the nearest-center postulate is invalid. Hubbard reviews a range of shopping studies, concluding that less than one-half and often only about one-quarter of shopping trips go to the nearest available outlet. The prevalence of multipurpose trips in certain kinds of travel obviously complicates this issue. So does uncertainty about whether people's perception of disutility corresponds to real distance, cost, or travel time. There is evidence

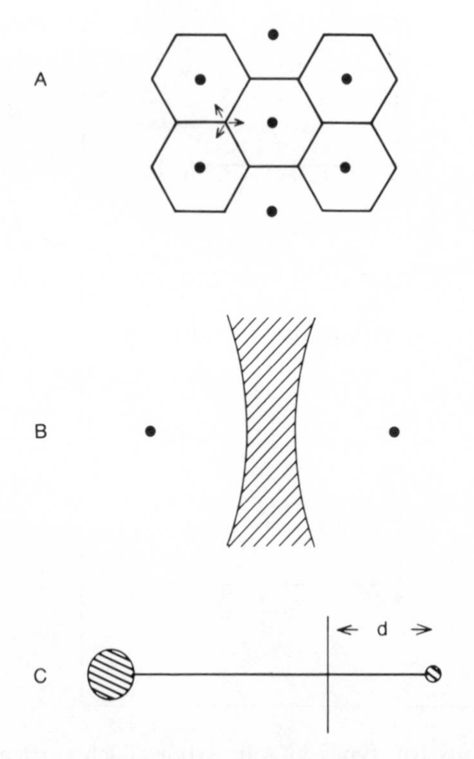

Figure 8-2. Trade area boundaries according to three models. (a) Equidistant boundaries between centers of equal size: central place theory; (b) Equidistant boundaries between centers of equal size, with zone of indifference (Devletoglou, 1971); and (c) Boundaries between centers of unequal size (Reilly, 1931).

that people more often minimize perceived travel time than real distance. (Cadwallader, 1975). Distance minimization is most plausible as a model of real behavior in societies in which the friction of distance is high.

A variant of the distance-minimizing principle assumes that people are only sensitive to significant differences in distance, differences that exceed a certain threshold of indifference. One early indifference model was by the economist Devletoglou (1971). Under some simplifying assumptions market area boundaries are geometric hyperbolas, separating zones of indifference. Within these zones, Devletoglou suggests, "fashion effects" (imitation) and other factors unrelated to distance may dictate destination choices. Yet another approach to delineating interaction breakpoints used a simple form of the gravity model to take into account the longer "pull" of larger places (Batty, 1978; Reilly, 1931). Figure 8-2 contrasts the implications of these assumptions for destination choice.

Minimization and breakpoint principles fail to account for one crucial process: the substitution between distance disutility and the attraction of destinations (the better the store, the farther one will go to it), and origins (one may be prepared to travel farther to work in exchange for a better residence and neighborhood). A unifying theme in many varied models of disaggregate travel is a concern to account for the interaction of site and distance attributes. Next we shall look at two different views of this trade-off: deductive utility models and inductive revealed preference analysis.

Deductive Theories of Distance Substitution

Many economists and regional scientists have developed theories in which consumers make trade-offs (substitutions) that maximize utility and which generate gravity-like (concave) profiles of interaction against distance. For example, Isard (1975) assumes that the trade-offs people make are inherently very simple. He assumes that trip intensity (for example, number of trips) is compared with distance traveled in a simple two-variable comparison, in such a way that a constant trade-off exists between the absolute change in number of miles traveled *(DX)* and percentage change in number of trips ($100 \cdot DT/T$), hence

$$100 \cdot \frac{DT}{T} = cDX \tag{2}$$

After mathematical manipulations and some assumptions on the form of utility functions, Isard shows that the number of trips or intensity of interaction with a site declines as a negative exponential function of its distance. Thus, in simplified notation,

$$T = A \exp(-Bd) \tag{3}$$

where T is interaction intensity and A and B are parameters (Isard, 1975, p. 29). It is a useful exercise to plot the form of this curve for different parameter values. This is easy to do with a calculator that handles the exponential function. The decreasing "value" of alternatives with increasing distance is called *spatial discounting.* A somewhat more general theory of spatial discounting is provided by Smith (1975, 1984). Some other key contributions were by Golob and Beckmann (1971), Golob, Gustafson, and Beckmann (1973), and Niedercorn and Bechdolt (1969).

Within the framework of deductive reasoning from highly idealized assumptions these (and similar utility-based theories) render the distance-decay regularity plausible at an individual level. It is certainly reasonable that the perceived value of activities in various places is discounted systematically over space. The normative, deterministic flavor of utility-oriented travel theories make them appealing to economists: The language and theoretical style connect travel behavior naturally to the main body of microeconomic theory.

Axiomatic theories such as Isard's are deductive: They start out not with observed choice data but with simplified assumptions. A different approach to the problem of site–distance trade-offs is discussed next. It starts out by looking at the actual (revealed or observed) choices people make in everyday travel.

Revealed Preference Theory

The basic premise of revealed preference theory is that, with sufficiently rich data, it is possible to make direct inferences about the trade-offs travelers make among site and distance attributes. The approach originated in economic consumer theory. Applications to destination choice were pioneered by Rushton and his associates (Girt, 1976; Rushton, 1969, 1976).

Revealed preference analysis begins by defining relevant attributes of potential destinations. In Rushton's earliest work the behavior studied was grocery shopping in rural Iowa. Destinations were towns. Two attributes were defined, size of town and its distance to the consumer. Population size of places was taken as an appropriate surrogate for retail attractiveness (such as variety of merchandise, which is strongly correlated with size). Places were cross-classified by size and distance into one of 30 locational types. Locations of the major grocery purchase were recorded for a random sample of rural Iowans.

The next stage of the analysis is to obtain revealed choices, preferably by finding out where consumers actually go. Sometimes subjects are asked to generate choices by selecting among hypothetical alternatives in the laboratory. Then preference analysis is applied to infer an indifference map showing the trade-offs subjects are making. This preference ordering can be defined over all possible combinations of attributes, whether observed or not, thus allowing the model to be generalized to new places or to hypothetical options, such as new centers proposed by planners.

A common preference scaling procedure is as follows (Rushton, 1976). First a similarity or proximity measure is defined between each pair of locational types. Similarity can be defined as the proportion of times one type is preferred to another by respondents who have a valid choice between the two types. (The closer this proportion is to one-half, the more "similar" the locational types are.) Second, using one of a number of different scaling methodologies, these proximity measures can be transformed into a unidimensional preference or utility scale. This is equivalent to constructing contours (indifference curves) of equal utility in the space defined by size versus distance. This indifference map provides an abstract summary of the preference structure of the group, or, if a homogeneous sample can be assumed, of any one household. Figure 8-3 shows the indifference maps for Rushton's original Iowa sample and another from a comparable study of grocery shopping conducted by Girt in Newfoundland. Each indifference curve joins combinations of size and distance that yield equal total utility. The curves clearly show the trade-off between size and distance: An increment in the size of a town induces consumers to travel farther while maintaining the same level of utility. For example, the curve marked 0 in Figure 8-3 shows that a town 20 miles away with 25 functions yields roughly the same utility as a town with 12 functions 10 miles away.

Lieber (1977) provides an application of revealed preference analysis in an intraurban setting. He studied recreational trips to bowling alleys in Buffalo, New York. After examining a number of attributes, Lieber found that the number of bowling lanes (a surrogate for size and other services) and distance provided two salient dimensions, that an indifference space comparable to those in Figure 8-3 could be constructed, and that preferences inferred from this indifference map provided good predictions of actual choices. The derived utility or preference scale could be written as:

$$U = .148 + (.0156)S + (.2378)D \tag{4}$$

where U is the derived preference scale of a combination of size (S) and distance (D). The numerical coefficients are produced by the scaling algorithm. They measure the partial contribution of each attribute to total utility.

The revealed preference approach has several advantages. One is its insistence on working from actual observed choice rather than from expressed or reported preferences; it is well known that people don't always do as they tell interviewers. Another is the ability of the method to generalize away from the particular configuration of destinations studied to produce an abstract description of potential choices over any possible set of alternatives. Yet the method has drawbacks. Because only one or two choices are typically observed for particular individuals or households, it is usually necessary to assume that different people share the same preference structure. To represent utilities in a form like equation 4 requires strong assumptions on the additivity and separability of attributes. Another more general problem is that people only reveal preferences at attainable alternatives. The elderly, minorities, the handicapped, the carless, or the otherwise disadvantaged may be forced to forgo certain kinds of trips or to travel to nearer, smaller, and perhaps more expensive alternatives such as many inner-city food stores. No amount of revealed preference analysis can reveal the desires and needs of such people, though it could well uncover the excess burden they bear in their actual travel patterns (Pirie, 1976). A critical discussion of revealed preference analysis is given in MacLennan and Williams (1980).

TRAVEL CHOICE MODELS

The previous section outlined some theory on the trade-offs and substitutions underlying destination choice. In their simplest forms both the deductive and revealed preference approaches make some dubious simplifications of the ways in which real travel decisions are made. Specifically, they assume that appropriate measures of distance and center size are the only factors affecting destination choice. It is clear from the previous chapter that many effects need to be incorporated into a realistic model of travel. Some of the principal factors that contribute statistically to explaining travel patterns are individual variables, such as age, ethnicity, income, sex, car ownership, level of education; site effects such as the price, quality, convenience, and variety available at shopping,

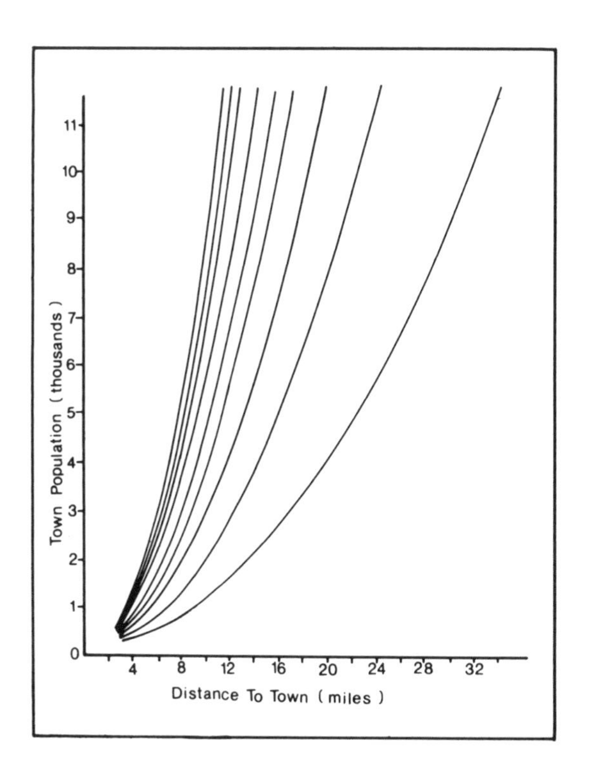

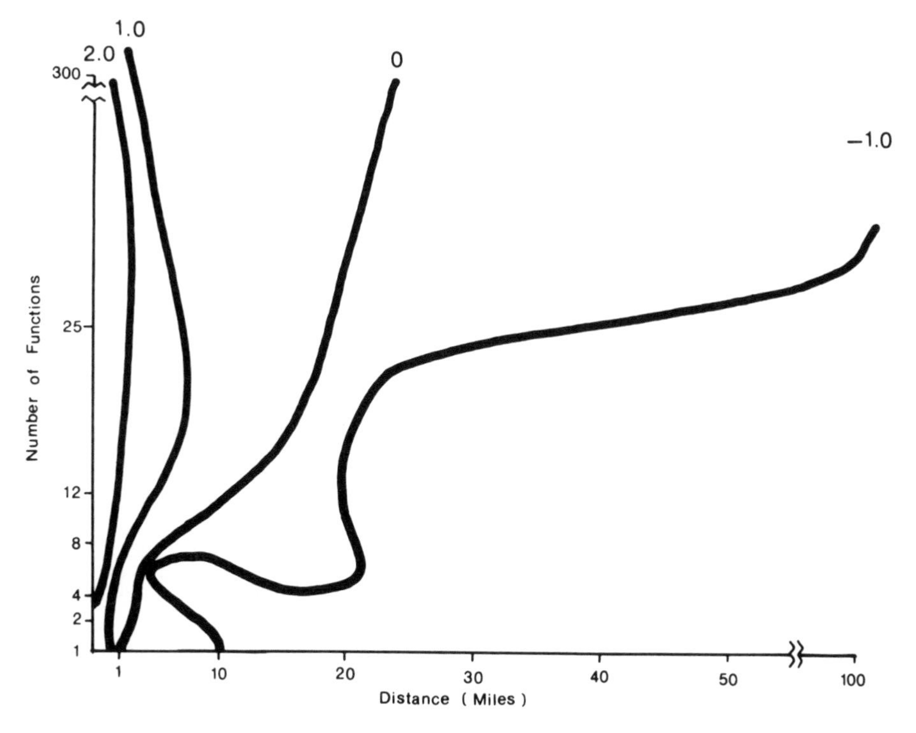

Figure 8-3. Revealed preference indifference maps. The contours (indifference curves) join combinations of size and distance that yield equal utility. From (a) "Analysis of Spatial Behavior by Revealed Preference Analysis" by G. Rushton, 1969, *Annals of the Association of American Geographers*, *59*, p. 398. Copyright 1969 by the Association of American Geographers. Reprinted by permission. (b) "Some Extensions of Rushton's Spatial Preference Scaling Model" by J. L. Girt, 1976, *Geographical Analysis*, *8*, p. 148. Copyright 1976 by Ohio State University Press. Reprinted by permission.

service, or recreational destinations; and the link or distance effects including the mileage, cost, travel time, and inconvenience of getting from *a* to *b*. Modern travel models in transportation planning and geography can incorporate explicitly a variety of such variables.

Modern travel models usually focus on the idea of choice. In early transport engineering and planning studies the mode choice for work travel was the main concern, though in the past ten years there has been an interest in destination, route, and other choices (for example, Lerman & Adler, 1976). From the first, geographers' interest focused on destination choice. Travel choice models are usually probabilistic. An early and particularly simple destination choice model will illustrate the principles of probabilistic choice theory.

A Simple Destination Choice Model

The model outlined here is an early version of an origin-constrained gravity model first proposed for retail applications of Huff (1962) and widely applied in this context (for example, Lakshmanan & Hansen, 1965). The model now exists in more general and complex forms (for example, Wilson et al., 1981, pp. 212, 243).

Consider choices made by *m* respondents over *n* homogeneous, competitive destinations such as shopping centers. Suppose that the respondents live in a localized area, so that the distances of the centers are essentially the same for everyone. Suppose we obtain from each respondent the name of the shopping center he or she uses most often. Let T_i be the total number of visits observed at the *i*th center. Huff's model is a one-parameter model calibrated on these observed choices, using various attributes of the centers as predictor variables. In the simplest form of his model Huff used two variables: S_i, the size of the *i*th center as measured by square feet of floor space, and D_i, the distance of the *i*th center which could be measured by travel time. Attractiveness or utility of the *i*th destination (A_i) was then represented as:

$$A_i = S_i D_i^{-b};\ A_i > 0;\quad i = 1, \ldots, n \qquad (5)$$

Here A_i is directly proportional to site attractiveness and inversely proportional to travel time, modulated by the unknown distance-decay parameter.

The probability of a trip to the *i*th center is defined as:

$$p_i = \frac{A_i}{\sum_{j=1}^{n} A_j};\quad i = 1, \ldots, n \qquad (6)$$

This probability is directly proportional to the utility of the *i*th center and inversely proportional to the sum of all utilities. The model is competitive in the sense that if A_j increases for $j \neq i$, then p_i decreases. This structure is particularly easy to deal with on a small computer or a good calculator. It is instructive to examine the effects of changing *b* for hypothetical values of *S* and *D*.

The calibration or fitting process is particularly simple in the Huff model, because there is only one free parameter *(b)*. The object is to choose a value of *b* which best replicates (fits) the observed T_i given the values of the variables D_i and S_i, $i = 1, \ldots,$ *n*. One common method, which can be used with much more complex, multiparameter models, is the method of maximum likelihood. It involves choosing the value of *b* which maximizes the probability of the data, taking into account all the observed values of S_i and D_i. It can be proved that this probability is proportional to:

$$\sum_{i=1}^{n} T_i \log p_i = \sum_{i=1}^{n} T_i \log \left\{ \frac{S_i D_i^{-b}}{\Sigma_j S_j D_j^{-b}} \right\} \qquad (7)$$

We choose *b* to maximize this expression. The resulting value, $\tilde{b}$, is called a maximum likelihood estimate. A numerical example is given in Table 8-1. Since this is a one-variable problem, it is particularly amenable to various maximization or search algorithms or to experimental trial and error to gain a

Table 8-1. The Huff model

The Huff model states that $P_i = (S_i d_i^{-b}) \cdot (\Sigma_j S_j d_j^{-b})^{-1}$ where P_i is the predicted choice probability of a store or shopping center, S_i and d_i are known size and distance measures, and b is a parameter to be estimated. Suppose that M consumers choose among N stores, $i = 1, \ldots, N$. Then the predicted number of trips to the ith store is $(M) \cdot (P_i)$. The following values show how the predicted number of visits T_i at the ith center varies with b. Let $N = 5$ and $M = 67$.

	S_i	d_i	I	II	III	IV
Store 1	17	1.00	17.25	15.43	13.61	15
Store 2	5	3.20	2.84	.44	.12	1
Store 3	9	0.80	10.21	12.77	14.07	13
Store 4	14	0.75	16.40	22.60	26.57	22
Store 5	21	1.10	20.31	15.76	12.63	16

I = predicted T_i with b = .5
II = predicted T_i with b = 2.0
III = predicted T_i with b = 3.0
IV = observed T_i

Worked example: Consider the first store with $b = 2$.

$$P_1 = (S_1 d_1^{-2}) \cdot (\Sigma_j S_j d_j^{-2})^{-1}$$
$$= (17/1^2)/(17/1^2 + 5/3.2^2 + 9/.8^2 + 14/.75^2 + 21/1.1^2)$$
$$= .2304$$

The predicted number of trips at site 1 is $(M \cdot P_1) = (67) \cdot (.2304) = 15.43$
If a set of visit frequencies T_i are known in advance, the model may be calibrated to estimate the value of b that best fits the data. One method is the method of maximum likelihood. The maximum likelihood estimate of b based on column IV above is b = 1.696.

Exercise: See how well the model predicts the data in column IV using $b = 1.696$. These are hypothetical data.

feel for the behavior of the nonlinear estimation problems that arise in travel models. (The mathematics can be simplified by using a negative exponential rather than a power function to represent distance-decay. In other words, replace D^{-b} with $\exp(-Db)$).

A model of essentially this form was applied by Lakshmanan and Hansen (1965) to examine possible distributions for large retail trade centers in the Baltimore area. Their study area was divided into origin and destination zones, and the dependent variable was the number of shopping goods trips, which is a good surrogate for retail sales. The model performed very well in replicating actual trips. For example, during the study period, 16,425 trips to the CBD were predicted by this model, and 17,466 were actually recorded by an origin–destination survey. A more advanced application of this kind of model is found in Baxter and Ewing (1981).

The Huff formulation contains an implicit model of choice. Study of the economic and psychological underpinnings of travel behavior led researchers to examine formal choice theory. We will now examine more abstractly two related models (of constant and random utilities) that came to dominate travel choice theory during the decade of the 1970s.

Formal Choice Theory

Consider a set of alternatives faced by a decision maker. They could be routes, modes, destinations, times of day, or any other well-defined choice objects. Let p_i represent the probability that the ith object is chosen. In the *constant utility* approach it is assumed that:

$$P_i = \frac{v_i}{\sum_j v_j} \tag{8}$$

The v values are utilities associated with the choice alternatives. Equation 8 states that choice probabilities are proportional to utilities. Utilities are not assumed to be maximized. To make this model useful, the v values have to be specified as functions of parameters and variables. One of the most useful specifications leads to the multinomial logit model (see below). The Huff model is a constant utility with the specification:

$$v_i = S_i \, D_i^{-b} \tag{9}$$

The constant utility model was formalized by Luce (1959). It has been very influential in the decision sciences (Luce & Suppes, 1965). Equation 8 is consistent with a controversial assumption called Luce's choice axiom or Independence from Irrelevant Alternatives (IIA). This axiom states that the relative choice probabilities (that is, the ratios p_i/p_j) depend only on the corresponding v values (that is, the ratios v_i/v_j) for any i and j. Thus in comparing a pair of objects, the presence or absence of other alternatives in the choice set is irrelevant to their relative worth. This assumption is compelling in many situations, but it does not necessarily apply in all transport choice contexts, as we shall see (for example, Stopher & Meyburg, 1976).

An alternative and more general approach in mathematical choice theory is the *random utility model*, which preserves the powerful idea of utility maximizing as a decision principle but assumes that utilities themselves vary. The abstract structure was developed in psychology by Thurstone in the 1920s (see, for example, Thurstone, 1927). One of the earliest economic–transportation applications was by McFadden (1973). Assume that each object is characterized by a utility U_i, defined as a random variable with density function f_i and distribution function F_i. The utility-maximizing principle then implies:

$$p_i = P\,(U_i \geq U_j;\ \text{all}\ _j) \tag{10}$$

Under the assumption of statistically independent utilities,

$$p_i = \int_{-\infty}^{\infty} f_i(t) \prod_{j \neq i} F_j(t)dt \tag{11}$$

Some subtleties arise in interpreting the U_i. They may represent variations in tastes, preferences, and myriad uncontrolled variables for a specific individual. Or they may represent variations in an otherwise homogeneous preference system across a sample population. Several radically different interpretations of the scale values are possible (for example, Pipkin, 1979). The assumption used in most transportation applications is that the stochastic variation describes variation in preferences over some sample population. This assumption is most often expressed by writing:

$$U_i = V_i + e_i \tag{12}$$

where V_i is constant and e_i is a random variable (for example, Horowitz, 1980).

Different forms of the models arise when different shapes ("probability distributions") are assumed for the utilities. By far the simplest form arises when the U_i have a Weibull probability distribution and the utility differences are logistic (Domencich & McFadden, 1975). In this special case the integral in 11 reduces to equation 8 under an exponential transformation implying that, despite its apparently different assumptions, the constant utility model is formally a special case of the random utility model. This version is called the *multinomial logit model* (MNL), and it has proved an extremely powerful tool in modeling travel choices (see Hensher & Johnson, 1981).

A formal statement of the MNL is given in Figure 8-4. The MNL builds into an explicit mathematical model parameters and variables characterizing decision makers (such as income, age, car ownership, sex, class), characteristics of the alternatives (modal level of service measures, distance, travel time, and site attributes), as well as stochastic variation in preferences across a population. The MNL satisfies the IIA assumption. This model has become the work horse of disaggregate

Model

P_i^h = probability that decision maker h chooses option i $= \dfrac{V_i}{\sum_{j=1}^{J} V_j}$

where $V_i = \exp\left\{\beta_{0i} + \sum_{k=1}^{K} \beta_k X_{ik}^h\right\}; \quad i \in J$

Notation

$h = 1, \ldots, H$ decision makers
$j = 1, \ldots, J$ alternatives
$k = 1, \ldots, K$ attributes of alternatives and decision makers

U_j^h = perceived utility of alternative j for decision maker h (unobserved)
β_{0i} = constant term (unspecified utility), to be estimated
β_k = coefficient for kth attribute (constant for all h), to be estimated
X_{jk}^h = value of kth attribute for alternative j and decision maker h (measured in data).

Assumptions

Utility distributions are constant over the population up to a stochastic error term; errors have identical, independent, constant variance Gumbel distributions; choices are utility maximizing. The U values are unobserved random variables. The V values are constants generating the U values with addition of an error term (see equation 12). Detailed examples of the multinomial logit model are provided in the next chapter.

Figure 8-4. *The Multinomial Logit Model.* Adapted from "Discrete Choice Theory, Information Theory and the Multinomial Logit and Gravity Models" by A. Anas, 1983, *Transportation Research B*, *17*, pp. 14–15. Copyright 1983 by Pergamon Press. Used by permission.

transportation theory. A more detailed discussion with worked examples is given in chapter 9. One of the earliest applications (to mode split) appears in Warner (1962). Stopher, Meyburg, and Brőg (1981) give a comprehensive summary of applications through the 1970s. A recent large-scale application to mode choice is reported by Southworth (1981). A sophisticated application to destination choice is provided by Kern, Lerman, Parcells, and Wolfe (1984), who study home-based shopping trips to regional shopping centers in Boston.

Recent Developments in the Mathematical Theory of Travel Choice

Because of its power and generality the MNL has come to be perhaps the principal tool in travel choice modeling. Its strengths and weaknesses are increasingly well understood (for example, Horowitz, 1981). Its domain of applicability has been steadily extended from binary mode choices to simultaneous choices over complex choice sets, with site, link, and individual variables included as predictor variables. Nevertheless, the MNL is only one particularly tractable form in the family of random utility models.

The IIA assumption has itself come under criticism. It appears compelling as a description only of simultaneous choices over genuinely comparable yet somewhat differentiated alternatives. Problems arise when elements that are essentially the same as existing ones are added to the choice set. For example, suppose that a consumer chooses between two supermarkets A and B, with constant utility values $v_A = 3$ and $v_B = 1$. (These correspond to the constant utilities of equation 8). Assume that they summarize a shopper's perception of the utility of the stores, taking into account all site and distance effects. Then from equation 8 $P_A = .75$ and $P_B = .25$. Now suppose that a new supermarket, C, is added to the choice set. Suppose that C is in the same store chain as B, with $v_C = 1$, and with B and C being essentially identical from the consumer's point of view. Intuition says that A ought to retain the same relative choice probability to B and C combined. But the probability of choosing A from the whole set diminishes to three in five (to see this, calculate P_A using equation 8). Because of this and other objections to the MNL, alternative choice theories have been explored.

The probit model, in which utilities and utility differences are normally distributed, has been shown to be computationally feasible in some cases (Daganzo, 1979). The situation of sequential or nested choices over routes, modes, times, and destinations (in which IIA need not apply between choices) can be modeled using structured logit models (Wilson et al., 1981, chapter 3). Another innovation is the dogit model, which is similar in form to the logit, but ratios of choice probabilities depend upon all alternatives (Gaudry & Dagenais, 1979). Tversky's (1972) elimination-by-aspects model has also been suggested as a way around IIA. Tversky proposes a sequential decision process in which significant dimensions of alternatives are considered in turn, with whole subsets of the choice set being eliminated until one option is left.

The MNL and its variants were originally calibrated using objectively measurable attributes of the alternatives and decision makers such as the size and distance of stores, the price of modes, and measures such as income and car ownership. However, both in transport science and geography there arose in the 1970s a conviction that less tangible attitudes, beliefs, and preferences were crucial determinants of travel behavior. The need to understand subjective elements in travel is well illustrated in research on travel time (for example, Hensher & McLeod, 1977; Levin, Louviere, Meyer, & Henley, 1979; Watson, 1974). Travel time can be measured objectively and is acknowledged to be an important factor in destination and mode choice decisions. But it turns out that (1) people often misjudge the actual time taken in travel, and (2) there are important qualitative differences in the disutility of time spent on different parts of a trip. Ten minutes spent waiting for the bus is more "costly" than ten minutes on the road. Thus the "objective" question of pricing travel time inevitably raises psychological questions about perceptions, attitudes, and beliefs.

Attitudinal and Cognitive Models of Travel Choice

Attitudes and subjective perceptions can be and have been built into mathematical choice models such as the MNL. But attitudinal and cognitive researchers have a different perspective than early proponents of random and constant utility models. Instead of accepting a straightforward rational and utility-maximizing theory of choice, researchers have asked whether respondents share the same choice sets, which attributes of alternatives are psychologically salient for people, how subjective trade-offs between price and comfort or time and quality are made, how intentions are formed, and whether, in fact, people's attitudes have much to do with their actual behavior (see Cadwallader, 1981). Levin defines attitudinal models in transportation research as "descriptions or predictions of behavior that stress the role of subjective perceptions, judgments, and evaluations" (1981, p. 173). Inventories and commentaries on this research in the transportation planning literature are provided by Charles River Associates (1978) and Stopher et al. (1981). In geography a wealth of cognitive–behavioral research relevant to repetitive travel has been conducted.

A wide variety of methods have been used in spatial and mode choice studies. The problem has several different components including identifying the salient dimensions of alternatives, modeling the ways in which these are weighed and combined by decision makers, and predicting actual choices.

The most desirable methods of identifying the relevant attributes are those in which individuals themselves implicitly or explicitly identify attributes that are important to them in judging alternatives. The "construct elicitation" process of personal construct theory is one increasingly popular method (for example, Bannister, 1970; Downs, 1976). Another indirect method uses multidimensional scaling (such as an application to grocery shopping trips, Spencer, 1980). It is more common, though, to identify a range of potentially relevant attributes in advance and allow subjects to estimate their interrelationships and relative importance.

The *semantic differential* is one such method. Two applications of the semantic differential to shopping center images are Downs (1970) and Nevin and Houston (1980). These studies are part of an enormous literature on the image dimensions of retail stores and shopping centers. Using this method, subjects are given several verbal descriptions of each hypothesized attribute arranged on a five- or seven-point evaluation scale. The 16 shopping center attributes used by Nevin and Houston are shown in Table 8-2. Factor analysis was used to identify underlying dimensions in the perception of attributes. Three were identified. The first was related to quality, selection, and variety and was identified as a retail diversity factor. The second factor was associated with amenities such as layout, parking, and restaurant facilities. The third dimension correlated with price, helpfulness of store personnel, and whether or not the center was perceived as conservative. Studies of retail attractiveness such as Nevin and Houston's have identified a confusing variety of dimensions of attractiveness. It has become clear that the traditional predictors of size of center, price, and distance are less important for many consumers than quality, cleanliness, layout, and level of service. Serious problems arise in correlating dimensions such as "quality" across different types of business (for example, clothing stores versus supermarkets). Another issue involves separation of the effects of brands, commodity types, stores, and the context of stores relative to other places that may be visited on multipurpose trips (see, for example, Patricios, 1978). It is clearly questionable whether shopping centers possessing a wide range of outlets can be said to have a single unitary "image" at all.

Table 8-2. Shopping center image attributes

Attribute	Anchor descriptors
Quality of stores	High-low
Variety of stores	Excellent-poor
Merchandise quality	Excellent-poor
Product selection	Excellent-poor
General price level	Fair-unfair
Special sales/ promotions	Attractive-unattractive
Layout of area	Convenient-inconvenient
Parking facilities	Adequate-inadequate
Availability of lunch/ refreshments	Adequate-inadequate
Comfort areas	Adequate-inadequate
Special events/exhibits	Attractive-unattractive
Atmosphere	Friendly-unfriendly
Store personnel	Helpful-not helpful
Easy to take children	Very easy-very hard
Great place to spend a few hours	Agree-disagree
A conservative center	Agree-disagree

Note: From "Image as a Component of Attraction to Intra-Urban Shopping Areas" by J. R. Nevin and M. J. Houston, 1980, *Journal of Retailing, 56,* p. 85. Copyright 1980 by Institute of Retail Management, New York University, *Journal of Retailing.* Reprinted by permission.

Once relevant dimensions of choice alternatives have been identified, it is necessary to model the way in which people combine them in forming overall evaluations and preferences (see, for example, Lieber, 1976). This is usually done by inferring a utility scale that is the sum of attribute values multiplied by coefficients measuring the relative importance of each attribute (as in equation 4 above). This is called an additive combination rule. Finally utilities are related to observed choices by correlating inferred utilities with visit frequencies. Once the coefficients have been estimated, predictions of unknown visit frequencies can be made.

A common method for this kind of work is *conjoint analysis.* Conjoint analysis is a scaling model that infers the contribution made to an overall preference ranking by various attributes of alternatives. It can be applied in revealed preference applications such as those described above. It is more commonly applied to reported preferences generated by subjects in ranking multiattribute alternatives. The technique has been widely used in geography, psychology, marketing research, and also in transportation modeling. Some have expressed reservation about the technique because of the lack of an explicit theory of errors and fit testing (Levin, 1981).

Schuler (1979) applied conjoint analysis as part of a larger study of grocery shopping in Bloomington, Indiana. His conceptualization of the choice process is shown in Table 8-3 and Figure 8-5. The sequence cognition, evaluation, preference ranking, and choice is implicit in many studies of consumer behavior. A larger sample of people were asked to rate 19 attributes relevant to grocery shopping. The variables included commodity traits, store attributes, and locational measures. Within-group rankings are shown in Table 8-3. Price and quality were the main commodity traits, while parking and service quality dominate perception of outlets. Interestingly, attractiveness of displays (on which retailers spend much effort) were apparently not very important. Among contextual factors proximity to the residence and the route to work were most important, reflecting the fact that grocery shopping is often a single-purpose, home-base trip but is also sometimes linked to the work place. The five most important variables indicated in the table were carried forward to the next phase of the analysis. Here, subjects ranked the supermarkets by name, ranked them separately across each attribute, and also ranked hypo-

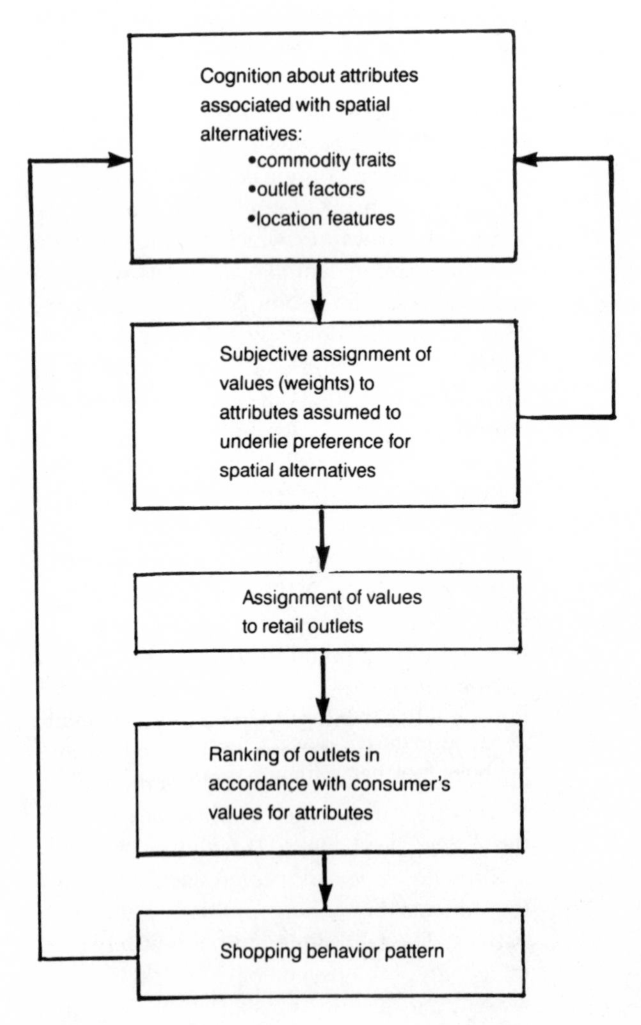

Figure 8-5. Schuler's conceptualization of destination choice. From "A Disaggregate Store-Choice Model of Spatial Decision-Making" by H. J. Schuler, 1979, *Professional Geographer, 31*, p. 147. Copyright 1979 by the Association of American Geographers. Reprinted by permission.

Table 8-3. The relative importance of factors in grocery shopping

Group class and attributes	Total points	Average	Within-group rank
A. *Commodity Traits*			
1. Price	1,083	5.791	1*
2. Quality	1,076	5.754	2*
3. Variety	910	4.866	4
4. Brand Range	985	5.267	3
B. *Outlet Factors*			
5. Friendly and courteous	1,028	5.487	3
6. 24-hour shopping	611	3.267	9
7. Cleanliness	906	4.844	5
8. Attractive displays	732	3.914	8
9. Pleasant surroundings	1,005	5.374	4
10. Quick service	1,079	5.770	2*
11. Internal organization	801	4.283	7
12. Nearness of parking	1,130	6.042	1*
13. Check cashing	876	4.684	6
C. *Locational Features*			
14. Proximity to residence	1,064	5.689	1*
15. Proximity to bank	436	2.331	4
16. Proximity to drugstore	587	3.139	3
17. Proximity to discount store	379	2.026	5
18. Proximity to bus stop	263	1.406	6
19. Proximity to a route you follow to or from work	695	3.716	2

*Selected for Detailed Analysis.

Note. From "A Disaggregate Store-Choice Model of Spatial Decision-Making" by H. J. Schuler, 1979, *Professional Geographer, 31,* p. 149. Copyright 1979 by the Association of American Geographers. Reprinted by permission.

thetical alternatives incorporating various levels of attributes. This yielded the following inferred utility scale. The coefficients indicate that the first two traits are most important.

$$U = .456\ (\text{Quality}) + .244\ (\text{Price}) + .119\ (\text{Distance}) + .092\ (\text{Speed of service}) + .086\ (\text{Nearness of parking}) \qquad (13)$$

Schuler's work illustrates several different ways in which such models can be tested. First, the validity of the conjoint model (and the additive combination rule) was tested by comparing the derived rankings of alternatives and the actual rankings available for each subject. In 63% of the cases the rankings correlated significantly. Then derived and observed scales aggregated across the sample were correlated ($r = .77$). Finally actual trip frequencies were found to be significantly correlated with the derived utility measures. These results show (1) that conjoint scaling adequately models the way multiple attributes are combined in overall evaluations, and (2) these evaluations are significant predictors of actual choices.

An alternative scaling method that some analysts feel has a firmer theoretical foundation than conjoint scaling, is *information integration theory.* Like conjoint scaling, the method seeks to uncover the subjective rules by which multiple attributes are combined in evaluations. The essence of information integration theory is the idea that three functions or tranformations are involved in making choices. A subjective filtering function transforms objective into subjectively perceived attributes. An integration function combines attributes into an overall evaluation. And behavior is a function of evaluations of the alternatives. Some applications of information integration theory to travel decisions are provided by Louviere (1976) and Louviere and Henley (1977). Timmermans (1982) applies the method to clothing shopping in Holland, while Lieber and Fesenmaier (1984) give a detailed demonstration of how the model was used in recreational trail management in the Chicago region.

The previous chapter raised the question of how the complexity of travel activity patterns could be modeled effectively. Which of the innumerable factors affecting travel could be built explicitly into models? It is obvious by now that much progress has been made. Early choice models such as Huff's dealt with site and distance attributes in a very simple way. Such approaches had no mechanism for dealing explicitly with variations in utilities (tastes and preferences) among the consumers themselves. More advanced methods such as the MNL are complicated but mathematically tractable models that can be applied to various kinds of travel choice (route, mode, time, destination). Many variables can be included explicitly, describing the choice alternatives and the people making the choices. Statistical error terms, at least in theory, account for individual variations in tastes. And attitudinal–cognitive methods allow some of the subjective and intangible aspects of choices to be quantified.

This progress notwithstanding, choice models as presently formulated are incapable of dealing with significant aspects of travel, a fact that has emerged from recent discussion and criticism.

Criticisms of Travel Choice Models

During the 1970s choice modeling dominated disaggregate travel research. Continual technical innovation meant that, in geography at least, it was hard to find two studies with identical methodology. In the past 6 or 7 years various criticisms of choice modeling have been expressed. In transportation planning these have focused on limitations of the variants of the MNL model and have generated a whole range of new approaches (reviews are given in Carpenter & Jones, 1983; and Stopher et al., 1981). In geography criticisms have often been voiced in evaluations of behavioral geography as a whole (Bunting & Guelke, 1979; Thrift, 1981).

Many technical problems have been identified in conventional choice models. One is the definition of the choice set. This problem is far from trivial because meaningful choices can only be made from known and evaluated alternatives, a set which differs among individuals and over time for any one person (Meyer, 1981). Aggregation of individuals is another question: It is evident from the preceding examples that revealed preference analysis, the MNL, functional analysis, and other techniques require strong assumptions about the homogeneity of preference structures among individuals. Another question is transferability: whether parameter estimates are constant over time and space, so that models calibrated in one setting can be used to make predictions in another (Burnett, 1981). Interaction models as a whole often show spatial structure effects, in which parameter estimates depend on the spatial configuration of a particular study area (Futheringham, 1983). The issue of functional form and combination rules is another serious problem in multiattribute preference models. We have encountered two such models here: revealed preference analysis and conjoint scaling. In each we assumed that individuals subjectively combine attributes by "adding them up," weighted by coefficients measuring the attributes' relative importance. This represents a linear function and an additive combination rule. The additive form is compensatory in the sense that low scores on one attribute can be offset by high scores on another. The assumption is more one of analytical convenience than of empirical necessity. Other rules are possible. For example, a multiplicative scale is not compensatory in the same sense as an additive one. We don't know much about which is really appropriate, although functional measurement in information integration theory can attack the problem (Lieber, 1976; Louviere, 1976; Timmermans, 1984).

Another enduring problem in attitudinal and cognitive research is attitude–behavior consistency. It is a cliché of attitude theory that the kinds of evaluations people express toward various alternatives may bear little relation to their actual behavior (Fishbein, 1967). A related issue is the mutual interdependence of attitude and behavior. A typical mode or destination choice model takes attitudes to be predictor variables and treats behavior as dependent. Yet it is clear that attitudes are themselves molded by behavior. For example, Charles River Associates (1978) provide statistical evidence of mutual dependence between evaluations of auto/bus modes and behavior.

Cognitive choice models, particularly within geography, have often been compelled to oversimplify attitude–behavior connections. One psychological theory of attitude–behavior connections is that of Fishbein and Ajzen (1975). Figure 8-6 shows an interpretation of this "reasoned action" theory in a transportation context by Desbarats (1983). Two key mediations appear, which are often omitted in geographic analyses. First, social norms, as well as individually determined attitudes, affect behavior. The individualistic approach of many cognitive–behavioral models assumes, wrongly, that attitudes are formed only in an individual learning transaction with reality, unaffected by expectations, peer pressure, customs, and social norms generally. Second, norms and attitudes do not determine behavior directly. Intentions intervene (Hensher & Louviere, 1979). Obviously in many travel contexts a high evaluation of an alternative does not imply an intention to travel there.

Another line of criticism questions the assumptions of choice and utility maximization common to all random utility theory and most cognitive–behavioral models. Probably most travel behavior is habitual, involving little conscious decision making of any kind.

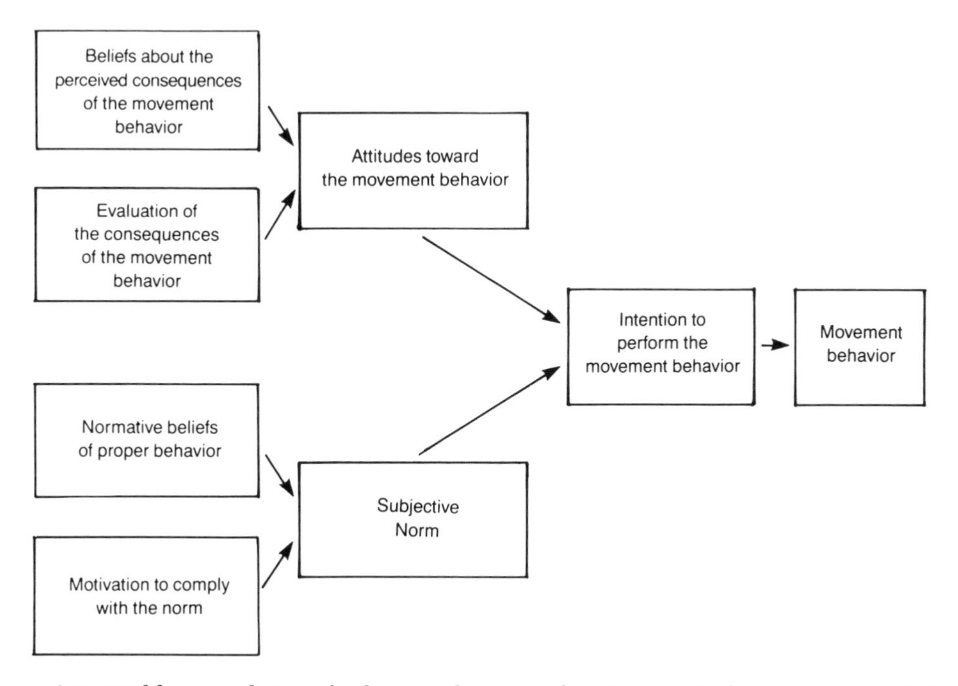

Figure 8-6. Fishbein and Ajzen's theory of reasoned action applied to movement behavior. From "Spatial Choice and Constraints on Behavior" by J. Desbarats, 1983, *Annals of the Association of American Geographers*, 73, p. 349. Copyright 1983 by the Association of American Geographers. Reprinted by permission.

Choices may be more concerned with meeting certain minimum needs ("satisficing") than with maximizing any utility, whether subjectively or objectively measured. In fact, many people have few genuine travel choices at all. It is easy for affluent, mobile, price-conscious researchers to project their values onto people with little mobility: the poor, the elderly, inner-city minorities, and home-bound suburbanites (often women) (Pirie, 1976; Spencer, 1980). In fact choice-based models sometimes tend to overpredict the effects on behavior of planned changes in policy variables, because behaviors are really more constrained than planners assume.

The dominance of utility and choice-based theory in travel studies can be viewed as a largely unconscious ideological slant (Sheppard, 1980). This diverts attention from the problems of groups who possess few choices. It contains strong and unexamined assumptions that people are the calculating egoists portrayed in elementary economic theory. And at a deeper level it dooms planners to propose improvements that have relatively small scope and impact in the context of what radicals see as the crisis of urbanism in late capitalist society (see, for example Dear & Scott, 1981; Scott, 1980). This and other lines of criticism suggest a need to refocus travel research on the constraints and the societal context of travel behavior.

A final (closely related) critique of conventional choice models follows from the space–time complexity of behavior described in the previous chapter. People's travel choices are constrained by time, money, and the activities of others in the same household. The rhythms of work and urban life enable and constrain opportunities in time and space. Destinations don't exist in isolation. Multi-purpose trips link visits to different functions in ways that could not possibly be predicted from the attributes of any one site alone (S. Hanson, 1980b). Household stocks of consumables and durables change in complicated and periodic ways. Sometimes we have use of a car, sometimes we don't. All these facts are hard to build into conventional models, which predict constant-parameter choices on single-purpose, home-based trips, using data from only one time period. A new generation of travel-demand models has grown up to address these problems (Burns, 1979; Carpenter & Jones, 1983; Jones, Dix, Clarke, & Heggie, 1983).

ACTIVITY MODELS OF TRAVEL BEHAVIOR

Data requirements on both travel characteristics and attributes of the urban environment are very demanding for activity models as indicated in the last chapter. Conventional preference scaling tasks are supplemented or replaced by travel diaries, unstructured interviews, gaming techniques, and participant observation (Jones, 1979). It is almost obligatory that data be longitudinal and that activities and trip lines be timed and geocoded. There is a new sensitivity to the biases of various survey instruments (Brőg & Meyburg, 1983). Stratification or segmentation is a critical issue, given evidence that different demographic, social, economic, and age groups have distinct travel patterns (for example, S. Hanson & P. Hanson, 1980).

Attempts to confront the complexity of behavior encounter formidable problems of definition and measurement and the lack of a theoretical background for explanation and models. The adaptability of human behavior provides an example of the problems encountered. Changes in transport policy (such as bus fares) or market conditions (gas prices) can lead to very complex adjustments in household behavior: rescheduling of trips, ridesharing, reassignment of tasks in the household, trip chaining (Adler & Ben Akiva, 1979), nontravel uses of leisure time, and so on. A change in one part of a household's activities is likely to have unexpected repercussions on travel by other people in the family. Later in this section, for example, we will outline a case in which a shift in school-bus hours in the morning caused other kinds of travel to increase in the evening.

One way of dealing with problems of complexity and interdependence in travel is to identify domains for different models. A domain is a subset of travel, defined by a specific kind of trip or type of traveler, for which a particular set of models is appropriate. Studies of model domains are important in bringing order to what might otherwise be futile attempts to confront the whole complexity of household behavior. They also suggest that conventional choice theories have a continuing place among travel models, though a narrower one than optimists originally believed. Heggie and Jones (1978) define four model domains. Their concern is with the adjustments households make in response to transport policies. Their model domains are essentially a classification of response patterns in space–time and interpersonal linkages. The domain of *independence* applies to individuals whose travel decisions are independent of other destinations (that is, they are single-purpose trips) and in which decisions such as mode and timing do not depend significantly on activities of other people. This domain is most applicable to young, childless adults. Conventional choice models are most relevant here. The domain of *spatio-temporal linkages* is relevant when trip purposes are interdependent in time and space but do not depend much on other people. (Multipurpose trips by single-person households fall in this category.) The domain of *interpersonal linkages* covers travel decisions dominated by person-to-person connections whether inside or outside the family. The domain of *full interdependence* allows for complex links between people and in space–time. These domains highlight the different modeling requirements of independent choices and of connections between people and activities in space and time.

Another approach to defining model domains focuses on the kind of behavior involved. Geographers have not devoted much effort to classification of behaviors in space and time. From a cognitive–behavioral point of view the issue of how "choice" behaviors can be defined is quite complicated (Pipkin, 1981a). Burnett and S. Hanson (1982) suggest a fourfold classification; for each kind a different modeling strategy is appropriate. *Habitual* behaviors are probably quite common in repetitive travel. People are not always making conscious price and attribute comparisons. Behaviors such as the weekly grocery shopping trip or the route followed to work may be so routinized as to be very resistant to change. Small policy changes such as modest bus fare cuts will hardly affect such behaviors at all. *Avoidance* behaviors involve avoiding courses of action that might seem rational. For example, many people avoid subways in large cities for fear of crime, noise, or dirt, even though the subway may be a cheap and efficient way to travel. *Institutionally constrained* behaviors are undertaken because of authority constraints imposed by corporations or government (such as certain kinds of work-related travel). *Choice* behaviors are the remaining kinds of behaviors, which involve real choices.

Neither Heggie and Jones nor Burnett and Hanson's taxonomies have been translated explicitly into operational sets of models, but it is becoming clear that both domains and types of behavior require different models. In developing new models there has been a renewed emphasis on *description* rather than *prediction*. Indeed the divisions between description, explanation, and prediction have become blurred. We will outline next a technique designed to model behaviors in the domain of full interdependence. It is used primarily to describe activity patterns but has been used successfully to predict likely outcomes of transport policies.

Activity Simulation Models

HATS (Household Activity–Travel Simulator) is one of the best known survey methods used to detail household activity patterns. It was developed by Dix and Jones of the Transport Studies Unit, Oxford, England (Dix 1981; Jones, 1979, 1981; Jones et al., 1983). HATS is a multistage procedure (Figure 8-7). Ideally it begins with completion of a conventional travel diary by the respondents. The HATS interview uses a display board showing a map of the study area and three parallel time scales labeled "Home Activities," "Travel," and "Out of Home Activities." The

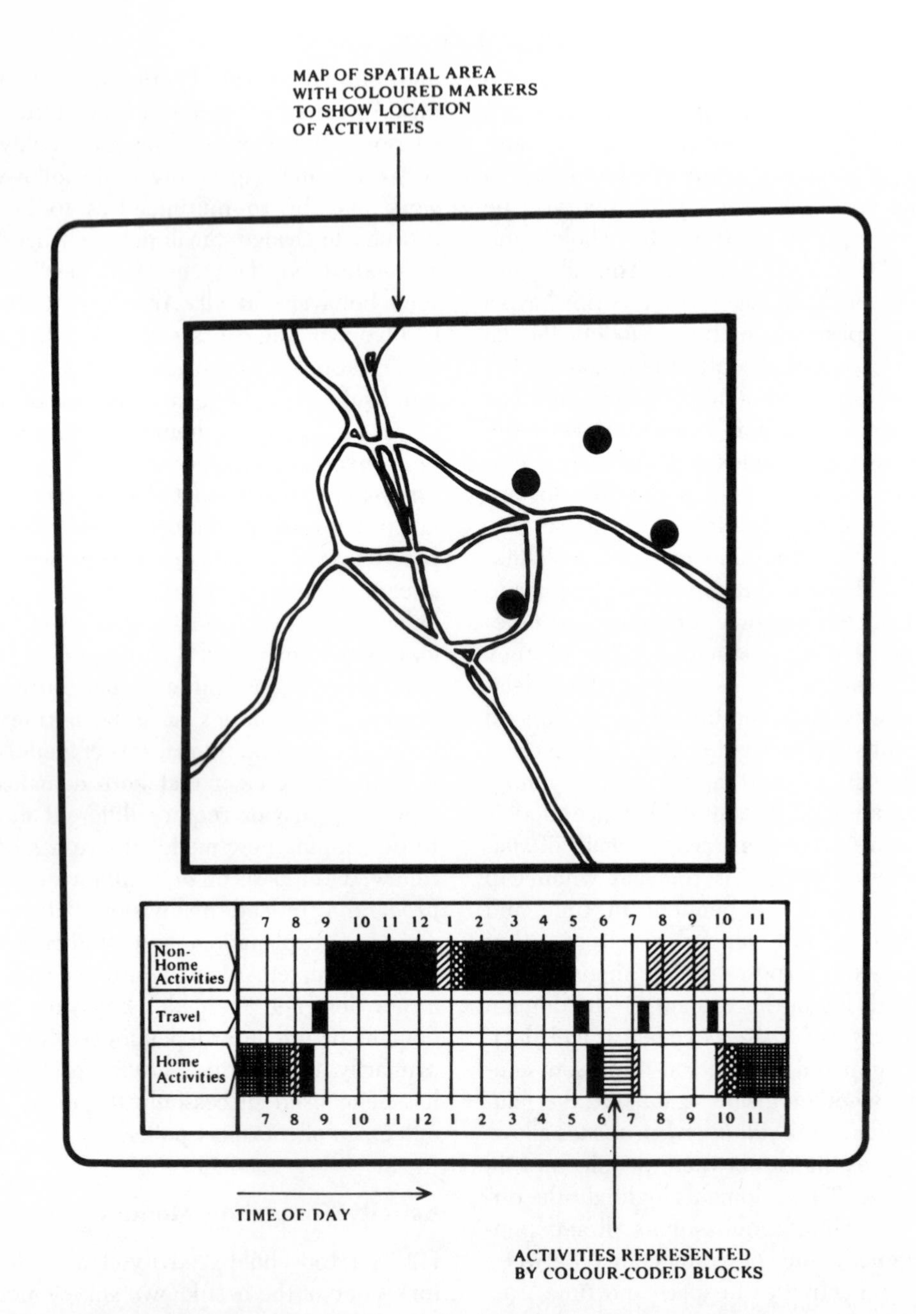

Figure 8-7. HATS gameboard. The top of the board displays a map of the study area. The three rows below represent parallel time scales for Out of Home, Home, and Travel Activities. Respondents block out the times they engage in these three activity types, taking care that responses are complete and consistent; that is, people must be somewhere at all times and never in two places at once! Typically time blocks on the Home and Out of Home scales must be linked by appropriate segments of travel. From "Structuring our understanding of travel choices: The use of psychometric and social science research" by M. C. Dix, pp. 89–109 in P. R. Stopher, A. H. Meyburg & W. Brog (Eds.), *New Horizons in Travel Behavior Research.* Lexington, Mass.: Lexington Books, p. 102. Copyright 1981, D. C. Heath and Co. Reproduced by permission.

interview begins as the respondent blocks out one day from the diary record on the display board. Different activities such as work and shopping are shown by different color blocks. Travel modes are indicated on the Travel scale. This simple representation of the day's activities forces respondents to be *consistent* (they cannot be in two places at once) and *comprehensive* (all times must be blocked out on one of the three rows). Boards must also be consistent across all participating members of the household. When the respondent is familiar with the display system, the interviewer discusses the observed behavior and explores the rationale and the constraints of time, cost, and family commitments that underlie it. The next stage in the process is to use the display and interview to explore and evaluate household reactions to proposed policy or market changes. "What if" questions are translated into hypothetical activity displays, representing household responses to changes in school times, bus availability, working hours, store opening times, and so on.

That HATS method and related gaming simulations have several advantages over conventional interview techniques. Their concrete, visual representation forces respondents to be thoughtful and consistent in deciding how they would respond to proposed changes in transport availability. For example, if a change in bus timetables and a lowering of bus fares were proposed, and if people were simply asked if they would be more likely to use the bus, they would probably say "yes." A technique such as HATS, on the other hand, might force respondents to look at the constraints of work, picking up children, shopping, and child-care needs leading to a conclusion that bus travel is not possible, no matter how low the fare. In this sense the predictions of the HATS method tend to be more conservative and realistic than some choice-based models.

Multi-instrument gaming (Burnett & Ellerman, 1980) has been proposed which combines a HATS-like display board with mapping of places, routes, and modes used by respondents. Burnett and Ellerman's method explores attributes of places and reasons for visits by card sorting and by analysis of tape recordings made while respondents think aloud about their travel choices. Their method also requires subjects to examine in detail the attributes, strategies, and decision rules underlying a particular discretionary trip. This kind of interview process is very time consuming and expensive to apply, particularly for large samples, but it provides insights into the rationale for travel behaviors that would be hard to obtain in any other way.

Reconciling Activity and Choice-Based Models

Although the activity approach differs in vocabulary, methodology, and philosophy from utility-based choice models, there have been several attempts to reconcile the two approaches. Considering the power of models such as the MNL, it certainly seems appropriate to generalize rather than discard them. One obvious strategy is to define choice sets and constraints in a more refined way. For example, Burnett and Hanson propose an abstract choice model as follows:

$$P_i(j) = P_i(j \epsilon A) \cdot P_i(j | j \epsilon A) \quad (14)$$

$P_i(j)$ denotes the probability that individual i chooses alternative j. $P_i(j \epsilon A)$ symbolizes a probabilistic model of choice set formation which would incorporate institutional and other constraints. "The principal constraints are deemed to be the characteristic distances to economic activities about the places of work and residence. . . . [They] also include characteristic sociodemographics defining the role complex (race or ethnic, class and family) of the person" (Burnett & Hanson, 1982, p. 91). The second term on the right side of equation 14 is a conventional choice model of the probability that j will be chosen given that it is in the choice set A. Burnett and Hanson regard this model as a relatively "open-ended" schematic, requiring "inductive exploration" and measurement of travel patterns and constraints, including classification of behaviors as habits, avoidance routines, or genuine constrained choice. They orient the model more to explanation than forecasting and proceed to discuss incorporation of complex behavior patterns represented as space–time diagrams as discussed in the previous chapter.

A practical illustration of the power of a combination of choice and activity models is presented by Van der Hoorn (1983). This study was conducted in Holland. Its objectives were (1) to construct a system to simulate (predict) the effects of changes in the transport system and the economy as a whole on travel behavior, and (2) to model individual trip generation as a function of daily activity schedules. The data base was a travel diary administered to 1,300 people for one week in 1975. Travel purposes, modes, and times were recorded. The population was stratified into several subgroups including working women, working men, and housewives. An appropriate classification of activities and locations was developed (see Table 8-4).

Van der Hoorn's model consisted, in part, of an interrelated set of multinomial logits describing transitions between activities and locations. Choice sets were defined to take into account constraints (for example, opening hours, duration of activity, out-of-home availability such as shopping), and the fact that some activities are mandatory and not discretionary (such as sleeping, eating). Location and activity choice models were calibrated. For example, Van der Hoorn describes an activity model of the form $p(A|L°,A°,T)$. This represents the probability that activity A will be chosen given a previous location of $L°$, a previous activity of $A°$, and a current time of day T. Joint probabilities of activities and locations were obtained by multiplying conditional probabilities of location and activity. Table 8-4 is a "validation table" comparing observations and predictions of the amounts of time working men spend in various activities and places. The fit is remarkably good.

Van der Hoorn's system of models also provided the basis for extensive simulation studies under hypothetical changes in variables ("scenario tests"). Table 8-5 shows the predicted results of giving working men Friday afternoon off. It appears that they would socialize more, help more with housekeeping, watch more television (but not sleep more!). Van der Hoorn outlines some other scenario studies including analysis of the effects of income changes and changes in working hours. This study has several limitations, including a very crude locational classification (home, in town, and out of town), neglect of interactions within the household, and an aggregation of the simulation over individuals mandated by computing feasibility (Van der Hoorn, 1983, p. 66). Nevertheless, this work illustrates the power of combining a constraint-oriented activity philosophy with the proven tool of logit analysis.

INTEGRATING TRAVEL MODELS WITH GENERAL SOCIAL THEORY

In the increasingly complex details of modeling and policy applications it is easy to lose sight of a basic fact about travel. It is a derived demand that can only be fully understood as part of a social and economic whole. For example, conventional travel choice models can make good predictions about the market shares obtained by existing shopping centers. They can also make useful estimates of the impact of new centers and of changes in policy and market conditions for travel. But standard travel models have nothing to say on the broader evolution of shopping centers and travel patterns as relatively new features in an evolving urban land system. This pattern is, in a broader view, a moment in an evolving system of production and reproduction. (For example, shopping travel can be seen as part of the "consumption sphere" while the work trip is determined by the spatial organization of homes and workplaces.) In the last analysis travel patterns are superficial products of deeper processes in which behavior, society, and urban space are mutually shaped.

It is very easy to make grand statements that "everything is related to everything else." It is true that a deeper understanding of

Table 8-4. Sample activity-location array

A Typical Validation Table (Working men, times in min. per day)

Activity	Time spent at home		Time spent in town		Time spent out of town		Total time spent		Number of trips	
	Predicted	Observed	Predicted	Observed	Predicted	Observed	Predicted	Observed	Predicted	Observed
Sleep	428	429	2	2	9	9	439	440	0	0
Personal hygiene	54	44	1	1	0	2	55	47	0	0
Eating	85	66	5	9	10	16	100	91	0.5	0.1
Work	25	25	197	197	202	203	425	425	2.5	2.4
Study	25	5	1	6	1	5	27	16	0	0.1
Housekeeping	34	47	0	4	0	1	34	52	0	0.2
Child-care	24	11	1	1	0	0	25	12	0	0
Private leisure	83	99	3	5	5	5	96	109	0.3	0.1
TV	82	75	0	1	0	1	82	77	0	0
Social	51	39	26	27	13	22	90	88	1.1	0.7
Entertainment	0	0	11	7	6	5	17	12	0.3	0.2
Recreation	0	0	15	9	5	4	20	13	0.4	0.1
Shopping (daily)	0	0	5	5	2	1	7	6	0.2	0.2
Shopping (durables)	0	0	5	7	2	3	7	10	0.2	0.3
Personal business	0	0	5	3	1	1	6	4	0.2	0.1
Participation[a]	19	4	10	14	5	6	34	24	0.2	0.3
Total	915	844	287	298	261	284	1,463	1,426	5.9	4.8

Number of trips per time-of-day

Period	Trip frequency per workday	
	Predicted	Observed
0.00– 7.00 h	0.4	0.2
7.00– 9.30 h	0.9	0.9
9.30–16.00 h	2.1	1.7
16.00–18.30 h	1.2	1.0
18.30–24.00 h	1.3	1.0
Total	5.9	4.8

Predicted and observed daily trip matrix

Origin		Destination		
		At home	In town	Out of town
At home	Predicted	—	1.4	0.6
	Observed	—	1.2	0.5
In town	Predicted	1.4	1.0	0.0
	Observed	1.1	0.5	0.1
Out of town	Predicted	0.5	0.1	1.0
	Observed	0.6	0.1	0.6

[a] Participation in neighborhood and community life and the like.

Note: From "Experiments with an Activity-Based Travel Model" by T. Van der Hoorn, 1983, *Transportation, 12*, p. 67. Copyright 1983 by Elsevier. Reprinted by permission.

Table 8-5. Simulation of the effects of Friday afternoon off for working men

Impacts of the Scenario Friday Afternoon Off (Working Men)

Activity group	Time spent (min. per day)	Percentage change from base	Trip frequency per day	Percentage change from base	Travel time (min. per day)	Percentage change from base
1. Sleep	439	0	0.02	+∞	0.54	+∞
2. Work	203	−52	2.23	−11	37.86	−10
3. Personal hygiene Eating	211	+33	0.58	+2	10.58	+8
4. Housekeeping Child care Shopping (daily)	76	+15	0.28	+22	3.20	+8
5. Social Entertainment Recreation Shopping (durables) Personal business Participation Study	276	+37	2.65	+13	38.65	+12
6. Private leisure TV	234	+35	0.32	+14	6.35	+22
Total	1,439	−2	6.08	+1	97.18	+2

Trips per time-of-day

Period	Trip frequency	Percentage change from base
0.00– 7.00 h	0.40	−18
7.00– 9.30 h	0.82	−4
9.30–16.00 h	3.02	+51
16.00–18.30 h	0.79	+52
18.30–24.00 h	1.12	−25
Total	6.15	+3

Time spent at three locations for nonwork activities

	Time spent (min. per day)	Percentage change
At home	1,070	+20
In town	106	+19
Out of town	63	+7
Total	1,239	+19

Predicted daily trip matrix (changes from base situation in parentheses)

	Destination		
Origin	At home	In town	Out of town
At home	—	1.58 (+14%)	0.63 (+7%)
In town	1.55 (+12%)	0.86 (−15%)	0.03 (−25%)
Out of town	0.60 (+5%)	0.07 (+0%)	0.84 (−10%)

Note. From "Experiments with an Activity-Based Travel Model" by T. Van der Hoorn, 1983, *Transportation 12*, p. 72. Copyright 1983 by Elsevier. Reprinted by permission.

travel behavior requires us to situate it in broader social processes. It is certainly true that the resolute behavioral and individualistic approach of travel demand modeling—particularly choice theory—tends to obscure these deep ties. It remains very hard to incorporate travel behavior into social theory in concrete ways. Yet some of the newer approaches in activity modeling, time–geography, and structural urban theory lead this way. Impetus has come from several directions.

First, social theorists have begun to recognize their traditional neglect of spatial structure. Giddens, for example, has called for integration of space and time, distancing, presence, and absence as basic ingredients of social theory (Giddens, 1979, 1984; Soja, 1983). As Burnett and Hanson put it:

> The appropriate definition and measurement of travel (or other) behaviors are of concern to any scientist who wishes to advance credible claims of studying human groups in society over the next 25 years. (1982, p. 88)

Second, the logic of activity models leads us to recognize that interpersonal and space–time linkages lock travel behavior inseparably into the structure of society as a whole. The same conclusion is implicit in recent holistic, materialist approaches to urbanism (for example, Dear & Scott, 1981; Scott, 1980).

Perhaps most important is the development of the concrete descriptive and explanatory languages of paths and projects in time–geography, which clarify the connection between individual and institutionally determined projects (for example, Hägerstrand, 1970; Pred, 1981a, 1981b). *Space–time autonomy* is one idea from time–geography that is potentially useful in travel analysis. It refers to the degree of free choice that individuals have in space and time, as measured on *space–time prisms* (see chapter 7). Burns (1979) provides a book-length study of space–time autonomy in the context of accessibility. Among many other issues, he discusses problems in quantifying the benefits of autonomy in space and time. Janelle and Goodchild (1983) give a detailed empirical evaluation of ways of measuring autonomy, using travel diary data from Halifax, Nova Scotia. A common concern of these and many other recent studies is to understand travel behavior in its total social and economic context.

Recently, then, travel-demand modeling has changed direction. The earlier disaggregate models moved away from aggregation in any form, into finer and finer disaggregation of trips, modes, and people. Powerful mathematical models like the MNL were obtained, which required strong independence assumptions. Every trip choice was modeled into isolation. More recently activity-based models have moved the other way—toward a recognition of the complex interdependencies and constraints underlying travel. The price has been disorder and uncertainty about the appropriate form of models. The potential gain is an end of the isolation of travel theory from the body of geographic and socioeconomic theory on the city.

References

Adler, T., & Ben Akiva, M. (1979). A theoretical and empirical model of trip chaining behavior. *Transportation Research B, 13*, 243–257.

Anas, A. (1983). Discrete choice theory, information theory and the multinomial logit and gravity models. *Transportation Research B, 17*, 13–23.

Bannister, D. (1970). *Perspectives in personal construct theory*. London: Academic Press.

Batty, M. (1978). Reilly's challenge: New laws of retail gravitation which define systems of central places. *Environment and Planning A, 10*, 185–219.

Baxter, M., & Ewing, G. (1981). Models of recreational trip distribution. *Regional Studies, 15*, 327–344.

Brőg, W., & Meyburg, A. H. (1983). Influence of survey methods on the results of representative travel surveys. *Transportation Research A, 17*, 149–156.

Bunting, T., & Guelke, L. (1979). Behavioral and perception geography: A critical appraisal. *Annals of the Association of American Geographers, 69*, 448–462.

Burnett, K. P. (1981). Spatial transferability of

travel-demand models. In P. R. Stopher, A. H. Meyburg, & W. Brőg (Eds.), *New horizons in travel behavior research* (pp. 623–636). Lexington, Mass.: Lexington Books.

Burnett, K. P., & Ellerman, D. R. (1980). *New methodologies for modeling the travel behavior of diverse human groups in American cities.* Bloomington, Ind.: Center for Urban and Regional Analysis.

Burnett, K. P., & Hanson, S. (1982). The analysis of travel as an example of complex human behavior in spatially-constrained situations: Definition and measurement issues. *Transportation Research A, 2,* 87–102.

Burns, L. D. (1979). *Transportation, temporal and spatial components of accessibility.* Lexington, Mass.: Lexington Books.

Cadwallader, M. (1975). A behavioral model of consumer spatial decision making. *Economic Geography, 51,* 339–349.

Cadwallader, M. (1981). Towards a cognitive gravity model: The case of consumer spatial behavior. *Regional Studies, 15,* 275–284.

Carpenter, S., & Jones, P. M. (Eds.). (1983). *Recent advances in travel demand analysis.* Aldershot, Hants., England: Gower.

Charles River Associates. (1978). *On the development of a theory of traveler attitude–behavior interrelationships.* Washington, D.C.: U.S. Department of Transportation, Research and Special Programs Administration.

Clark, W.A.V., & Rushton, G. (1970). Models of intra-urban consumer behavior and their implications for central place theory. *Economic Geography, 46,* 486–497.

Cox, K. R., & Golledge, R. G. (Eds.). (1981). *Behavioral problems in geography revisited.* New York: Methuen.

Daganzo, C. (1979). *Multinomial probit: The theory and its applications to demand modeling.* New York: Academic Press.

Dear, M., & Scott, A. (Eds.). (1981). *Urbanization and urban planning in capitalist society.* London: Methuen.

Desbarats, J. (1983). Spatial choice and constraints on behavior. *Annals of the Association of American Geographers, 73,* 340–357.

Devletoglou, N. E. (1971). *Consumer behaviour.* London: Harper and Row.

Dix, M. C. (1981). Structuring our understanding of travel choices. The use of psychometric and social science research. In P. R. Stopher, A. H. Meyburg, & W. Brőg (Eds.) *New horizons in travel behavior research* (pp. 89–109). Lexington, Mass.: Lexington Books.

Domencich, T. A., & McFadden, D. (1975). *Urban travel demand: A behavioral analysis.* New York: North Holland.

Downs, R. M. (1970). The cognitive structure of an urban shopping center. *Environment and Behavior, 2,* 13–39.

Downs, R. M. (1976). Personal constructions of personal construct theory. In G. T. Moore & R. G. Golledge (Eds.), *Environmental Knowing* (pp. 72–87). Stroudsburg, Pa.: Dowden, Hutchinson and Ross.

Fishbein, M. (Ed.). (1967). *Readings in attitude theory and measurement.* New York: John Wiley.

Fishbein, M., and Ajzen, I. (1975). *Belief, attitude, intention and behavior.* Reading, Mass.: Addison-Wesley.

Fotheringham, A. S. (1983). A new set of spatial interaction models: The theory of competing destinations. *Environment and Planning A, 15,* 15–36.

Gaudry, M.J.I., & Dagenais, M. G. (1979). The dogit model. *Transportation Research B, 13,* 105–111.

Giddens, A. (1979). *Central problems of social theory.* London: Macmillan.

Giddens, A. (1984). *The constitution of society.* Berkeley and Los Angeles: University of California Press.

Girt, J. L. (1976). Some extensions of Rushton's spatial preference scaling model. *Geographical Analysis, 8,* 137–152.

Golledge, R. G., & Rushton, G. (Eds.). (1976). *Spatial choice and spatial behavior.* Columbus: Ohio State University Press.

Golob, T. F., & Beckmann, M. J. (1971). A utility model for travel forecasting. *Transportation Science, 5,* 79–90.

Golob, T. F., Gustafson, R. L., & Beckmann, M. J. (1973). An economic utility theory approach to spatial interaction. *Papers of the Regional Science Association, 30,* 159–182.

Hăgerstrand, T. (1970). What about people in regional science? *Papers of the Regional Science Association, 24,* 7–21.

Hanson, S. (1980). Spatial diversification and multipurpose travel: Implications for choice theory. *Geographical Analysis, 12,* 245–257.

Hanson, S., & Hanson, P. (1981). The travel-activity patterns of urban residents: Dimensions and relationships to sociodemographic characteristics. *Economic Geography, 57,* 332–347.

Heggie, I. G., & Jones, P. M. (1978). Defining domains for models of travel demand. *Transportation, 7,* 119–125.

Hensher, D. A., & Johnson, L. W. (1981) *Applied discrete choice modeling.* New York: Croom Helm.

Hensher, D. A., & Louviere, J. (1979). Behavioral intentions as predictors of very specific behavior. *Transportation, 8,* 167–182.

Hensher, D. A., & McLeod, P. B. (1977). Towards

an integrated approach to the identification and evaluation of the transport determinants of travel choices. *Transportation Research, 11*, 77–93.

Horowitz, J. (1980). Random utility models of urban nonwork travel demand. *Papers of the Regional Science Association, 45*, 125–137.

Horowitz, J. (1981). Identification and diagnosis of specification errors in the multinomial logit model. *Transportation Research B, 15*, 345–360.

Hubbard, R. (1978). A review of selected factors conditioning consumer travel behavior. *Journal of Consumer Research, 5*, 1–21.

Huff, D. L. (1962). *Determination of intra-urban retail trade areas*. Los Angeles: University of California Real Estate Research Program.

Isard, W. (1975). A simple rationale for gravity model type behavior. *Papers of the Regional Science Association, 35*, 25–30.

Janelle, D. G., & Goodchild, M. F. (1983). Transportation indicators of space–time autonomy. *Urban Geography, 4*, 317–337.

Jones, P. M. (1979). "HATS": A technique for investigating household decisions. *Environment and Planning A, 11*, 59–70.

Jones, P. M. (1981). Activity approaches to understanding travel behavior. In P. R. Stopher, A. H. Meyburg, & W. Brőg (Eds.), *New horizons in travel behavior research* (pp. 253–266). Lexington, Mass.: Lexington Books.

Jones, P. M., Dix, M. C., Clarke, M. I., & Heggie, I. G. (1983). *Oxford studies in transport: Understanding travel behavior*. Aldershot, Hants., England: Gower.

Kern, C. R., Lerman, S. R., Parcells, R. J., & Wolfe, R. A. (1984). *Impact of transportation policy on the spatial distribution of retail activity: Final report on DOTRC92024*. Washington, D.C.: U.S. Department of Transportation.

Lakshmanan, T. R., & Hansen, W. G. (1965). A retail market potential model. *Journal of the American Institute of Planners, 31*, 134–143.

Lerman, S. R., & Adler, T. J. (1976). Development of disaggregate trip distribution models. In P. R. Stopher & A. H. Meyburgh (Eds.), *Behavioral travel-demand models* (pp. 125–139). Lexington, Mass.: Lexington Books.

Levin, I. P. (1981). New applications of attitude measurement and attitudinal modeling techniques in transportation research. In P. R. Stopher, A. H. Meyburg, & W. Brőg (Eds.), *New horizons in travel behavior research* (pp. 171–188). Lexington, Mass.: Lexington Books. Books.

Levin, I. P., Louviere, J. J., Meyer, R. J., & Henley, D. (1979). *Perceived versus actual modal travel times and costs for the worktrip*. Iowa City: Institute of Urban and Regional Research, University of Iowa, Technical Report 120.

Lieber, S. R. (1976). A comparison of metric and nonmetric scaling models in preference research. In R. G. Golledge & G. Rushton (Eds.), *Spatial choice and spatial behavior* (pp. 191–208). Columbus: Ohio State University Press.

Lieber, S. R. (1977). Attitudes and revealed behavior: A case study. *Professional Geographer, 29*, 53–58.

Lieber, S. R., & Fesenmaier, D. R. (1984). Modeling recreational choice: A case study of management alternatives in Chicago. *Regional Studies, 18*, 31–43.

Louviere, J. J. (1976). Information processing theory and functional form in spatial behavior. In R. G. Golledge & G. Rushton (Eds.), *Spatial choice and spatial behavior* (pp. 211–246). Columbus: Ohio State University Press.

Louviere, J. J., & Henley, D. (1977). Information integration theory applied to student apartment selection decisions. *Geographical Analysis, 9*, 130–141.

Luce, R. D. (1959). *Individual choice behavior*. New York: John Wiley.

Luce, R. D., & Suppes, P. (1965). Preference, utility and subjective probability. In R. D. Luce, R. R. Bush, & E. Galanter (Eds.), *Handbook of mathematical psychology III* (pp. 249–410). New York: John Wiley.

McFadden, D. (1973). The measurement of urban travel demand. *Journal of Public Economics, 3*, 303–328.

MacLennan, D., & Williams, N. J. (1980). Revealed preference theory and spatial choices: Some limitations. *Environment and Planning A, 12*, 909–919.

Meyer, R. (1981). *Utility, uncertainty and spatial adaptation: A behavioral theory of destination choice set formation*. Iowa City: Institute of Urban and Regional Research, University of Iowa, Working Paper 39.

Nevin, J. R., & Houston, M. J. (1980). Image as a component of attraction to intraurban shopping areas. *Journal of Retailing, 56*, 77–93.

Niedercorn, J. H., & Bechdolt, B. V. (1969). An economic derivation of the "gravity law" of spatial interaction. *Journal of Regional Science, 9*, 273–282.

Okabe, A. (1977). Spatial aggregation bias in trip distribution probabilities: The case of the opportunity model. *Transportation Research, 11*, 197–202.

Openshaw, S. (1977). Optimal zoning systems for spatial interaction models. *Environment and Planning A, 9*, 169–84.

Patricios, N. N. (1978). Consumer images of spatial choice and planning of shopping centers.

South African Geographical Journal, 60, 103–120.

Pipkin, J. S. (1979). Respondent heterogeneity and alternative interpretation of scale values in destination choice models. *Professional Geographer, 31,* 16–24.

Pipkin, J. S. (1981a). The concept of choice and cognitive explanations of spatial behavior. *Economic Geography, 57,* 315–331.

Pipkin, J. S. (1981b). Cognitive behavioral geography and repetitive travel. In K. R. Cox & R. G. Golledge (Eds.), *Problems in behavioral geography revisited* (pp. 145–181). New York: Methuen.

Pirie, G. H. (1976). Thoughts on revealed preference and spatial behavior. *Environment and Planning A, 8,* 947–955.

Pred, A. (1981a). Social reproduction and the time geography of everyday life. *Geografiska Annaler B, 63,* 5–22.

Pred, A. (1981b). Everyday practice and the discipline of human geography. In A. Pred (Ed.), *Space and time in geography.* Lund, Sweden: Gleerup.

Reilly, W. J. (1931). *The law of retail gravitation.* New York: Pilsbury.

Rushton, G. (1969). Analysis of spatial behavior by revealed preference analysis. *Annals of the Association of American Geographers, 59,* 391–400.

Rushton, G. (1976). Decomposition of space preference functions. In R. G. Golledge & G. Rushton (Eds.), *Spatial choice and spatial behavior* (pp. 119–133). Columbus: Ohio State University Press.

Schuler, H. J. (1979). A disaggregate store-choice model of spatial decision-making. *Professional Geographer, 31,* 146–156.

Scott, A. J. (1980). *The urban land nexus and the state.* London: Pion.

Sheppard, E. S. (1980). The ideology of spatial choice. *Papers of the Regional Science Association, 45,* 197–213.

Smith, T. E. (1975). An axiomatic theory of spatial discounting. *Papers of the Regional Science Association, 35,* 31–44.

Smith, T. E. (1984). Testable characteristics of gravity models. *Geographical Analysis, 16,* 74–94.

Soja, E. W. (1983). Redoubling the helix: Space–time and the critical social theory of Anthony Giddens. *Environment and Planning A, 13,* 1267–1272.

Southworth, F. (1981). Calibration of multinomial logit models of mode and destination choice. *Transportation Research A, 15,* 315–325.

Spencer, A. H. (1980). Cognition and shopping choice: A multidimensional scaling approach. *Environment and Planning A, 12,* 1235–1251.

Stopher, P. R., & Meyburg, A. H. (1976). *Behavioral travel-demand models.* Lexington, Mass.: Lexington Books.

Stopher, P. R., Meyburg, A. H., & Brőg, W. (Eds.). (1981). *New horizons in travel behavior research.* Lexington, Mass.: Lexington Books.

Thrift, N. (1981). Behavioral geography. In N. Wrigley & R. J. Bennett (Eds.), *Quantitative geography: A British view* (pp. 352–365). London: Routledge and Kegan Paul.

Thurstone, L. L. (1927). A law of comparative judgment. *Pyschological Review, 34,* 273–286.

Timmermans, H. (1982). Consumer choice of shopping center: Information integration approach. *Regional Studies, 16,* 171–182.

Tversky, A. (1972). Choice by elimination. *Journal of Mathematical Psychology, 9,* 341–367.

Van der Hoorn, T. (1983). Experiments with an activity-based travel model. *Transportation, 12,* 61–77.

Warner, S. L. (1962). *Stochastic choice of mode in urban travel.* Evanston, Ill.: Northwestern University Press.

Watson, P. L. (1974). *The value of time: Behavioral models of modal choice.* Lexington, Mass.: Lexington Books.

Webber, M. J. (1980). A theoretical analysis of aggregation in spatial interaction models. *Geographical Analysis, 12,* 129–141.

Wilson, A. G., Coelho, J. D., McGill, S. M., & Williams, H.C.W.L. (1981). *Optimization in locational and transport analysis.* New York: John Wiley.

Yuill, R. S. (1967). Spatial behavior of retail consumers: Some empirical measurements. *Geografiska Annaler B, 49,* 105–115.

Zipf, G. K. (1949). *Human behavior and the principle of least effort.* Cambridge, Mass.: Addison-Wesley.

9 EXAMPLE
MODELING CHOICES OF RESIDENTIAL LOCATION AND MODE OF TRAVEL TO WORK

JOEL L. HOROWITZ
University of Iowa

INTRODUCTION

In this chapter, the application of the theoretical concepts discussed in chapter 8 will be illustrated by using them to develop an empirical model of an important class of choices: residential location and mode of travel to work (bus, automobile) by single-worker households in the Washington, D.C., area. The model that will be developed is a simplified version of one developed by Lerman (1976, 1979). The purpose of the model will be to explain and predict the outcomes of the choices under consideration in terms of independent (or explanatory) variables such as housing and transportation costs, indicators of the quality of service provided by automobiles and buses, indicators of neighborhood quality, and attributes of households (such as income). The model will be a mathematical formula that enables a household's residential location and mode of travel to work to be predicted if the values of the explanatory variables pertaining to that household are known.

There are several reasons for concentrating on choices of residential location and mode of travel to work in this chapter. First, these choices include both spatial and nonspatial aspects, and they illustrate the importance of including both types of considerations in explanations of behavior. For example, residential location is clearly a spatial choice, but it is influenced by nonspatial factors such as the income and size of the household making the choice. It is also influenced by spatial factors such as the quality of houses and schools in different neighborhoods. Second, the example of choice of residential location and work-trip mode illustrates the interdependence of choices that may seem separate. For instance, people who live far away from bus stops are unlikely to choose to go to work by bus, so the choice of residential location influences the choice of mode of travel to work. On the other hand, some people may deliberately choose to live near bus stops because they want to be able to travel to work by bus. In this case, the choice of mode of travel to work influences the choice of residential location. Thus, there is not a clear direction of causality from one choice to the other. Causality may run both ways simultaneously, thereby making the choices interdependent.

Finally, choices of residential location and mode of travel to work have important implications for public policy. For example, traffic congestion during peak hours is a major problem in many cities. One possible way of reducing congestion is to improve the quality of transit service for commuters. This possibility gives rise to a large number of questions related to choices of residential location and mode of travel to work. For example, if buses are made to run faster so that bus travel time is reduced by 10%, how many automobile drivers will switch to the bus? Will the improvement in bus service tend to benefit mainly high-income travelers, low-income ones, or both more or less equally? Will improved bus service tend to make more people want to live in neighborhoods that are near bus routes, thereby giving rise to the possibility that property values will rise in these neighborhoods and that new residential construction will take place in them? Later in this chapter, two examples will be presented that illustrate how policy questions such as these can be answered with the model that will be developed.

DEFINING THE PROBLEM TO BE SOLVED

It takes only a moment's thought to realize that the choices of residential location and mode of travel to work are enormously complex. In addition to depending on each other, these choices are likely to be influenced by related choices such as whether to own or rent one's dwelling, whether to live in a house or an apartment, how many cars to own, and what kinds of cars to own. For example, a decision to live in an apartment restricts one's choices of location to areas where there are apartment buildings, so the choice of whether to live in a house or an apartment influences the choice of residential location. Similarly, a worker in a household that chooses to own only one car may be forced to rely heavily on transit for travel to work (for example, so that the household's automobile will be available for use by other family members), thereby making it necessary to live near a bus line. Thus, the choice of the number of automobiles owned can influence both the choice of mode of travel and the choice of residential location. Of course, causality also can operate in the other direction. A two-car household that chooses to live near a bus line may find that it can dispose of one of its vehicles; in this case the choice of residential location influences the choice of the number of cars to own.

Another source of complexity is the large and detailed set of variables that are likely to influence the choices. For example, the choice of residential location is likely to be influenced not only by relatively major and easily measured factors such as housing costs, but also by difficult-to-quantify and, possibly, highly personal factors such as the appearance of a neighborhood, the style of architecture there, and whether one is likely to be able to make friends with people who already live there. Similarly, mode choice depends not only on easily measured variables such as the fare and the time required to travel from home to work by bus and automobile, but also on highly personal factors such as whether one plans to go directly from work to the theater tonight or whether another person in the household needs the family car today.

A further source of complexity is the definition of the term "residential location." It could sensibly refer to the exact house or apartment that a household chooses, but it also could refer equally sensibly to such things as the block, neighborhood, or political jurisdiction that is chosen.

It is simply not possible to treat all these complexities in a single model. Even if the processes by which people make choices of residential locations and travel modes were understood well enough to make such a treatment possible in principle, it would be impossible in practice. A model that attempted to incorporate all the details and complexities of choice would be too large and cumbersome to use, and the cost of fulfilling its data requirements would make it inaccessible to even the best-financed users. Therefore, the first step in building a model of choice of residential location and mode of travel to work (and, in fact, the first step in building a model of any highly complex process) is to define what is to be modeled in a way that is simple enough to be tractable while retaining enough of the complexity of the full problem to make the model useful. In other words, it is necessary to reach a compromise between tractability and detail or realism.

One compromise that will be made in the model discussed here concerns the spatial scale of the model or, equivalently, the definition of "residential location." For the purposes of the model, a household's residential location will be defined as the census tract in which the household lives. A census tract is a geographical unit that has an area of about 0.25 square mile and that may contain several hundred to over 1,000 dwelling units. A model that operates at the census-tract level cannot resolve the details of individual houses or apartments. However, census tracts provide adequate spatial detail for dealing with policy questions such as how transit improvements will affect choices of residential location. Moreover, it is not necessary to model the highly complex process through which houses are bought and sold when the lowest level of spatial resolution is a census tract.

Another compromise that will be made here is to model choices of residential location and mode of travel only by households that have chosen to own one car and to live in owner-occupied, single-family dwellings. This simplifies the model by eliminating the need to treat interactions between residential location and mode choices on the one hand and choices of automobile ownership, tenure (own or rent), and dwelling type (for example, single-family or apartment) on the other. Of course, this means that the model will have nothing to say about the behavior of households who choose to own no automobiles or several automobiles or to live in rental housing or apartments. It also means that the model will correctly predict the effects of policy measures only if the measures have little effect on automobile ownership levels and choices of tenure and dwelling type. Despite these limitations, the model will provide useful insights into behavior and the possible consequences implementing a variety of public policy measures. (The model developed by Lerman, of which the model presented in this chapter is a simplified version, treats choices of tenure, housing type, and automobile ownership in addition to residential location and mode of travel to work. Readers interested in seeing how this broader class of choices can be included in a model are referred to Lerman, 1976, 1979.)

The travel modes that will be represented in the model are single-occupant automobile and bus. The model will not be able to treat travel to work by such modes as bicycle, taxi, and walking, but only a small proportion of work trips in the Washington area use these modes, so that their omission is not serious. The omission of the carpool mode is more serious, since it is widely used. The carpool mode is not essential to the purposes of this chapter, however, and its omission helps to minimize the complexity of the model.

A final compromise that must be made to develop the model concerns the set of explanatory variables that will be included. It is not possible to include all variables relevant to choices of residential location and mode of travel to work because many are difficult or impossible to measure and some may not even be known to modelers (for example, some variables may influence households' choices subconsciously). Even if these problems were not present, any attempt to include all variables relevant to choice in a single model would result in a model that would be far too large and cumbersome to use. Therefore, it is possible to include in a practical model only a very small subset of the variables that are known or likely to influence the choices being modeled. In deciding which variables to include in a model, it is necessary to take account of the behavioral and mathematical structure of the model, the intended uses of the model, and the data that are available for developing and applying the model. Accordingly, further discussion of the selection of variables for inclusion in the model of residential location and mode choice will be postponed until these other matters have been discussed.

BEHAVIORAL AND MATHEMATICAL FOUNDATIONS

Choice is an expression of preferences. Given a set of available alternative residential locations and travel modes, a household will choose the location and mode that it prefers. Preferences depend on the attributes of the available alternatives and of the household that is making the choice. For example, other things being equal, a household with several children is likely to prefer a neighborhood with large houses to a neighborhood with small ones, whereas a household with no children may prefer a neighborhood with small houses (such as a townhouse development). In this case preferences depend on house sizes (an attribute of the alternative locations among which the household chooses) and on the number of children in the household making the choice (an attribute of the households).

Suppose that there are J different combinations of residential locations and modes of travel to work available to a household. For example, combination 1 may consist of living in census tract number 1 and traveling to work by automobile, combination 2 may consist of living in census tract number 1 and traveling to work by bus, and so forth. Each of these combinations is called an "alternative." Let x_{ih} denote the attributes of alternative i and household h that are relevant to choices among the alternatives. For example, x_{11} might refer to the average size of the houses in census tract 1, the time required to travel by automobile from census tract 1 to the workplace of the employed member of household 1, and the number of children in household 1. Similarly, x_{13} might refer to the average size of the houses in census tract 1, the time required to travel by automobile from census tract 1 to the workplace of the employed member of household 3, and the number of children in household 3. The fundamental behavioral assumption of the model that will be developed in this chapter is that there is a mathematical function U, called a *utility function*, such that household h prefers alternative i to alternative j if and only if $U(x_{ih})$ (the value of the utility function corresponding to the attributes of alternative i and household h) exceeds $U(x_{jh})$ (the value of the utility function corresponding to the attributes of alternative j and household h). In other words, alternative i is preferred to alternative j by household h if and only if

$$U(x_{ih}) > U(x_{jh}) \qquad (1)$$

Since the household chooses the alternative it most prefers among those that are available, it follows that alternative i is chosen if $U(x_{ih})$ exceeds the value of U for every other alternative. In other words, alternative i is chosen by household h if

$$U(x_{ih}) > U(x_{jh}) \text{ for all alternatives } j \neq i \qquad (2)$$

Inequality 2 implies that the choice of household h could be predicted if the utility function U and the values of the attributes x were known. This ability to predict choice is illustrated by the following example.

Example 1: Suppose for simplicity that the only choice that must be made is the mode of travel to work. Let the relevant attributes of the alternatives (single-occupant automobile and bus) be travel time and cost, and let the relevant attribute of households be annual income. Let the utility function be

$$U(T,C,Y) = -T - 5C/Y \qquad (3)$$

where T is door-to-door travel time in hours, C is door-to-door travel cost in dollars, and Y is household income in thousands of dollars per year. Suppose the travel times and costs of the available alternatives are:

Alternative	Travel Time (T)	Travel Cost (C)
Automobile	0.50 hr.	$2.00
Bus	1.00	0.75

Then for a household whose income is \$40,000 per year, the values of the utility function corresponding to the two alternatives are $U(\text{automobile}) = -0.50 - 5(2.00)/40 = -0.75$ and $U(\text{bus}) = -1 - 5(0.75)/40 = -1.09$. Since $U(\text{automobile}) > U(\text{bus})$, this household is predicted to choose automobile as the mode of travel to work. In contrast, the utility values for a household whose income is \$10,000 per year are $U(\text{automobile}) = -0.50 - 5(2.00)/10 = -1.5$ and $U(\text{bus}) = -1 - 5(0.75)/10 = -1.38$. Therefore, this household chooses bus.

Inequality 2 and Example 1 show how choice can be predicted if the utility function and all of the relevant attributes, or variables, are known. In fact, equation 3 constitutes a simple model of mode choice. As was discussed in the previous section, however, it is not possible in practice to include all relevant attributes in a model. The most important consequence of omitting variables relevant to choice from a model is that it is no longer possible to predict the outcomes of choices with certainty. (To see why this is so, suppose that it were possible to predict mode choice with certainty using a model that did not include the variable "reliability of transit service." Then, transit reliability could have no influence on mode choice and, therefore, would not be relevant to choice. Otherwise, mode choice would vary according to the level of the omitted reliability variable and could not be predicted with certainty.) Rather, it is necessary to predict choices probabilistically. A probabilistic prediction of choice is a statement of the probabilities that each of the available alternatives will be chosen. A model that relates these probabilities to the values of a set of explanatory variables is called a probabilistic choice model.

In order to make an operational probabilistic choice model, it is necessary to specify the form of the relation between the probability that an alternative is chosen, and the explanatory variables included in the model. A specification that is very convenient in practice is the multinomial logit specification. In this specification, the probability P_{ih} that household h chooses the ith of the available alternatives is given by

$$P_{ih} = \frac{\exp[V(z_{ih})]}{\sum_{j=1}^{J} \exp[V(z_{jh})]} \tag{4}$$

In this equation $z_{jh}(j = 1 \ldots, J)$ denotes the attributes of alternative j and household h that are both relevant to the choice being considered and are included in the model (such as travel time, travel cost, and annual income in the case of a model of work trip mode choice), V is the utility function U averaged over the omitted attributes (such as reliability and comfort), and the sum extends over all alternatives available to the household. Exp denotes the exponential operator—given any number a, exp $(a) = \epsilon^a$, where $\epsilon = 2.71828 \ldots$ is the base of natural logarithms. (The multinomial logit specification can be derived mathematically by treating the omitted attributes of alternatives and households—that is, the elements of the x_{jh}'s that are not elements of the z_{jh}'s—as random variables with an appropriate probability distribution. Since this derivation requires the use of advanced mathematics, it will not be given here. The derivation is presented in Domencich & McFadden, 1975.) The use of the multinomial logit specification in choice modeling is illustrated by the following example.

Example 2: Assume as in Example 1 that the only choice that must be made is travel mode and that the available modes are single-occupant automobile and bus. Let the function V be

$$V(T,C,Y) = -T - 5C/Y \tag{5}$$

where T, C, and Y are as defined in Example 1. Then according to the multinomial logit specification, the probability that automobile is chosen by a household is

$$P_{\text{auto}} = \frac{[\exp(-T_a - 5C_a/Y)]}{[\exp(-T_a - 5C_a/Y) + \exp(-T_b - 5C_b/Y)]} \tag{6}$$

where the subscripts a and b, respectively, denote variables that are evaluated for automobile and bus. The probability that bus is chosen is

$$P_{bus} = \frac{[\exp(-T_b - 5C_b/Y)]}{[\exp(-T_a - 5C_a/Y) + \exp(-T_b - 5C_b/Y)]} \quad (7)$$

Notice that $P_{auto} + P_{bus} = 1$, reflecting the fact that since the only alternatives are automobile and bus, one of these must be chosen. Notice, also, that the probabilistic choice model of equations 6 and 7 differs from the deterministic choice model used in Example 1, although the variables of both models are the same. This is because it was assumed in Example 1 that T, C, and Y are the only variables relevant to choice, whereas it is now assumed that additional variables not included in the model also are relevant.

Suppose the values of T and C are as in Example 1. Then for a household whose income is \$40,000 per year ($Y = 40$), the values of the choice probabilities are $P_{auto} = \exp[-0.5 - 5(2.00)/40] / \{\exp[-0.5 - 5(2.00)/40] + \exp[-1 - 5(0.75)/40]\} = 0.59$ and $P_{bus} = \exp[-1 - 5(0.75)/40] / \{\exp[-0.5 - 5(2.00)/40] + \exp[-1 - 5(0.75)/40]\} = 0.41$. For a household whose income is \$10,000 per year, the choice probabilities are $P_{auto} = \exp[-0.5 - 5(2.00)/10] / \{\exp[-0.5 - 5(2.00)/10] + \exp[-1 - 5(0.75)/10]\} = 0.47$ and $P_{bus} = \exp[-1 - 5(0.75)/10] \{\exp[-0.5 - 5(2.00)/10] + \exp[-1 - 5(0.75)/10]\} = 0.53$. These probabilities can be interpreted as meaning that on the average, 59% of households with incomes of \$40,000 per year and whose travel times and costs are as in Example 1 will choose automobile, whereas 41% will choose bus. Similarly, on the average 47% of households that earn \$10,000 per year and whose travel times are as in Example 1 will choose automobile, whereas 53% will choose bus. Households with the same income, travel times, and travel costs do not all make the same choices because of the influence of the variables relevant to choice that are not included in the model. In other words, the omitted variables cause some households to favor automobile and others to favor bus, even when the values of the included variables are the same for all households.

In the multinomial logit specification, a change in a variable that causes an increase in the value of V for a particular alternative also causes an increase in the probability that that alternative is chosen. The probabilities that other alternatives are chosen decrease. A change in a variable that causes a decrease in the value of V for a particular alternative has the opposite effect. Thus, for example, in the case of choice between automobile and bus travel with V given by equation 5, a decrease in bus travel time, which increases the attractiveness of the bus relative to the automobile, increases the value of V for bus travel. Accordingly, the probability that the bus is chosen increases, and the probability that the automobile is chosen decreases. If bus travel time increases, then the probability that the bus is chosen decreases and the probability that the automobile is chosen increases. This dependence of the bus and automobile choice probabilities on transit travel time is illustrated in Figure 9-1 for the case in which T_a, C_a, and C_b have the values used in Example 1, Y is \$40,000 per year, and P_{auto} and P_{bus} are given by equations 6 and 7.

Choice models based on the multinomial logit specification are called multinomial logit models or, simply, logit models. Example 2 illustrates how a logit model can be used to make probabilistic predictions of choices once the V function is known. In practice, however, the function V is not known *a priori*. It must be inferred from observations of people's choices. The most frequently used method of inference is called the *maximum likelihood method*. Although the details of this method involve the use of advanced statistical techniques, its basic concepts are easy to understand. (See Domencich & McFadden, 1975, for a detailed discussion of the use of the maximum likelihood method for the development of multinomial logit models.) Essentially, the method consists of choosing the V function so as to maximize the likelihood

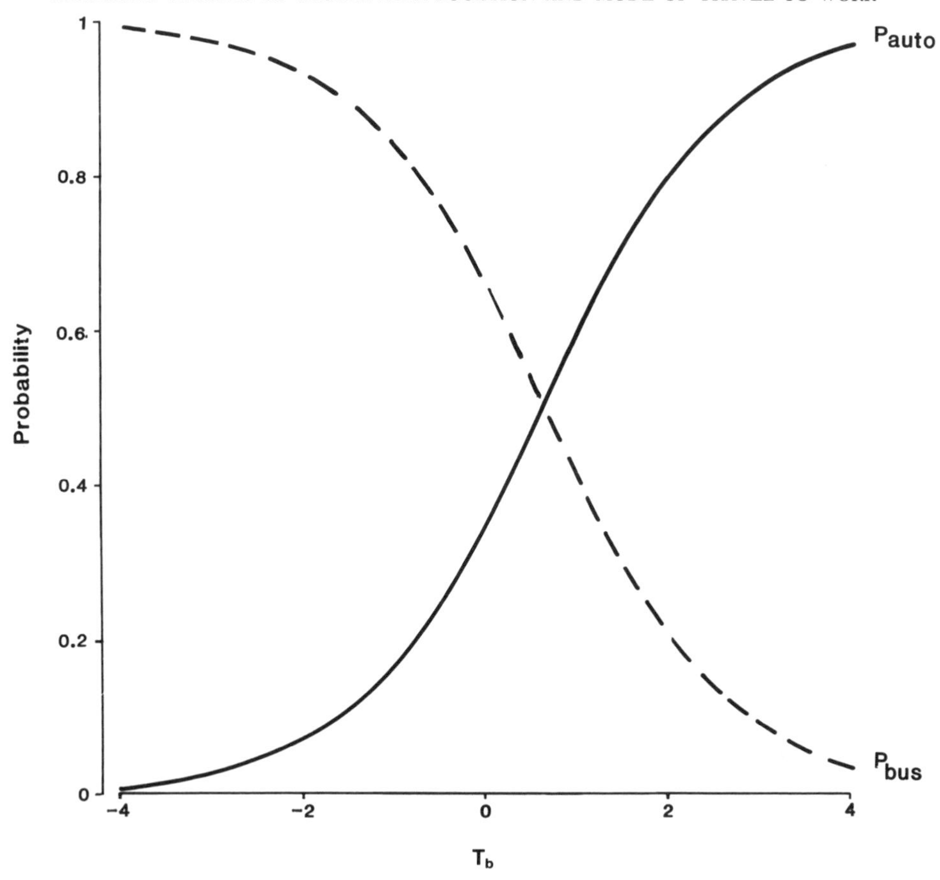

Figure 9-1. An example of the dependence of P_{auto} and P_{bus} on bus travel time, T_b in the mode choice model of equations 6 and 7.

(or probability) according to the model being developed of observing the choices that actually have been observed.

More specifically, suppose it is desired to develop a multinomial logit model of commuters' choices between automobile and bus. Suppose, also, that the variables to be included in the model are the time required to travel from home to work by automobile and bus, the costs of travel by these modes, and the annual income of the traveler's household. Then it is necessary to infer the dependence of V on the variables T, C, and Y, where T, C, and Y, are defined as in Examples 1 and 2. To carry out this inference by the method of maximum likelihood, it is necessary to have the following data:

1. Observations of the mode choices (automobile or bus) actually made by a random sample of commuters. This sample is called the estimation sample. Typically, the estimation sample should contain observations on at least several hundred and, possibly, over 1,000 individuals.

2. The values of T_a, T_b, C_a, C_b, and Y corresponding to each individual in the sample. As in Example 2, the subscripts a and b denote variables pertaining to automobile and bus, respectively. Note that it is necessary to have data on the T and C values for both the mode chosen by an individual and the mode not chosen. This is because in the multinomial logit model, the probability that a particular alternative is chosen depends on the values of the variables associated with all alternatives, not just on the values of the variables associated with the chosen alternative.

It is also necessary to specify tentatively the form of the functional dependence of V on the variables of the model. For example, V might be specified to have the form

$$V(T,C,Y)=a_1T+a_2C/Y \tag{8}$$

where a_1 and a_2 are constant coefficients whose numerical values remain to be determined. Since the tentative specification of the form of V usually must be based on past experience or considerations of mathematical and computational tractability, rather than on specific knowledge of the behavior of the individuals in the estimation sample, there can be no assurance that the selected specification is correct. In fact, even after a model is completely developed, there can be no assurance that it is correct. However, a variety of statistical test procedures are available that frequently make it possible to detect serious errors in the specification of V and thereby help one to avoid the unwitting development of highly erroneous models.

The tentative specification (equation 8) does not include the numerical values of the coefficients a_1 and a_2. Rather, the maximum likelihood method determines the values that are best in the sense of causing the model to assign the highest possible probability to the choices observed in the estimation sample. (The maximum likelihood method is analogous in some respects to the ordinary least squares method frequently used to estimate the coefficients of linear regression models. In ordinary least squares, the coefficients are assigned the values that minimize the regression model's mean square prediction error. Ordinary least squares is not applicable to the models discussed in this chapter.)

The process of determining the numerical values of the coefficients is illustrated by the following example.

Example 3: Suppose that in a logit model of choice between automobile and bus the function V is specified as

$$V(T)=aT \tag{9}$$

where T is travel time and a is a constant whose value must be estimated using the method of maximum likelihood. Suppose, in addition, that the estimation sample consists of observations of the choices of three travelers. (A sample of size three is far too small to be useful for developing a real model, but it is convenient for illustrative purposes because it permits all the necessary computations to be performed with a desk calculator. Parameter estimation with a sample of realistic size requires the use of a digital computer.) Let the choices and T values of the individuals in the estimation sample be

Individual	Choice	T_a	T_b
1	Auto	0.50	1.00
2	Bus	0.40	0.65
3	Bus	0.40	0.60

T is in units of hours. The probabilities of the observed choices according to the model being developed are

Individual 1:
$$P(\text{Auto})=\frac{[\exp(0.50a)]}{[\exp(0.50a)+\exp(1.00a)]}$$

Individual 2:
$$P(\text{Bus}) = \frac{[\exp(0.65a)]}{[\exp(0.40a)+\exp(0.65a)}$$

Individual 3:
$$P(\text{Bus}) = \frac{[\exp(0.60a)]}{[\exp(0.40a)+\exp(0.60a)]}$$

Equivalently, the probabilities are

$$\text{Individual 1: } P(\text{Auto})=\frac{1}{[1+\exp(0.50a)]}$$

$$\text{Individual 2: } P(\text{Bus}) =\frac{1}{[1+\exp(-0.25a)]}$$

$$\text{Individual 3: } P(\text{Bus}) =\frac{1}{[1+\exp(-0.20a)]}$$

The probability of observing the entire set of observations in the estimation sample is the product of the probabilities of the individual observations. Hence

$$P(\text{sample})= \frac{1}{[1+\exp(0.50a)][1+\exp(-0.25a)][1+\exp(-0.20a)]} \tag{10}$$

The maximum likelihood method chooses a so as to maximize P(sample). Although this usually requires the use of a digital computer, it can be done graphically in this case. Figure 9-2 shows the graph of P(sample) in equation 9 as a function of a. It can be seen that the maximum occurs at $a = -0.28$. Thus, the maximum likelihood estimate of a is -0.28.

Probabilistic choice models generally and logit models in particular make it possible to develop useful choice models that do not include all the variables that influence the choice being modeled. This does not imply, however, that a model based on any subset of the influential variables will be useful. On the contrary, there are certain types of variables that must be included to obtain a useful model. These classes are:

1. Policy Variables: One of the most important uses of choice models is predicting the effects of policy measures. For example, a transportation planner may want to predict what changes will occur in bus ridership or, possibly, choices of residential location if bus fares change or if bus travel becomes faster. Such questions can be answered only if the model includes explanatory variables, called policy variables, that represent the policy measures being considered. Thus, in the example just given, policy variables such as bus fare and bus travel time must be included in the model to make it useful for predicting the effects of policy measures that would change fares or travel times. In the models discussed in Examples 1 and 2, T (travel time) and C (travel cost) are the policy variables. In the model of Example 3, T is the policy variable. There is no variable that reflects travel costs in the model of Example 3, so that model is not useful for predicting the effects of changes in bus fare.

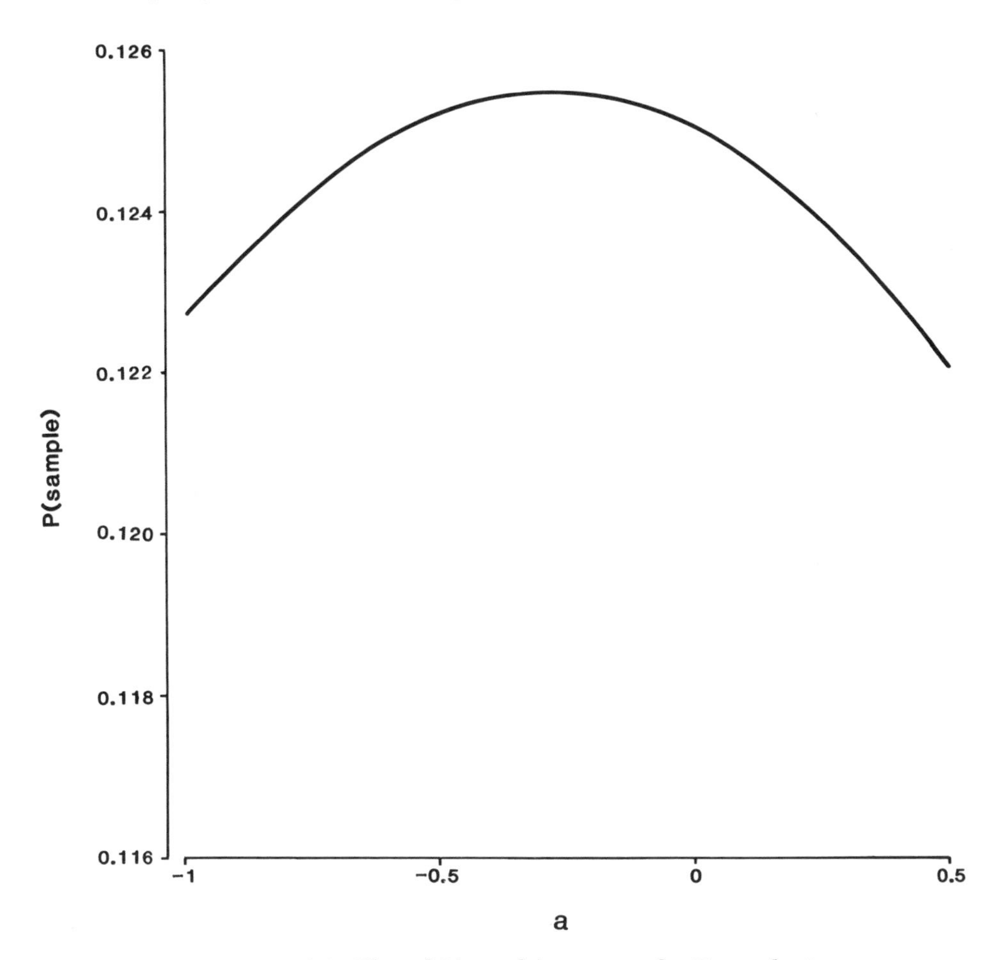

Figure 9-2. Plot of P(sample) versus a for Example 3.

2. *Variables That Identify Population Groups:* Often, it is important to be able to predict the effects of policy measures on different groups in the population. For example, it may be important to know whether increasing bus fares will be particularly burdensome to low-income travelers or whether a certain improvement in transit service will succeed in attracting members of multicar households to transit. A model can answer questions such as these only if it includes variables that permit the effects of policy measures on different population groups of interest to be differentiated. In the example just given, the variables income and automobiles owned would serve this purpose. The models discussed in Examples 1 and 2 include the variable Y (income) and therefore are capable of differentiating among the effects of changes in travel time (T) and travel cost (C) on different income groups. The model discussed in Example 3 does not include Y and therefore cannot differentiate among income groups.

3. *Variables That Influence Choice and Are Correlated with the Policy or Grouping Variables:* Frequently, it is possible to identify variables that are themselves not of interest but that affect choice and are correlated with one or more of the policy or grouping variables in a model. Such variables also must be included in the model, or else the model will give incorrect predictions of the effects of the policy and grouping variables. For example, suppose that travel time (a policy variable) and income (a grouping variable) are included in a model of mode choice. Suppose also that the number of automobiles owned by a traveler's household is not of interest in a particular study but that multicar households tend to have higher incomes and to live farther from their workplaces than do single-car and non-car-owing households. Since automobile ownership undoubtedly has an independent effect on mode choice, apart from its association with travel time and income, a model that does not include automobile ownership as a variable will give incorrect estimates of the effects of travel time and income on mode choice. Such a model's estimates of the effects of travel time will reflect not only the true effects of travel time but also the effects of changes in automobile ownership that are associated with changes in travel time through the tendency of households with large travel times to own many cars. In other words, the travel time variable will operate, in part, as a surrogate for automobile ownership and therefore will not correctly describe the true effects of changes in travel time alone. Similarly, the model's estimates of the effects of income on mode choice will reflect both the true effects of income and the effects of changes in automobile ownership that are associated with changes in income through the tendency of high-income households to own many cars.

In addition to the foregoing considerations, the selection of variables for inclusion in practical models is virtually always constrained by the availability of data. Data acquisition is usually a difficult and expensive process, so that there are strong incentives in practice to rely on existing data sets for use in model development. Reliance on existing data sets limits the variables that can be included in models to ones for which suitable data already are available. As a result, it is not necessarily possible in practice to include in a model all the variables that consideration of the foregoing three classes might suggest are desirable. Nonetheless, knowledge of the classes provides valuable guidance for deciding which variables to include in a model.

DATA, VARIABLES, AND SPECIFICATION OF THE MODEL

The model presented in this chapter is based on data obtained from three sources:

1. The Washington, D.C., area transportation survey.
2. The 1970 U.S. census.
3. Estimates of tract-to-tract travel times and costs according to mode of travel (automobile or bus).

The Washington, D.C., area transportation survey was carried out in 1968 and was designed to provide data on the travel behavior and socioeconomic characteristics of some 30,000 surveyed households. Data available for each household include the census tracts in which home and the place of work are located, the type of structure in which the home is located (for example, single-family detached, high-rise apartment, and so on), whether the household owns or rents its home, the number of automobiles owned, the household's size and income, and the mode of travel used by each worker in the household to travel to work on the day of the survey. The census data provide information on neighborhood attributes of census tracts (such as average costs of houses, racial composition, and residential densities according to tract).

The variables of the model are shown in Table 9-1. (As discussed in the introduction to this chapter, the model that is presented here is based on that of Lerman, 1976, 1979. The variables listed in Table 9-1 are the variables of Lerman's model that are relevant

Table 9-1. Variables of the model

Variable	Type	Definition
DCAR	Transportation	Dummy variable equal to 1 for alternatives in which a car is driven to work and 0 otherwise.
OUT/DIST	Transportation	Two-way out-of-vehicle travel time (in minutes) divided by two-way travel distance (in miles).
TOTIME	Transportation	Total two-way travel time (in minutes).
DCITY	Transportation	Dummy variable equal to 1 for alternatives in which a car is driven if the workplace is downtown. Equal to 0 otherwise.
ln(Z)	Household	Natural logarithm of remaining income.
AALD	Household	Number of automobiles owned (always 1 in this chapter) divided by the number of licensed drivers in the household for alternatives in which a car is driven to work. Equals 0 for other alternatives.
GPTINV	Locational	Transit accessibility of tract to shopping locations.
R1	Locational	Automobile accessibility of tract to shopping locations.
FBFORW	Locational	Fraction of nonwhite households in tract for choices by whites; 0 for nonwhites.
FBFORB	Locational	Fraction of nonwhite households in tract for choices by nonwhites; 0 for whites.
DENSITY	Locational	Residential density of tract (in households per acre).
SCHOOL	Locational	Per pupil school expenditure for households with children and tracts in jurisdictions outside of D.C.; 0 for households without children and for tracts in D.C.
INCDIFF	Locational	Square of difference between choosing household's income and average income of households in tract if this difference exceeds 0; 0 otherwise.
DOC	Locational	Dummy variable equal to 1 for tracts in D.C. and 0 for other tracts.
ln(NL)	Locational	Natural logarithm of the number of dwelling units in tract.

to the choices being treated in this chapter.) The variables belong to three broad categories:

1. *Transportation Variables:* These variables indicate the time and cost involved in traveling from home to work by automobile and transit. There is a separate variable for out-of-vehicle travel time (that is, time spent walking to and from parking lots or bus stops and time spent waiting for buses) because, as explained in chapter 8, experience has indicated that travelers find out-of-vehicle travel time more onerous than in-vehicle travel time. Out-of-vehicle travel time is divided by total trip length, since travelers may find out-of-vehicle travel time less onerous on long trips than on short ones. The variable *DCITY* accounts for the possibility that commuters may perceive automobile travel to the downtown area to be more difficult than automobile travel to other locations owing to the effects of congestion in center city. The variable *DCAR* accounts for the average effects of all unobserved attributes of automobile and transit travel that affect mode choice. Travel cost does not appear as an explicit variable of the model. Rather, it is subsumed in the variable "remaining income," which is included in the next class of variables.

2. *Household Variables:* These measure attributes of households that affect choices of residential location or mode of travel to work. Remaining income is the money (in 1968 dollars) a household has left from its income after paying taxes, housing costs, and the costs of owning a car and commuting to work. Other things being equal, alternatives that cause remaining income to be high are preferred over alternatives that cause remaining income to be low. Remaining income subsumes the variable "travel cost," because an increase in travel cost causes a corresponding decrease in remaining income. The variable "automobiles per licensed driver" *(AALD)* enables the model to represent the effects of competition among household members for the use of the household's automobile. In particular, increases in the number of licensed drivers in a household, which cause decreases in the value of *AALD*, make it less likely that the household's automobile will be available for commuting purposes.

3. *Locational Variables:* These measure attributes of census tracts that are relevant to choice of residential location. *GPTINV* and *R*1 measure the accessibility of tracts to shopping locations. These variables are defined such that larger values of *GPTINV* and smaller values of *R*1 are associated with better accessibility. *FBFORW* and *FBFORB* measure the racial composition of neighborhoods. These variables capture the tendency of neighborhoods in the Washington area to be racially homogeneous. *DENSITY* measures the effects of residential density on choice of residential location. *SCHOOL* provides an indication of school quality. *INCDIFF* captures the tendency of households to prefer not to live in neighborhoods where the average income level is substantially below their own. *DOC* represents the mean value of unobserved attributes of census tracts that affect the choice as to whether to live in the District of Columbia (the city of Washington) or the suburbs. *NL* accounts for the fact that other things being equal, a household is more likely to live in a census tract where there is a large number of dwelling units than in a tract where there is a small number of units.

The policy variables of the model are the components of travel time and travel cost. These are the only variables among those listed in Table 9-1 that can easily be influenced by public policy makers. (Recall that the travel cost variable is subsumed in remaining income and, therefore, is not an explicit variable of the model.) Many of the other variables are needed in the model because they are likely to be correlated with the policy variables. For example, census tracts with high densities, low incomes, and high proportions of black residents tend to be located near the center of the Washington area and therefore to have shorter travel times to work than other census tracts. In addition, several of the other variables make it possible to study the effects of policy changes on different population groups. For example, the variables of the model permit the effects of policy changes to be differentiated according to such criteria as income, the number of licensed drivers in a household, race, and whether the residence is located in the center city or the suburbs.

The model has the multinomial logit functional form. In other words, the probability that alternative i is chosen is given by equation 4, where the alternatives consist of combinations of the census tract of residence and mode of travel to work. The function V is specified as

$$V = a_1 DCAR + a_2(OVTT/DIST) + a_3 TOTIME + a_4 DCITY + a_5 \ln(Z) + a_6 AALD + a_7 GPTINV + a_8 R1 + a_9 FBFORW + a_{10} FBFORB + a_{11} DENSITY + a_{12} SCHOOL + a_{13} INCDIFF + a_{14} DOC + a_{15} \ln(NL) \quad (11)$$

Thus, V is evaluated by multiplying each variable (*DCAR*, *OVTT/DIST*, and so on) by its coefficient (a_1, a_2, and so on) and then summing the resulting products.

ESTIMATION RESULTS

Table 9-2 shows the estimated values of the coefficients of the model. The coefficients were estimated by the maximum likelihood method using the data described in the previous section.

Table 9-2. Estimated values of the model's coefficients

Coefficient	Corresponding variable	Estimated value of coefficient
a_1	*DCAR*	0.433
a_2	*OVTT/DIST*	−0.0570
a_3	*TOTIME*	−0.00831
a_4	*DCITY*	−0.437
a_5	ln(Z)	1.07
a_6	*AALD*	0.964
a_7	*GPTINV*	2.92
a_8	*R*1	−1.35
a_9	*FBFORW*	−2.18
a_{10}	*FBFORB*	1.95
a_{11}	*DENSITY*	−0.00557
a_{12}	*SCHOOL*	0.000442
a_{13}	*INCDIFF*	−0.0123
a_{14}	*DOC*	−0.00993
a_{15}	ln(*NL*)	0.492

Note. Adapted from Lerman, 1976, 1979.

The signs of several of the estimated coefficients are worthy of attention. The negative signs of the coefficients of *TOTIME*, *OVIT/DIST*, and *DCITY* indicate that other things being equal, travel alternatives with high travel times and that involve driving to the center city tend to be less preferred than alternatives that have low travel times and do not involve driving to the city. The positive coefficient of ln(Z) implies that residential location and travel alternatives that permit a large remaining income are preferred, other things being equal, to alternatives that cause remaining income to be small. In other words, households prefer less costly housing and transportation, other things being equal. The positive coefficient of *AALD* implies that a car is more likely to be driven to work by members of households where there are few licensed drivers than by members of households where there are many licensed drivers, other things being equal. The negative coefficient of *FBFORW* and the positive coefficient of *FBFORB* indicate that white households tend to live in predominantly white neighborhoods and nonwhite households tend to live in predominantly nonwhite neighborhoods. The positive coefficient of ln(*NL*) indicates that households are more likely to live in census tracts where there are many dwelling units than in tracts where there are few. All of these findings are consistent with intuition and expectation.

The coefficient estimates provide insight into the relative importance that households place on different attributes of housing and travel alternatives. Consider, for example, the relative values assigned to in-vehicle and out-of-vehicle travel time. Let *IVTT* denote in-vehicle travel time, and note that total travel time is the sum of in-vehicle and out-of-vehicle travel time:

$$TOTIME = IVTT + OVTT \quad (12)$$

Suppose the values of *IVTT* and *OVTT* change by the amounts $\Delta(IVTT)$ and $\Delta(OVTT)$, respectively. Then the corresponding change in *TOTIME* is

$$\Delta(TOTIME) = \Delta(IVTT) + \Delta(OVTT) \quad (13)$$

If these travel time variables are the only variables of the model whose values change, the corresponding change in utility is

$$\Delta U = a_2\Delta(OVTT)/DIST + a_3[\Delta(IVTT) + \Delta(OVTT)] \quad (14)$$

Thus, the change in utility is the change in *OVTT/DIST* multiplied by a_2, the coefficient of *OVTT/DIST*, plus the change in *TOTIME* multiplied by a_3, the coefficient of *TOTIME*. Equivalently,

$$\Delta U = (a_2/DIST + a_3)\Delta(OVTT) + a_3\Delta(IVTT) \quad (15)$$

The value of $\Delta(IVTT)$ that exactly counterbalances $\Delta(OVTT)$, so that ΔU is zero, is obtained by setting $\Delta U = 0$ in equation 15 and solving for $\Delta(IVTT)$. The result is

$$\Delta(IVTT) = -[a_2/(a_3DIST) + 1]\Delta(OVTT) \quad (16)$$

The change in *IVTT* that exactly counterbalances a 1-minute change in *OVTT* is obtained by setting $\Delta(OVTT) = 1$ in equation 16. The result is

$$\Delta(IVTT) = -a_2/(a_3DIST) - 1 \text{ minutes} \quad (17)$$

Since the estimated value of a_2/a_3 is 6.86, equation 17 implies that

$$\Delta(IVTT) = -6.86/DIST - 1 \text{ minutes} \quad (18)$$

Consider a trip whose length (round trip) is 15 miles, which is the approximate average length of work trips in the Washington area. For such a trip, equation 18 implies that $\Delta(IVTT)$ is -1.46 minutes. Therefore, a 1.46-minute decrease in in-vehicle travel time is needed to compensate exactly for a 1-minute increase in out-of-vehicle travel time. In other words, travelers whose round trips to work are 15 miles long consider out-of-vehicle travel time to be 1.46 times more onerous than in-vehicle travel time.

It is also possible to estimate the money value of travel time to a commuter—that is, the amount of money a commuter would have to be paid to provide exact compensation for an increase in in-vehicle or out-of-vehicle travel time. Consider in-vehicle travel time. The change in utility that occurs when *IVTT* changes by $\Delta IVTT$ and $\ln(Z)$ (the natural logarithm of remaining income) changes by $\Delta\ln(Z)$ is

$$\Delta U = a_3\Delta IVTT + a_5\Delta\ln(Z) \quad (19)$$

The corresponding change in Z, ΔZ, is related to $\Delta\ln(Z)$ by

$$\Delta\ln(Z) = \ln(Z + \Delta Z) - \ln(Z) \quad (20)$$

That is, the change in the logarithm of Z is the logarithm of the new value of Z minus the logarithm of the original value. Equivalently,

$$\Delta\ln(Z) = \ln(1 + \Delta Z/Z) \quad (20a)$$

Therefore, the change in Z that exactly compensates for the change $\Delta IVTT$ in in-vehicle travel time (that is, that causes ΔU to be zero) is given by

$$0 = a_3\Delta IVTT + a_5\ln(1 + \Delta Z/Z) \quad (21)$$

The solution to this equation is

$$\Delta Z = Z\{\exp[(-a_3/a_5)\Delta IVTT] - 1\} \text{ dollars} \quad (22)$$

Since the estimated value of a_3/a_5 is -0.0078, equation 22 implies that

$$\Delta Z = Z[\exp(0.0078\Delta IVTT) - 1] \text{ dollars} \quad (23)$$

Suppose that $\Delta IVTT = 1$ minute. Then equation 23 together with the fact that $\exp(0.0078) = 1.0078$ yields the result

$$\Delta Z = 0.0078Z \text{ dollars} \quad (24)$$

This value of ΔZ is the increase in remaining income needed to compensate a commuter exactly whose remaining income is Z dollars per year for a 1-minute increase in the in-vehicle travel time required for the commuter's daily trip to work. If the commuter's taxes and costs of housing, automobile ownership, and commuting remain unchanged, this also is the total amount of money the commuter must be paid to provide exact compensation. Thus, the value of Z given by equation 24 is the value (in 1968 dollars) of a 1-minute increase in daily commuting in-vehicle travel time.

The value per minute of in-vehicle travel time can be obtained by dividing ΔZ by the number of work days per year. Since there are approximately 250 work days per year, the result is $3.12 \times 10^{-5}Z$. Multiplying this number by 60 yields the value of commuting in-vehicle travel time in dollars per hour. The result is 0.0019Z. Thus a commuter whose remaining income is $10,000 per year (1968 dollars) considers in-vehicle travel time to be worth $19 per hour according to the estimated model.

POLICY ANALYSIS WITH THE MODEL

In this section, two examples are presented illustrating the use of the estimated model to predict the effects of transit service improvements on households' travel behavior and choices of residential location.

It is important to recognize, when using the model to predict changes in households' behavior, that there are great differences between the lengths of time households need to adjust their travel behavior to changes in transit service quality and the lengths of time needed to adjust residential locations. Changes in the mode of travel to work can be made relatively easily and quickly. If the change does not entail the purchase of a vehicle, it can be made in a few days. Even if a vehicle must be bought, most households are likely to be able to accomplish a change of mode in a period of weeks. Thus, changes in transit service quality are likely to have noticeable effects on mode choice quickly. In contrast, changes of residential location usually are difficult and costly. As a result, they occur infrequently. The mean time between moves for homeowning households in the U.S. exceeds 10 years, and the mean time between moves for all households is about 5 years. Therefore, adjustments in residential location in response to transit service changes are likely to occur over periods of years or possibly even decades.

Because of the large differences between the time periods required for changes in mode choice and changes in residential location to occur, it is useful to distinguish between the "short run" and the "long run" in making predictions of the effects of changes in transit service quality. In the case of the discussion in this chapter, the short run refers to a time period in which mode choice conditional on residential location has adjusted to the service quality changes but residential location has not adjusted significantly. Thus, the short run consists of a time period that may begin a few days or weeks after the change has occurred and last for several years. The long run refers to a time period during which both mode choice and residential location have adjusted to the change in service quality. Thus, the long run consists of a time period that begins several years after the change has occurred and lasts indefinitely. The short-run effects of a change in transit service quality are investigated in Example 4, and the long-run effects are investigated in Example 5.

Example 4: This example illustrates the use of the estimated model to predict the short-run effects on mode choice of improvements in transit service quality. Since residential locations remain fixed in the short run, it is necessary to be able to compute the probabilities that the automobile and bus transit modes are chosen conditional on a given, fixed, residential location. These probabilities can be obtained from the estimated model by applying rules of conditional probability. According to these rules, the probability $P(m|i)$ that mode m is chosen conditional on residential location i is given by

$$P(m|i) = \frac{P(m,i)}{\sum_{\text{modes } m'} P(m',i)} \qquad (25)$$

where $P(m,i)$ is the probability that mode m and residential location i are chosen (that is, the probabilities given by the estimated model), and the sum in the denominator is over all available modes (automobile and bus in this case). Recall that in the estimated model, $P(m,i)$ has the multinomial logit form with the V function given by equation 11. The model $P(m|i)$ also has the multinomial logit form, but its V function contains only the variables of the original model that have different values for different modes. This is because variables whose values are the same for all modes (that is, variables that depend only on the residential location) cancel in the quotient on the righthand side of equation 25. Thus, the V function for the model $P(m|i)$, which will be denoted by V^* to distinguish it from the V function of the full model, is

$$V^* = a_1 DCAR + a_2(OVTT/DIST) + a_3 TOTIME + a_4 DCITY + a_5 \ln(Z) + a_6 AALD \qquad (26)$$

where the a coefficients have the values shown in Table 9-2. The complete specification of the model $P(m|i)$ is

$$P(m|i) = \frac{\exp[V^*(m)]}{\sum_{m'} \exp[V^*(m')]} \qquad (27)$$

where $V^*(m)$ denotes the value of V^* corresponding to mode m, m denotes either automobile or bus, and the sum in the denominator of equation 27 is over these two modes.

In this example, two households will be considered, one that lives relatively near to its workplace and one that lives farther away. Both households have incomes of $15,000 per year (1968 dollars), and both have two licensed drivers (so $AALD = 0.5$ for the automobile alternative, since the estimated model treats only households that own one car). Each household has one worker, and both workers have downtown workplaces (so $DCITY = 1$ for the automobile alternative). The near household lives 3.5 miles from its workplace, and the far household lives 7.5 miles from its workplace. Therefore, the round-trip distances to work $(DIST)$ are 7 miles and 15 miles for the near and far households, respectively. It is assumed that taxes, housing costs, and the costs of automobile ownership are 50% of total income. The round-trip bus fare is $0.70 per day or $175 per year for a year that has 250 work days. Therefore, if the bus mode is chosen, remaining income (Z) for both households is $[0.5(15,000) − 175] or $7,325 per year. The cost of driving to work is $0.05 per mile. Therefore, if the automobile mode is chosen, the near household spends $0.05(7) (250) = $87.50 per year for commuting, and the far household spends $0.05(15) (250) = $187.50 per year. Thus, if the automobile mode is chosen, the remaining income of the near household is $[0.5(15,000) − 87.50] = $7,412.50, and the remaining income of the far household is $[0.5(15,000) − 187.50] = $7,312.50.

The values of the variables of the mode choice model prior to implementation of transit service improvements are given in Table 9-3. The tabulated values of $\ln(Z)$ are based on the Z values just derived. Note that bus travel is considerably more time consuming than automobile travel. The tabulated differences between bus and automobile travel times are typical for U.S. cities.

The values of V^* for the two modes and households can be obtained by substituting the values of the variables shown in Table 9-3 and the coefficient values shown in Table 9-2 into equation 26. Thus, for example, V^* (automobile) for the near household is

$$V^*(\text{automobile}) = 0.433 - 0.0570(1.429) - 0.00831(35) - 0.437 + 1.07(8.911) + 0.964(0.5) \qquad (28a)$$

$$= 9.641. \qquad (28b)$$

The complete set of V^* values is

	Near	Far
Automobile	9.641	9.503
Bus	8.696	8.660

The mode choice probabilities are obtained by substituting these values of V^* into equation 27. Thus, for example, for the near household

$$P(\text{auto}|\text{near location}) = \frac{\exp(9.641)}{[\exp(9.641) + \exp(8.696)]} \quad (29a)$$

$$= 0.720. \quad (29b)$$

The complete set of choice probabilities is

	Near	Far
Automobile	0.720	0.699
Bus	0.280	0.301

Thus, among a large group of workers from households identical in terms of the variables shown in Table 9-3 to the near household, 72.0% would travel to work by automobile and 28.0% would travel to work by bus. Among a large group of workers from households identical in terms of the tabulated variables to the far household, 69.9% would travel to work by automobile and 30.1% would travel to work by bus.

Now suppose that improvements in transit service cause a 50% reduction in transit out-of-vehicle travel time and a 20% reduction in transit in-vehicle travel time. Then transit out-of-vehicle travel time for both households is reduced to 15 minutes. *OVTT/DIST* for transit is reduced to 2.143 for the near household and 1.0 for the far one. Transit in-vehicle travel time is reduced to 32 minutes for the near household and 48 minutes for the far household. Therefore, *TOTIME* for the bus mode is 47 minutes and 63 minutes for the near and far households, respectively. The values of the variables associated with the automobile mode and of Z do not change.

The values of V^* for the two modes and households after the improvement in transit service are

	Near	Far
Automobile	9.641	9.503
Bus	9.009	8.941

The corresponding mode choice probabilities are

	Near	Far
Automobile	0.653	0.637
Bus	0.347	0.363

Relative to the situation before transit was improved, the probability of bus usage has increased by 23.9% for the near household, that is, 100(.347 – .280)/.280% for the near household and by 20.6% for the far household. Thus, the improvements in transit service quality have had a substantial effect on bus usage.

Example 5: The long-run effects of the improvement in transit service discussed in Example 4 now will be investigated. Since the long-run effects involve households' choices of residential locations as well as

Table 9-3. Values of variables for base case of Example 4

	Near household		Far household	
Variable	Auto	Transit	Auto	Transit
IVTT[a]	25	40	45	60
OVTT[a]	10	30	10	30
DIST[a]	7	7	15	15
Z[a]	7412.50	7325.00	7312.50	7325.00
DCAR	1	0	1	0
OVTT/DIST	1.429	4.286	0.667	2.00
TOTIME	35	70	55	90
DCITY	1	0	1	0
ln(Z)	8.911	8.899	8.897	8.899
AALD	0.5	0.0	0.5	0.0

[a]Not a variable of the model but used in computing one or more variables of the model.

choices of modes of travel to work, the analysis in this example makes use of the full estimated model.

The analysis is carried out for a household whose income is $15,000 per year, has two licensed drivers, and whose only worker has a downtown workplace. Thus, the attributes of the household are identical to those of the households considered in Example 4. The household is assumed to choose between the near and far residential locations described in Example 4 as well as between driving and traveling by bus to work. Table 4 shows the values of the model's variables for the four location/mode alternatives available in this example. Note that the values of the transportation and household variables are the same as they were in Example 4. (If this were not the case, then the results of this example would describe the long-run effects of a set of transportation changes different from those considered in Example 4.) The values of the locational variables have been chosen so that the near and far locations are identical except in terms of accessibility and density. The near location has a higher density and better accessibility to both the workplace and shopping locations than does the far location.

Substitution of the values of the variables shown in Table 9-4 and the values of the coefficients shown in Table 9-2 into equation 11 yields the following values of the V function for the location/mode alternatives:

	Near	Far
Automobile	10.500	8.919
Bus	9.555	8.076

The corresponding probabilities of choices are obtained by substituting these values of V into equation 4. The results are:

	Near	Far
Automobile	0.594	0.122
Bus	0.231	0.053

Thus among a large group of households identical in terms of the model's variables to the one being considered in this example,

Table 9-4. Values of variables for base case of Example 5

	Near location		Far location	
Variable	Auto	Transit	Auto	Transit
IVTT[a]	25	40	45	60
OVTT[a]	10	30	10	30
DIST[a]	7	7	15	15
Z[a]	7412.50	7325.00	7312.50	7325.00
NL[a]	500	500	500	500
DCAR	1	0	1	0
OVTT/DIST	1.429	4.286	0.667	2.00
TOTIME	35	70	55	90
DCITY	1	0	1	0
ln(Z)	8.911	8.899	8.897	8.899
AALD	0.5	0.0	0.5	0.0
GPTINV	1.0	1.0	0.5	0.5
*R*1	4.0	4.0	4.0	4.0
FBFORW	0	0	0	0
FBFORB	0	0	0	0
DENSITY	5	5	2	2
SCHOOL	700	700	700	700
INCDIFF	0	0	0	0
DOC	0	0	0	0
ln(*NL*)	6.215	6.215	6.215	6.215

[a] Not a variable of the model but used in computing one or more variables of the model.

59.4% would choose to live at the near location and travel to work by automobile, 23.1% would choose to live at the near location and travel to work by bus, 12.2% would choose to live at the far location and travel to work by automobile, and 5.3% would choose to live at the far location and travel to work by bus.

The probabilities that the household chooses each location regardless of mode, chooses each mode regardless of location, and chooses each mode conditional on location (the base case results of Example 4) also can be obtained from the model. The probability of living at the near location regardless of mode is the probability of living at the near location and choosing automobile plus the probability of living at the near location and choosing bus. This probability is 0.594 + 0.231 = 0.825. Similarly, the probability of living at the far location regardless of mode is 0.122 + 0.053 = 0.175. Thus, among a large group of households identical in terms of the model's variables to the one considered in this example, 82.5% would choose to live at the near location and 17.5% would choose to live at the far location.

The probability of choosing to travel to work by automobile regardless of location is the sum of the automobile choice probabilities for the two locations or 0.716. The probability of choosing bus regardless of residential location is the sum of the transit choice probabilities for the two locations or 0.284. Thus, among a large group of households identical in terms of the model's variables to the one considered in this example, 71.6% would use the automobile for travel to work, and 28.4% would use bus.

The probability of choosing automobile conditional on living at the near location can be obtained by applying the conditional probability equation (equation 25) to the probabilities of choosing the location/mode alternatives. The resulting conditional probabilities are:

	Near	Far
Automobile	0.720	0.699
Bus	0.280	0.301

These probabilities are identical to those given for the base case in Example 4, as must happen if the model is to be internally consistent.

Now suppose that the transit improvements described in Example 4 are implemented. Then as discussed in Example 4, *OVTT/DIST* for bus is reduced to 2.143 and 1.0 at the near and far locations, respectively. *TOTIME* for bus is reduced to 47 minutes and 63 minutes at the near and far locations. The values of the other variables remain unchanged. The resulting values of V are

	Near	Far
Automobile	10.500	8.919
Bus	9.868	8.357

The choice probabilities are

	Near	Far
Automobile	0.539	0.111
Bus	0.287	0.063

The probability of living at the near location regardless of the mode of travel to work is now 0.826, compared to a value of 0.825 before the improvement in transit service. The probability of living at the far location regardless of the mode of travel to work is 0.174, compared to a preimprovement value of 0.175. Thus, in the long run, the model predicts that the improvement in transit service does little to bring households to the near location. (Although the changes in choices of residential location are very small in this example, they are realistic. Improvements in transit service usually have very little impact on households' choices of residential location in U.S. cities.)

The probability of choosing automobile for travel to work regardless of residential location is now 0.650 compared to 0.716 before transit service improved. The probability of choosing bus for travel to work regardless of residential location is now 0.350 compared to 0.284 before the service improvement. Thus, among a large group of households identical in terms of the model's variables to the one considered in this example, the transit service improvement would increase bus ridership by 100[(0.350 − 0.284)/0.284]% or 23%. It can be shown that virtually all of this increase in bus ridership occurs in the short run. This is because, as shown above, there is virtually no change in residential location as a result of the improvement in transit service.

The probabilities of choosing automobile and bus conditional on residential location after the improvement in transit service can be computed from equation 25 and are

	Near	Far
Automobile	0.653	0.637
Bus	0.347	0.363

These are identical to the conditional probabilities obtained in Example 4 as must happen if the model is internally consistent.

SUMMARY

This chapter has described a model of households' joint choices of residential location and mode of travel to work. The model is based on a theory of individual choice called utility theory and has been implemented empirically using data from the Washington, D.C., metropolitan area. The model has been used to predict the effects of transit service improvements on the residential location and mode of travel to work of a hypothetical household. It has been found that the transit improvements have relatively large effects on mode choice. However, their effects on choice of residential location are very small.

References

Domencich, T. A., & McFadden, D. (1975). *Urban travel demand: A behavioral analysis.* New York: American Elsevier.

Lerman, S. R. (1979). Neighborhood choice and transportation services. In D. Segal (Ed.), *The economics of neighborhood.* New York: Academic Press.

Lerman, S. R. (1976). Location, housing, automobile ownership, and mode to work: A joint choice model. *Transportation Research Record, 610,* 6–11.

SECTION III
POLICY CONCERNS

10 TRANSIT IN AMERICAN CITIES

GORDON J. FIELDING
University of California, Irvine

Transit is a minor contributor to mobility in American cities. Only 2.8% of all person trips were made by transit in 1980 compared to 73% made by auto. More important to mobility, however, were the 6.2% using transit for their journey to work. Urban thoroughfares are heavily congested during the peak commuting hours, and the contribution that transit makes to lessening congestion is a reason why governments have sought to assist the industry. Proportionate use of transit for work trips is geographically varied. Highest use occurs in the older and most congested urban areas, whereas it falls to 2 or 3% in cities of the Southwest. Explanations for these spatial variations will be provided in this chapter to provide a link with earlier chapters on urban form and the modeling of movement patterns. There will also be a definite political emphasis. Federal, state, and local governments have molded transit assistance to reflect policy objectives, so that any geography of transit also provides a case study in political geography. For this reason, the chapter has an American orientation. Students from other countries may find the description of the elements of transit and the techniques for performance analysis useful for understanding local systems, but the political context differs. In American cities transit exists in a transportation policy environment that is dominated by highway, trucking, and automobile interests. Considerable academic attention has been given to public transit during the last two decades, and billions of dollars have been provided by government agencies to help revitalize transit; yet, its contribution to mobility is important in only a few locations: in the congested areas of the central city, the older suburbs, and along arterial corridors between the new suburbs and the central city.

SPATIAL DISTRIBUTION

Transit use (measured in passenger miles) is concentrated in the largest cities. Although all metropolitan areas, most small cities, and many rural areas have some form of public transit, 86% of all transit use occurs in the 21 urbanized areas with 50% of the national population (Figure 10-1). The New York–New Jersey and Chicago urbanized areas to-

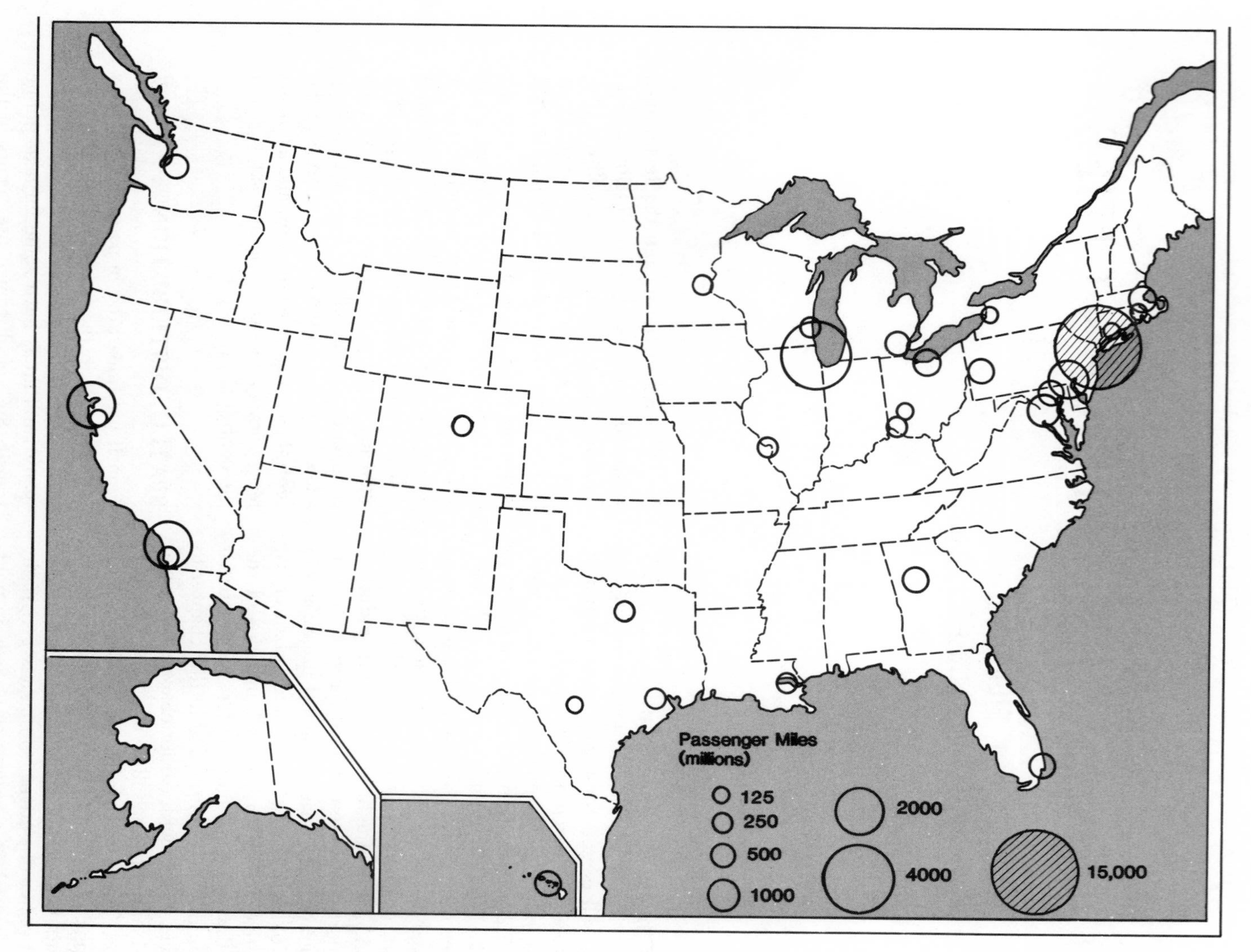

Figure 10-1. Transit use in urbanized areas, 1982, in annual passenger miles. Areas with less than 100 million passenger miles use 10% of the total transit, but are not represented. From *National Urban Mass Transportation Statistics*, by Urban Mass Transit Administration, 1982, Washington, D.C.: U.S. Department of Transportation.

gether use 52% of all passenger miles consumed.[1] Although the journey to work is the most important trip in all these areas, transit also serves other trip purposes, especially in the central cities where many short-distance trips are made by transit. Only 10% of all passenger miles are consumed in the smaller urbanized areas not shown in Figure 10-1. Most transit patrons in these places have no other reasonably priced alternative: they are transit dependent.

New York is overwhelmingly important. With 11% of the nation's population, the New York region, which includes northeastern New Jersey, uses 41.4% of the transit. Thirty-four agencies operate public transit service in the urbanized area. The two largest agencies, New York Metropolitan Transportation Authority and New Jersey Transit, operate six systems and about three-quarters of all service. The other 29 agencies are independent. Many are small, private companies operating under contract to public agencies. In addition, there are many commuter bus companies operating with subsidies whose statistics are not included in the national figures. It has been estimated that 550 private express commuter buses enter Manhattan every day, operating without public subsidy and at a profit (Orski, 1982). In terms of service use New York dominates American transit and its policy. Many of the national programs discussed in subsequent sections were promulgated to assist New York but have benefited other cities as well. New York is not distinctive in its organizational complexity. All urbanized areas represented in Figure 10-1 are characterized by at least one dominant transit agency operating alongside many smaller agencies.

[1] Unless otherwise cited, statistics come from two sources: the *National urban mass transportation statistics: 1982 Section 15 Annual Report* (U.S. Department Transportation, 1983) and *The status of the nation's local public transportation: Conditions and performance*, Report of the Secretary of Transportation to the U.S. Congress (U.S. Department of Transportation, Section 310 Report, 1984). Material for this chapter is based upon the 336 agencies who have reported under Section 15 requirements. No reliable data are available on the other operators.

Although passenger miles is the ideal measure of transit consumption, reliable statistics are difficult to obtain. National statistics have been available in the United States only since 1979, and for many nations comparative statistics are unavailable. For this reason transportation geographers prefer to use passenger trips as a measure of consumption in comparative or time-series research. Each time a person boards a vehicle that traveler is counted separately *(unlinked trips)* even though several trips might be made in one journey.

Where Transit Is Effective

Work trips continue to be the mainstay of transit use. They accounted for 59% of all weekday transit trips in 1980, down from 70% in 1970. The percentage of workers using transit also declined from 9.0 to 6.2% in the same period. A variety of factors account for this decline, including increased competition from automobiles, dispersal of workplaces, higher rates of unemployment, and increased government investment in highways. These factors vary regionally: the greatest decline, from 19 to 14%, occurred in the Northeast whereas transit use increased slightly in the West from 4.6 to 5.0%. A combination of declining employment opportunities and reduced transit service affected Northeastern metropolitan areas like Buffalo, New York, where transit use by workers declined from 10.5 to 6.6% in the decade before 1980. In comparison, use increased from 7.1 to 9.6% over the same period in the Seattle SMSA, where both population and jobs have increased and transit service has been modernized and aggressively promoted.

An example of what good transit management can accomplish is provided by Minneapolis–St. Paul. In a region where transit use has been declining, it increased slightly from 8.5% to 8.7% of all workers between 1970 and 1980. This was accomplished by connecting sheltered pedestrian walkways to transit malls in both downtown areas of the Twin Cities and increasing express bus service from suburban areas. Service was also improved: on-time departures and passenger assistance have been emphasized in driver training and recognized by merit awards; quality-circle techniques were used in maintenance shops to emphasize the importance of dependable vehicles; and organized labor cooperated with management to reduce employee absenteeism. These personnel and service development programs have increased the reliability of transit service. Operating cost has increased as a result, and fares have had to be increased, but patrons have responded through continued use.

A market where transit use has not declined is the work trip to the CBD (Table 10-1). Even in the Northeast and North Central regions where transit use declined the most between 1970 and 1980, CBD-destined trips have been retained. Slight declines are evident for cities like Boston and Philadelphia; others like Chicago remain stable, whereas transit's share has increased in Washington, D.C., where commuters have benefited from construction of a rapid rail system. More spectacular increases have been achieved in the West where a combination of local, state, and federal funds has been used to revitalize and expand bus systems as well as new rail systems. More than half the CBD-oriented work trips in San Francisco and Oakland were made by transit in 1980 and almost one-quarter were in Los Angeles.

Table 10-1. Transit's share of CBD-bound work trips

Area	1970	1980
Boston	60.6%	58.5%
New York	60.6	60.2
Washington, D.C.	36.5	42.6
Chicago	74.9	74.1
St. Louis	29.7	26.7
Dallas–Fort Worth	19.9	19.2
Houston	12.9	15.0
San Francisco–Oakland	47.9	52.4
Los Angeles–Long Beach	21.4	23.8
Anaheim–Santa Ana–Garden Grove	1.1	3.4
San Diego	8.7	13.7

Note. From *The Status of the Nation's Local Public Transportation* by the Secretary of Transportation, 1984, Washington, D.C.: U.S. Department of Transportation.

Los Angeles could not function if it were not for the contribution of transit in reducing congestion on arterial routes leading to the CBD. A transit strike causes intolerable delays and a substantial decline in CBD commerce. On August 12, 1974, the Los Angeles area public transit system operated by the Southern California Rapid Transit District was shut down for nine weeks by a labor dispute. Decline in retail sales by major stores in the CBD resulted in losses of between 20 and 30% compared to the same period in previous years. Even sales in suburban centers declined, because in one-auto households, the car was needed for commuting and was therefore unavailable for midday shopping trips. Commuters adapted in many ways: carpooling increased, taxi ridership increased by 25%, and used car sales rose between 10 and 15% during the strike. But the worst hardships were suffered by transit-dependent persons. Many were unable to get to doctors' appointments or to nutritional centers. Their condition was epitomized by one elderly woman who stated, "Transportation is as important to me as water or electricity. No one would think of cutting off all these services for two months" (Crain & Flynn, 1975, p. 11).

Other Markets

Another important market is transporting college and high school students. Fourteen percent of all transit trips are school related. This can cause operational problems when school travel coincides with work-related demands. Demand concentrated into peak periods is very costly to accommodate because labor and equipment are difficult to schedule for just a few hours at both ends of the normal work day.

Medical and shopping-related trips account for 11% of transit travel. Transit is not a convenient way to travel for these purposes, and it is used primarily by persons from low-income or autoless households. Often these are the same (Table 10-2). Seventy-one percent of all trips are made by persons from households with less than $20,000 per year, and 40% had less than $10,000. Many of these trips originate from homes with women as heads of household. On average they had an income less than half that of traditional households, and one in three female-headed families is beneath the poverty line. These families are dependent upon transit for both nonwork and work trips. This, in part, explains why 62% of riders are women (Table 10-3).

Women's use of transit has declined as more women have joined the work force. In 1970 only 43% of the women over 16 years of age were working, and many homemakers used transit in the midday for shopping and social purposes. By 1980, 52% of women were working. This has decreased the demand for midday transit travel by women but not their use of transit for commuting.

Successful transit markets exist for work trips to the CBD and to schools and colleges. Another market exists for travel by persons from low-income households. This suggests that transit does best in high-density neighborhoods (where low-income individuals and students dwell) and along radial corridors connecting the suburbs with the CBD or other commercial megacenters. Pushkarev and Zupan (1977) summarized this argument in *Public Transit and Land Use Policy*. Using comparative statistics from several urban areas, they demonstrated how transit use increases sharply above a density of seven dwelling units per acre. They also indicate how the demand for various modes of transit is related to nonresidential floor space in the CBD, which is used as a surrogate for employment opportunities. Transit systems require both density of supply (potential users) from residential areas and density of demand (potential employment opportunities) for routes to be successful. Few areas possess these characteristics, so that transit systems must be sustained with governmental assistance.

Forecasting Demand

Geographers and civil engineers use knowledge about transportation generators and attractors to forecast the demand for new transit systems. The techniques used are adequately summarized in previous chapters, but a note of warning is appropriate. The temptation to overestimate demand for new transit systems is strong because the work is completed under contract to agencies who want to construct these facilities. Trip generation can be inflated either by overestimating growth in population and employment or by forecasting more frequent trip making by households. Another source of error comes from forecasting a higher mode split for transit by assuming highway congestion and faster comparative travel times on transit.

An example of optimistic forecasting is the ridership estimates prepared for the Miami rapid transit system. Disaggregate demand models were used during the planning phase, and patronage of 100,000 trips per day was forecast for the first segment. When the system opened in 1984, daily ridership was only a paltry 9,500—about 10% of what was predicted! Outcomes like this jeopardize the professional credibility of geographers and engineers. And this is not the only example of overly optimistic estimates.

The Bay Area Rapid Transit (BART) system planned for San Francisco both underestimated the technical difficulty of implementing a new rail system and overestimated the ability of rail transit to compete with autos in the commuter market (Webber, 1976). Planning studies conducted in 1967 estimated an average weekday ridership of 253,400 for the planned 71-mile system. The system opened in 1972, but construction delays and problems with automated equipment precluded operation of the system as originally planned. Even after two years the best weekday ridership was 126,500 passengers—not quite one-half of what highly regarded transportation planning firms had predicted. It has taken 12 years to achieve reliable operation for BART, with peak-hour headways of 3.8 minutes in the CBD segments of San Francisco and Oakland. Almost as much time was required to coordinate bus services with BART. None of these obstacles were anticipated by planners. Average weekday ridership in 1984 was 211,000, still below what had been projected, but far better than the decade earlier.

Travel demand forecasts for the rail system in Atlanta, however, were accurate. The 12 miles of east-west rail line in Atlanta were projected in 1971 to attract 17.6 million riders annually. This estimate was exceeded by 200,000 riders in 1980, the first year of operation (Giuliano, 1984). Reliable forecasts can be produced using the methods outlined in previous chapters, but caution is recommended. The results can easily be distorted to achieve political objectives.

Why is it, then, that in some instances geographers and engineers can provide accurate travel demand forecasts, but in others miss by a factor of two or three? The answer involves politics. Cities compete with each other for federal funds, and professionals are under pressure to come up with results favorable to desired outcomes. A real ethical quandary is presented to transportation professionals (Wachs, 1982). They recognize that the output from travel demand models depends heavily upon the demographic and employment assumptions that they use. Relationships between travel time and cost and choice of destination and mode are also relevant. Slight changes in either assumptions or relationships result in substantial changes in predicted outcomes. Yet, they also recognize that the government agency that has hired them will benefit from the jobs and prestige that will result from obtaining federal assistance for construction. An illusion of objectivity is provided by the computer-based models, but the core assumptions and the relationships upon which the models depend are chosen subjectively. Professional objectivity can easily be subverted in these circumstances, especially when the geographer is employed by the agency whose prestige will be enhanced through construction of a rapid transit system. To whom does the individual owe loyalty: to the profession or the employer? Planning schools are beginning to give attention to this dilemma, and geographers who desire employment in public agencies should become acquainted with the issues (Wachs, 1985).

Forecasting travel demand presents an ethical problem for the geographer–planner. Use of mathematical models and complex travel assignments compels decision makers to rely upon the results provided. Too often these are presented as predictable outcomes rather than as estimates with some probability of occurrence under specific assumptions. In most cases examined estimates have been found to exceed actual use by a factor of two or three (Hamer, 1976). Errors of such magnitude create financial hardship for the operating agency as well as prejudice against geographers and planners. The very prospects for transit have been diminished by the exaggerated forecasts of what transit might accomplish.

AUTOS VERSUS TRANSIT

Too much has been expected from transit. During the 1960s limited capital assistance was provided to transit agencies, so that they might contribute to urban redevelopment. But in the 1970s America's attention shifted to new problems associated with the adverse impact of the automobile. Transit investment was advocated as a solution to air pollution and highway congestion; as a means to conserve energy and reduce urban sprawl; and to provide mobility to the poor, the elderly, and the handicapped (Fielding, 1983a).

Growth of the transit program was overwhelming. Transit assistance from all government sources increased almost tenfold, from only $540 million in 1973 to $5.2 billion in 1978. Funding was made available not only for planning and capital assistance but also for operating subsidies and personnel development. Increases of such magnitude were surprising because transit's direct constituency was relatively small. However, support for assistance had broad political appeal. The explanation, say Altshuler, Womack, and Pucher (1979), lies in the fact that transit proved to be a "policy for all perspectives" on the urban problem.

> Whether one's concern was the economic vitality of cities, protecting the environment, stopping highways, energy conservation, assisting the elderly and handicapped and poor, or simply getting other people off the road so as to be able to driver faster, transit was a policy that could be embraced. This is not to say that transit was an effective way of serving all these objectives, but simply that it was widely believed to be so. Additionally, because the absolute magnitude of transit spending was so meager at the beginning of this period, it was possible to obtain credit for rapid program growth with quite modest increase in the absolute magnitude of expenditures. (p. 36)

Portland, Oregon, demonstrates how the geography of cities was altered by these politically inspired investments. The CBD was redesigned around a transit mall complete with brick and cobblestone pavements, futuristic bus shelters, and antique lighting. A new maintenance facility and administrative tower were constructed; articulated buses were introduced for express service; and planning and construction were begun for a new light rail line. Bus routes were redesigned, increasing miles of service by 178%, which resulted in a doubling of ridership between 1970 and 1976. Portland was regarded as one of transit's "success stories." Local government had aggressively sought federal funding to improve transit and revitalize their city. And yet, in this same period, operating expenses increased more than fivefold, and revenues fell to less than 40% of operating cost. By 1980 the ratio of total operating revenues to operating cost had fallen to 31%, and the transit agency had placed a moratorium on service expansion. Local officials now realized that they had used federal funds to create an organization far larger than either the patrons or the city were prepared to sustain. Ridership had doubled between 1970 and 1980, but this was not reflected by corresponding increases in revenues because of the low fares. And the city had not solved its auto congestion problems. Highways remained congested during peak hours, and the proportion of workers using public transit had only increased from 6.0 to 6.2%.

What Went Wrong?

Transit has not been a panacea for auto-related problems. Space does not permit a thorough discussion of reasons why, and the interested reader should refer to either Altshuler et al. (1979) or Meyer and Gomez-Ibanez (1981). The explanation will not surprise geographers who are accustomed to seeking explanations in terms of human behavior: time, cost, and convenience are all involved.

For only a few types of trips can transit compete with auto travel. Transit is competitive for journeys to the CBD where roadways are congested and parking costly and difficult to find; for short trips from neighborhoods to the CBD or to suburban shopping malls and colleges, and for trips within higher-density neighborhoods where transit service is frequent and stops are easily accessible. For most other trips, transit takes too long. Buses on neighborhood streets average 13 to 15 miles per hour. The average vehicular work trip in metropolitan areas is 9 miles, which means about 40 minutes each way in travel time by bus. Additional time must be allowed for walking to the transit stop and waiting for the vehicle. Let us add 10 minutes. How many workers are willing to allocate 1 hour and 40 minutes a day to traveling by transit? Only those who travel shorter distances than the average or those who have access to faster (express bus or rail) service will do so. Of course, there are those who do not have access to an automobile (primarily women) or who have little money but adequate time (the unemployed) or have flexible work schedules (college students). These are the predominant transit users (Tables 10-2 and 10-3).

Availability of an automobile is a real deterrent to transit use. Whereas 17.5% of the

Table 10-2. Transit trips by household income of rider

	Percent of transit trips	Percent of total population, 1980
Less than $10,000 per year	40.3	25.1
$10,000 to $20,000 per year	30.9	26.1
$20,000 to $30,000 per year	15.4	20.3
Greater than $30,000	13.3	28.4

Note. From *The Status of the Nation's Local Public Transportation* by the Secretary of Transportation, 1984, Washington, D.C.: U.S. Department of Transportation.

households in the United States did not own an automobile in 1970, the proportion has declined to 15.9% in 1980. And the cost of purchasing an automobile increased at only one-half the rate of consumer prices over the decade. More American households could afford to purchase automobiles in 1980, and they were using them instead of transit despite the real increase in gasoline prices. Few people consider the full cost of operating an automobile. Depreciation, insurance, and licensing are commonly regarded as "sunk costs." As a car is needed for shopping and recreation, the concept of leaving the car at home when not needed seldom occurs to Americans unless automobility is seriously constrained by congestion, parking costs, or scarcity of gasoline. Auto ownership has become increasingly affordable and once available, is used for every purpose even when transit alternatives exist. The auto's relatively low variable cost and great convenience make it the overwhelming mode of choice. Like it or not, people prefer to be prosperous: to own cars, to live in the suburbs, and not use transit. Rising personal wealth has made this life-style possible.

Table 10-3. Transit trips by age and sex of rider

Age class	Percent of transit trips	Percent of population, 1980
Less than 18	15.2	28.0
18 to 25	20.3	13.3
26 to 55	43.2	37.9
56 to 65	11.3	9.6
Greater than 65	10.0	11.3
Male	38.4	48.7
Female	61.6	51.3

Note. From *The Status of the Nation's Local Public Transportation* by the Secretary of Transportation, 1984, Washington, D.C.: U.S. Department of Transportation.

Availability of Gasoline

Hostile action in the Middle East interrupted the international distribution of petroleum first in 1973 and again in 1978. Both events created gasoline shortages and price increases even though only an 8% reduction in supply occurred in the United States. Transit systems were overwhelmed during both interruptions, but people soon returned to their cars and accepted the higher prices for gasoline when available. Although the possibility of future shortages is often used as a reason for assisting transit, it is unlikely that either scarcity or price of gasoline will force Americans out of their cars and onto transit in the foreseeable future.

Although petroleum is a scarce resource that will one day be exhausted, proven reserves are of such magnitude that the primary issues will revolve around price. Increased domestic production from 55% of demand in 1980 to 70% in 1983, together with strategic reserves, will help insulate Americans against supply interruption. Real price increases for gasoline have also aided conservation. Now the Department of Energy is forecasting a 17% decrease in national petroleum consumption between 1980 and 1995 despite projected increases in both population and automobiles. The trend toward smaller cars and projected increases in overall automobile fuel efficiency should make continued automobile use possible without disruption.

Transit is also not an effective way to save energy. Other strategies have proved more beneficial. As transit only provides a miniscule 3.2% of all urban trips, even a doubling of transit use is not going to conserve much gasoline.

Although a full bus is more energy efficient than an auto with an average occupancy of 1.3 persons, the figures can be deceptive. First, buses are full only at rush hours. Average occupancy over the whole day is only about 15 persons. Even at this level, a bus is still more energy efficient than an auto with one or two occupants. Autos with more than three occupants and vans with ten are, however, more energy efficient than buses. Therefore, ridesharing strategies that achieve higher occupancies for automobiles can be more energy efficient than transit.

Bus transit is used for comparison because almost 83% of all transit service is provided by buses. Rail transit can be more energy efficient. When available, it contributes real energy savings in comparison to the auto. However, this cannot be used as a rationale for constructing rail systems, because the energy resources consumed in construction may take 45 years or more to recover through energy savings.

Other Reasons for Transit

Transit's contribution to reducing air pollution merits less support than its role as an energy conserver. Again, limited use, especially in those urban areas most affected by automobile-generated air pollution, reduces transit's contribution. Moreover, bus transit is itself a source of pollutants. Exhaust from diesel buses is not only nauseating, it is also harmful. It is heavily laden with hydrocarbons and particulates, and the engines produce more oxides of nitrogen than comparable gasoline engines. It is the complex interaction of these reactive emissions with other pollutants and sunlight that creates the components of what is commonly called smog. More stringent control of emissions from diesel engines is required before buses can claim to reduce air pollution.

Only in terms of reducing traffic congestion during peak periods can transit claim to be effective. Arterial roads would be more congested and commute times far longer and more troublesome if transit were not attracting some commuters. And when buses and trains are filled, transit can also claim to be both an energy saver and a reducer of air pollution.

Transit programs expanded rapidly during the 1970s because transit was regarded as a solution to a host of urban dilemmas. Ridership did increase and has now leveled off again (Figure 10-2). The optimism that transit would regain a share of ridership proportionate to that enjoyed in the electric streetcar era described in chapter 2 was misplaced. Transit policies were promoted as a palliative to the automobile without critical evaluation of the cost and convenience advantages that the automobile offered. Too often, transit was developed for someone else to use by interest groups who had other objectives.

INTERNATIONAL COMPARISONS

Transit trends in other industrialized countries are similar to those described for America. Until the mid-1960s, the demand for public transportation was so great that most agencies managed to operate without external financial assistance. But a number of factors, including increased auto ownership, have changed this situation. Transit has survived because of shifts in attitudes: public transit has come to be regarded as a social service rather than a commercial enterprise. Government assistance has increased to the level of two-thirds of cost in most industrial countries (62% in the United States), and officials have now begun to question whether transit is providing social benefits equivalent to its cost.

Bly, Webster, and Pounds (1980) provide an excellent summary of these trends for 15 nations. When the effects of fare subsidies and increased service have been removed, transit ridership has decreased in Australia, the United Kingdom, New Zealand, and the United States. Continental European nations like Norway, the Netherlands, France, and

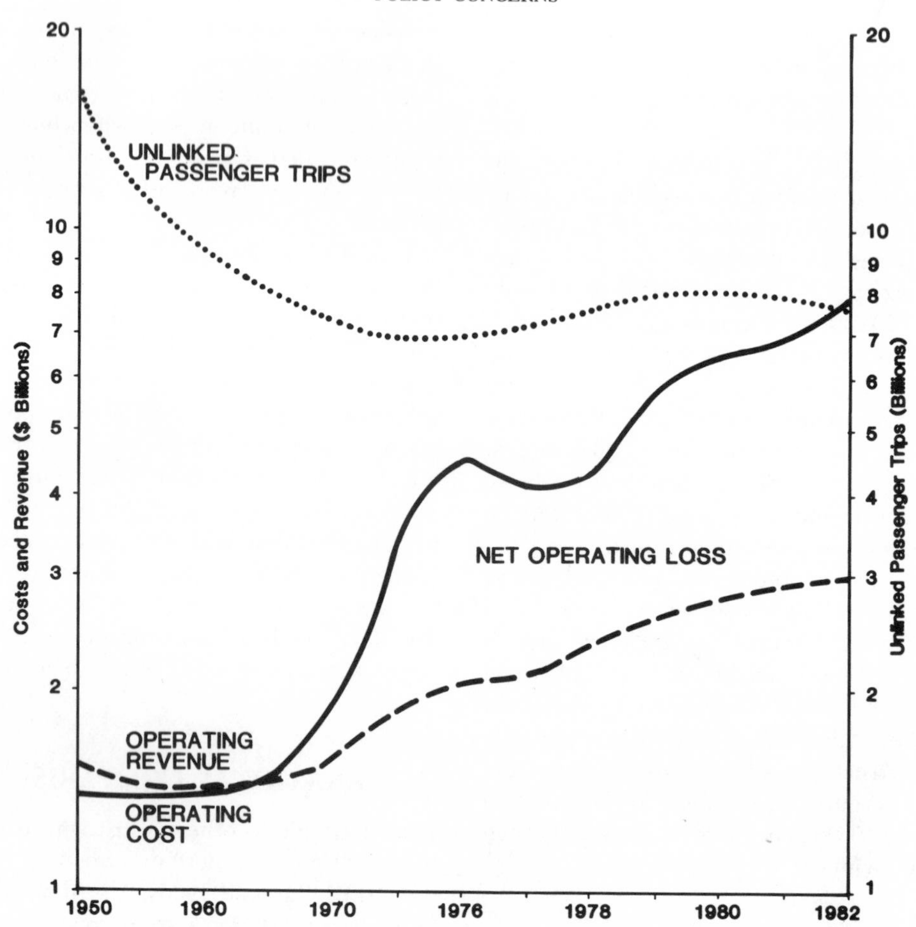

Figure 10-2. Relationships between transit passengers, operating cost and operating revenue, 1960–1982. Revenues covered operating costs until 1963. Marked divergence occurs after 1970 when first state and then federal operating assistance was made available. From *APTA Fact Book* 1981, and *National Mass Transportation Statistics*, 1981 and 1982.

Sweden have maintained patronage which Bly, Webster, and Pounds ascribe to their more compact urban spatial structure.

Canada presents an interesting example which does not fit well with this explanation. Ridership has remained fairly constant when the effects of fare subsidies and increased service have been removed, but there are sharp regional differences between the cities. In Toronto, for example, many residents and public officials fear the "Manhattanization" of their city and have used transit to disperse development without reliance on the automobile. Local government has implemented a host of land use and zoning controls designed to shape and regulate growth. The overarching goal is to encourage the development of suburban, transit-oriented, residential and employment nodes. Currently 75 to 80% of workers travel to the Toronto CBD by transit (bus and rail). Current policies, which explicitly tie zoning regulations to transit use, are aimed at a 50% mode split for suburban travel. Parking controls are used to achieve this objective.

Although auto ownership rates in Canada are roughly equivalent to those in the United States, most individuals in Toronto own and use a transit pass. While transit ridership has declined within the CBD, it has increased in the Toronto suburbs. This increase is the result, in part, of an image of public transportation in Canada that is very different from that in the United States: transit is perceived as being a service for the community as a whole, not just for the transit dependent.

POLITICS OF TRANSIT

As previously indicated, the program of federal transit assistance in America began as a minor element of the urban renewal program and grew rapidly in the 1970s to the largest discretionary program in the domestic budget. In part this was a reaction to the negative impact of the automobile and the disruption caused by freeway construction, and in part it was the result of clever politics by interest groups who stood to gain. The impress of the political influence is apparent in every urban area. Sleek, new, rapid transit vehicles move commuters in San Francisco, Washington, D.C., Atlanta, Baltimore, and Miami; streetcars have been reintroduced in San Diego, Portland, and Buffalo; and new buses are apparent in every urban area and many rural areas as well.

Transit provides a fascinating illustration of the way in which politics influences how we use resources. Seattle has received four times as many federal dollars for capital projects per person than the average for all-bus cities (Briggs, 1980). When bus projects are combined with rail, Washington and Atlanta lead the nation. In all three cities, the skill with which local administrators prepared projects and the help of influential congressional leaders representing these cities, not need, has been responsible for the federal largesse.

Federal transit assistance was initiated in 1961 at the behest of the big city mayors (Smerk, 1974). Eastern cities and Chicago were alarmed by legislation passed in 1958 that reduced the regulatory constraints on railroads that wished to abandon commuter services. They sought capital assistance to purchase and renovate rail systems but achieved only a small loan program because they could not obtain the support of organized labor and cities without rail systems.

A broader constituency was assembled to support the legislation that became the Urban Mass Transportation Act of 1964. Grants for capital assistance were made available for the purchase of ailing systems as well as for the construction of new systems, and organized labor was given guarantees against adverse impact from federal assistance.

Organized labor supported federal assistance only after Section 13(c) of the act guaranteed that employees of private transit systems would suffer no "worsening of their positions with respect to their employment" when the systems were transferred to public ownership. And Section 13(c) has been used by labor unions to obstruct changes in work rules and capital investments that could have increased labor efficiency. The concern of labor was genuine. Support for modernization of public transit had been promoted in cities like San Francisco and Miami on the basis of claims that new capital investment would reduce the dependency upon labor. Automated transit systems would only require capital funding and, once operating, would be self-supporting out of the fare box because of their low operating costs. Not only were funds to be made available for construction of rapid transit facilities, but they were also to be available to help communities purchase failing private bus companies. Labor's concern over the continuation of collective bargaining rights and the maintenance of the level of transit employment was warranted, given the plans that were being promulgated.

The political muscle of organized labor was apparent in the early attempts to obtain federal assistance. Organized labor had refused to support earlier legislation and only after labor had endorsed the legislation in 1964 did President Johnson place the Urban Mass Transportation Act on his "must pass" agenda (Smerk, 1974). Without labor's endorsement, transit assistance would not have passed, and labor has played a similar role in subsequent amendments to the 1964 act. Seldom do transit officials recognize the pivotal role of labor when they criticize organized labor for declining productivity in the industry. Without labor's support, transit would not have survived outside the major metropolitan areas.

Throughout the 1960s the mayors and transit managers in the largest cities together with organized labor gradually enlarged funding for capital projects. Many private systems in small and medium-sized cities passed from private to public ownership assisted by federal grants. This expanded the transit constituency in Congress because representatives from these cities realized tangible benefits for their constituents.

Even traditionally conservative areas like Orange County, California, saw the opportunity to obtain federal funds. A transit district was created by referendum in 1970 to purchase the eight-bus private system that was failing. It began with a single employee and a $300,000 budget but expanded quickly to take advantage of federal and state funds. By 1975 the transit district was operating a fleet of 60 sleek, orange and white buses out of a modern garage with 80% of the cost paid for by a $10 million federal grant. By 1985 it had a fleet of 434 buses, three operating garages, and an annual operating budget of $107 million. It had even purchased a right-of-way for a future rail connector to Los Angeles. All of these changes were accomplished with federal and state subsidies.

Environmental and Energy Concerns

Although transit's direct constituency was relatively small, the big city mayors, transit managers, and labor leaders took advantage of public opinion to increase financial support for transit. In the early 1970s, America's attention had shifted to new problems—dwindling energy supplies, air pollution, traffic congestion, urban sprawl, and the needs of the increasingly vocal elderly and handicapped sectors of the population. Transit adopted new goals. It would reduce pollution and congestion, save energy, curb urban sprawl, and provide essential mobility to the elderly and handicapped as well as the poor. Transit advocates used these concerns not only to increase funding but also to obtain federal operating assistance for transit after 1974. Even small cities and rural areas that had no transit service began systems. Funds to cover 80% of capital costs and 50% of operating costs could be obtained from the federal government with most of the balance available from the state transit programs. Many systems recovered as little as 10% of their operating costs from fares.

Congress backed into a deeper and deeper commitment to mass transit without reaching agreement on the objectives to be achieved and without anticipating the eventual cost of federal involvement. Federal subsidies were approved without systematic inquiry into the ability of local governments to shoulder the cost of subsidy without federal assistance. And federal aid was universally available without any standard of need or measure of merit.

The operating loss increased sharply after the introduction of federal operating assistance (Figure 10-2). Service miles operated increased by 5% between 1975 and 1980 and passengers by 13%, but meanwhile fares were declining and operating costs increasing. The ratio between operating revenue and operating cost per mile declined from 0.53 to 0.41 (Fielding, 1983a). Transit systems have been modernized as a result of governmental intervention, and ridership has responded to increased service and lower fares, but the major beneficiary has been transit labor. Pickrell (1983) estimates that between 1970 and 1980, the cost of operating a vehicle mile of transit increased from $1.89 to $2.89 (in 1980 dollars). Labor expenses accounted for 46% of the increase—18% for declining productivity and 28% as increased labor prices.

Organized labor has been an effective lobbyist for transit programs, and their members have been the principal beneficiaries. If this bothers you, then consider the price supports that agricultural lobbyists have won for farmers. The crop support payments for cotton and tobacco are paid so that there will be no worsening of conditions for farmers. Transit subsidies have had a similar benefit for transit labor. Both are income transfers that have had social benefits.

New Federalism

Responding to public concern over the rising cost of federal programs, the Reagan administration unveiled a policy of New Federalism in 1981. The policy emphasized local control of public programs and expenditures, fiscal prudence and accountability, elimination of costly federal regulations, and the increased involvement of the private sector in public service provision. Like many other social programs, transit became a vulnerable target for administration cost cutting.

But after the first four years of the Reagan administration not much had changed because of strenuous opposition in Congress. Nearly every urban area in the nation and many rural areas as well have benefited from federal assistance. An intense and successful lobbying effort was conducted by labor and management to prevent reductions. Only regulatory reform was achieved because this could be accomplished administratively. Other policies were stalemated. Federal operating assistance continued, but the increases sought by transit agencies and the phase-out sought by the administration were stalled by congressional opposition until an infusion of new money provided the opportunity to restructure the program.

The Surface Transportation Act of 1982 changed the nature of transit assistance. For the first time, transit obtained a portion of the federal tax on gasoline. Limits were placed on operating assistance, but transit agencies had more money for capital projects, and they could still use state and local funds for operating assistance. Restraints placed on operating assistance have had beneficial effects. Transit management and labor organizations have begun to emphasize the need to manage resources, especially labor resources, efficiently.

This austere financial outlook provides geographers with an opportunity to become involved in transit planning and management. Just as the geographer's skill in analyzing patterns of development in terms of the causal processes has helped forecast transportation demand and provide an understanding of how politics has influenced the development of public transit, these same skills can help managers plan for "hard times."

IMPROVING TRANSIT PERFORMANCE

In order to improve transit performance, directors, managers, and consultants must understand the elements of transit and how these elements are combined to produce service under different circumstances. The geographer's systematic approach as well as the technique of regional analysis can be helpful for these purposes.

Using the *National Mass Transportation Statistics* for 1980 and the statistical techniques of factor analysis, geographers have identified the seven factors that best represent the dimensions of transit performance (Fielding, Brenner, & Babitsky, 1985a). Three dimensions were identified: *inputs* (labor, capital, and fuel) used to produce service; *outputs* (miles and hours of service produced); and *consumption* (passengers, passenger miles, and revenue used). The relationship between inputs and outputs measures *cost efficiency;* between outputs and consumption, *service effectiveness;* and between inputs and consumption, *cost effectiveness.* Highest scoring variables on each factor were chosen as indicators for concepts associated with the three dimensions (Table 10-4). These seven indicators provide markers for a broad range of statistical data presented in the national statistics.

The performance of any transit agency can be described using these indicators. Achievements can be monitored over several years with the confidence of knowing that these indicators capture changes in a wide range of vital transit functions. Managers and planning analysts using these seven indicators are better able to detect trends and to recommend changes that will improve performance.

Comparisons over time for one agency can be more effective if achievements can be compared with other agencies operating under similar conditions. Here the geographic

Table 10-4. Performance dimensions and indicators

Dimension and concept	Best indicator
Cost efficiency	
Output per dollar cost	Revenue vehicle hours per dollar operating expense *(RVH/OEXP)*
Labor efficiency	Total vehicle hours per employee *(TVH/EMP)*
Vehicle efficiency	Total vehicle miles per peak vehicle *(TVM/PVEH)*
Maintenance efficiency	Total vehicle miles per maintenance employee *(TVM/MNT)*
Service effectiveness	
Utilization of service	Passenger trips per revenue vehicle hour *(TPAS/RVH)*
Safety	Total vehicle miles per reported accident *(TVM/ACC)*
Cost effectiveness	
Revenue generation per operating expense	Operating revenue per dollar of operating expense *(CORV/OEXP)*

Note. From Performance evaluation for bus transit by G. J. Fielding, M. E. Brenner, & T. T. Babitsky, 1985a, *Transportation Research,* 19A; 73–82.

concept of regional typologies can be helpful in transportation management. Using the same national transportation statistics for 1980, geographers have shown how transit agencies can be assigned to systematic groups using the statistical techniques of cluster analysis (Fielding, Brenner, & Faust, 1985b). Twelve peer groups for bus transit were differentiated by size (measured by number of vehicles required for peak operations and total revenue miles reported) and characteristics of operation (measured by average speed and the relative emphasis given to peak-period service).

These geographical techniques have many applications for transportation analysis. Performance indicators are used by the departments of transportation in California, New York, and Pennsylvania to analyze whether service is being provided efficiently and used effectively (Miller, 1979). Congress mandated in 1982 that triennial audits be conducted for every transit agency receiving federal financial assistance, and the Urban Mass Transportation Administration is using both the indicators and the peer group concept to assess the "vital signs" of each agency and to compare its performance with other agencies in the same peer group. Consulting firms hired to analyze such issues as the costs of producing transit, the effect of capital investments on future operating costs, and whether private firms can operate transit service more efficiently than public providers are using these same geographical concepts.

Case Study

The Niagara Frontier Transportation Authority, New York, operates a fleet of 483 buses in the Buffalo urbanized area. Some 369 buses are required for their peak afternoon schedule, and 119 buses, one-third of peak demand, for the midday. This gives a peak-to-base requirement of 3.10 and a vehicle speed of 11 miles per hour, which is representative of transit systems serving both suburban and central city locations. Using the techniques of cluster analysis, Buffalo was placed in a peer group with similar systems. The 32 other systems in the group are primarily located in the eastern half of the United States and serve medium-sized metropolitan areas or are in the suburbs of major metropolitan areas (Figure 10-3). These transit systems are characterized by a higher-than-av-

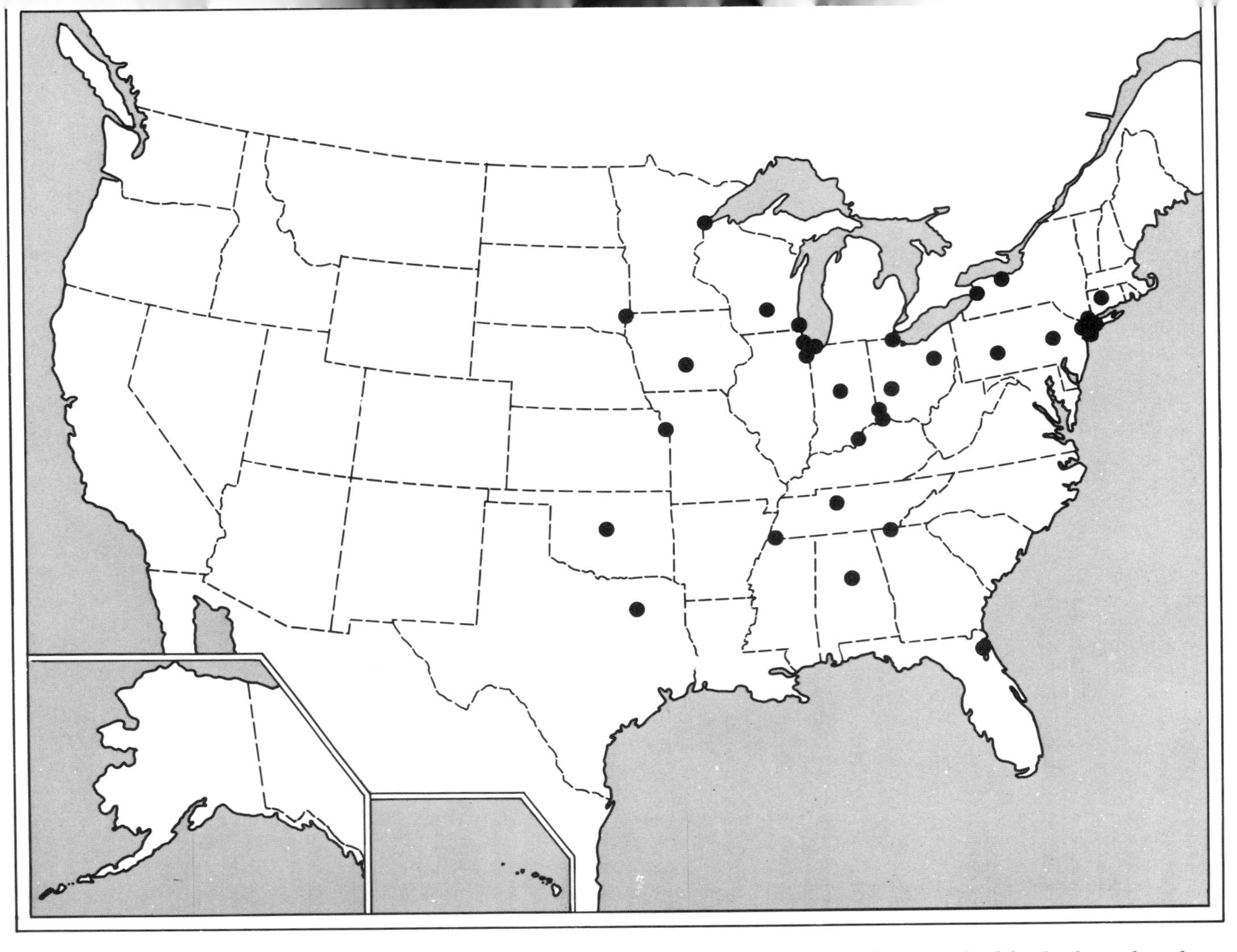

Figure 10-3. Distribution of transit agencies in Peer Group 8. These agencies are distinguished by high peak-to-base ratio, resulting from an orientation to commuter service.

erage peak-to-base ratio as service is designed to accommodate commuters. Buffalo operates more peak vehicles than is characteristic for this peer group, and the performance indicators must be interpreted by taking the size of the agency into consideration because there are no economies of scale in transit.

The Transportation Authority has been deeply involved in developing a light rail system and felt it had neglected the bus system. Therefore, the governing board decided to examine how the bus system was performing in terms of the major dimensions of transit performance; not only over the three years from 1980–1982, but also in comparison to the achievements of the other 32 agencies in the same peer group.

A statistical technique was used to normalize the data calculated for the seven performance indicators plus two additional indicators. The *standard score* (z score) for each indicator was computed by subtracting Buffalo's performance from the mean score for the group and dividing by the standard deviation. The statistics for Buffalo can then be graphed as falling above or below the group mean (Figure 10-4). Each financial year is then shown for all indicators. Nine

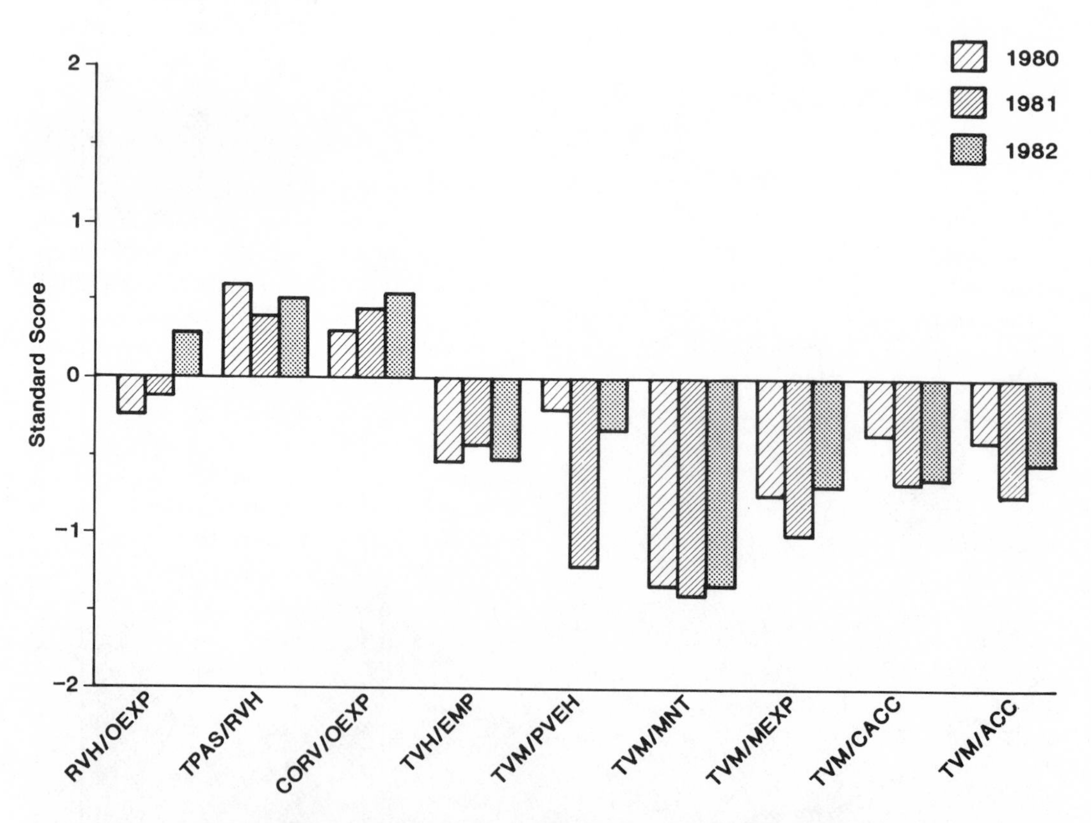

Figure 10-4. Transit performance for Buffalo, New York, 1980–1982. Indicators are expressed so that higher standard scores are better than lower. The 0 represents the group mean so that positive scores represent performance better than the group mean while minus scores lower on that indicator. Each financial year is shown separately so that change over time is apparent. Two additional indicators are included alongside those listed in Table 4. These provide additional measures for maintenance efficiency *(TVM/MEXP)* and safety *(TVM/CACC)*.

indicators were used embracing the seven dimension of performance including two indicators for maintenance and safety.

Performance for Buffalo was better than the peer group mean on the first three indicators in 1982. Considering the size of the Buffalo system, this is better than what might normally be expected. Annual improvement in cost efficiency *(RVH/OEXP)* and cost effectiveness *(CORV/OEXP)* is especially noteworthy. Maintenance efficiency, measured by the two indicators *(TVM/MNT* and *TVM/MEXP)*, warrants concern. It requires more employees and greater cost per vehicle mile to maintain the vehicles in Buffalo than is normal for this peer group. Buffalo operates service out of three depots, and the facilities in two of them are outmoded. They are paying a high price for their inefficient maintenance facilities, and the directors are considering capital investments that will correct the problem. In this manner, the result of geographical research can be helpful to policy makers. It helps focus attention on key indicators of performance over time as well as facilitating comparison with other transit systems operating under similar conditions.

The exceptionally low performance on vehicle use efficiency *(TVH/PVEH)* in 1981 is an anomaly. It occurred because 115 vehicles were added to peak-hour service with little use for them during the midday. The consequences for vehicle efficiency are apparent on the graph. Peak-period service was reduced in 1982 when 103 of the 115 buses were removed.

Safety indicators were also a concern. More accidents, both collision accidents *(TVM/CACC)* and total accidents *(TVM/ACC)*, occurred in Buffalo than would normally be expected for the peer group. Considering the adverse driving conditions in winter in Buffalo, however, the high accident rate is understandable.

Similar profiles have been created for all 335 bus transit agencies in America that receive federal operating assistance. Explanations for these profiles provide an example of how geographical techniques can be used to improve transportation management. Other geographers have used similar techniques to explain the effects of climate on bus durability and maintenance costs (Arlinghaus & Nystuen, 1985).

CONCLUSION

After two decades of expansion, the American transit industry has begun to consolidate its rejuvenation and emphasize more efficient production of service. Transit has not been able to restructure cities as it did in the first decades of this century. Nor has transit been especially effective in conserving energy or reducing air pollution. However, transit does play an important role in reducing peak-hour congestion in the central city and along arterial corridors between the CBD and suburbs. Peak-hour commuter travel is a market that transit has retained and is now expanding to accommodate travel to and from the regional megacenters emerging in the outer suburbs.

Transit is apparent in all American cities whether it be offered by rapid rail, streetcar, bus, or paratransit modes. In almost every instance it represents the impress of politics on the geography of metropolitan areas. Geographers can play a role in consolidating these political accomplishments. Managers, planners, and consultants are needed who can understand the role that transit can play in urban development and how the elements of transit might be analyzed and compared in a regional framework so that transit agencies produce service efficiently and ensure that it is used effectively.

References

Altshuler, A. A., Womack, J. P., & Pucher, J. R. (1979). *The urban transportation system: Politics and policy innovation.* Cambridge: MIT Press.

American Public Transit Association (APTA). (1981). *Transit fact book.* Washington, D.C.: APTA Statistical Department.

Arlinghaus, S. L., & Nystuen, J. D. (1985). *Climatic effects on bus durability.* Ann Arbor, Mich.: University of Michigan, Transportation Research Institute.

Bly, P. H., Webster, F. V., & Pounds, S. (1980). The effect of subsidies on urban public transport. *Transportation, 9,* 311–331.

Briggs, R. (1980). The impact of federal local public transportation assistance upon travel behavior. *Professional Geographer, 32,* 316–325.

Crain, J. L., & Flynn, S. D. (1975). *Southern California Rapid Transit District 1974 strike impact study.* Sacramento, Calif: California Department of Transportation.

Fielding, G. J. (1983a). Changing objectives for American transit. Part I. 1950–1980. *Transport Reviews, 3,* 287–299.

Fielding, G. J. (1983b). Changing objectives for American transit. Part II. Management's response to hard times. *Transport Reviews, 3,* 341–362.

Fielding, G. J., Brenner, M. E., & Babitsky, T. T. (1985a). Performance evaluation for bus transit. *Transportation Research, 19A,* 73–82.

Fielding, G. J., Brenner, M. E., & Faust, K. (1985b). Typology for public transit. *Transportation Research, 19A,* 269–278.

Giuliano, G. (1984). *Standard transportation forecasting techniques: How they fail.* Staff Paper No. ITS-UCI-SP-84-5. Irvine, Calif.: University of California, Institute of Transportation Studies.

Hamer, A. M. (1976). *The selling of rail rapid transit: A critical look at urban transportation planning.* Lexington, Mass.: D. C. Heath.

Meyer, J. R., & Gomez-Ibanez, J. A. (1981). *Autos, transit and cities.* Cambridge: Harvard University Press.

Miller, J. H. (1979). An evaluation of allocation methodologies for public transportation operating assistance. *Transportation Journal, 19,* 40–49.

Orski, C. K. (1982). The changing environment of urban transportation. *Journal of the American Planning Association, 48,* 309–314.

Pickrell, D. (1983). *The causes of rising transit operating deficits.* Final Report No. UMTA-MA-11-0037. Cambridge: Harvard University, School of Government.

Pushkarev, B. S., & Zupan, J. M. (1977). *Public transportation and land use policy.* Bloomington: Indiana University Press.

Smerk, G. M. (1974). *Urban mass transportation: A dozen years of federal policy.* Bloomington: Indiana University Press.

Urban Mass Transportation Administration. *National urban mass transportation statistics: Section 15 reporting system.* Washington, D.C.: U.S. Department of Transportation. Annual.

Wachs, M. (1982). Ethical dilemmas in forecasting for public policy. *Public Administration Review, 46,* 562–567.

Wachs, M. (Ed.). (1985). *Ethics in planning.* New Brunswick, N.J.: Transaction Books.

Webber, M. M. (1976). The BART experience—what have we learned? *The Public Interest, 35,* 79–108.

11 LAND USE IMPACTS OF TRANSPORTATION INVESTMENTS: *HIGHWAY AND TRANSIT*

GENEVIEVE GIULIANO
University of California, Irvine

INTRODUCTION

The role of the transportation system in fostering growth and affecting land use structure has been of great interest to academics, planners, investors, and politicians. The historical record demonstrates that land use changes and transportation investments go hand in hand. Indeed, the evolution of urban form appears to be so clearly linked with transportation technology that Muller formulates his explanation of urban form in Chapter 2 in terms of transportation eras. On a more mundane scale, transportation investment decisions are the subject of much debate and lobbying, particularly at the local government level. Transportation investments are advocated as economic necessities. Public transit investments are pursued as the key to economic vitality in central cities, while in the suburbs new freeways are pursued as the means for continued growth and development. Similarly, transportation system constraints, which are manifested in extreme levels of traffic congestion, are frequently cited as a cause of urban deterioration or stagnation.

The ability of transportation investments to shape or influence urban structure is also a widely held conviction. Thus transit investments are also advocated for their ability to promote high-density development, whereas freeways are criticized for their role in causing urban sprawl. Among those who prefer low-density development, the modal preferences are reversed, but the belief is the same: transportation investments are *perceived* as having a significant impact on urban structure. The purpose of this chapter is to explore this idea; to examine the theoretical basis for expecting that transportation influences land use and to determine the extent to which such impacts can be documented.

The chapter begins with a brief description of factors that must be considered when examining land use impacts of transportation investments and of the mutual dependence between land use and transportation changes. In the second section, the conceptual relationship between transportation and land use is described in the context of accessibility. The third section reviews the major land use/transportation theories and models that have been developed and presents an example of how one model has been used in transportation policy analysis. The final section summarizes the findings of empirical studies of highway and transit impacts and draws conclusions based on these findings.

Overview of Transportation/Land Use Relationships

There are four factors that must be considered when examining transportation and land use impacts. First, land use must be distinguished from growth. This is difficult, because most of the land use changes we observe in conjunction with transportation improvements tend to be manifestations of economic growth. These manifestations can take many forms, however. For example, commercial development can occur in the form of garden office developments as well as multistory skyscrapers. Our emphasis here is on the *form* these land use changes take, rather than on the economic implications of these changes.

The second factor to be considered is the level of intensity of the transportation investment. Improvements that represent significant technological change, such as the construction of electric streetcar systems in the late nineteenth century, can be expected to have much greater impact than changes that are more incremental. The scale of the investment relative to the existing system is also important. For example, the first 10 freeway miles constructed within an urban area should have a greater impact than the last 10 miles, all other things being equal, because the relative influence of those first 10 miles will be so much greater.

A third and related factor is the level of analysis. In a large metropolitan area, even a very large project in monetary terms, such as a new freeway or transit system, would have a very limited impact. If for example a new freeway segment carries 100,000 work trips per day in an area with 3 million daily work trips, its impact is hardly perceptible from a regional perspective. This is not to say, however, that significant changes might not take place in the vicinity of the new system. In examining land use impacts, then, the level of analysis must be defined, and generally both micro (local) and macro (regional) impacts are of interest.

The fourth factor is the longevity and durability of urban structure. The residential and commercial structures that make up the urban landscape are valuable and long-lived investments. Once an area is built up, changes in the capital stock tend to be marginal, because the cost of reconstruction is usually high. Thus the location of the transportation investment is important, since the potential rate of change (in terms of land use) is much lower in developed areas than in undeveloped areas.

Mutual Dependence of Land Use and Transportation Changes

It is particularly difficult to examine the land use impacts of transportation because land use and transportation are mutually dependent. A simple illustration of this relationship is presented in Figure 11-1. The characteristics of the transportation system determine accessibility, or the ease of moving from one place to another. Accessibility in turn affects the location of activities, or the land use pattern. The location of activities in space affects activity patterns which result in travel

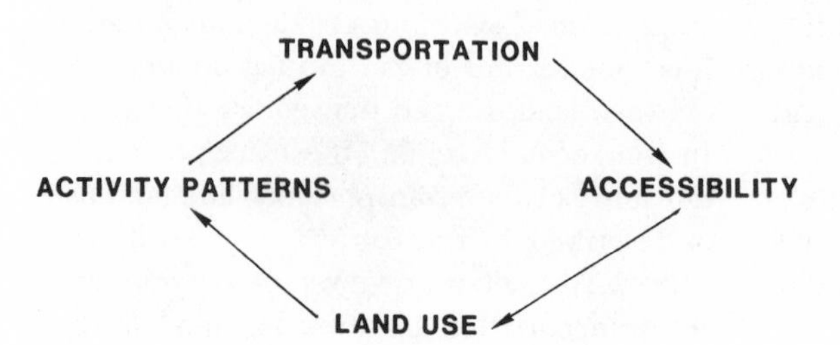

Figure 11-1. The transportation–land use relationship.

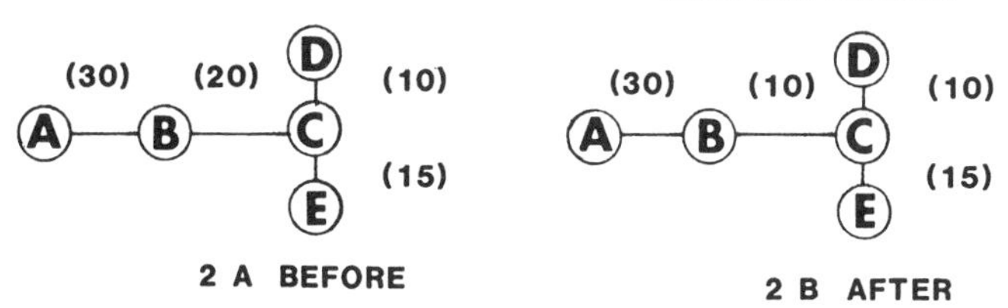

Figure 11-2. Accessibility in a simple network.

patterns (a set of trips within the region). These in turn affect the transportation system. The model in Figure 11-1 shows that a change in land use will affect transportation, just as transportation affects land use. Note that this model does not imply anything about the strength of these relationships. It serves only to illustrate the interdependence of empirically observed land use and transportation changes. It is quite difficult to isolate the impact of transportation on land use, as it requires examining one half of the figure, holding the other half constant. The land use system is dynamic, however, and what we observe in empirical analysis is the totality of relationships expressed in Figure 11-1.

TRANSPORTATION IN THE CONTEXT OF ACCESSIBILITY

The basic concept underlying the relationship between land use and transportation is accessibility. In its broadest context, accessibility refers to the ease of movement between places. As movement becomes less costly—either in terms of money or time—between any two places, accessibility increases. The propensity for interaction between two places increases as the cost of movement between them decreases. Consequently, the structure and capacity of the transportation network affect the level of accessibility within a given area.

Accessibility also includes the concept of attractiveness: the opportunities or activities that are located in a given place. Thus the ease of movement between places, as well as the attractiveness of these places as origins or destinations, is expressed in accessibility.

More specifically, we can define accessibility as the attractiveness of a place as an origin (how easy is it to get *from* there to all other destinations) and as a destination (how easy is it to get *to* there from all other destinations). Note that these two measures are not symmetrical. In the first case, the emphasis is on accessing opportunities located in other places; in the second case the emphasis is on the opportunities located in that place. It is important to incorporate attractiveness in the concept of accessibility because of the unique function of transportation as the mechanism for spatial interaction. Trips are usually made only in order to engage in some other activity such as working or shopping, and therefore the opportunities made available by the transportation system are the important factor.

How does a change in the transportation network affect accessibility? We can illustrate this with a simple network example, as illustrated in Figure 11-2. Each node represents a possible origin/destination, and the numbers above each link represent travel times. We begin with the network in Figure 11-2a, and make an improvement such that travel time between B and C is reduced by one-half, as shown in Figure 11-2b. Network accessibility can be measured by calculating the travel time from each node to every other node and summing over each node, as in Tables 11-1a and 1b. Each row in the matrix corresponds to travel times from one to every other node. The row sums are the accessibility measure for each node. (Since we have used travel times, lower numbers mean greater accessibility.) Comparing the row sums of Tables 11-1a and 1b, we see that the improvement made in the network not only improves the accessibility between B and C;

Table 11-1. Accessibility matrix for networks shown in Figure 11-2: internode travel times

(a) Network A							(b) Network B							Percent change in travel time
From/To	A	B	C	D	E	Σ	From/To	A	B	C	D	E	Σ	
A	0	30	50	60	65	205	A	0	30	40	50	55	175	−14%
B	30	0	20	30	35	115	B	30	0	10	20	25	85	−26
C	50	20	0	10	15	95	C	40	10	0	10	15	75	−21
D	60	30	10	0	25	125	D	50	20	10	0	25	105	−24
E	65	35	15	25	0	140	E	55	25	15	25	0	120	−14

it also benefits the entire network. This simple example demonstrates how an improvement in a network (say a new freeway link between two places) affects both the accessibility of the two directly connected places and the accessibility of the entire network. And as accessibility increases, the level of spatial interaction increases, because travel has in effect become less costly.

How does this change affect land use? As more interaction occurs, more activities will locate in those places in response to their increased accessibility. In our example, nodes B, D, and C have benefited most from the improvement, and we would expect the greatest land use changes to occur at these nodes. We can think of this process in terms of regional growth. As population and employment increase, their relative location will be affected by the transportation system. In a theoretical world where capital is completely mobile, then, transportation system changes would result in observable shifts in activity location.

The concept of accessibility can also be used to predict changes resulting from a deterioration in the transportation system. Suppose now that the link between two places has reached capacity, and travel speeds have deteriorated. Accordingly, the level of accessibility deteriorates, generating incentives for activities to shift away from these two places.

These examples illustrate how transportation investments are justified. Traditionally, transportation investments are evaluated with some form of benefit/cost analysis. Like other public investments, transportation projects are deemed worthwhile if the benefits outweigh the costs. Two types of benefits are generally considered: traveler benefits and local economic benefits. In the case of congested central cities, transportation investments are viewed as necessary in order to avoid a loss of economic activity and allow continued growth. Subways and other high-capacity transit systems are promoted to avoid gridlock and maintain economic vitality. In the suburbs, transportation investments (usually highways) are advocated as a means to attract economic growth by increasing access to skilled labor markets and affluent consumers. Even a very costly investment such as a subway system might be considered worthwhile if the anticipated benefits are sufficiently large.

THEORIES OF LAND USE/TRANSPORTATION INTERACTION

Research on the relationship between land use and transportation has generated numerous theories and models. This section discusses theoretical expectations regarding transportation and land use changes.

Theories of land use/transportation attempt to explain the effect of transportation costs on location decision making. Theories posit a set of behavioral or causal relationships that describe location decisions in terms of the factors determining such decisions. Models are to be distinguished from theories. Models are based on observed correlations between factors related to location, rather than causal relationships. Location theories have been developed at both the disaggregate and aggregate levels of analysis. Disaggregate theories attempt to explain individual location behavior, whereas aggregate theories focus on regional outcomes. Regional outcomes are derived from disaggregate theories by generalizing individual location behavior; however, as was pointed out in earlier chapters, aggregate models cannot be "broken down" to the individual level. Location theories typically focus on only one type of land use (for example, residential), one type of transportation (goods movement or passenger), and one mode of transportation (Meyer & Gomez-Ibanez, 1981). Though individually incomplete, these theories describe the determinants of location and help us to understand the factors involved in the location process. Theories of residential and employment location will be reviewed as they are most relevant to the subject of this chapter.

Residential Location Theory

Theories of residential location were developed in the 1960s by Alonso (1964), Mills (1972), and Muth (1969). Residential location is modeled as a utility maximization problem. Location choice is predicted given the following assumptions: (1) the total amount of employment is fixed and located at the center of the city, (2) each household has one worker, (3) housing is homogeneous (that is, location is the distinguishing factor), and (4) transportation cost is a function only of distance in all directions. Each household chooses the best combination of housing, transportation, and all other goods for its given level of income.

The unit cost of housing must decline with distance from the center of the city, since transportation (commute) costs increase with distance. (If this were not true, all households would locate at the center of the city.) The best location for a given household is the point at which the marginal savings of housing are equal to the marginal cost of transportation, or the savings in housing are just offset by the increase in transportation cost. The particular location of any given household depends on its relative preferences between housing and transportation. It is generally assumed that housing preferences are the stronger of the two, and therefore the higher the income, the more housing will be consumed, even at the cost of additional commuting.

What does the theory predict in response to a change in transportation cost? If commuting costs are reduced, the theory predicts movement away from the center, as shown in Figure 11-3. The curve $-\Delta H_d$ is the savings on housing costs and T_{d1} is the original transportation cost curve. The intersection of the two curves at point D_1 is the equilibrium location. When transportation cost is reduced, as denoted by the curve T_{d2}, the equilibrium location shifts to D_2.

Commuting costs can be measured in terms of both time and money, and therefore they can be reduced in two ways. Either actual monetary costs can be reduced, such as by a reduction in the price of gas, or travel speed can be increased. Different types of travel-cost reductions are likely to have differing impacts across income groups. A simple price

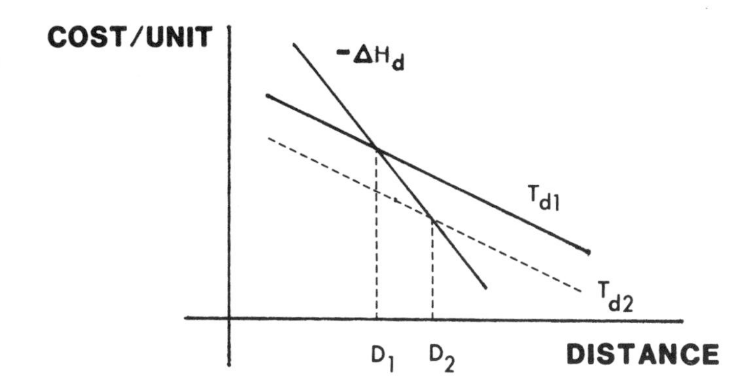

Figure 11-3. Effect of reduction in transportation cost on residential location.

reduction will affect all income groups in the same way: more income would be available for housing, leading to a more distant location choice. However, travel-time reductions tend to have a greater impact on higher-income groups, because saving time is worth more to higher-income groups than to lower-income groups (all other things being equal). Consequently, the decentralizing effect of a travel-time reduction is greater for higher-income groups.

Suppose that a transportation improvement is accompanied by a price increase, say by the construction of a commuter rail system that improves travel time and increases transit fares. The price increase could offset the value of time saved for the low-income household, but the opposite would occur for the high-income household. Thus the benefits of this investment would accrue disproportionately to higher-income households.

Location theory also suggests that land values will be affected by a change in transportation cost because of the shifts in housing demand. As more households seek locations farther from the center, the relative price of land closer to the center will decline. The degree to which these changes take place depends, however, on the relevant cross-elasticities, or the rate at which demand for one factor changes in response to a change in the price of another factor.

On a regional scale, residential location theory predicts declining residential densities. Lower-income households would consume less housing and locate closer to the center, whereas higher-income households would consume more housing and locate farther from the center. Furthermore, the theory predicts that transportation improvements that reduce the cost of commuting lead to greater decentralization. In a very general way, the empirical evidence tends to support residential location theory. Population densities tend to decline with increasing distance from the city center (Clark, 1951; Guest, 1975). The density gradients of central cities have also declined over time in conjunction with transportation system improvements.

Employment Location

Employment location theories generally deal with certain types of firms or economic activities. It is useful to distinguish between market-sensitive and non-market-sensitive activities. Market-sensitive activities (such as sales and services) depend on accessibility to consumers, whereas non-market-sensitive activities include industrial or manufacturing activities.

Retail and Service Activity Location

The classical theory of market-sensitive activities is central place theory, formulated by W. Christaller and A. Lösch. More recently, gravity model formulations of consumer shopping behavior have been developed (Huff, 1964; Lakshmanan & Hansen, 1965). These models are similar in form to the trip distribution models described in Chapters 5 and 8. Shopping destination choice is modeled as a function of the attractiveness of each possible alternative and its associated travel cost. Berry (1967) has shown that these formulations are consistent with central place theory, so the two approaches are complementary. Since the distance that consumers are willing to travel varies for different types of goods, the theory gives rise to a hierarchical pattern of market centers. According to this theory, a reduction in transport cost would result in larger, more dispersed centers, because consumers would be willing to travel farther to shop, and an increase in transport cost would result in smaller, more concentrated centers, all other things being equal.

Only remnants of central place structure remain in contemporary metropolitan areas. However, empirical evidence supports the behavioral concepts of the theory. Market centers have enlarged in conjunction with transportation improvements and population decentralization (Berry, 1967; Muller, 1981).

Industrial Location

Industrial location theory was developed originally by A. Weber and expanded by W. Isard, L. Moses, and others. The theory predicts that the firm locates so as to minimize total transport costs, or the transport costs of both inputs and outputs, assuming an optimal level of production. In contrast to the location theories previously discussed, the focus here is on goods movement rather than passenger movement. In fact, workforce availability is assumed. The importance of transport costs in determining industrial location has declined in the U.S. manufacturing sector as a result of technological shifts and a highly developed goods movement network. Consequently, transport-cost location models are no longer adequate predictors of industrial location, implying of course that a change in transportation would not have a significant impact on industrial location. Rather, agglomeration economies, economies of scale, linkage patterns, infrastructure availability, and labor-force availability have been identified as important industrial location factors (Norcliffe, 1975).

Business Location Theory

A third type of employment location theory is essentially an extension of the residential location theory described above. In this case, however, the location of employment is not fixed; rather, employment location is a function of commuter transportation costs and land values. How does commuting cost affect employment location? If commuting cost is reduced, then the net wage of the employee increases. Residential location theory predicts that reduced commuting cost will decentralize housing location. As this occurs, central-area land values decline relative to land values farther out. The drop in relative value of the central locations makes them more attractive for the employer. Furthermore, in a competitive market an employer should be able to pay a lower wage when commuting cost is reduced, enhancing the firm's competitive advantage. That is, an employee will accept lower wages in return for lower commuting cost, since the net wage remains the same. The net effect of a reduction in commuting cost then is *centralization* of employment as well as *decentralization* of residences. Conversely, an increase in commuting cost has just the opposite effect: relatively higher land values at the center, decentralizing employment, and centralizing residences. Table 11-2 summarizes these effects. The extent to which these changes occur depends on the supply-and-demand elasticities, in other words, the degree of change in the supply or demand of a good in response to a change in its price.

Business location theory has one obvious shortcoming: its failure to account for market-sensitive firm location. These are the firms that depend on close proximity to the population. Decentralization of population would also imply decentralization of these activities. Consequently, employment associated with market-sensitive firms would also decentralize. Moreover, the theory predicts conflicting incentives. If employment becomes more centralized, more workers must locate within commuting distance of the jobs, and popu-

Table 11-2. Impact of commute cost changes on residences and employment predicted by location theory

Commuting cost	Land value	Household response	Employment response
Reduction	Increase in suburbs, decrease in central area	Decentralization; longer commute traded for more housing	Centralization; take advantage of lower land value and lower wage rates in central area
Increase	Increase in central area, decrease in suburbs	Centralization; housing traded for shorter commute	Decentralization; shift away from higher land values and wage rates in central area

lation concentration should increase. Location theory therefore implies conflicting incentives in response to transportation changes. Meyer and Gomez-Ibanez (1981) conclude that it is probably impossible to predict the impact of changes in transportation on urban form on the basis of location theory alone.

Theoretical Simplicity versus Real World Complexity

The development of theories and models of complex phenomena requires the use of simplifying assumptions. These assumptions are what make the problem manageable. It would be practically impossible, for example, to develop a theory of residential location that takes into account all the factors influencing location choice. (Residential location choice models have been estimated with up to 30 or more independent variables; yet we still cannot accurately predict individual choices.) Moreover, such a model would be so complex that the most important factors might be obscured.

In the previous section we discussed land use responses to changes in transportation costs predicted by various theories. Next we discuss some of the major assumptions associated with these theories and examine how relaxation of these assumptions affects our ability to predict.

Residential Location

The assumption of a monocentric city (all employment located at the center) is unsupported in contemporary urban areas. Rather, as Muller (chapter 2) and others have pointed out, urban areas today are polycentric. Although the downtown CBD generally remains the location of the greatest employment density and highest land values within the region, the central city is surrounded by competing employment subcenters. For example, Figures 11-4 and 11-5 illustrate the employment and population densities for the greater Los Angeles region. The area includes five counties (four SMSA's) and encompasses 9,740 square miles. The highest peak represents one very small zone within the CBD containing 18,600 jobs and is somewhat misleading. The entire CBD area—about 7 square miles—contains 308,000 jobs, or about 8% of the total for Los Angeles County (3,980,000 jobs). Thus, although the highest concentration of jobs remains downtown, downtown jobs make up a small proportion of the total employment within the metropolitan area. The population density surface is much more irregular, with the highest density peaks located along the coastline and in the northeast sector of the region. Associations between the employment and population distributions are not immediately apparent.

Residential location choice models are of the form, $D=f(P_H,P_T,P_G)$, where D is housing demand, P_H is the housing price, P_T is transportation cost, and P_G is the price of all other goods and services the household consumes. By assuming identical preferences among all households, the model becomes an aggregate model of residential location. Household preferences are not identical, however, even among households with similar socioeconomic and demographic characteristics. As a result, the demand for housing services (particularly with respect to other goods and services) is not the same for all households. In economic terms, housing demand elasticities are different among households, meaning that relative preferences between transport and housing are not necessarily identical. It is difficult to predict the impact of transportation changes in the absence of identical preferences, as the response depends on the relevant elasticities.

The P_G term in the residential location model (representing all other goods and services) also has a location component. Public goods like schools, parks, and air quality exhibit a high degree of spatial variation. If households have strong preferences for such goods, location choice will be affected. The quality and availability of public goods is an important consideration in housing choice for many households, and increased transport costs might willingly be incurred in order to obtain them.

Figure 11-4. Employment density in the greater Los Angeles region, 1984. From Southern California Association of Governments, 1985, *Travel Forecast Atlas, 1984, Base Model*, Los Angeles, Calif.

Figure 11-5. Population density in the greater Los Angeles region, 1984. From Southern California Association of Governments, 1985, *Trend Forecast Atlas, 1984 Base Model*, Los Angeles, Calif.

Residential location theory defines housing in the broadest terms: the bundle of all housing-related services, so public services and site amenities can be conceptualized as being contained within this housing bundle. However, the theory also assumes that housing is uniformly available throughout the region, but the supply of housing and public services is in fact quite uneven. Housing stock, public goods, and neighborhood characteristics exhibit a high degree of spatial variation.

There are also a number of hard-to-qualify factors that affect location choice. These may include ethnic preferences, racial biases, family loyalty to specific neighborhoods, and preferences for specific housing styles. Last, but no means least, the cost of moving, both financial and psychological, is so high that a large amount of inertia is associated with location choice. Even when a significant change in income or transport cost occurs, household response may be minimal.

These issues imply that location choice is much more complicated than the simple trade-off between housing, transport, and all other goods that location theory posits. On the contrary, work-trip cost is just one among many factors considered. It should be no surprise then that the theoretical model is seldom observed in reality. Even when households move, recent research indicates that the move is not necessarily closer to work. Rather, the probability of moving closer to work increases with distance from the workplace (Clark & Burt, 1980).

Demographic and market changes that have occurred during the past decade make the traditional model even less relevant. The recent inflation in housing prices and interest rates has dramatically increased the price of housing and reduced housing mobility. The increase in the number of two-worker households creates a different problem: which work trip is minimized? A minimum-cost solution for two different work trips can generate an infinite number of trip combinations; therefore, residential location is indeterminant for the two-worker, two-workplace household. Though many other examples might be discussed, these suffice to illustrate the limited empirical applicability of residential location models.

Employment Location

The problems associated with predicting employment location have already been identified. To review, market-sensitive employment is dependent upon the population distribution, and, as we have seen, household location is dependent upon many factors other than workplace location. Non-market-sensitive location has little relationship to transportation costs because for most firms these costs have become a relatively small part of total costs. The empirical evidence suggests that labor force and infrastructure availability, agglomeration economies, and location preferences of corporate leaders are the key factors in contemporary industrial location decisions.

The Supply Side

A number of factors associated with the supply of land in urban areas also affect theoretical expectations. In urbanized areas, land use is regulated by local jurisdictions. Municipalities can influence land use through zoning codes, development restrictions and requirements, and the provision of infrastructure and services.

A major concern for municipalities is tax revenues. Land uses that contribute positively to the tax base are likely to be encouraged, while uses that may require services may be discouraged. In most metropolitan areas, local municipalities compete for activities that are perceived as economically beneficial. This competition may range from booster efforts to the provision of development infrastructure. On the other hand, some municipalities may wish to preserve a suburban or rural atmosphere and use stringent development restrictions to do so. The likelihood of realizing theoretical expectations regarding land use and transportation is reduced to the degree that the public sector can control or regulate the land market.

A second important supply side factor is the nature of the contemporary land market.

The scale of land development has increased dramatically in the past few decades. Residential development has evolved from individual houses to gradually larger tract developments to entire planned communities. In the commercial sector, the trend has been from individual buildings to office parks to multiuse centers. The consequences of these changes are twofold. First, land use change is most likely to occur when large tracts of land are available, and second, land use changes are determined by fewer decision makers. Under these conditions, we might expect to see a very close relationship between transportation and land use decisions when major public and private actors have common objectives, and no relationship at all when the consensus is lacking. To sum up, then, the complexity of land use and location decisions limits the applicability of conventional land use/transportation theories.

PREDICTING LAND USE IMPACTS OF TRANSPORTATION INVESTMENTS

Our discussion of land use/transportation theory suggests that transportation investments have a significant but difficult-to-predict impact on land use. Because transportation investments affect relative accessibility within the region, long-run changes in land use (such as shifts in travel origins and destinations) should occur in response. These changes should be a key consideration in the evaluation of alternative transportation investments. Surprisingly, however, land use impacts are not considered in the conventional planning process outlined in chapter 3. Rather, future population and employment distributions are assumed (usually on the basis of regional plan projections) and used as input to the four-step transportation planning model; this process was described for the case of Atlanta in chapter 6. Thus, as Sheppard pointed out in chapter 5, there is an implicit assumption of no relationship between transportation and land use, and the evaluation of alternative transportation investments is based on how well anticipated (fixed) travel demand is served.

Methods for predicting land use impacts of transportation investments have been developed and used in policy analysis applications, however. The most recent generation of predictive models consists generally of two types: economic equilibrium models based on the location theories discussed in the previous section and gravity-type simulation models.[1] The economic equilibrium models predict residential location as a function of employment location and characteristics of the transportation system. The Chicago Area Transportation/Land Use Analysis System, or CATLUS, developed by Anas (1983) is an example of this type of model. It is a recursive model that simulates supply and demand in the land market over several time periods. Employment location is exogenous, that is, determined outside the model. Transportation system changes are also exogenous and can be input at any time period. The model uses this information to simulate land market transactions during each time period, yielding an equilibrium residential distribution. Residential location is modeled as a function of work-trip travel cost, housing price, and housing availability. The CATLUS model flow chart is shown in Figure 11-6.

Gravity-type land use simulation models were developed in the mid-1960s. They are based on the gravity model formulation described in chapter 5. In this case zonal population is modeled as a function of attractiveness (usually measured by the location of employment and housing supply) and (generalized) travel cost. An example of a gravity model for residential location is as follows:

$$T_{ij} = \frac{E_i H_j C_{ij}^{-\beta}}{\sum_j H_j C_{ij}^{-\beta}} \quad (1)$$

where T_{ij} represents trips from work zone i to residential zone j, E_i is the number of jobs

[1] For a more complete summary, see S. H. Putman, "Urban land use and transportation models: A state of the art summary." *Transportation Research*, 9 (1975) 187–202.

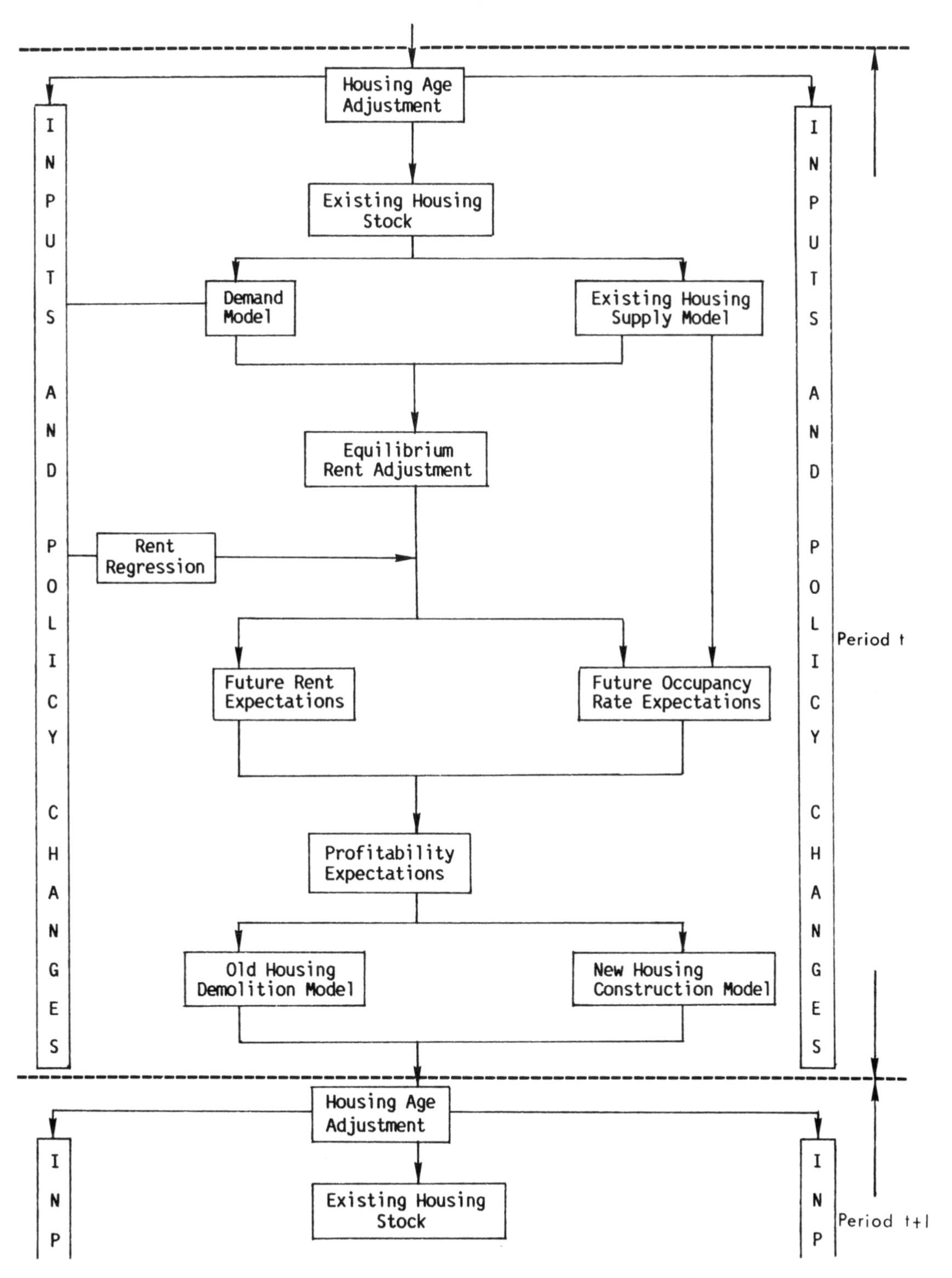

Figure 11-6. Flow chart for the CATLUS model. From *The Effects of Transportation on the Tax Base and Development of Cities* by A. Anas, 1983, Final Report No. DOT/OST/P-30/85/005, U.S. Department of Transportation, Evanston, Ill.: Northwestern University.

in zone i; H_j is a measure of residential attractiveness of zone j, C_{ij} is the generalized travel cost between i and j, and β is the travel impedance (distance decay) parameter. The equation states that the number of workers employed in zone i and residing in zone j is directly related to each zone's attractiveness and indirectly related to the cost of traveling between them. This particular example is a constrained model form. The term in the denominator assures that the total number of residences located is equal to the total number of jobs.[2] A flow chart for a gravity-type land use model is illustrated in Figure 11-7.

Predictive models have had limited applicability. Empirical applications reveal that these models tend to be nontransferable; they can be calibrated to *replicate* land use and transportation characteristics in a given area, but they do a poor job if applied to other areas. The lack of transferability results from a weak theoretical basis. The behavioral relationships that determine land use/transportation interaction have not been adequately identified, as discussed in the previous section.

Owing to the complexity of the problem, many simplifying assumptions must be employed in predictive models. The most serious shortcoming in the context of this chapter is the treatment of travel costs. Location is determined *given* travel costs between zones. However, travel costs are in fact a function of location, as the distribution of activities determines travel demand, which in turn affects travel speeds. Ideally, then, a predictive model should simultaneously solve for both activity distribution (employment and population) and travel costs. Models of this type are currently in the development stages and have very limited empirical applicability (Boyce, 1980; Fernandez & Friesz, 1983; Putnam, 1983). Despite their shortcomings, however, models like these do enable us to predict land use impacts of transportation investments in a general way.

[2]Note that for the sake of simplicity we are equating workers and population. In practice, a ratio of workers per household is assumed, and zonal population is extrapolated from this ratio.

A Gravity Model Application: The Buffalo LRRT Project

The following example shows how a gravity-type model was used to predict land use impacts for a proposed light rail transit system in Buffalo, New York.

Project Background

Buffalo, New York, is one of many eastern "snowbelt" cities that have suffered a loss of employment and population as a result of a declining economy and out-migration. The Buffalo metropolitan area population has remained stable at about 1 million since 1960, while city population has declined from 533,000 to 357,000. Over the same period, employment increased by 3.6% in the metropolitan area and decreased 5.5% in the city.

As a way of revitalizing downtown Buffalo, the city proposed constructing a light rail transit (LRRT) line from the CBD to the area of the campus of the State University of New York–Buffalo on the northeast side of the city, as illustrated in Figure 11-8. The route is 6.4 miles long and was estimated to cost about $430 million. In conjunction with the LRRT, the major downtown shopping street was to be reconstructed as a pedestrian mall. The project was approved for 80% federal funding in 1975. Construction began in 1979, and the system began operation in 1985. The U.S. Department of Transportation commissioned a research project to determine the impacts of the LRRT system on downtown revitalization (Berechman & Paaswell, 1983). The research was to evaluate the impact of the LRRT system on downtown employment, accessibility, retail sales, and potential for private investment and development. (It should be emphasized that the purpose of the study was to *predict* impacts rather than to document the actual outcomes of the LRRT project.)

It was anticipated that the LRRT would

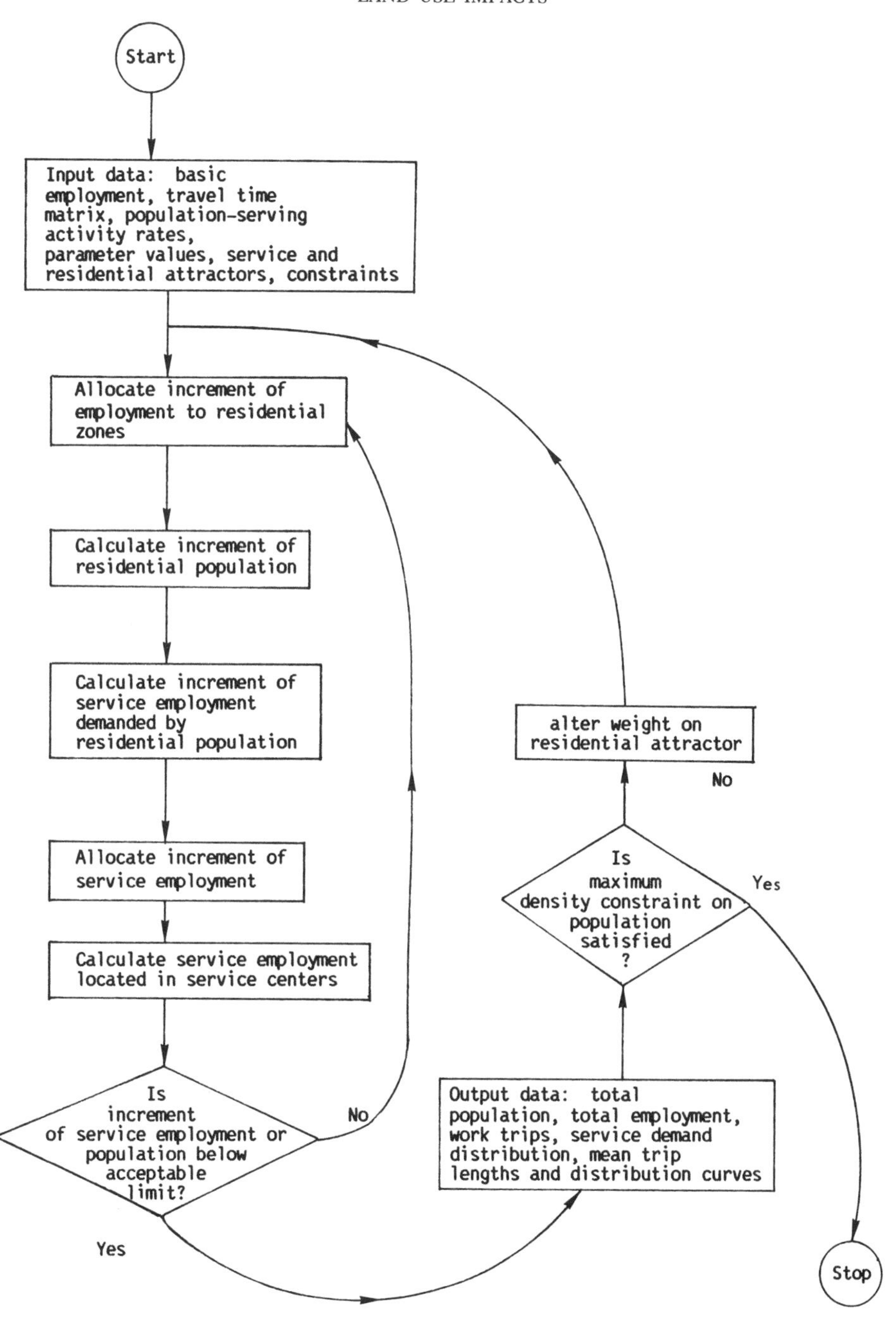

Figure 11-7. Flow chart for the Garin-Lowry (a gravity-type land use) model. Adapted from *Urban Modeling: Algorithm, Calibrations, Predictions,* Batty, 1976. Cambridge: Cambridge University Press. Copyright 1976 by Cambridge University Press. Reprinted by permission.

affect downtown activity by increasing accessibility to downtown. The rail line, together with anticipated bus service improvements, would significantly reduce transit travel time to and from downtown, making it relatively more accessible. Thus even in a stagnant economic situation, some economic activity would be expected to shift to downtown, all

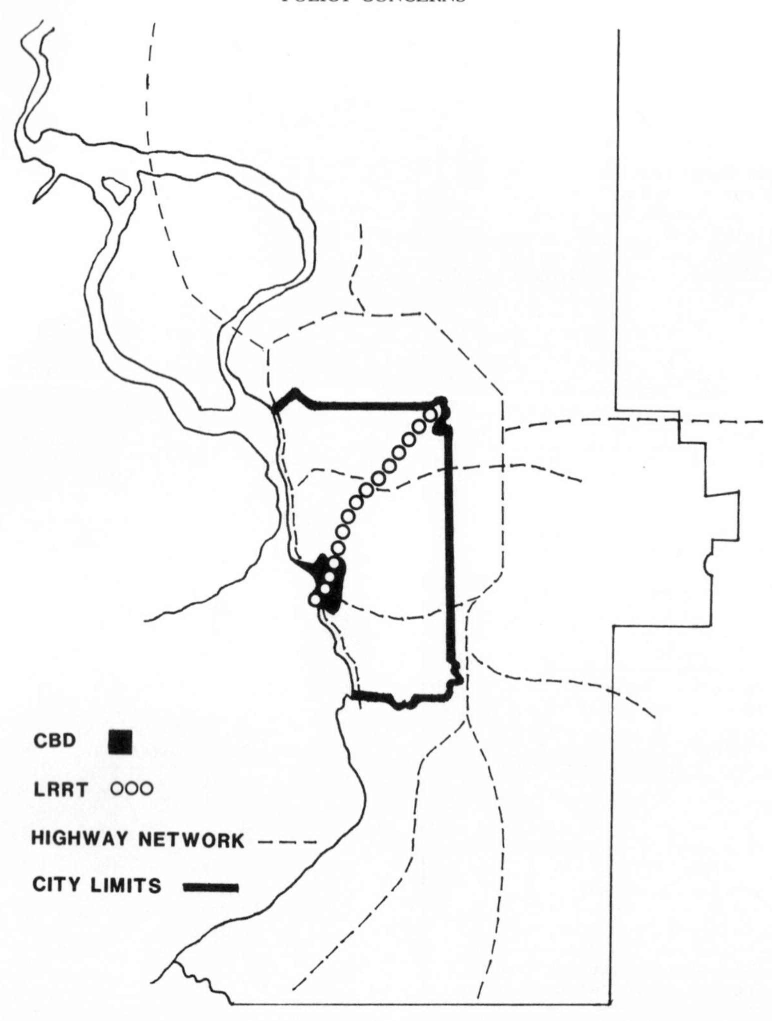

Figure 11-8. Buffalo light rail transit line and metropolitan area. From *An Analysis of Rapid Transit Investments* by R. Paaswell and J. Berechman, 1981, Final Report No. NY-11-0022, Urban Mass Transit Administration, Buffalo, N.Y.: State University of New York at Buffalo.

other things being equal. The researchers used the Garin-Lowry land use model (Figure 11-7) to estimate land use impacts resulting from the transportation system changes. Impacts on retail activity, downtown accessibility, and economic growth were evaluated using the model results.

In order to estimate land use impacts, the Garin-Lowry model was calibrated on base-year data (1976), and then estimated for the target year (1985) with the LRRT project. The zonal travel-cost parameters (the C_{ij} terms in equation 1) were adjusted to reflect the target-year improved transportation system, which included the LRRT, planned bus system improvements, and a new major highway. Basic employment generated by the $430 million LRRT investment was also included in the target-year data. The Garin-Lowry model then estimated the project-generated additional nonbasic (that is, service) employment and its distribution within the Buffalo metropolitan area. The analysis was based on a comparison of base-year versus target-year results.

Results of the LRRT project analysis showed that the rail system would have very little impact on the region as a whole for three reasons. First, the highway system in the Buffalo region provides a very high level of service, even during peak periods. Because of the loss of population and jobs in the central area, there is little traffic congestion and no parking shortage. Thus, although the LRRT project significantly improves *transit* accessibility to the CBD, there is little shift in transit market share because the competitive advantage of auto travel does not materially change. The analysis predicted a change in transit mode share from 12% to 14% during the peak and from 5% to 7% during the rest of the day. Second, the process of suburbanization is expected to continue because of favorable investment conditions, available land, and a good highway system. A large proportion of the new service employment generated by the LRRT project will shift to the suburban area, where residential growth will continue to take place. Third, suburban competition for retail activities will continue because of the greater attractiveness of several large suburban shopping malls. The authors conclude that the LRRT will have a slightly positive economic impact in the downtown area because the project is a large public investment and because redevelopment money has been concentrated in the same area.

The results of this study show that the single most important aspect of the LRRT project was the large sum of money invested in the Buffalo economy. The multiplier effect of the LRRT investment created more jobs and disposable income. It should be noted that any investment of similar magnitude, such as government financing provided to locate new industry in the area, would have the same effect. The land use impacts of such an investment could have been at least as significant, because the attractiveness of the downtown area as a work location would have significantly increased. Moreover, the analysis did not take into account the additional subsidy burden generated by the increase in transit service. To the extent that the increased burden is supported by local tax funds, the economic benefits of the LRRT project will be offset accordingly.

EMPIRICAL EVIDENCE OF LAND USE IMPACTS OF TRANSPORTATION INVESTMENTS

Empirical documentation of the land use impacts of transportation investments has been of interest to researchers for many decades. Studies of specific projects in the U.S. were conducted as early as 1930, and more than a dozen have been performed since 1960. Despite this large and growing body of empirical research, there is little consensus on the results or the conclusions that might be drawn from it. Indeed, the available evidence provides conflicting results for both highway and transit investments. Before reviewing a few of the more comprehensive studies, however, a discussion of the problems involved in tracing the impact of transportation changes and some important factors affecting impacts is in order.

Problems of Identifying the Effect of a Transportation Investment

There are two problems involved in attempting to identify land use impacts of transportation investments. First, transportation changes take place in a highly dynamic system. The change in the transportation system is just one of many occurring at the same time. Furthermore, this process of change continues after the investment takes place, making it very difficult to attribute observed changes correctly. That is, land use changes that may be observed following a given transportation investment do not necessarily result from that investment. Second, the interaction of land use and transportation takes place over time. Because of the longevity of capital infrastructure, the market response to changes in transportation may take place only after many years — say 10 or 20. The longer the time span, however, the more other factors like general economic conditions, regional employment, and demographic characteristics change as well. It therefore becomes increasingly difficult to hold everything constant as the time frame of the analysis increases.

The ideal way to trace the land use impacts of a transportation investment is to compare them with a "no-project" scenario: How different is the land use distribution we observe from what would have occurred without the given project if all other relevant factors had remained unchanged? This approach is of course impossible in empirical studies. Instead, either time-series or comparative studies must be conducted.

Impact studies using the time-series approach focus on one area and study the changes that take place over a given time period. If lead time is sufficient and data are available, land use changes within the area of impact (however defined) are traced from before the project took place to some period after completion. In order to isolate transportation-related impacts, all other factors thought to be relevant are incorporated into the analysis. For example, the regional rates of housing and/or office construction would be considered in the impact assessment.

The time-series approach may not control for the distributional shifts in land use activities that transportation investments may generate if the area of analysis is limited. The presence of a given transportation investment can attract land use change that might otherwise have located elsewhere in the region. The investment might have had no net impact at the regional level, even though very significant local changes might have taken place.

Studies using the comparative approach attempt to make comparisons between similar areas as an approximation of the "no-project" comparison. For example, comparisons might be made between cities of similar size, employment, and income characteristics with and without rail transit systems. However, differences between apparently similar metropolitan areas are sufficiently great that it is difficult to ascribe much validity to the results of a comparative study unless the observed impacts are highly significant.

Several comparative impact studies have been performed at the subregional level in order to make the comparison as appropriate as possible. Different sectors within a region are the subjects of study. For example, rail transit impact studies are often performed at the corridor level. Land use changes in a rail corridor are compared to those in a similar nonrail corridor. This type of comparison has another pitfall: If transit investments cause a shift in activity location because of improvements in relative accessibility, such a comparison will exaggerate the extent of the impact. The combined effect of increased accessibility in the rail corridor and decreased accessibility in the nonrail corridor is actually being measured.

Finally, neither of these approaches allows us to draw any conclusion on the direction of causality between land use and transportation investments. Even when significant land use changes occur in conjunction with transportation investments, there is no way to determine whether the investment *caused* these changes, or whether the changes created the demand for the transportation improvement. The fact that the development community is an active proponent of transportation investments, and in recent years has become an increasingly important financial contributor to such investments, lends credence to the latter interpretation. In any event, land use and transportation decisions are so closely tied together that it has been impossible so far to separate their effects.

The Contemporary Context of Urban Highway Investments

There are several aspects of the context of urban highway investments that can affect the extent of land use impacts. First, it is important to realize that any single highway investment is but a part of a much larger urban transportation system. Highway investments are marginal; they add some increment of accessibility to the area. In an area that already enjoys a high level of accessibility, we would not expect a new investment to have much of an impact. In an area with limited accessibility, the same investment would be expected to have a much greater impact. Most U.S. urban areas fall into the former category, however.

Second, the availability of developable land must be a key consideration. Land use change is more likely to occur in the form of new construction rather than reconstruction because of the high cost and longevity of most residential and commercial infrastructure. Available land is of course a prerequisite for new construction. All other things being equal, then, the land use impacts of highway investments in areas with vacant land available should be greater than in areas where development has already taken place. This is not to say that highway investments may not promote redevelopment, but rather that significant change is more likely when developable land is available.

Even when developable land is available, local public policy may not be favorable to development. Local zoning is a third critical factor in impact assessment, because land use change cannot occur unless local zoning permits it. Consequently, transportation investments may not result in significant land use changes if there is strong local opposition to such changes. Major transportation investment decisions are generally made at the regional level, and local constituencies may not support these decisions.

A fourth factor to consider in studying land use impacts is the state of the regional economy. In a stagnant or declining region, little new investment takes place. More disinvestment occurs; residential areas deteriorate, factories close down, and the general level of economic activity declines. Under these circumstances, we would not expect a highway investment to have much impact. If it did have an impact, it would most likely be at the expense of other areas within the region.

Finally, the scale of analysis must be considered. It is relatively straightforward to determine whether highway investments have an impact within their immediate vicinity. Indeed, any observer of the urban scene can recount changes observed after a new freeway opened or an interchange was constructed. All that appears to be necessary to determine whether and how impacts occur is a series of before and after studies. The question is more complicated if the region as a whole is of interest. Regional impacts are more difficult to observe because transportation investments are incremental. More importantly, however, observation of local changes does not support the hypothesis that highway investments generate land use impacts. In order to test such a hypothesis validly, we must study the region as a whole. Only then can we determine whether the land use changes taking place within the vicinity of a given highway investment are significantly different from those occurring outside that vicinity.

Highway Impacts

During the decade following World War II, the automobile became the overwhelmingly dominant choice for personal transportation. By the mid-1970s, the only significant market left to public transit was CBD-destined work-trip travel. A similar shift from rail to trucking occurred in goods movement, with trucking dominating all but the large bulk/long-distance market.

Research on highway impacts was stimulated by the highway-building boom of the 1950s and early 1960s. Several studies were commissioned to identify the social and economic benefits of highways and to identify future highway needs. Highway-related environmental impacts became the subject of concern by the late 1960s, however. Transportation policy interest subsequently shifted toward public transit at the national level, and no major highway impact studies were conducted again until the mid-1970s.

Considering that our urban highways carry over 90% of all person trips and a large proportion of goods movement, it seems reasonable to expect that highway investments would generate significant changes in land use in the surrounding area. As discussed earlier, a major highway investment (say construction of a new freeway link within a metropolitan area) changes accessibility within the area. By reducing travel time between the origins and destinations linked by the new facility, these points become more accessible relative to the other origins and destinations in the area. More activities will be attracted to these areas as a result of the increased level of accessibility, some of which may have shifted from formerly favored locations.

Early studies of highway impacts tended to validate these theoretical expectations. (Adkins, 1959; Mohring, 1961). These studies used land value as the measure of impact. Increases in accessibility are reflected in the land price developers or investors are willing to pay.[3] These studies indicated that land values near the highway increased significantly. It was not clear, however, whether these increases represented a net gain within the region as a whole or whether they represented a distributional shift; that is, whether land-value increases near the highway were offset by decreases in other areas within the region.

More than a decade passed before additional research on highway impacts appeared. In the interim the interstate highway system was effectively completed, and freeway systems had been constructed within all the major cities in the United States. The rate of auto ownership more than doubled, and the shift of both population and commercial activity growth to the suburbs had occurred (chapter 2). More recent studies of highway impacts, then, have taken place in a vastly different setting. It is not surprising, therefore, that the results of these studies tell a somewhat different story.

The DOT Beltway Study

One of the most comprehensive recent highway impact studies was performed for the U.S. Department of Transportation (Payne-Maxie Consultants, 1980). This study examined the land use and related impacts of beltways (circumferential limited-access highways). Originally conceived as a means of diverting through traffic away from congested, central-city areas, beltways have increasingly become integral parts of the intrametropolitan highway system in areas where they have been constructed.

The beltway study consisted of three different phases of analysis. First, a review of all previous related research was conducted. Second, a statistical analysis was performed on a set of 54 U.S. cities, of which 27 had beltways. Third, in-depth case studies of eight beltway cities were conducted. Some examples of beltway and nonbeltway cities used in the study are illustrated in Figure 11-9.

The statistical analysis attempted to take into account all of the quantifiable factors that affect land use decision making. Measures of economic and population growth; measures of residential, manufacturing, retail, and other business activity; and measures of regional accessibility were used. The statistical analysis revealed that the land use impacts of beltways have been largely insignificant. Growth effects as manifested by regional population and employment changes during the period of study (1960–1977) were related to the change in manufacturing employment and city age. That is, employment changes in other sectors were positively related to changes in the manufacturing sector. The city-age relationship was negative, indicating that newer cities were growing more rapidly. The existence of a beltway, the beltway's relative location, and its length had no consistent effect on growth. Selected beltway study results are presented in Table 11-3.

Distributional impacts, measured by relative changes in population and employment for the central city and the SMSA showed mixed results. The presence of a suburban beltway (a beltway located outside the central city) was associated with less growth in central-city manufacturing employment but more growth in central-city service employment. The suburban beltway was also associated with a smaller percentage of workers residing in the suburbs and working in the central city. Since suburban beltways are generally located in the largest cities, how-

[3]This concept has been well documented in land-rent density studies. The downtown CBD has historically been the point of highest land value and is the most accessible point as well.

Table 11-3. Selected beltway study results

Measure	Significant factors: Beltway	Significant factors: Other
Growth Impacts		
SMSA population growth, 1960–1970	Slight (+)	Change in manufacturing employment (+) City age (−)
SMSA population growth, 1970–1977	No significant effect	Change in manufacturing employment (+) City age (−)
Wholesale employment growth, 1972–1977	Beltway distance from CBD (+)	Change in manufacturing employment (+) City age (−) Arterial mileage (−)
Distributional Impacts		
Central city population growth, 1960–1970	No significant effect	Change in central city manufacturing employment (+) City age (−)
Central city wholesale employment, 1972–1977	Interchanges per beltway mile (−)	Change in central city population (+) Change in central city manufacturing employment (+)
Central city service employment, 1972–1977	Interchanges per beltway mile (+) Beltway distance from CBD (+)	Change in central city employment (+) wholesale

Note. Adapted from *The land use and urban development impacts of beltways,* Final Report No. DOT-0S-90079, U.S. Department of Transportation and Department of Housing and Urban Development, by Payne-Maxie Consultants, 1980, Washington, D.C.: Payne-Maxie Consultants.

ever, these results are more a function of city size than of beltway effects.

The statistical analysis indicated that for manufacturing and retail activity location, the key factors were land availability and economic vitality. Where developable land was available and the regional economy was growing, these activities tended to locate near beltways. In cities without beltways, however, location patterns were similar (for example, the distribution of employment growth between central city and suburb was not significantly different).

Distributional impacts in the housing market were also examined. Table 11-4 gives the ratio of SMSA to central-city rate of change in new housing units from 1960 to 1970. The central-city rate of increase was greater in cities with the beltway located within the central cities and slower in cities surrounded by a suburban beltway. The authors point out that these differences result from development patterns. Beltways were usually built on the urban fringe, where new development would occur in any case. It is also noteworthy

Table 11-4. Beltway presence and ratio of SMSA to central-city rate of change in new housing units, 1960–1970

	Ratio
Location of beltway (all)	1.50
Central-city location	.76
Mixed jurisdiction location	.69
Suburban location	2.28
No beltway	3.15

Note. From *The land use and urban development impacts of beltways,* Final Report No. DOT-0S-90079, U.S. Department of Transportation and Department of Housing and Urban Development, by Payne-Maxie Consultants, 1980, Washington, D.C.: Payne-Maxie Consultants, p. 60.

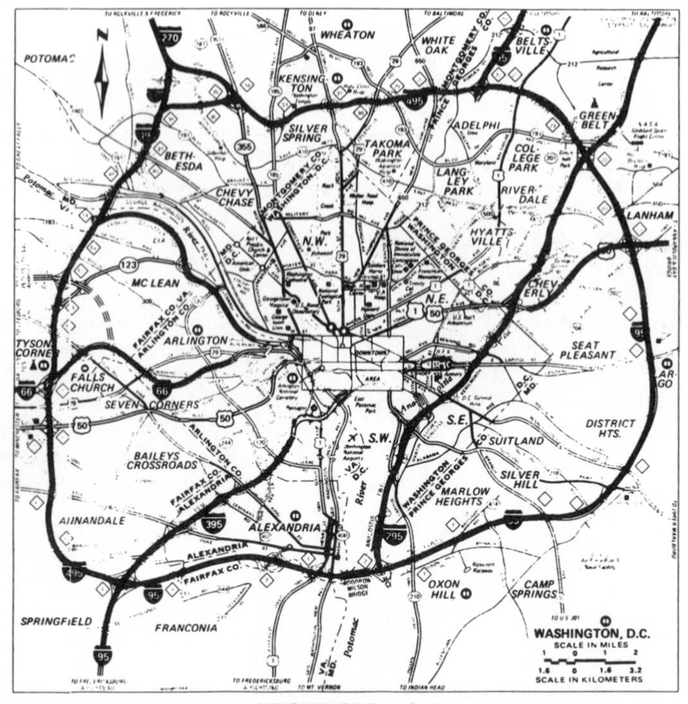

WASHINGTON, D.C.

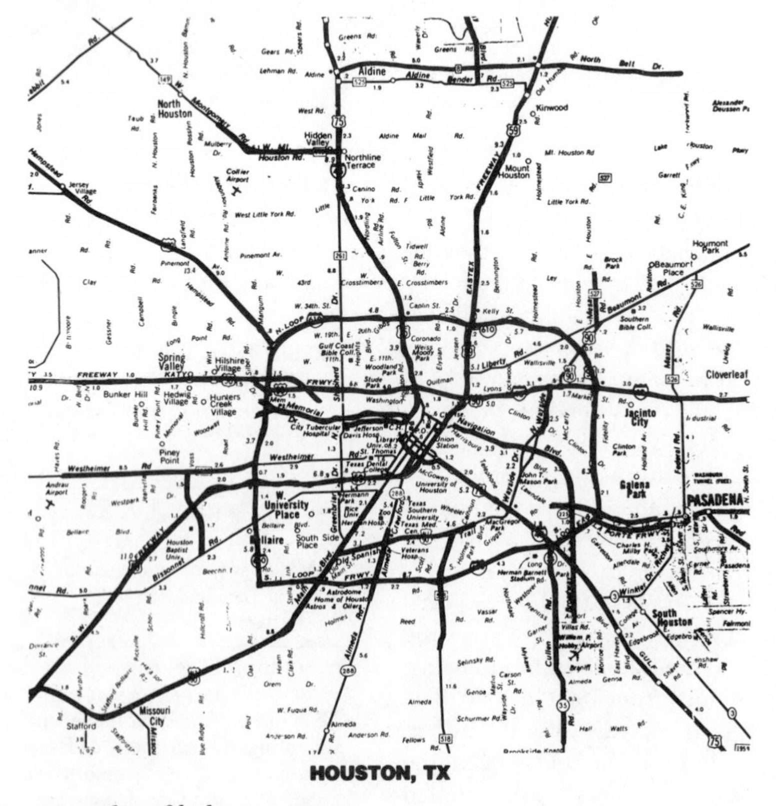

HOUSTON, TX

Figure 11-9a. Examples of beltway cities.

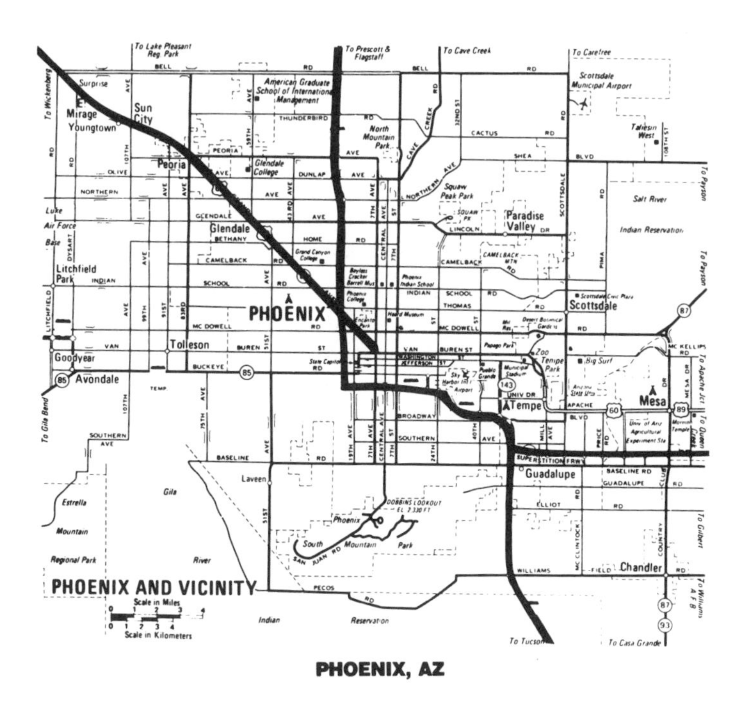

Figure 11-9b. Examples of nonbeltway cities. Figures 11-9a and 11-9b were constructed by the U. C. Irvine publications department and were based on street maps.

that the greatest proportion of suburban residential growth occurred in cities without beltways. The analysis also showed that low-density residential development tended to occur away from beltway areas, while medium- and high-density development was observed near beltways.

The absence of any consistently significant impact of beltways was corroborated in the case studies. In each city, portions of the beltway area underwent significant development, while other portions experienced no change, even though the time frame of the study encompassed nearly two decades. Areas that experienced development were those that (1) were located near medium- or high-income residential areas, (2) had developable land available, and (3) had local zoning policies favorable to development. Furthermore, the extent of development that occurred in each area seemed to be largely a function of the rate of economic growth of the area. The study authors conclude that the extent of land use impacts is determined largely by market conditions: When conditions are favorable, land use change will occur; when conditions are not favorable, impacts are unlikely.

The case studies also documented the absence of any significant distributional impact of beltway investments. In a few specific instances (such as Atlanta and Louisville) one-time "shifts" of office activity location from central city to beltway were observed, but these were attributed to unique regional circumstances.

The results of the beltway study are not surprising. Given the high level of accessibility that exists in U.S. urban areas, the impact of any single facility will be marginal. In most urban areas the supply of accessible, developable land is relatively plentiful. As the beltway study demonstrated, a variety of other factors affect location decisions.

Transit Impacts

A renewed interest in public transportation emerged in the 1960s. Public transportation systems in several major cities were in serious financial trouble, and the federal government was called upon to provide capital for rebuilding the nation's transit industry (chapter 10). At the same time, public awareness of the adverse environmental and social impacts of highway building in urban areas was growing. A broad consensus developed in support of federal assistance for transit, with the improvement and expansion of public transportation services justified as a means to reverse the trend of increasing reliance on the automobile, to reduce air pollution and save energy, and to revitalize central cities. Advocates argued that better transit would improve central-city accessibility, which would in turn lead to higher densities in central areas and halt urban sprawl. The federal government responded with a massive capital grants program that eventually led to the construction of the "new generation" rail transit systems.

These rail systems proved to be some of the most costly public projects ever constructed. In contrast to highway projects, which are largely funded by gasoline tax revenues, transit capital projects were funded from general revenue sources. That is, whereas highway projects were supported by user fees, rail projects were not. Therefore, it became very important to justify these projects on the basis of indirect benefits. Land use benefits were the subject of particular interest, because if transit projects increased land values, that increase in value would generate additional tax revenue, offsetting the cost of the system. Not surprisingly, then, the federal government sponsored impact studies for all three major new systems: BART in San Francisco, Washington D.C.'s METRO, and MARTA in Atlanta.

Theoretical Expectation of Rail Transit Impacts

Before reviewing the empirical evidence, it is useful to examine what land use impacts we might expect on the basis of theory. Rail transit systems generate changes in accessibility only in the immediate vicinity of the rail line itself. The construction of a rail transit system should improve accessibility along the rail line corridors and increase the relative advantage of rail corridors compared to areas that are not served by the rail system. All other things being equal, then, activity location should shift toward the rail corridors, and this shift should be reflected in increased land values. In addition, since rail systems are focused on the CBD, the position of the CBD as the most accessible point in the area should be enhanced, leading to an increase in activities and land values in the CBD area.

Proponents of rail transit systems made two rather fundamental errors in predicting land use impacts. First, they assumed that the introduction of a rail system would significantly change accessibility. In fact, rail service usually replaces existing bus transit, and its impact on accessibility is actually very slight. Even when this it not the case, however, rail system impacts tend to be inconsequential in the context of the transportation system as a whole, because the transit market share is so small. Second, rail-transit proponents assumed that rail system–induced changes in accessibility would result in an increase in density. Our review of the theory, however, pointed out that a reduction in travel cost should lead to decentralization, as workers trade more housing against transportation costs. Theory suggests that the impact of a rail system should not be much different from that of a highway system. The difference should be in degree: The land value gradient should be steeper for rail than for highway. The following case study of the Philadelphia-Lindenwold line illustrates these points.

The Philadelphia-Lindenwold High-Speed Line

The High-Speed Line is the first link of a rail mass transit system planned for the Philadelphia region. It is unique among modern rail systems in that it was the first to be built without federal funding by the Delaware River Port Authority.[4] The High-Speed Line runs from downtown Philadelphia to the suburban community of Lindenwold, New Jersey, a distance of approximately 14.5 miles (see Figure 11-10). It is located in a previously developed corridor consisting mainly of low-density suburban housing. It was built to service suburban commuters and reduce congestion on highway routes to the downtown area. The six suburban stations of the line have park-and-ride facilities. Since it began operation in 1969, the line has enjoyed a constantly increasing ridership.

A University of Pennsylvania research team sponsored by the Department of Transportation conducted a long-term analysis of the High-Speed Line (Boyce, Allen, Mudge, Salter, & Isserman, 1972). One of their objectives was to determine whether nonuser benefits could be attributed to the line. The research team reasoned that if existence of the line improved accessibility, then the value of this improvement should be capitalized into residential housing values within the corridor. Land values were examined to test this idea. Residential sales prices in the corridor were compared with prices in a control corridor — a similarly built-up corridor in the South Jersey region slated for a rail link in the future. The period of study was extended back to 1965, in order to capture any speculation effect that might have occurred in anticipation of the opening of the High-Speed Line. The research team also hypothesized that the improved accessibility within the corridor should lead to higher-density land use patterns in areas undergoing development within the corridor. In-depth case studies of two municipalities located near the outlying stations were performed in order to examine development patterns.

In order to test the land value hypothesis, the research team developed a Savings Model

[4]The San Diego Trolley, a light rail line constructed in San Diego, California, is the second system to be built without federal funds. The 16-mile system began operation in 1982.

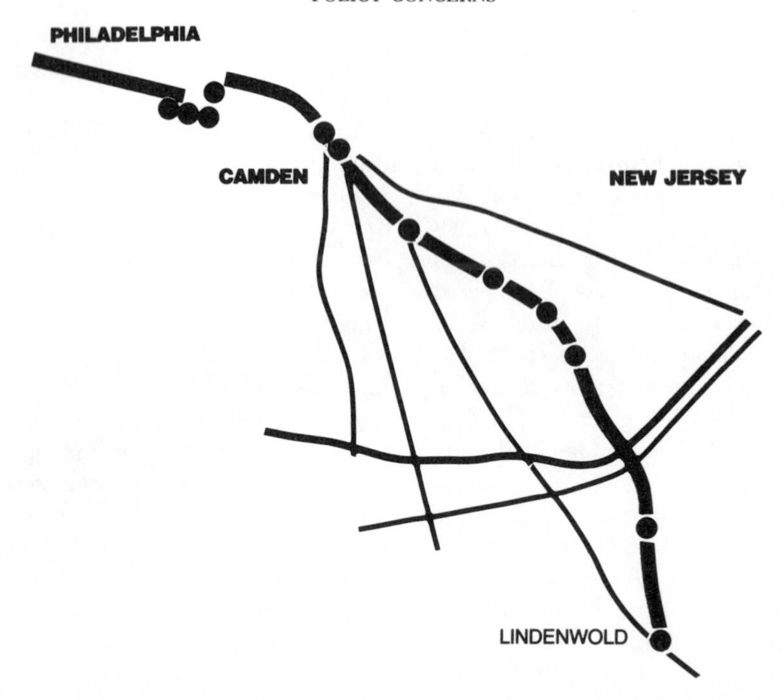

Figure 11-10. The Philadelphia-Lindenwold High-Speed Line. From *Land use impacts of rapid transit: Implications of recent experience*, by R. Knight and L. Trygg, 1977, Final Report No. DOT-TPC-10-77-29, U.S. Department of Transportation. San Francisco: De Leuw Cather and Co.

of Property Value Impact based on earlier highway impact research (Mohring & Hartwitz, 1962). The model was used to identify market areas for the High-Speed Line corridor and to develop market areas for a simulated transit line in the control corridor. Market areas are determined by the (generalized) cost of using the transit line compared to the costs of driving a car. Relative cost is a function of driving cost, location with respect to the nearest transit station, and transit fare. The market area encompasses all points where using the transit line is less costly than driving. The mathematics of the model indicate that the market area of each station can be partitioned by a series of curves that trace increasing levels of transit line savings, as illustrated in Figure 11-11. These curves bend inward as savings increase. Savings increase as distance to the station decreases within the market area, because the cost of getting to the station declines. Maximum

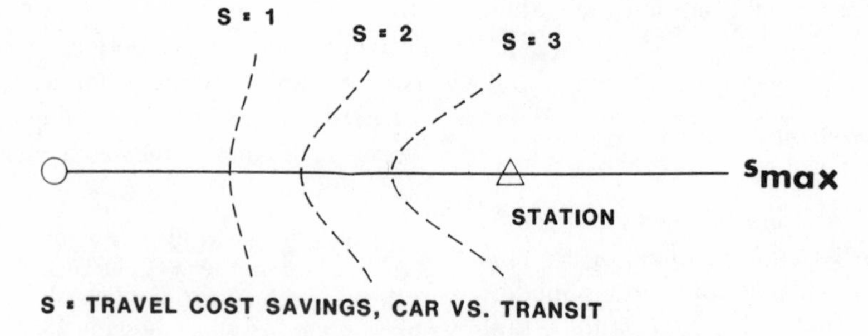

Figure 11-11. Illustration of the savings model for one station.

savings occur on a line extended along the axis of the curves (that is, along S_{max} in Figure 11-11).

This model illustrates two important points. First, market areas around stations are not concentric rings or parallel bands as is usually assumed, and second, the greatest savings from the transit line accrue to locations most distant from the destination. The implication is that commuters may choose even more distant housing locations in order to take advantage of these savings.

Using the savings model, the authors developed market areas for each of the stations of the High-Speed Line and for a simulated system in the control corridor. Real estate sales data for the corridor were compared with data from the control corridor. It was hypothesized that land values near the stations (within the market area) would be higher than the density/land value gradient based on distance from the CBD would predict. In other words, households should be willing to pay more for locations near the line, all other things being equal.

Sales price was modeled as a linear combination of site variables, neighborhood variables, locational variables, and impact variables. The Lindenwold corridor impact variable coefficient was found to be small and only marginally significant in the statistical analysis. The estimated value of rail line accessibility amounted to between 0.5 and 1% of the average residential sales price within the corridor. The authors concluded that some capitalization within the Lindenwold corridor had occurred. Although statistically insignificant, the control corridor impact variable was negative, meaning that location in the control corridor had a negative impact on the sales price. The authors took this to indicate that some shifting had occurred.

Later studies of the line's impact on residential property values also corroborated these findings (Boyce, Allen, & Tange, 1976). Line-related property value impacts were positive and a function of the distance from the transit line. Moreover, results showed that the greatest absolute increases in property value occurred at the most distant locations (the outer suburbs), just as the savings model would predict.

The Lindenwold Line studies support the two points made earlier. Land use impacts of transit operations are small because the relative change in accessibility generated by the system is small. Second, rail-system land use impacts are decentralizing, because the greatest improvement in accessibility is realized at the most distant locations.

The case study research of residential development in two communities within the rail corridor also yielded interesting results. In one community, substantial apartment development occurred, while in the other almost none occurred. Similar findings were observed for office and commercial development (Gannon & Dear, 1972). New development was a continuation of suburbanization trends already in progress before construction of the line, and developments were as likely to locate near existing major highways as near the line. The authors conclude that while the line may have been a factor in the location decisions of developers, land availability, perceived demand for new housing, existing zoning practices, and the attitudes of local political leaders were more important.

The San Francisco Bay Area Rapid Transit (BART) System

The BART system was the first of the new federally funded rail systems to be constructed. The first line began operation in 1972, and the 71-mile system was completed by 1974 (see Figure 11-12). The system was beset by operating problems, however, and did not achieve its designed service frequency until a decade later. Since BART has the longest operating history of the new Urban Mass Transit Administration–funded rail systems, it has the greatest potential for providing evidence on land use impacts. Unfortunately, however, the BART impact study was completed in 1978, too soon to expect significant land use changes to have taken place. The study does, however, summarize BART's short-term impacts. The study indi-

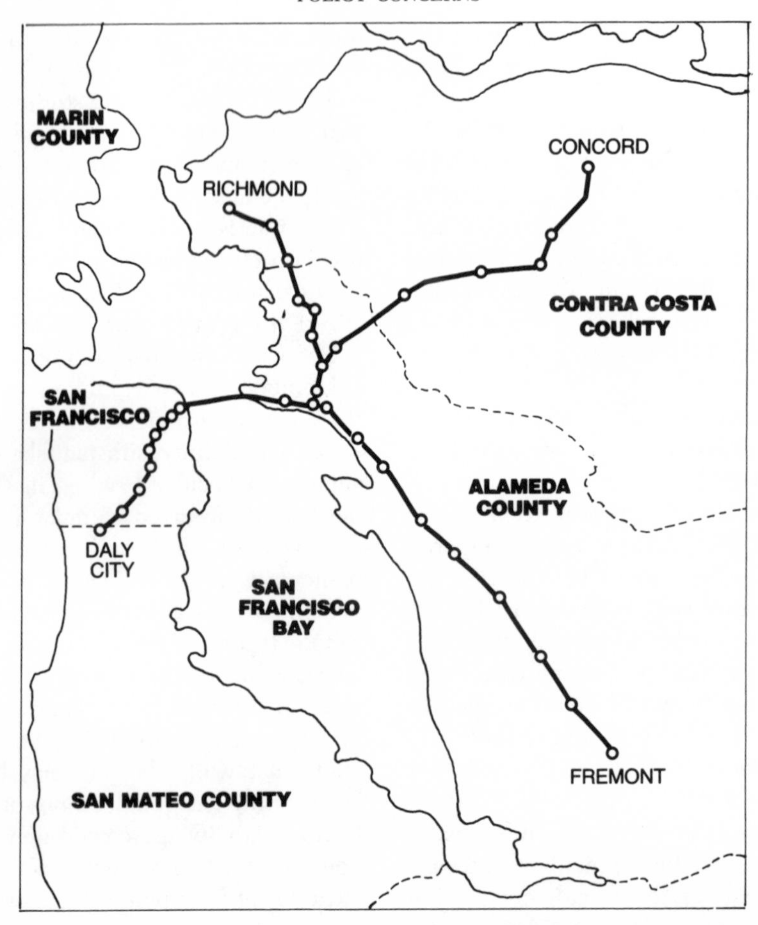

Figure 11-12. The San Francisco Bay Area Rapid Transit (BART) system. *From Land Use and Urban Development Impacts of BART*, by M. Dvett, D. Dornbusch, M. Fajans, C. Falcke and V. Gussman, & Merchant, 1979, Final Report No. DOT-P-30-79-09. U.S. Department of Transportation and Department of Housing and Urban Development. San Francisco: Blayney Assoc. and Dornbusch and Co.

cated that BART has had little effect on land use and urban development (Dvett, Dornbusch, Fajans, Falcke, Gussman, & Merchant, 1979).

As discussed earlier in this chapter, the effect of any transportation improvement depends on its impact on accessibility. BART had a very limited impact on accessibility, despite its great scale. Operation problems affected service frequency and reliability, and although BART improved transit travel times to the major employment centers, auto travel times to these cities remained on average 35% shorter than transit. As a consequence of BART's negligible impact on accessibility, significant shifts in travel pattern failed to occur.

The BART study also examined location decisions of workers, households, and employers. Survey results indicated that transportation factors were seldom a factor in job location choice, and access to BART is a minor consideration in household location choice. Housing type, general access to the workplace (such as highway access), and neighborhood characteristics were the key factors in household location decisions. BART was also rarely a factor in employer location decisions. The most important employer location considerations were site availability, price, and proximity to other firms.

The study team concluded that, with few exceptions, BART's impact on land use after its first five years of operation was insignificant. The study indicated that determining factors in land use change are land availability, local market demand, neighborhood preferences, and the general trend toward continued suburbanization.

Vacant land is, of course, limited along the BART routes, because the system was built primarily to serve the developed portion of the Bay Area. Thus land-intensive development, like low-density housing, has occurred in the outlying areas where land is relatively plentiful and cheap.

The anticipated redevelopment to high densities around station areas has also not yet occurred. Neighborhood opposition resulted in zoning changes around nine stations that reduced allowable density of development in order to prevent redevelopment. However, high-density redevelopment has not occurred even when zoning has remained favorable. The reasons cited were lack of demand, continued reliance on the automobile, and the short time span of BART's operation.

The most significant impact attributed to BART was the shift in downtown San Francisco office construction to the Market Street area. A major redevelopment project and favorable new zoning provisions for the area were implemented in conjunction with BART, and these were identified as the key factors in the shift. Moreover, the (relatively) low price of land (the area was deteriorated), constraints to growth in other areas of the downtown, and good access even without BART made the Market Street area a prime target for new office construction. The study authors conclude, "At a regional scale, BART has not had a measurable impact on population and employment growth, but development in BART-served corridors and downtown is somewhat greater than it would have been had BART not been built" (Dvett et al., 1979, p. i).

The general trends identified in the BART impact study are corroborated by census data presented in Table 11-5 and illustrated in Figure 11-13. The San Francisco metropolitan region is shown in Figure 11-13. The area is divided into large analysis areas called superdistricts. Superdistricts served by BART are denoted as BART zones, although only a small portion of the larger zones are actually within a BART corridor. Table 11-5 gives

Table 11-5. Worker residence and job locations, San Francisco region, 1970–1980

Workers	BART zones	Non-BART zones	BART zones share
By residence:			
1970	724,843	1,078,641	40%
1980	849,661	1,598,319	34%
Percent change	+17.2%	+48.2%	
Highest absolute change	25,743 (#21)	46,654 (#10)	
Highest percent change	92.6% (#16)	205.5% (#23)	
By job location			
1970	871,922	931,562	48%
1980	1,044,504	1,403,476	43%
Percent change	+19.8%	+50.6%	
Highest absolute change	61,487 (#1)	132,393 (#9)	
Highest percent change	93.8% (#22)	174.3% (#23)	

(#) = Zone number (see Figure 11-13)

Note. Developed by the author from data provided by the Metropolitan Transportation Commission, Oakland, Calif.

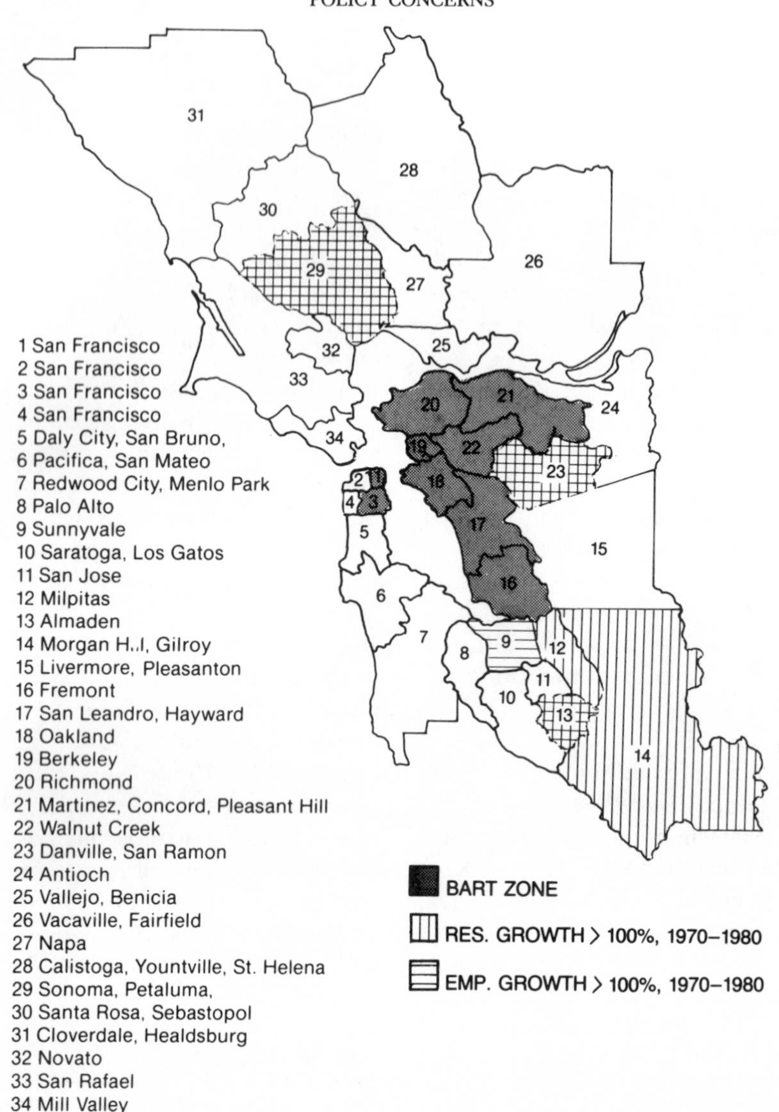

Figure 11-13. Population and employment growth, greater San Francisco region, by superdistricts. Developed by the author from data provided by the Metropolitan Transportation Commission, Oakland, Calif.

residence and job location for all workers, 1970 and 1980, by superdistrict. The suburbanization of residences is quite evident. Growth in non-BART zones is greater than in BART zones. The most rapid growth in BART zones has occurred in the outermost zones, but growth in suburban non-BART zones has been even greater. Job location trends are similar, with the suburban non-BART zones experiencing the most rapid growth. The greatest absolute increase in employment within BART zones took place in downtown San Francisco, but this increase is far surpassed by the growth in Zone 9, the heart of the Bay Area's famed Silicon Valley.

Conclusions on Rail System Impacts

The evidence provided by impact studies indicates that rail systems have had little influence on urban structure. The most significant impacts have occurred in Toronto,

Canada, where stringent land use controls have been employed to direct land use changes. In the United States, these controls do not exist, and there is frequently a lack of local support for increases in development density. Even when market conditions are favorable, the local community may prevent such changes from taking place. Specific instances of downzoning around transit stations in response to local opposition have been documented in Washington, D.C., and Atlanta, as well as in the BART area, and the absence of pro-development policies has been identified as one cause of the lack of transit-related development in Chicago and Cleveland (Knight & Trygg, 1977b; Lerman, Damm, Lam, & Young, 1978; USEPA, 1975). Other consistently identified constraints are regional development trends and availability of developable land.

Finally, as noted earlier, any transportation improvement promotes residential decentralization. Thus the population and employment shifts observed in the BART region, for example, are just as expected: the greatest BART-generated change in accessibility occurred at the outermost areas of the system.

Conclusions on Land Use Impacts

Empirical evidence on the land use impacts of both highways and transit indicates that transportation investments do not have a consistent or predictable impact on land use. The beltway study found that within the same metropolitan area some beltway locations experienced significant development while others did not. Similarly, land development has occurred near some rail stations but not near others. In Atlanta rapid growth has occurred along the East Line, but not along the West and South lines. The evidence clearly shows that land use change does not necessarily follow transportation investments, even when the dollar value of these investments is large. Rather, availability of developable land, favorable economic conditions, and local political support have been identified as key factors in most studies.

These conclusions are not surprising. Land use–transportation theory suggests that transportation changes generate conflicting location incentives. A transportation improvement has a decentralizing incentive for households but a centralizing incentive for businesses. Centralizing businesses, however, create an incentive for households to locate closer to the center. At the same time, market-dependent firms must follow the population. A transportation improvement enlarges the market area for such firms, allowing greater decentralization. The inconsistencies observed empirically are therefore just what we would expect, given the conflicts in the theory.

Two other points deserve mention in conclusion. First, metropolitan areas throughout the United States have very highly developed transportation systems. Even the largest investments can effect only minor changes in the general level of accessibility within the region; consequently we should not expect them to have much impact. The historical record provides ample evidence in the case of highways. The studies conducted in the late 1950s and the early 1960s almost invariably found increases in land values and development associated with highway development. The two most recent studies (1975 and 1980 respectively) show no consistent relationship. During the earlier period, highways generated a substantial improvement in the area's accessibility and usually made more land available for development. By the 1970s this was no longer the case. Transportation considerations had become relatively less important in land use decisions.

Second, the existing transportation system is quite flexible. There is no longer a lower bound constraint: no urban area suffers from a scarcity of developable land because of a lack of transportation. An upper bound constraint also remains to be discovered. To date, development has never been halted because transportation system capacity was reached. Development moratoriums based on transportation system-capacity concerns have been imposed, but these seem to be the result of local perceptions. If moratoriums were a response to real constraints, they would occur in the cities with the most serious transportation problems. Downtown areas continue to attract new office and commercial activity (albeit at much lower rates than suburban areas), for example, whether or not they are served by high-capacity rail transit systems, and some of the most congested downtowns, namely New York and Los Angeles, are experiencing the greatest growth. Land use change is incremental; a single project will have an imperceptible impact on the transportation system, unless the project is extremely large. Given the apparent flexibility of urban transportation systems and the number of other factors that affect location decision, it is understandable that the linkage between transportation and land use in contemporary metropolitan areas is rarely clear.

References

Adkins, W. G. (1959). Land value impacts of expressways in Dallas, Houston, and San Antonio, Texas. *Highway Research Board,* Bulletin 227, 50–65.

Alcaly, R. (1976). Transportation and urban land values: A review of the theoretical literature. *Land Economics, 52,* 42–53.

Alonso, W. (1964). *Location and land use.* Cambridge: Harvard University Press.

Anas, A. (1983). *The effects of transportation on the tax base and development of cities.* Final Report No. DOT/OST/P-30/85/005, U.S. Department of Transportation. Evanston, Ill.: Northwestern University.

Arnott, R. (1979). Optimal city size in a spatial economy. *Journal of Urban Economics, 6,* 65–89.

Batty, M. (1976). *Urban modelling: Algorithm, calibrations, predictions.* Cambridge: Cambridge University Press.

Berechman, J., & Paaswell, R. E. (1983). Rail rapid transit investment and CBD revitalization: Methodology and results. *Urban Studies, 20,* 471–486.

Berry, B. J. L. (1967). *Geography of market centers and retail distribution.* Englewood Cliffs, N.J.: Prentice-Hall.

Boyce, D. E. (1980). A framework for constructing network equilibrium models of urban location. *Transportation Science, 14,* 77–96.

Boyce, D., Allen, W. B., Mudge, R., Slater, P., & Isserman, A. (1972). *Impact of rapid transit on suburban residential property values and land development: Analysis of the Philadelphia-Lindenwold High-Speed Line.* Final Report, U.S. Department of Transportation. Philadelphia: University of Pennsylvania.

Boyce, D., Allen, W. B., & Tang, F. (1976). Impact of rapid transit on residential property sales prices. In M. Chatterji (Ed.), *Space, location and regional development* (pp. 145–153). London: Pion.

Clark, C. (1951). Urban Population Densities. *Journal of the Royal Statistical Society, 114A,* 490–496.

Clark, W. A. V., & Burt, J. (1981). The impact of workplace on residential relocation. *Annals of Association of American Geographers, 70,* 59–67.

Dvett, M., Dornbusch, D., Fajans, M., Falcke, C., Gussman, V., & Merchant, J. (1979). *Land use and urban development impacts of BART.* Final Report No. DOT-P-30-79-09, U.S. Department of Transportation and Department of Housing and Urban Development. San Francisco: Blayney Assoc. and Dornbusch and Co.

Fernandez, J. E., & Friesz, T. L. (1983). Equilibrium predictions in transportation markets: The state of the art. *Transportation Research, 17B,* 155–172.

Gannon, C., & Dear, M. (1972). *The impact of rapid transit systems on commercial office development: The case of the Philadelphia-Lindenwold Line.* Final Report No. PA-11-0011, Urban Mass Transportation Administration. Philadelphia: University of Pennsylvania.

Guest, A. M. (1975). Population suburbanization in American metropolitan areas. *Geographical Analysis, 7,* 267–283.

Huff, D. L. (1964). Defining and estimating a trading area. *Journal of Marketing, 28,* 37–38.

Knight, R., & Trygg, L. (1977a). *Land use impacts of rapid transit: Implications of recent experience.* Final Report No. DOT-TPI-10-77-29, U.S. Department of Transportation. San Francisco: De Leuw Cather and Co.

Knight, R., & Trygg, L. (1977b). Evidence of land use impacts of rapid transit systems. *Transportation, 6,* 231–248.

Lakshmanan, T. R., & Hansen, W. G. (1965). A retail market potential model. *Journal of the American Institute of Planners, 31,* 134–143.

Lerman, S., Damm D., Lam, E. L., & Young, J. (1978). *The effect of the Washington Metro on urban property values.* Final Report No. UMTA-MA-11-0004-79-1, Urban Mass Transportation Administration. Cambridge: MIT.

Meyer, J. R., & Gomez-Ibanez, J. A. (1981). *Autos, transit and cities.* Cambridge: Harvard University Press.

Meyer, M. D., & Miller, E. J. (1984). *Urban*

transportation planning: A decision-oriented approach. New York: McGraw-Hill.

Mills, E. S. (1972). *Studies in the structure of the urban economy.* Baltimore: Johns Hopkins University Press.

Mohring, H. (1961). Land values and the measurement of highway benefits. *Journal of Political Economy, 79,* 236–249.

Mohring, H., & Hartwitz, M. (1962). *Highway benefits.* Evanston, Ill: Northwestern University Press.

Muller, P. (1981). *Contemporary suburban America.* Englewood Cliffs, N.J.: Prentice-Hall.

Muth, R. (1969). *Cities and housing.* Chicago: University of Chicago Press.

Norcliffe, G. B. (1975). A theory of manufacturing places. In L. Collins & D. F. Walker (Eds.), *Locational dynamics of manufacturing.* New York: John Wiley.

Paaswell, R., & Berechman, J. (1981). *An Analysis of rapid transit investments.* Final Report No. NY-11-0022, Urban Mass Transportation Administration. Buffalo, N.Y.: State University of New York at Buffalo.

Payne-Maxie Consultants (1980). *The land use and urban development impacts of beltways. Final* Report No. DOT-OS-90079, U.S. Department of Transportation and Department of Housing and Urban Development. Washington, D.C.: Payne-Maxie Consultants.

Putman, S. H. (1975). Urban land use and transportation models: A state of the art summary. *Transportation Research, 9,* 187–202.

Southern California Association of Governments. (1985). *Travel forecast atlas, 1984 base model,* Los Angeles, Calif.

U.S. Environmental Protection Agency. (1975). *Secondary impacts of transportation and wastewater investments: Research results.* Report No. EPA-600/5-785-013. Washington, D.C.: USEPA Office of Research and Development.

12 TRANSPORTATION AND ENERGY

GEOFFREY J. D. HEWINGS
University of Illinois

Many writers have commented about the inextricable link between the transportation system and location decision making. In fact, both elements of the metropolitan system may be seen as the two sides of the same coin, since a location decision implies a transportation decision and vice versa. This chapter explores the link between transportation, location, and energy. The link has received attention only relatively recently in response to the sharp increase in energy prices beginning in 1973. In the years immediately following, a large literature appeared on the topic, portraying the likely effects that increased energy costs would have on the *form* of the urban system and the patronage of transportation systems within urban areas. Furthermore, an attempt was made to examine the role that conservation practices in transportation systems could have on reducing the total amount of energy consumed in the nation.

Many of the analyses were flawed, often mistakenly confusing normative and positive analysis. The distinction between the two forms of analysis may be summarized as follows. Attempts to interpret actions within the economic system in terms of what should occur (for example, a certain policy prescription) may be interpreted as *normative* in nature. On the other hand, attempts to describe the system characteristics with little or no statement on the expectation of the results or any prescription about what should occur may be considered to be examples of *positive* analysis. In actuality, the distinction is not always quite so clear-cut, since many forms of positive analysis often rely very heavily on assumptions (about market or individual behavior) which may be construed as normative in nature. The net results of increased energy prices have been far more subtle than the normative studies predicted; for the most part, the structure of the urban system has not changed to a marked degree in response to the expected outcomes of a sharp rise in energy prices. In the interim, many other factors have combined to produce more dramatic changes in urban systems, for example, higher interest rates, industrial restructuring, and changes in tastes and life-styles. As a result, the expected centripetal influences created by rising energy prices, which would see urban densities increasing once again as individuals moved back toward the center of cities, did not occur although some movement back into the central cities has been observed. These movements have been characterized as gentrification processes, and although rises in energy costs may have contributed to their occurrence, the influence of the cost of energy was probably small in comparison to changes in tastes and life-styles and a reduction in household size for those members of the population in the 20 to 30 age bracket.

We shall explore some of the analyses that were undertaken in an attempt to seek some insights into the processes of modeling the relationships between transportation and energy in metropolitan areas. We shall begin by examining some of the macroeconomic relationships between transportation and energy consumption within the United States. Thereafter, we shall explore these relationships at the individual level before shifting the focus of attention back to a more macro level of analysis. At that level again, we will explore the energy effects of shifting funds from one form of transportation to another and the effect of changes in patronage of transportation mode. Finally, energy issues will be explored in the context of urban areas in developing countries.

TRANSPORTATION AND ENERGY: PRELIMINARY CONSIDERATIONS

We noted earlier that there was an expectation that after the oil price shock of the early 1970s the impact on the transportation system would be a profound change in demand for different modes; there was a longer-run expectation that even the form and character of the metropolitan area would change. However, these expectations were made without careful consideration of the nature of the demands for transportation energy. In this section, we shall explore some of the basic relationships between transportation and energy over the last decade or so. Table 12-1 shows the relationships between the primary consumption of energy by major end uses in 1983. The percentage allocation of energy by end use has remained rather stable over the last decade: the minor changes that have occurred have been a slight increase in transportation demand and a slight decrease in industrial demand. One striking, rather obvious feature of Table 12-1 is the dominant role transportation plays in the consumption of petroleum energy (over 60%). Hence, while transportation energy consumes only a little over one-quarter of total energy consumed in the United States, it is the form of that energy consumption which has created the major concerns and perhaps provided the best opportunities for energy conservation.

Table 12-2 shows the way in which the transportation energy was consumed by mode in 1981. Here we find that highway modes account for nearly three-quarters of total energy consumed in the transportation sector but about 96% of the gasoline consumption. Within this category (highway), the dominant role of automobiles is evident. However, note

Table 12-1. Total primary consumption by end use, 1983

	Coal	Nat. Gas	Petroleum	Electric	Other[a]	Total energy
Residential/commercial	1.2	40.7	7.8	63.8	0	36.2
Industrial	15.3	38.7	25.8	36.1	0.9	36.8
Transportation	0.0	3.3	61.3	0.1	0	27.0
Electric utilities	83.5	17.3	5.1	—	100.0	—
Total	100.0	100.0	100.0	100.0	100.0	100.0

Numbers may not add up due to rounding.

[a]Other category includes hydroelectric, nuclear, and geothermal.

Note. From *Transportation Energy Data Book, Edition 7* by M. C. Holcomb and S. Kolsky, 1984, Oak Ridge, Tenn.: Oak Ridge National Laboratory.

Table 12.2. Transportation energy use by mode, 1981

	Percentage of total	Percentage gas consumption
I Highway	73.3	96.4
Automobiles	45.3	68.5
Trucks	27.2	27.3
Motorcycles	0.1	0.2
Buses	0.7	0.4
II Nonhighway	23.2	1.7
Air	7.5	0.5
Water	7.9	—
Pipeline	4.6	—
Rail	3.2	—
Other[a]	0.1	3.1
III Military operations	3.4	
Total	100	

Average annual percentage change in transportation energy use

1973–1978	2.1%
1978–1983	−1.6%

[a]Includes recreation, state and local government

Note. From *Transportation Energy Data Book, Edition 7* by M. C. Holcomb and S. Kolsky, 1984, Oak Ridge, Tenn.: Oak Ridge National Laboratory.

that in the period 1978–1983, the average annual percentage change in transportation energy use *declined* by 1.6%, in contrast to the preceding five years in which the use increased by 2.1% per year.

The expectation of many analysts following the 1973 price rise in energy was that (1) energy used for transportation would decrease; (2) substitution of modes of transportation would take place, particularly in urban areas, and (3) the form of urban areas would begin to change in response to higher energy prices. One of the major problems in conducting analysis of this kind is the issue of differentiating the effect of an increase in the price if a flow commodity (which is primarily used for immediate consumption) on the spatial demand for a number of stock commodities (such as housing). The distinction between "stocks" and "flows" is not always based on the magnitude of the expenditures, although comparing a household's energy expenditures and housing costs, the size differential is obviously large. Of greater importance in this case is the view of the consumption of one commodity (energy) within a short time frame (such as a year) in contrast to housing expenditures for which individuals usually negotiate loans that are amortized over 15- to 30-year time periods. In judging the price rises of the energy goods, the individuals consumer will be comparing these increases with the possibly much greater costs that would be incurred if the household changed its residential location.

Thus, when the stocks are large and far more expensive than the flow commodities, the expectation that rather sudden or dramatic changes might occur ought to be very low. In addition, the role of expectations becomes very important; in this case, the U.S. consuming public adjusted rather quickly to higher energy prices (although less easily to energy shortages) and, as a consequence, many of the expected effects did not materialize. Futhermore, although the number of automobile miles driven has risen almost continuously in the last decade, the data in Table 12-2 reveal that energy consumption has decreased. Clearly, technological changes in the form of lighter, more fuel-efficient automobiles have allowed individuals to retain their attachment to automobiles at rather small sacrifices of space, comfort, and size.

In the next section, we will explore the impacts of the transportation/energy interactions at the individual level. At this level of analysis, we will be in a position to explore some of the trade-offs facing an individual and to begin the task of identifying some of the factors that served to retard major locational reorientations within the North American city.

TRANSPORTATION AND ENERGY: THE INDIVIDUAL DECISION-MAKING FRAMEWORK

The best-known formal model demonstrating the choice of location for an individual within an urban area was initially developed by Alonso (1964) but has subsequently been modified and expanded by a large number of authors (Anas, 1982; Mills, 1972; Muth, 1984). This model was also introduced in chapter 11 on the land use impacts of transportation investments. Let us take a simple version of this model. Assume a city that is located on an isotropic (that is, a uniform, featureless) plain. All employment opportunities are located in the center of the city. An individual with income Y is faced with the choice of a location decision. Essentially, the choice entails trading off expenditures on (1) land (which we will assume to include either rent for an apartment or some mortgage payment for a house or condominium), (2) transportation costs involved in getting to and from work, and (3) all other goods and services, which are usually lumped together into something called a composite good. If we assume that land costs per unit fall with increasing distance from the city center (see Figure 12-1a) and that transportation costs rise at a decreasing rate from the city center (Figure 12-1b), then we may develop the cost functions shown in Figure 12-2a. If we use the procedures of classical consumer equilibrium analysis to derive the optimum location for an individual, then we need to specify the individual's utility function. This function essentially provides us with information on the degree of satisfaction a consumer receives from the consumption of a set of goods. Assume for the moment that there are only two goods available; the consumer is thought to be able to calculate how much satisfaction would be received from alternative combinations of the two goods. For any given level of satisfaction (utility), we may develop a set of curves that provide the different combinations of the two goods contributing to this satisfaction level. This trade-off between two goods may be shown in terms of an "indifference curve." An example is shown in Figure 12-2b in the case of the two goods land (q) and the composite good (z). In most cases, the indifference curves are convex to the origin; there will also be a whole family of these curves, each one representing a different level of utility.

Consumers usually wish to maximize their satisfaction; in other words, they seek to choose combinations of goods that will place them on indifference curves as far away from the origin as possible. They are not able to move to the very highest levels in most cases because their incomes are limited. Thus, we use the budget constraint to derive the optimal trade-off between the three "goods." However, the shapes of the indifference curves for these goods do not all appear to be similar

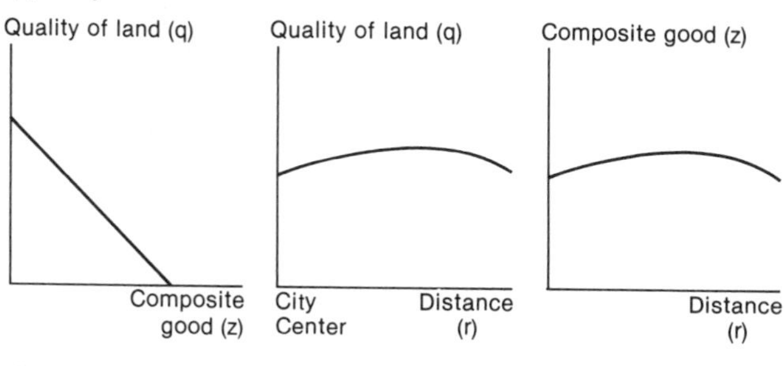

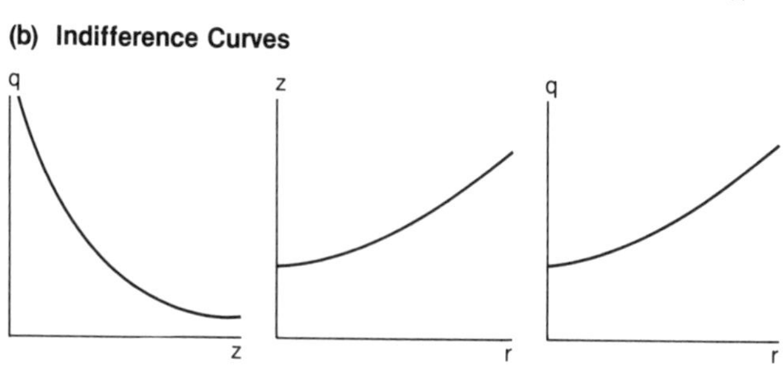

Figure 12-1. For the individual, (a) Isoquants, and (b) Indifference Curves. From *Location and Land Use* by W. Alonso, 1964, Cambridge: Harvard University Press. Copyright 1964 by Harvard University Press. Reprinted by permission.

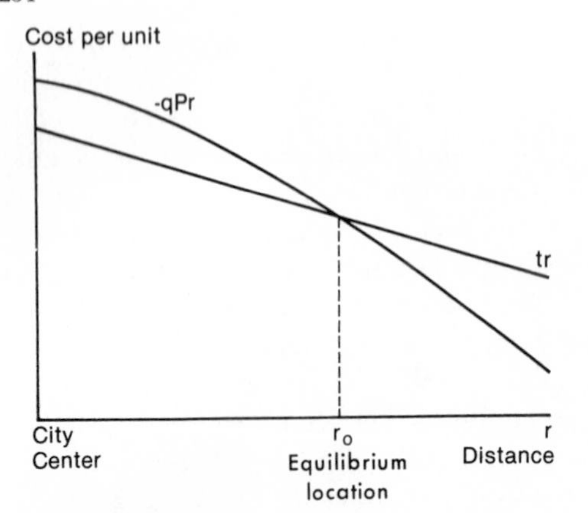

Figure 12-2. Spatial equilibrium for the individual. From "Energy-Price Effects on Metropolitan Spatial Structure and Form" by M. C. Romanos, 1978, *Environment and Planning A, 10,* pp. 93–94. Copyright 1978 by Pion Ltd. Reprinted by permission.

(see Figure 12-2b). For example, since transportation costs may be regarded as a "negative good" (in the sense that less will be preferred to more, other things being equal), then an individual would have to be compensated with more land as transportation costs increase (that is, as the individual moves away from the city center). Hence, the indifference curves between land and transportation and between transportation and the composite good are shaped a little differently from those between land and the composite good.

If we assume that the individual chooses to maximize utility with respect to the budget constraint, the optimum allocation of expenditures among the three goods will be provided. In addition, the *location* of the individual within the city will be known. Now suppose that energy prices rise, as they did in 1973. What will happen to the optimal location for the individual? Following Romanos (1978), we may examine the consequences.

Let us define the individual's utility function as:

$$U=U(z,q) \tag{1}$$

where z represents the composite good and q the purchase of land and housing. The individual's budget constraint may be shown as:

$$g=y-(z+p_r q=t_{r,y}) \tag{2}$$

where p_r is the price of housing at a distance r from the central city and t is the transport cost which we assume will be a function of distance r and the individual's income, y.

Maximization of utility yields:

$$L=U-\lambda g \tag{3}$$

where λ is the Lagrangian multiplier. This multiplier serves to provide a way of determining the degree to which the constraints imposed by the individual consumer's budget constrain the allocation of income among the goods available for purchase. In equilibrium, the partial derivatives of L with respect to all the parameters will be zero. Thus, we assume that the individual has found an optimum allocation of income such that there would be no incentive to change the combination of purchases. We can consider that at maximum utility, the consumer has reached a peak in the utility function; further adjustments can only be made in the combination of the goods purchased by the individual moving down from the peak and thus lowering the level of utility attained. In particular, we are interested in the partial derivative with respect to distance since our major interest is in tracing the impacts of price changes on location:

$$\frac{\delta L}{\delta r}=-\lambda\left[q\frac{\delta p}{\delta r}+\frac{\delta t}{\delta r}\right] \tag{4}$$

Setting this equal to zero yields:

$$-qp_r=t_r \tag{5}$$

How should this be interpreted? The right-hand side is the marginal transportation cost of moving away from the city center. As we noted earlier, the transportation cost function exhibits a form in which the marginal costs of going an additional mile diminish as the number of miles traveled increase. This may be seen by reference to Figure 12-1b; although the total cost per unit increases with distance, the marginal addition to that cost from going the extra mile becomes smaller. Hence, t_r will exhibit a negative slope (see Figure 12-2). The lefthand side of equation 5 reveals the reductions in housing costs as we move away from the center. Again, drawing on the shapes of the functions in Figure 12-1, we would expect the savings in housing costs to diminish as one moves away from the center. In all likelihood, these savings will fall at a faster rate than the marginal increases in transportation costs. In Figure 12-2, we have the equilibrium location, r_0, determined.

Next, Romanos (1978) considered the effect of an energy price increase. He explored the impacts of these price increases on two parts of the model, first, the effect on housing prices and then on transportation costs. In the former case, we may expand the definition of p_r as follows:

$$p_r = p_1(r) + p_2(r) \tag{6}$$

where $p_1(r)$ represents all purchases associated with housing except energy, and $p_2(r)$ represents energy purchases (heating, cooling). Romanos assumed that p_2 would increase with distance, since one would expect to find larger houses with higher heating/cooling costs with increasing distance from the city center. Hence, we may now write:

$$-qp'_r = -q\frac{\delta p_1 + \delta p_2}{\delta r} = \frac{\delta t}{\delta r} \tag{7}$$

Similarly, transportation costs may be divided into energy and nonenergy costs per unit distance:

$$t(r) = t_1(r) + t_2(r) \tag{8}$$

where t_1 are all costs except fuel and t_2 are the fuel costs. In this case, both components of transportation cost increase with distance. Linking the two divided cost components, we have:

$$-\text{q}\frac{\delta p_1 + \delta p_2}{\delta r} = \frac{\delta t_1 + \delta t_2}{\delta r} = t'_r \tag{9}$$

The effects are shown in Figure 12-3. The increased energy costs for housing would result in the $-qp_r$ curve shifting downward to $-qp'_r$; at increasing distance from the city center, the marginal savings in housing costs would be lower. This would occur because the increase in energy costs would affect larger houses (located farther away from the city center) proportionately more than smaller dwellings. On the other hand, energy price increases in transportation would shift the t_r curve upward to t'_r. The new equilibrium location for the household would now shift from r_0 to r_1 closer to the city center.

However, this did not happen in the sense that we have not witnessed a massive return toward the center of U.S. cities nor has the density of population exhibited an increase within the urban areas. The reasons for this are many as Muth (1984) has articulated.

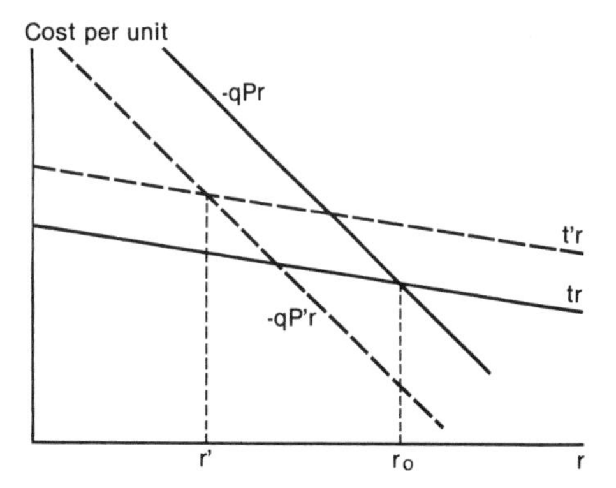

Figure 12-3. Spatial equilibrium of the individual after a rise in energy prices. From "Energy-Price Effects on Metropolitan Spatial Structure and Form" by M. C. Romanos, 1978, *Environment and Planning A, 10*, pp. 93–104. Copyright 1978 by Pion Ltd. Reprinted by permission.

First, marginal analysis of the kind undertaken in the Alonso-type framework ignores the *level* of expenditures of the alternative basket of goods (land, transportation, and the composite good). Second, the analysis takes no account of the time framework, particularly with reference to expectations and the discounted present value of assets held by the individual. In this case, we are attempting to provide a way of estimating the value during a specific year of a commodity (housing) that is being purchased and consumed over a much longer time horizon. Third, the analysis does not allow for technological developments.

Let us deal with these issues in turn. A rise in the price of gasoline in the United States from 36 cents to 60 cents and then to $1.00 per gallon would appear to be rather dramatic increases in percentage terms. However, the percentage increases mask the changes in *real* terms (that is, accounting for the effects of all price changes) and the percentage of the individual's income that would normally be devoted to gasoline purchases. Figure 12-4 shows that although gasoline prices have increased sharply from 1973 through 1984 in *current* dollar terms, the price rise in *real* dollars (dollars adjusted for inflation) reveals an increase of only 14 cents over this period. To calculate the effect this might have had on the average household (even using current dollars), assume, for example, that for work trips an individual drives 20 miles a day (10 miles each way). Over a 250-workday year, 5,000 miles would be logged. Assume further that the automobile gets 12 miles to the gallon. At 36 cents per gallon, the annual cost would be $150; at 60 cents per gallon, the cost would be only $100 more per year. The likelihood that the increased gasoline prices would induce individuals to move closer to work would depend on the effect that an additional $100 would have on other expenditures that would have to be sacrificed, assuming income remains constant. Clearly, for most people, the effect was more initial shock and adjustment to higher gasoline prices than a mad scramble to relocate to save a few hundred dollars in gasoline costs. At the same time that energy costs were rising, so were the costs of many other goods as well as incomes. In particular, the costs of borrowing money to purchase houses increased rather dramatically in the middle to late 1970s completely dwarfing the impact of gasoline price increases.

In essence, with real incomes rising over

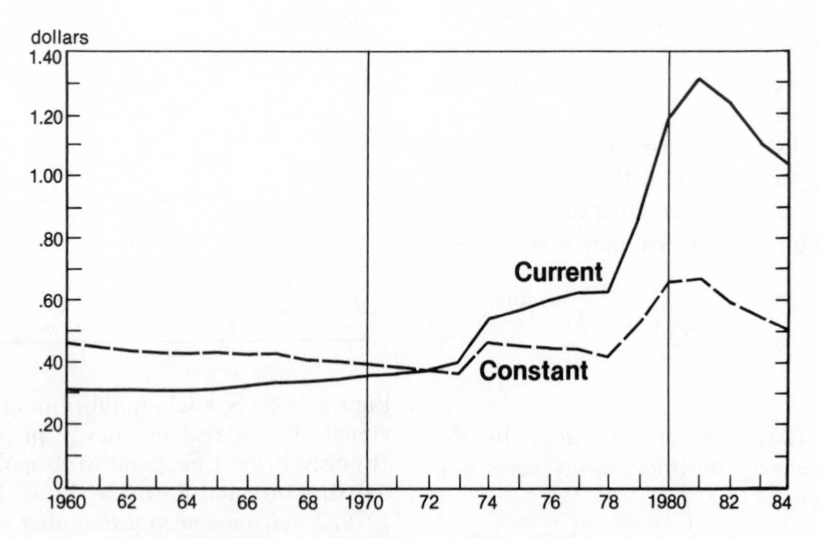

Figure 12-4. Constant and current prices of gasoline, 1960–1984. Adapted from *Annual Energy Review 1984* by the Energy Information Administration, 1985, Washington, D.C.: U.S. Government Printing Office.

Table 12-3. Consumer price indexes, selected years, 1973–1980

Item[a]	1973	1976	1980	1980/1973
All items	133.1	170.5	246.8	1.85
New automobiles	111.1	135.7	179.3	1.61
Gasoline	118.1	177.9	419.3	3.55
Fuel and other utilities	126.9	182.7	278.6	2.20
Deflated by all items				
New automobiles	83.5	76.6	72.6	0.87
Gasoline	88.7	104.3	106.9	1.92
Fuel and other utilities	95.3	107.2	112.9	1.18

[a] 1967 = 100.

Note. From "Energy Prices and Urban Decentralization" by R. F. Muth, 1984, in *Energy Costs, Urban Development and Housing*, ed. A. Downs and K. L. Bradbury, Washington, D.C.: Brookings Institution. Copyright 1984 by the Brookings Institution. Reprinted by permission.

the last decade, individuals have had to expend a greater share of their *marginal propensities to consume* for gasoline than in the previous two decades. In this regard, we make a distinction between expenditures made from what we may refer to as a base or average income (average propensities to consume) and those expenditures made from additions to income. Rarely do we find that individuals expend the same proportion of an additional dollar on a good as they do from an average dollar. Usually, greater expenditures will be devoted to luxury-type goods from the marginal dollar. Thus, the price rises of energy curtailed the degree of choice that the consumer had in allocating these marginal dollars.

This is borne out by the data in Table 12-3, developed by Muth (1984) showing that when the price rises for gasoline are compared with the all-items index, the real rise over the period 1973 to 1980 has been a doubling in cost only. Hence, we find that the individual price rise for gasoline from 1973 to 1980 has been 3.55, but only 1.92 when consideration is taken of price rises in all other items. Note, too, that the real cost of automobiles has actually decreased during this same period, even though the price index for new automobiles alone has increased by 1.61. During this period, the price effects for all items revealed a 1.85 increase (that is, prices almost doubled). One important issue that cannot be discerned from the table is the degree to which increases in energy prices lead to increases in the prices of other goods. In other words, what was the *indirect* contribution of increased energy prices in the economy?

The effect has been to increase the real cost of operating an automobile from 1973 to 1980 by 16%. This increase itself probably overstates the true increase, since a comparably sized automobile that obtained only 12 miles to the gallon in 1973 would be getting 20 to 25 miles per gallon in the early 1980s. Technological improvements in the efficiency of engines and the substitution of aluminum for steel in large parts of the automobile frame have served to produce more efficient automobiles without sacrificing too much space in the cabin. Muth (1984) also showed that when the other operating costs associated with running an automobile (depreciation, repair and maintenance, taxes, and so on) are included with gasoline and oil costs, the costs per mile (in constant 1973) dollars) have increased only about 5% between 1973 and 1980. (See Table 12-4.) Here we see that the total per mile operating costs for automobiles was 13.34 cents in 1980 and 6.63 cents in 1973; when the 1980 figures are converted to 1973 dollar equivalents, we find that the cost in 1980 was only 7.19 cents. A 5% rise over that period of time should not be expected to create many changes in

Table 12-4. Ten-year average cost of operating a 1976 subcompact automobile (cents per mile)

Year and deflator	Gasoline and oil	Various operating costs	Total
1973	1.19	5.64	6.63
1976	1.80	6.89	8.69
1980	4.24	9.10	13.34
1980 in 1973 dollars	2.29	4.91	7.19
1980/1973 (1973 dollars and 1980 auto prices)	1.92	0.87	1.05
1980/1973 (1973 dollars and 1973 auto prices)	1.92	1.00	1.16

Note. From "Energy Prices and Urban Decentralization" by R. F. Muth, 1984, in *Energy Costs, Urban Development and Housing*, ed. A. Downs and K. L. Bradbury, Washington, D.C.: Brookings Institution. Copyright 1984 by the Brookings Institution. Reprinted by permission.

consumer behavior. Furthermore, between 1973 and 1983, the average number of miles per gallon achieved with the vehicles on the highway rose from 13.1 to 16.7; hence, an increase in efficiency of automobiles meant that the fuel consumption per car in 1983 was only 75% of that 10 years earlier (Energy Information Administration, 1985). The issue, of course, was that so much of that change involved a set of large increases in gasoline prices, a component of automobile operating costs that individual consumers are made aware of on a regular basis. In the final analysis, though, these changes, while larger in comparison to prices increases elsewhere in the economy, were still relatively modest in terms of their effects on individual consumer behavior.

Hence, it should come as no surprise that there has not been a massive population shift back to the city center. The rise in gasoline prices certainly created some disruptive effects, but after the hyperbole is discounted, the net effects have proven to be negligible. The analysis was also flawed by the fact that the theoretical models operate on the assumption that all employment activities are located in the central city. In fact Romanos (1978) felt that the increasing suburbanization of job opportunities would see the city evolve into a series of nuclei surrounded by high population densities, with a far less demonstrable overall density gradient originating in the CBD. Anas (1982) noted that in the Chicago SMSA the percentage of jobs in the CBD declined from 20 to 15% of the SMSA total between 1970 and 1980. Thus, even if the price rises in gasoline had been significant in terms of household budgets, the fact that job opportunities were suburbanizing created centrifugal forces. The relocation of jobs away from the central area of cities provided greater attractions for residential locations in closer proximity to suburban business centers. Even if the data revealed that the overall length of the average commuting trip to work had declined over the last decade, it is doubtful that one could attribute more than a small percentage of the reason for this to rising energy costs.

THE ROLE OF TRANSPORTATION AND ENERGY CONSUMPTION

The micro-level analysis described above does hide some very important macro-level concerns about the relationship between transportation and energy. Since the early 1960s, when U.S. domestic consumption exceeded domestic production, there has been growing concern about the problems of increasing dependence on imported oil and the possibilities for energy conservation. Figure 12-5 shows the relationships between energy consumption, domestic production, and imported energy in the last 30 years. During

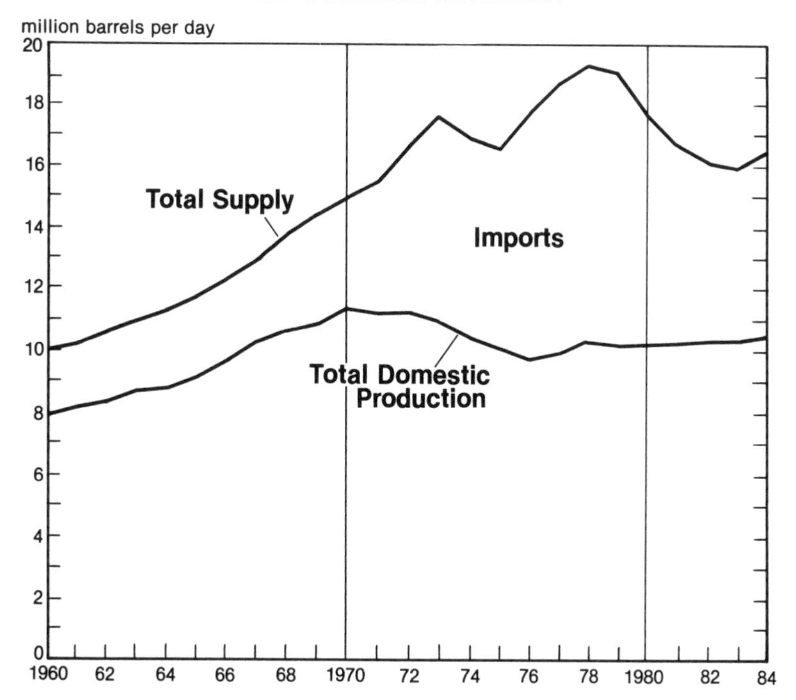

Figure 12-5. Petroleum production and imports, 1960–1984. From *Annual Energy Review 1984* by the Energy Information Administration, 1985, Washington, D.C.: U.S. Government Printing Office.

the late 1950s–early 1960s, the percentage share of domestic production in the provision of total energy needs began to decline, and a period of increasing import dependency set in. As we noted earlier, more than 25% of annual energy consumption in the United States occurs in the transportation sector, and over half of this is consumed by the private automobile. Though there have been some attempts to develop alternative fuels (such as gasohol or electric cars), the major fact remains that the possibilities for substitution in this sector are limited. In addition, the consumption of gasoline in *running* the automobile represents only a part of the total energy consumed in *producing* the automobile.

The concept of *direct* and *indirect energy consumption* is important. We consider energy used to run an automobile as an example of *direct* energy consumption. The construction of the automobile itself also requires energy—in the extraction of the iron ore and coal to make the steel, the production of the steel and the succeeding fabricated metal that is ultimately transformed into the automobile body. Energy will be required to make the glass for the windows, rubber for the tires, and so forth. Thus, the automobile delivered to the retail showroom will contain energy costs embodied in it, what we refer to as *indirect* energy consumption. Bezdek and Hannon (1974) examined this idea in the context of the possible impacts of changes in energy consumption and employment resulting from changes in the patronage of transportation modes within cities and between cities. They were interested in discovering the impacts of changes of this kind on some important attributes of the economy, especially employment. Would a redirection of investment funds away from highways produce massive unemployment even though there may be considerable energy savings? They considered the possibilities in terms of a redirection of $5 billion of the 1975 funds accumulated in the Highway Trust Fund away from its federally mandated use for highway

Table 12-5. The impact of energy consumption and employment of a $5 billion investment (1975 dollars) in seven federal programs. Five billion 1975 dollars are equal to $3.65 billion in 1963 and $3.165 billion in 1958. Data calculated from *(1, 3)*. No attempt was made to correct for the technological impact on energy use efficiency between 1963 and 1975. It is generally expected that 1975 technology will be more energy intensive.

	Energy consumption			Employment demand		
Federal program	Requirement per 1963 dollar of program (Btu)	Total requirement (10^9 Btu)	Decrease* (%)	Jobs per $100,000 of program (1975)	Total no. of jobs	Increase* (%)
Highway construction	112,200†	409.53		8.1	256,180	
Railroad and mass transit construction	43,100	157.32	+61.6	8.4	264,430	+3.2
Water and waste treatment facilities construction	65,400	238.71	+41.7	8.2	259,490	+1.3
Educational facilities construction	70,600	257.69	+37.1	8.5	268,980	+4.7
National health insurance	40,400	147.46	+64.0	13.4	423,220	+65.2
Criminal justice and civilian safety	118,500	432.53	−3.4	12.4	393,520	+53.6
Personal consumption expenditures (tax relief)	86,000‡	313.90	+23.4	8.7	275,120	+7.4

*Percent changes are relative to highway construction program.

†As in all programs this number is for a technology of estimated efficiency. The actual energy intensity of all highway construction in 1963 was 98,000 Btu per dollar *(3, 4)*. Similar construction (Army Corps of Engineers) varied from 92,000 to 146,000 Btu per dollar, the highest of all government programs.

‡Includes direct energy purchases and the energy and labor required for trade and transportation margins.

Note. From "Energy, Manpower and the Highway Trust Fund" by R. Bezdek and B. M. Hannon, 1974, *Science, 185*(4152), pp. 669–675. Copyright 1974 by American Association for the Advancement of Science. Reprinted by permission.

construction to other forms of investment (listed in Table 12-5). The base year of analysis was 1963 (the latest year for which a national input-output model for the United States was available in the early 1970s), and all dollar values were deflated to 1963 dollars. In addition, no attempt was made to correct for any improvements in energy-use efficiency between 1963 and 1975. Some years earlier, Leontief (1966) had performed a similar analysis in connection with a compensated cut in defense expenditures (by this we mean that the savings in defense expenditures would be redirected to other forms of investment or consumption), but his analysis considered only the effects on output and employment. In conducting their analysis, Bezdek and Hannon (1974) modified the usual input-output model in the following way:

$$M = z(I - A)^{-1}p_q \qquad (10)$$

where M is an $n \times p$ matrix showing the employment generated by and within each industry by a specified expenditure distribution (highway construction, railroads and mass transit, and so on), z is an $n \times n$ diagonal matrix of employment/output coefficients (the numbers employed per $1 million of output in each industry), and $(I-A)^{-1}$ is the Leontief Inverse Matrix; it details the direct and indirect requirements that each industry places on all other industries in the process of producing output. P_q is the product of an $n \times p$ matrix whose coefficients show the direct requirements for the outputs of each industry generated per dollar of expenditure on each activity, and q is a diagonal matrix whose elements show the expenditures allocated to each activity.

The energy requirements can be generated by:

$$e = [Q(I - A)^{-1} + T]P_q \qquad (11)$$

where Q is a matrix of energy sales of each energy sector (coal, natural gas, oil, and so on) to each industry per unit of output in that industry and T is a diagonal matrix of energy type sold to the final demand activities.

The results of the analysis, in terms of employment and energy impacts, are shown in Table 12-5. The columns marked "Decrease" and "Increase" should be read in terms of the relation of the alternative programs with the highway construction mandate of the Highway Trust Fund. Thus, we find that a redirect of funds from highways to railroads and mass transit construction would decrease energy consumption by 61.6% relative to that amount of energy used for the highway program. Similarly, employment would show an increase of 3.2% as a result of this redirection; again the increase is over and above the numbers employed as a result of the highway investment program. Note that all the alternatives, except the criminal justice and civilian safety programs would save energy, and all would generate more jobs than the highway construction program, although the number of additional jobs created would be small for many of the programs. The reason for the increase in energy consumption associated with the criminal justice program stems from the fact that the expenditure pattern for this program would be heavily oriented toward increased numbers of uniformed officers and many more patrol cars—hence, increases in direct and indirect energy consumption.

A few comments should be made here. First, the analysis assumes that there would be no disruptive effects in shifting from one program to another. Clearly, some industries, especially those producing intermediate goods, might be able to accomplish this rather easily. For others, the transition might not be possible, resulting in many firms going out of business. Second, there is no guarantee that the jobs created would be placed in the same geographic regions as those lost. Finally, the analysis does not discriminate among the skill categories of jobs lost and created. Bezdek and Hannon (1974) did extend the analysis to consider occupational categories. It should be obvious that construction workers may be shifted fairly easily from highway to mass transit construction but less easily from construction to a program focusing on national health care (although a major component of this program might involve the construction of additional hospitals and clinics).

The analysis does suggest that the substitution of railroad and transit for highway construction would save substantial amounts of energy. What needs to be addressed now is the problem of (1) calculating the energy savings associated with a shift in patronage from highway to rail and transit, and (2) estimating the probability that individuals would switch from private automobiles to mass transit. The former analysis is less difficult to undertake; the latter analysis is an almost impossible task!

Table 12-6 shows the energy impacts of switching patronage within and between cities. To provide some basis for comparison, a standardized flow, namely 1 million passengers, over a unit mile is used. No consideration is given to economies of scale, congestion costs, the fact that transportation rate structures are nonlinear (increase at a decreasing rate) or the fact that for some transportation systems (such as air), the terminal costs are very high. The load factors used in Table 12-6 reflect reasonable expectations: the results would change rather dramatically if, for example, transit load factors could be increased. Once again, the differences in energy consumption by mode shown in this table are rather dramatic. Note, too, the differences in the direct and indirect energy consumption. Recall from the earlier discussion that the former includes the energy required to move the vehicle whereas the latter includes the energy involved in construction of the vehicle and any support systems (such as, electrical transmission lines in the case of the electric commuter system). Thus, the energy cost of running an automobile represents only about 58% of the total energy costs within an urban area.

These estimates vary somewhat with those shown in Table 12-7 for a slightly larger set of transportation modes. In this case, the energy used to manufacture vehicles does not include the indirect energy requirements (those necessary to manufacture the components and assemble the automobile), and the

Table 12-6. Selected results on the total dollar, energy, and labor impacts of consumer options in transportation

	Requires			
To move a million passengers one mile by . . .	Load factor	Thousands of dollars (1971)	Million BTU of energy	Jobs
Intercity				
Car	2.2 people	55	7,000	3.7
Plane	50% full	58	9,800	3.8
Bus	50% full	38	2,500	3.2
Train	37% full	44	4,000	7.3
Urban				
Car	1.9 people	69	8,900	4.8
Bus	12.0 people	105	5,300	8.3
Bicycle	1.0 people	26	1,300	N/A
To move one million tons of freight one mile by . . .	Circuity			
Air	1.1	200	80,000	37.0
Rail	1.2	16	1,600	1.4
Truck	1.2	80	4,100	2.4
Barge	1.7	3	1,600	0.6

Note: From "Energy, Employment and Transportation" by B. M. Hannon, 1974, Urbana, Ill.: Center for Advanced Computation, University of Illinois. Copyright 1974 by the University of Illinois. Reprinted by permission.

Table 12-7. Middle estimates of basic components of operating energy intensiveness and line-haul energy by urban transportation modes (British thermal units (BTUs) per vehicle-mile)

Mode	Propulsion energy	Average number of occupants	Station and maintenance energy	Construction energy	Vehicle manufacturing energy	Total energy per capita
Single-occupant automobile	11,000	1.0	2,000	125	1,100	14,225
Average automobile	11,000	1.4	2,000	125	1,100	10,160
Carpool	11,000	3	2,000	125	1,100	4,741
Vanpool	14,000	9	2,000	200	2,000	2,022
Dial-a-ride	15,500	1.6	2,000	200	2,000	12,312
Heavy rail (old)	61,000	24	9,000	3,000	1,500	3,104
Heavy rail (new)	75,000	21	15,000	4,000	1,500	4,547
Commuter rail	105,000	40	7,000	1,200	2,500	2,892
Light rail transit	75,000	20	7,000	1,700	2,000	4,285
Bus	30,000	11.5	900	370	1,200	2,823
Personal rapid transit	11,000	2	5,000	300	1,000	8,650
Group rapid transit	20,000	6	6,000	600	1,000	4,600
Shuttle loop transit	23,000	10	7,000	600	1,000	3,160

Note. From *Urban transportation and energy: The potential savings of different modes* by Congressional Budget Office, 1977, Washington, D.C.: U.S. Government Printing Office.

energy costs have been amortized over the life of the automobile (that is, we assume that the average automobile might be run for 100,000 miles during its lifetime). Note again the variations in operating intensiveness on a per capita basis. Finally, for comparative purposes, note the energy efficiency of different modes in terms of fuel used per passenger mile shown in Figure 12-6.

Addressing the second question, how one might be able to persuade people to move from automobile to mass transit systems, opens up a complex literature in which the basic conclusions appear to be that the probabilities are not very high. Why? A number of plausible hypotheses can be advanced. First, we might make use of the notion of consumption theory based on characteristics rather than mode. Quandt (1972), using Lancaster's reformulation of consumer theory, developed the notion of an abstract mode. Individuals choose to patronize one form of transportation or another within an urban area on the basis of its characteristics, not whether it is a bus, car, or train per se. Thus, a consumer interested in utility maximization would choose the mode whose characteristics (speed, comfort, convenience, safety, cost) could be matched within the framework of the budget constraint and the nature of the trip purpose. The individual does not necessarily value or weight each of these characteristics equally; thus, a strong preference for convenience might completely dominate minor differences in costs among various competing modes. Second, the valuation of time and costs may vary substantially by mode. Consumers in general tend to underestimate substantially the real cost of driving a private automobile. Since the costs are

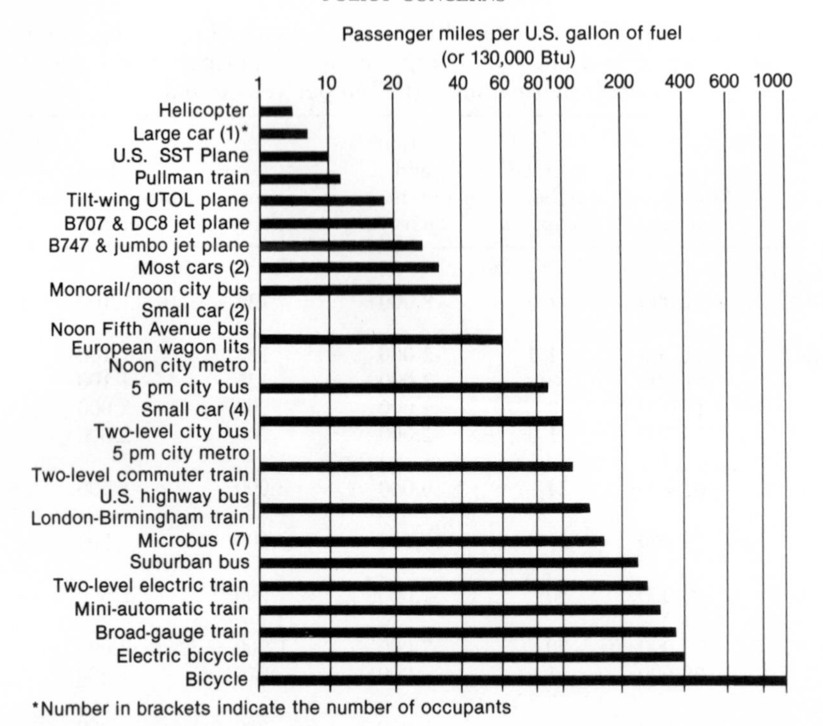

Figure 12-6. Energy efficiency of various passenger transport systems in the United States. From "The Impact of the Motor-Car on Oil Reserves" by G. Leach, 1973, *Energy Policy, 1*, pp. 195–208. Copyright 1973 by Butterworth Scientific Ltd. Reprinted by permission.

often incurred outside the context of the work trip (gasoline may be purchased at the weekend, insurance premiums are usually paid only once or twice per year), the consumer thus compares the apparent "free" automobile ride with the fact that transit fares have to be paid at the time the system is used. Thus, a 10 or 20% increase in transit fares would tend to have a large impact on transit riders, whereas a similar increase in gasoline prices might have little impact on automobile patronage.

To illustrate these difficulties, Anas (1982) investigated gas and parking tax policies to finance public transit in Chicago. He estimated in 1981 that the accumulated deficit (reduction in operating revenue and federal and state subsidies) for the public transportation system in that city amounted to about $100 per household. If fares on the system were set to zero, the tax necessary per household would rise to $245. Would free fares have any impact upon transit patronage? This is doubtful, concludes Anas (p. 222), since "in the U.S. the travel cost elasticity of ridership demand is too low to achieve such results." By this we imply that changes in the costs of public transit would be unlikely to yield significant changes in patronage. Would an alternative revenue-raising proposition have any impact? Anas considered a per gallon tax: he estimated that a tax of about 21 cents per gallon would be required (raising the average 1981 price in the Chicago SMSA by 14%) to cover the cost of the operating deficit. However, the effect on automobile commuting would have been negligible, a decline of only 0.9%! On the other hand, if the deficit in transit operations were covered by raising the fares, then ridership on rapid transit, bus, and commuter rail into the CBD would decline by 5% each while suburban rapid transit ridership would drop by 29%. If the transit fares are set to zero and the operating costs borne by an increase of 48 cents per gallon in gasoline, then automobile ridership would decline by 10%. If the revenue were raised instead by increasing parking charges in the CBD by $3.26 a day, automobile commuting to the CBD would decline by 16%.

This analysis has interesting welfare implications, since it would decrease the welfare of automobile owners, who tend to be from higher-income groups, and increase the welfare of transit and bus commuters, who tend to come from lower-income groups. The major point is, however, that the degree of elasticity in transit use exhibited with respect to the cost of automobile patronage is very small. Unless energy costs increase at an alarming rate over a very short period (especially in comparison to other price and income changes), it is unlikely that we will witness (1) major shifts in location of households, and (2) major shifts in the use of modes of transportation within urban areas.

TRANSPORTATION, ENERGY, AND URBAN FORM

In this section, we explore some issues in the context of the relationship between energy consumption for transportation and the form of the urban or metropolitan area. There is clear evidence that the per capita consumption of energy in the United States is much greater than almost anywhere else: our penchant for automobile transportation and the direct and indirect embodiment of energy in the goods we consume are obviously major factors contributing to this high consumption. However, the form of the U.S. city is often overlooked in terms of its contribution. Cities with lower population densities, other things being equal, will probably use more energy than more highly compact cities. An exploration of the effects of urban form on energy consumption was made in a very provocative report *The Costs of Sprawl* (Real Estate Research Corporation, 1975). Unfortunately, the analysis was conducted without any real data and, as Vining (1979, p. 74) has noted, "the basic conceptual problem of defining and specifying a settlement pattern has yet to be resolved, much less how one measures the energy costs of maintaining a settlement pattern at a given level of material well-being." He then goes on to comment on some of the difficulties with intuition. Higher-density settlements might produce higher environmental costs (not to mention the need for higher cooling costs in summer because of the heat retained within the built environment; offsetting this, though, might be lower heating costs because of the "heat island" effect of the city in winter). The problem of reorganizing structures in which over 80% of the current U.S. population live might prove to be a somewhat daunting proposition.

One interesting attempt to explore the effect of urban form on energy consumption for transportation was made by Edwards (1977). He used a Lowry-type model and associated components to generate trips, mode choice, and route assignment to produce land use patterns and characteristic behavior in several hypothetical cities with different urban forms (see Figure 12-7). The cities were assumed to have automobile and transit available and, by fixing the spatial distribution of trips, Edwards was able to calculate the effect that form had on energy consumption for transportation. He used data for Sioux Falls, South Dakota, and Baton Rouge, Louisiana. The basic conclusion for this analysis would appear to be the strong correlation between energy consumption and the average length of the work trip, which in turn was a function of the form of the city. The graph in Figure 12-3 illustrates the relationship between total energy and regional accessibility to population for the set of experiments Edwards performed with the hypothetical city forms and the real data. Regional accessibility was defined as:

$$\text{Regional Accessibility} = P_j F_{ij} \qquad (12)$$

where P_j is the population residing in the jth zone and F_{ij} is a measure of the work-trip friction-of-distance parameter. As one would expect, the more compact spatial forms use less energy; experiments numbered 1 through 13 assume dependence on the automobile

1	2	3	4	5	6	7	8	9	10	11	12	13
14	15	16	17	18	19	20	21	22	23	24	25	26
27	28	29	30	31	32	33	34	35	36	37	38	39

.5 mile

(a) **PURE LINEAR SHAPE**

---- Spinal auto/transit artery

2 Zone number

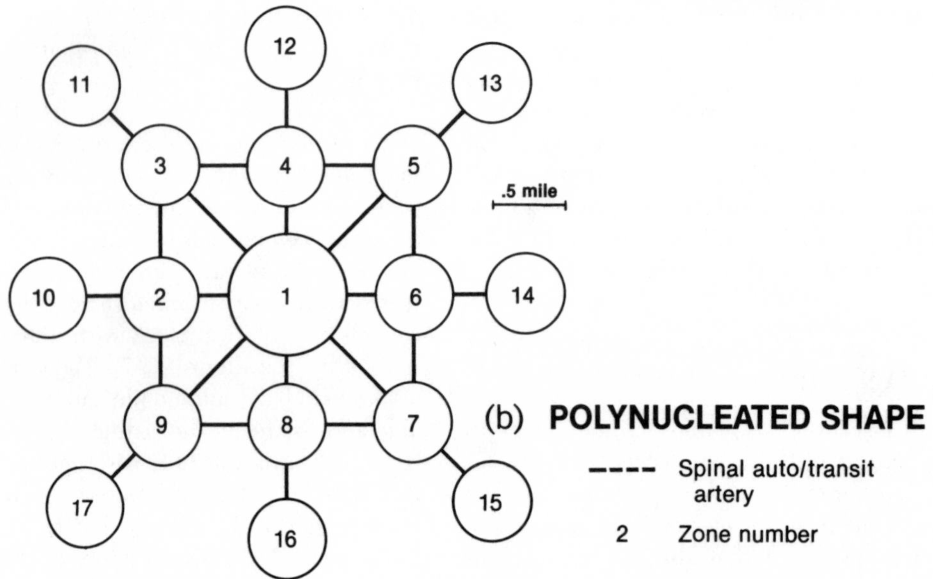

(b) **POLYNUCLEATED SHAPE**

---- Spinal auto/transit artery

2 Zone number

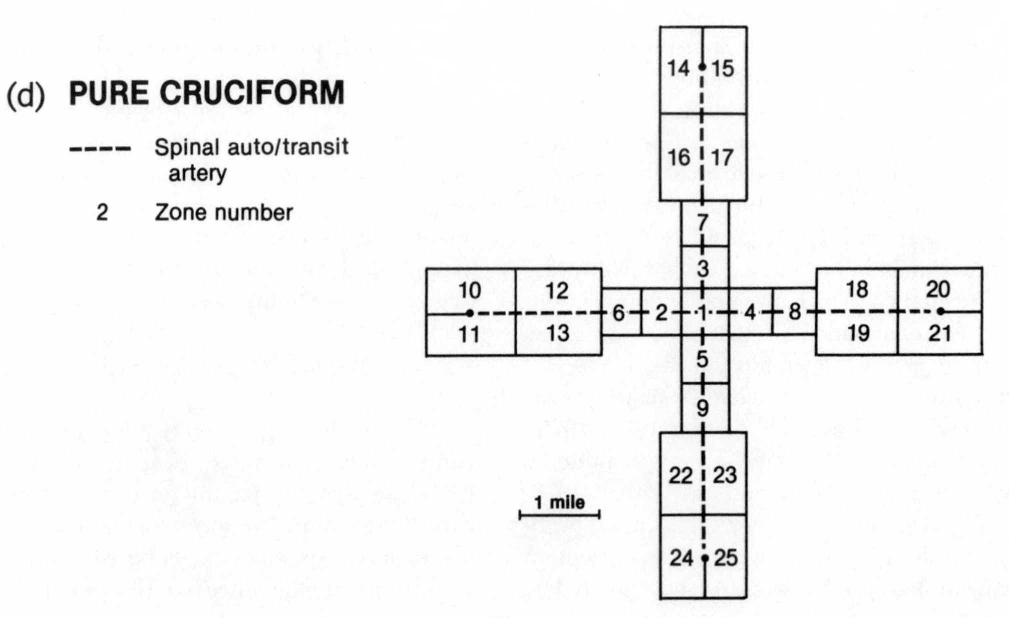

(d) **PURE CRUCIFORM**

---- Spinal auto/transit artery

2 Zone number

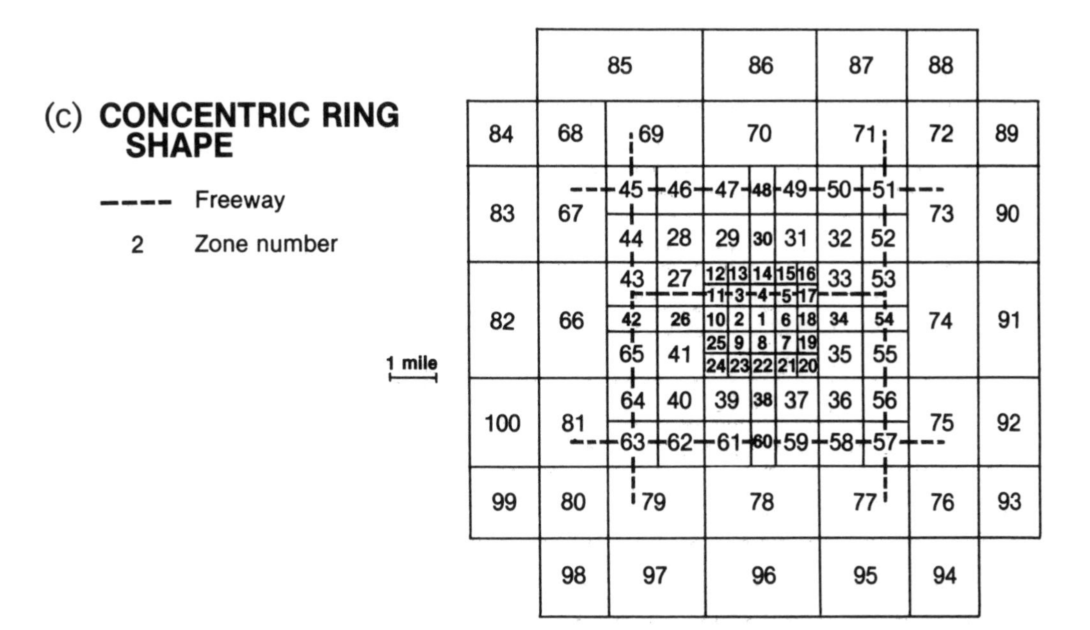

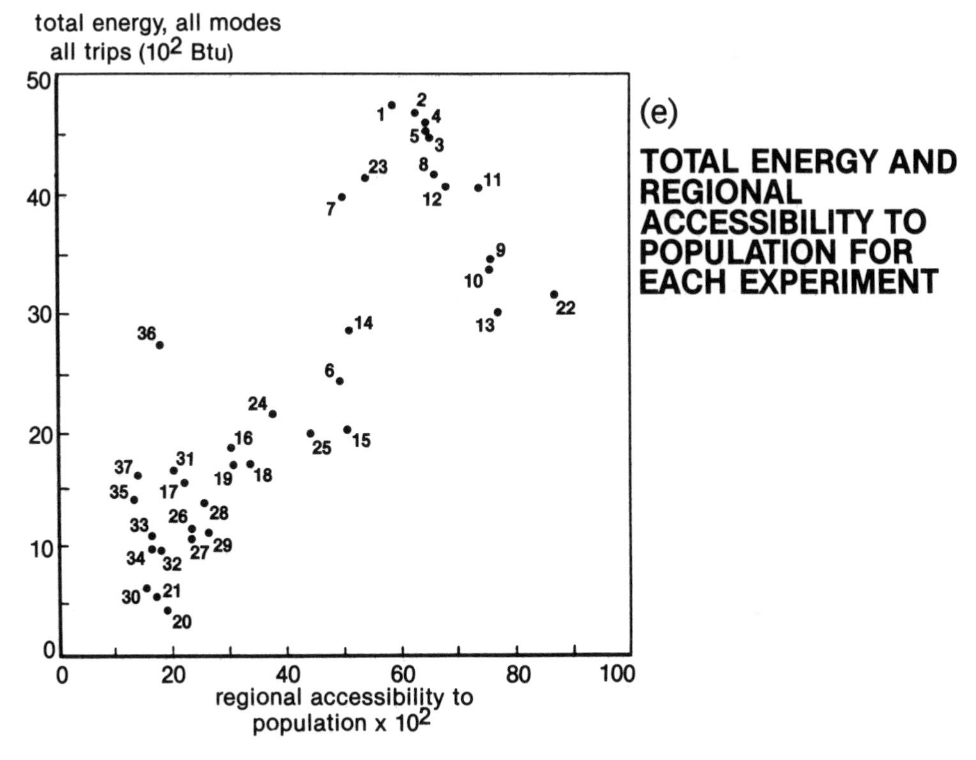

Figure 12-7. Experiments relating energy and urban form. (a) Pure linear shape, (b) Polynucleated shape, (c) Concentric ring shape, (d) Pure cruciform. (e) Total energy and regional accessibility to population for each experiment. From "Use of a Lowry-Type Spatial Allocation Model in an Urban Transportation Energy Study" by J. L. Edwards, 1985, *Transportation Research, 11*, pp. 117–126. Copyright 1985 by Pergamon Press, Inc. Reprinted by permission.

alone. Edwards found that the total energy consumed by all modes was highly correlated with any variable revealing variation in work travel, such as the average length of the work trip by automobile or automobile miles traveled to work. By using artificial data, Edwards was not able to address the problem of the degree of sensitivity of the relationship between those three attributes nor was he able to answer some of the questions discussed earlier such as the degree to which energy price changes might give rise to differences in journey-to-work patterns resulting from individuals changing jobs, residential location, or both.

TRANSPORTATION, ENERGY, AND CITY GROWTH IN DEVELOPING COUNTRIES

The discussion thus far has focused on U.S. metropolitan areas. In this concluding section, some comments will be made about the relationship between energy costs and urban growth in developing countries. Kelley and Williamson (1984) have developed a prototype computer general equilibrium (CGE) model to address the question of what drives third world city growth. Drawing on data for a sample of 40 developing countries, they produced an elegant model and subjected it to various assessments of possible impacts. The CGE model structure is rather complex and not easily tractable; as the name implies, it operates on an equilibriating assumption in various markets, such as the product and factor markets. Product markets are those in which the trade and exchange of goods and services take place, while the factor markets concentrate on the supply and demand for labor and capital inputs into the production process.

Table 12-8 shows the elasticities in the model relating certain city growth attributes to changes in the price of imported fuels and raw materials. The elasticities show how a 1% change in the price of imported fuel and raw materials would affect various attributes of third world city growth. The major impact would appear to be a slowing of the rate of city growth (the elasticity here is −0.89) and a lowering of the immigration rate (elasticity here is −1.77). Thus, for every 1 percent increase in fuel prices, we might anticipate a 0.89% decrease in the rate of city growth, but a larger decrease, 1.77%, in the level of immigration into the city. However, these effects are much smaller than the effect on the prices of manufactured goods and agriculture (see columns two and three of Table 12-8). The conclusion must be tempered with the fact that the model was calibrated for a period in which imported fuel prices remained relatively constant. Thus, the elastic-

Table 12-8. Impact of changes in imported fuel and raw materials on third world city growth (1970 statistics) [Elasticities]

	Prices		
	Imported fuel	Manufacturing	Agriculture
Quantity fuel	−1.15	1.76	−0.38
% urban population	−0.04	0.54	−0.32
City growth rate	−0.89	11.13	−6.51
In-migration rate	−1.77	22.23	−13.01
Squatter house rents	−0.31	3.57	−1.58
% urban land in squatter settlements	−0.11	1.33	−0.90
Cost of living differences urban-rural	−0.12	1.51	−1.11

Note: From *What Drives Third World City Growth* by A. C. Kelley and J. G. Williamson, 1984, Princeton, N.J.: Princeton University Press. Copyright 1984 by Princeton University Press. Reprinted by permission.

ities reported in Table 12-8 may seriously underestimate the responses of city growth to change in fuel costs. The experience since 1973 would suggest that at the aggregate level, the effect has been to retard the acceleration trends rather than cause a decline in the rates of growth themselves (Kelley & Williamson, 1984, p. 129).

Unfortunately, Kelley and Williamson were not able to relate their analysis to the land use and transportation patterns in the cities studied, since their model operated at such an aggregate level. However, this work does provide some stimulus for attempting to link more national or international level modeling efforts with the activity responses within urban areas. Such macro-to-micro modeling might prove useful in providing insights into internal structural changes in cities in developed as well as developing countries.

CONCLUSIONS

The ideas presented here suggest that the most important relationship between energy and transportation may be at the macro level. The evidence of the last decade has not revealed a great deal of sensitivity between energy costs and patronage of transportation modes and certainly very little evidence for changes in internal location of households within metropolitan areas.

The analysis reviewed in this chapter suggests that one needs to be very careful about drawing conclusions about individual behavior on the basis of changes in macro-level variables (such as energy prices). It is also clear that the distinction between changes affecting what we may refer to as marginal expenditure effects cannot be transferred to effects on average expenditures. The price rise in energy costs were dramatic: the short-run adjustment problems centered less on issues of demand but more on issues of supply (scarcity and rationing problems). Once the prices rises had been accepted as permanent features of the marketplace by consumers, the adjustment process proceeded rather rapidly. In fact, the effect of energy price increases had a far more profound effect on the *characteristics* of the private automobiles produced than on the *patronage* of automobiles vis-à-vis public transportation. Analysts predicting massive returns to public transportation patronage failed to appreciate that although the energy price rises were large, they still did not accelerate rapidly enough to change the elasticities of demand for public transport. Given the small percentage of total income allocated to transportation energy expenditures in the average household, even large increases in energy costs failed to provide significant incentive for households to relocate within the metropolitan area.

Adjustment to energy price increases has taken the form of substitution of automobiles with high fuel efficiencies for those with low fuel efficiencies. In the short run, this substitution effect had a deleterious effect on the U.S. automobile industry but provided little evidence of visible change in the commuting patterns or the structure or form of urban areas. In the longer run, it would be fair to state that in comparison to the major structural changes taking place in the U.S. economy, for example, the increasing dominance of the service sector, the effects of energy price increases are not going to exert very much influence. The normative claims provoked by analysis conducted in the mid-1970s appear to have been refuted by the recent empirical evidence.

References

Alonso, W. (1964). *Location and land use.* Cambridge: Harvard University Press.

Anas, A. (1982). *Residential location markets and urban transportation.* New York: Academic Press.

Bezdek, R., & Hannon, B. M. (1974). Energy, manpower and the Highway Trust Fund. *Science, 185*(4152), 669–675.

Congressional Budget Office (1977). *Urban transportation and energy: The potential savings of different modes* Washington DC: US Government Printing Office.

Edwards, J. L. (1977). Use of a Lowry-type spatial allocation model in an urban transportation energy study. *Transportation Research, 11,* 117–126.

Energy Information Administration. (1985). *An-*

nual Energy Review 1984. Washington, D.C.: U.S. Government Printing Office.

Hannon, B. M. (1974). Energy, employment and transportation. Urbana, Ill.: Center for Advanced Computation, University of Illinois.

Holcomb, M. C., & Koshy, S. (1984). *Transportation energy data book edition 7*. Oak Ridge, Tenn.: Oak Ridge National Laboratory.

Kelley, A. C., & Williamson, J. G. (1984). *What drives third world city growth*. Princeton, N.J.: Princeton University Press.

Leach, G. (1973). The impact of the motor-car on oil reserves. *Energy Policy, 1*, 195–208.

Leontief, W. W. (1966). *Input-output economics*. London: Oxford University Press.

Mills, E. S. (1972). *Studies in the structure of the urban economy*. Baltimore: Johns Hopkins Press.

Muth, R. F. (1984). Energy prices and urban decentralization. In A. Downs and K. L. Bradbury (Eds.), *Energy costs, urban development and housing*, pp. 110–142. Washington, D.C.: Brookings Institution.

Quandt, R. (1972). *The demand for transportation*. Lexington, Mass.: D.C. Heath.

Real Estate Research Corporation. (1975). *The costs of sprawl*. Washington, D.C.: U.S. Government Printing Office.

Romanos, M. C. (1978). Energy-price effects on metropolitan spatial structure and form. *Environment and Planning A, 10*, 93–104.

Vining, D. R. (1979). The president's national urban policy report: Issues skirted and statistics omitted. *Journal of Regional Science, 19*, 69–77.

13 SOCIAL IMPACTS OF URBAN TRANSPORTATION DECISIONS: *EQUITY ISSUES*

DAVID HODGE
University of Washington

The evolution of the American city has resulted in a distinct spatial patterning of social areas which are remarkably consistent from city to city. Significantly, urban transportation systems have played a major role in shaping these patterns. As described in chapter 2, the transition from pedestrian to rapid transit to streetcar and finally to the automobile city has been accompanied by a massive restructuring of urban social and economic patterns. For the most part improvements in transportation have resulted in ever greater mobility and prosperity for urban areas and their residents. But the impacts of these transportation and land use changes are not always positive, nor do they accrue to all social groups equally. In fact, much of the activity related to urban transportation creates distinct sets of winners and losers.

Analysis of social impacts of urban transportation decisions is complicated by the wide range of potential impacts that could be evaluated. For example, the construction of an urban freeway may improve access to a CBD and in so doing increase land value (and therefore profit) for CBD landowners. The freeway reduces effort to reach work thereby reducing wage demands of commuting employees. Increased land values and employment mean increased taxes for the central city, which increases the city's ability to fund social projects for the disadvantaged. Construction of the freeway, however, usually involves the destruction of housing and sometimes the physical division of communities. Daily auto traffic increases air pollution and noise pollution levels throughout the central city, posing health dangers to inner-city residents. Each of these impacts benefits or harms a different set of social groups, and in so doing they raise the issue of equity or fairness. Are urban transportation decisions creating costs and benefits that are fairly distributed among urban social groups?

As noted above, a complete answer to this question is virtually impossible given the enormous variety of ways in which urban transportation affects the city. In this chapter, therefore, the discussion is limited to a number of the most important impacts by focusing first on issues related to the automobile, especially problems associated with the construction of highways and with pollution generated by auto use, and, second, on issues related to mass transit, especially the problems of transit dependents and the issue of who pays and who benefits from public subsidy of mass transit.

SOCIAL IMPACTS OF THE AUTOMOBILE

Urban transportation systems are both a cause and a consequence of the separation of land uses in the American city. Transportation improvements have especially encouraged the separation of home and work. The widespread use of the electric streetcar in the first half of this century, which led to real estate developments at the periphery, gave way after World War II to the automobile. Automobiles, combined with a plentiful and inexpensive supply of mortgage funds, sustained massive suburbanization. Given that resident suburbanization was primarily composed of middle- and upper-income whites who required urban highways to connect their residences with their white- and blue-collar jobs concentrated in the central city, it follows that major benefits have gone to them while the negative impacts of highway construction and use have fallen most heavily on lower-income, and often minority, inner-city neighborhoods, which were politically powerless to prevent such impacts.

Impact of Highway Construction on Urban Communities

The destruction of housing units in the central city as a result of highway construction has been considerable. Between 1957 and 1968 a total of 335,000 housing units were destroyed, their households displaced (*Building the American City*, 1969). In the overwhelming number of cases, the displaced households were disproportionately from lower economic and social strata. Throughout most of the major highway construction period of the 1950s and 1960s relocation assistance was negligible; displaced households were simply forced to absorb fiscal and psychic costs. These impacts are now minimized through more realistic compensation programs, but these programs were not initiated until after the major construction era.

The disproportionate impact of urban highway construction on poor and minority households has been a function not only of the concentration of such households in the central city, but also of their lack of political power. Given the severe impacts of dislocation, planners typically have opted for the line of least effort, or in this case the route of least political opposition, which may or may not coincide with the best route from an economic point of view. At times such decisions have had devastating consequences for the affected communities. A classic example of such impact occurred in Nashville, Tennessee, in the construction of I-40 (Seley, 1970). For reasons never made clear by the Tennessee State Highway Department, the original plan for the routing of I-40 was changed in such a way that Jefferson Street, the heart of the Nashville black business community was destroyed. Prior to construction of I-40, the Jefferson Street corridor of North Nashville was home to 80% of all black-owned businesses in Nashville. Displaced businesses not only received no relocation assistance (only two businesses received loans) but also faced the additional problem of limited acceptance of black-owned businesses in other parts of the city. The construction also destroyed 650 homes and 27 apartment buildings, further disrupting the black community in North Nashville. In addition, the new route cut between the campuses of three major black educational institutions—Fisk University, Meharry Medical College, and Tennessee A and I University—restraining expansion, increasing isolation, and decreasing the capacity for cooperative programs.

The example of the Nashville I-40 case is important not only because of the inequitable social impact of the location decision but also because of flagrant inequity in the decision process. Through a number of subterfuges, such as posting notices of public meetings only outside the affected neighborhoods, the

Tennessee State Highway Department took advantage of the black community's lack of political power to impose its decision and its harsh social impacts on them with virtually no compensation. The black community had hardly any voice in the decision process.

While the Nashville I-40 case is perhaps extreme, it does serve to point out the real social impacts of urban highway construction especially during the 1960s. Almost every major U.S. city bears the scars of communities split apart by the nearly impenetrable barrier of concrete. The political process that made much of this possible systematically worked against those with little political power, namely the poor and minorities, and their neighborhoods bore the brunt of these impacts. Urban highway construction has slowed dramatically for a variety of reasons including higher costs, a more open and difficult political process, the fact that the major systems have already been completed, and the inability of CBDs to accomodate larger numbers of autos. Fewer and fewer households are now being displaced and fewer communities disrupted, but the legacy of those past decisions remains in both the physical and social landscapes.

Air and Noise Pollution

The ongoing problems of air and noise pollution associated with urban highways are perhaps the most important, and socially inequitable, legacies of urban transportation and land use changes from the 1950s and 1960s. In spite of more recent improvements in many cities, high levels of air and noise pollution persist in most cities. Although it is difficult to isolate the effect of automobile use from other sources of urban pollution, there is general agreement that transportation is the major source of most types of urban pollution. Once again we find that low-income and minority populations, whose residences are concentrated in the central city where pollution levels are typically highest, are most at risk. An example of the socially unequal burden of air pollution is portrayed graphically in Figure 13-1. In most of the cities identified, lower-income households have greater exposure to total suspended particulates while the differences for blacks are even more dramatic. This pattern is also true for elderly households, who are typically concentrated in inner-city neighborhoods where pollution levels are highest (Greenland & Yorty, 1985).

It is difficult to assess the full long-term effects of these higher levels of exposure to air and noise pollution. Increasing numbers of studies are finding evidence linking noise and air pollution to an even wider set of physical and mental problems as well as learning disabilities that may be systematically reducing achievement levels in affected youth. While the debate over the total effect continues, the unequal impact of air and noise pollution associated with the automobile continues to discriminate against low-income and minority social groups.

SOCIAL IMPACT OF MASS TRANSIT

The major social and equity issues associated with mass transit differ significantly from those associated with the automobile. In part this is due to a difference in goals, in part to a difference in physical impact on the city, and in part to a difference in funding. Although both highways and mass transit operations can be thought of broadly as public goods, that is, they are both financed by governments, their underlying motives and support differ sharply. They both have a general goal of helping individuals move from place to place, but there the similarity ends. Highways are funded largely by user taxes on gasoline and annual vehicle license taxes and simply provide the opportunity for private transportation. Mass transit is funded largely by nonuser taxes. Community support of mass transit is generally considered appropriate, since, in addition to moving people, mass transit is designed to reduce the negative impacts of the automobile on urban life as well as to provide services to those unable to use autos.

One of the major benefits often claimed

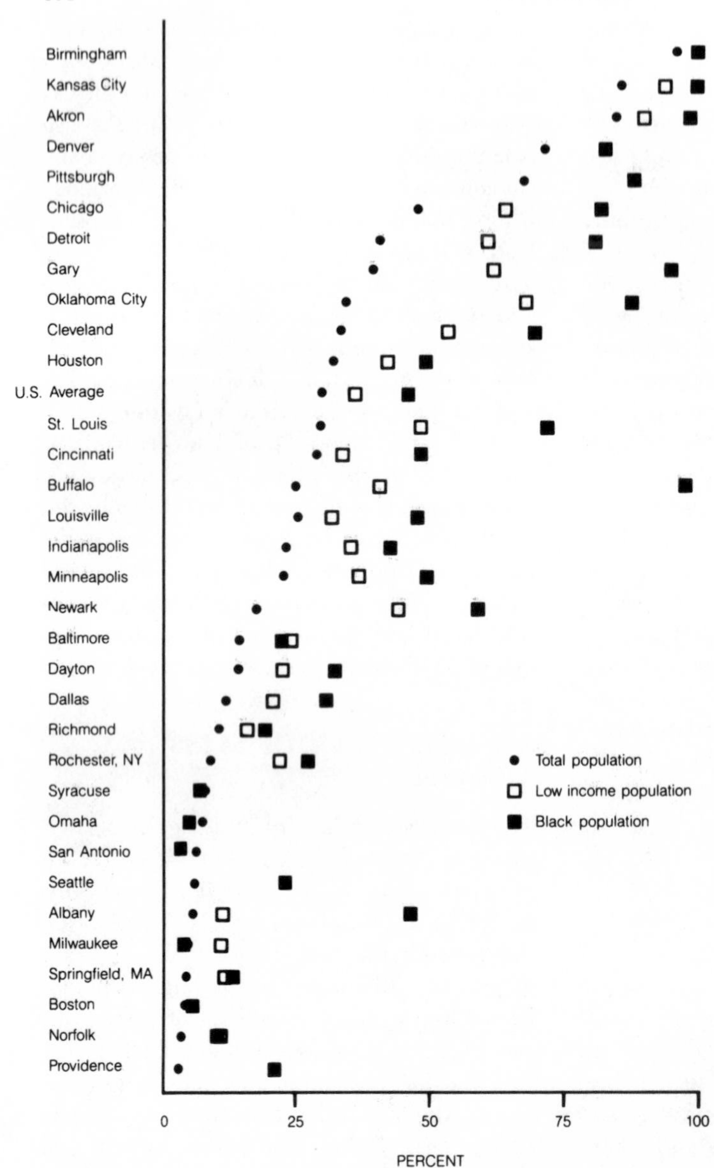

Figure 13-1. Percent of total population, low income persons, and black populations exposed to annual average TSP concentrations above the national primary standards, 1973. From *Urbanization and environmental quality* by T. R. Lakshmanan & R. Chatterjee, 1977, Washington, D.C.: Association of American Geographers.

for mass transit is the potential for reducing air pollution. It is ironic that one of the main forces behind increased use of mass transit in the CBD comes not from the Urban Mass Transportation Administration (UMTA) but from the Environmental Protection Agency (EPA) whose requirements for reducing air pollution have forced many major cities to encourage transit use into and within the city center. By lowering air pollution levels in the city center, mass transit addresses one of the key negative externalities of the automobile and, in so doing, balances some of the inequity imposed by auto use in downtowns. In general, those who stand to benefit most from reduced concentrations of air pollution in the CBD are likely to be poor and/or a racial minority. According to the American Public Transit Association (APTA) use of mass transit saves 40 million barrels of oil each year while reducing carbon monoxide pollution by about 150,000 tons and other air pollutants by another 55,000 tons (*Transit Fact Book*, 1981). However, as pointed out by Fielding (chapter 10) and Schou (1979), potential reductions in fuel consumption and air pollution through increased mass transit use are minor relative to total urban transportation levels and fall far short of improvement that would result from even small increases in the average number of passengers per auto or increased gas mileage of automobiles.

A more direct social impact of mass transit involves the provision of transportation services to those who have no alternative to automobiles because of fiscal or physical limitations. Unlike the support of highway construction and the use of automobiles, one of the major goals of mass transit operations is to assure basic transportation services to all urban dwellers. This goal contributes much to the justification for nonuser subsidies of public transit. As such, reduced fares for elderly and special equipment for the handicapped are subsidized as part of social rather than transportation policy. For these people, who are wholly dependent on mass transit for their transportation, the quality of mass transit service determines what urban resources they can or cannot use.

THE ISSUE OF EQUITY

The issue of equity in mass transit depends upon three important questions. First, we must address the question of which groups of people constitute an important social distinction. Obviously not everyone will benefit from mass transit equally or contribute equally to its operation, but the inevitable individual inequities become important only when, in the aggregate, they reflect a systematic discrimination against an identifiable and important group. Through various acts of legislation and relevant court decisions, it is clear that race constitutes the most important category for careful inspection. Transit systems may not discriminate on the basis of race although, as we shall see later, finding acceptable definitions of discrimination is a difficult task. In addition, various pieces of legislation have required transit systems to provide fair access to handicapped individuals, which poses its own particular definition problems, and to all persons regardless of sex. Beyond these groups there are no social categories that hold a legal right to nondiscrimination. However, there are other social groups that maintain a political claim to their fair share of the resources. Most notable are the distinctions based on income and age. In particular are low-income and/or older persons being treated fairly by transit? In addition to social groups the "who" question includes one other important distinction, spatial groups. As we shall soon see, there are clear patterns of subsidy and performance by geographic area that are in many ways an inevitable outcome of the spatial organization of cities. To some extent the spatial patterns of benefits and costs are reflected in social patterns because of the segregated nature of our cities. But the spatial patterns have an additional measure of importance when they coincide with jurisdictional boundaries. In most metropolitan areas, transit authorities cover several jurisdictions, each naturally intent on getting its fair share of the resources. The issue thus is both a political and legal question. When Seattle METRO was formed, for example, the enabling legislation (from the state government) dictated that operating deficits be distributed in proportion to the amount of sales tax generated in each jurisdiction. In this particular case only two jurisdictions were identified, inside the central city of Seattle and the rest of King County, but smaller geographic entities are no less vigorous in pressing their claims for improved service or in protesting reduced service to their areas.

The second major issue involves the question of what to measure, that is, what aspects of the transit operations should be subject to scrutiny for equity. There are four major groups of measures that can be considered: (1) nonuser revenues, (2) expenditures, (3) outputs, and (4) outcomes. Perhaps the most obvious measure involves the huge quantities of nonuser (tax) revenues. Who pays these revenues? Are they the same people who use the system? Is there a net shift of income between social groups as a result of this tax revenue? The second measure concerns the distribution of expenditures, that is, how are expenditures being distributed among the users of the system? What proportion of costs do riders from different social and spatial groups pay? A third measure involves the assessment of outputs, which may be defined as the level of service, including quality of service equipment and patterns of accessibility. A final set of measures, referred to as outcomes, are the most difficult to conceptualize and to operationalize. Basically these are indicators of the effect of providing transit service and as such are strongly related to the conditions of users. For example, providing transit service to a relatively wealthy individual may be a convenience and possibly a cost savings, but for a lower-income or autoless individual the availability of transit may mean the difference between having a job or not having one. Unfortunately, such an assessment is very difficult to determine and will not be dealt with in this chapter.

The third and final major issue involves setting the rules for evaluating the distribution of whatever measure we are examining. It is one thing to determine that Group A has a certain level of benefit and Group B another level, but it is a more difficult task to evaluate the degree to which that distribution of resources is fair. Consider, for example, a rule based on equality in the distribution of benefits in contrast to a rule based on a progressive scale. In other words, is the system equitable when riders of all income groups receive the same level of benefits (equality) or when lower-income groups receive proportionately more than higher-income groups (a progressive distribution)? The issue of equity rules is related to the type of measure under consideration, the social group of interest, and ultimately, to the policy objectives of the transit providers. In the studies to be examined in the following pages we will see a variety of measures and associated equity definitions that will give an understanding of the enormous complexity of this type of decision making.

Non-Fare-Box Revenues

In response to the rapid deterioration of many private transit systems being taken over by local public agencies in the early 1960s the federal government began a program of capital grants as part of the 1976 Urban Mass Transportation Act with the federal government paying two-thirds of the cost of acquiring private equipment or purchasing new equipment. By 1970 the federal subsidy for capital acquisition was $133 million. It climbed steadily and rapidly throughout the decade reaching $2.7 billion by 1980 with the federal share of capital expenses raised to 80% of costs. In general the distribution of capital costs is considered progressive because of the large portion paid by the federal government and the generally progressive federal tax structure.

Nonuser support of operating expenses has also become necessary because of sharp declines in revenue passengers and even sharper increases in operating costs (see Figure 10-2). Operating expenses are covered by a more complex set of subsidies, consisting of local and state taxes, which are generally considered to be regressive, and federal taxes, which are generally considered to be progressive. Like capital subsidies, nonuser operating subsidies became necessary in the 1960s and rose rapidly in the 1970s in response to a 60% increase (constant dollars) in the operating cost per passenger, the result of low ridership and increasing expenses. Recognizing the importance of mass transit to urban areas, the federal government began offering operations assistance in 1975 with grants totaling $301 million. Although the absolute dollar value of federal subsidies continues to increase, its relative contribution has declined from a peak of 30.9% of operations subsidies to 30.2% in 1980 with continued declines expected (*Transit Fact Book*, 1981, pp. 29–30). The greatest share of operating expenses continues to be shouldered by local governments with a sizable subsidy from state governments as well (Table 13-1). As a result, the issue of the distri-

Table 13-1. Distribution of 1978 transit tax burden (taxes as a percentage of total money income) by level of government and income group

	Dollar subsidy (millions)	Percent of total	Under $6,000	$6,000–$10,000	$10,000–$15,000	$15,000–$20,000	$20,000–$25,000	$25,000 and over
Capital subsidies								
Federal gov't	2,078	80	.070	.110	.145	.151	.181	.230
State & local gov't	531	20	.058	.059	.055	.052	.050	.040
Total	2,069	100	.127	.169	.198	.208	.230	.270
Operating subsidies								
Federal gov't	739	28	.025	.039	.052	.056	.065	.082
State & local gov't	1,920	72	.208	.214	.201	.187	.180	.143
Total	2,659	100	.234	.253	.255	.244	.246	.225
Total subsidies								
Federal gov't	2,817	53	.095	.149	.197	.214	.246	.312
State & local gov't	2,450	47	.266	.273	.256	.238	.230	.183
Total	5,268	100	.361	.422	.453	.452	.476	.495

Data refer to "baseline assumptions" that fall approximately midway between the most regressive and most progressive assumptions of the tax rates.

Note. Compiled and computed from "Distribution of the Tax Burden of Transit Subsidies in the United States" by John Pucher and Ira Hirschman, 1981, *Public Policy, 29.*

bution of costs among income groups is confused.

The best attempt to estimate the incidence of tax burden by income group is the work of Pucher and Hirschman (1981), who computed the incidence of tax burden under a varying set of most progressive to most regressive assumptions. The results of their analysis (Table 13-1) document the expected progressive nature of federal support and regressive nature of state and local taxes. As a result capital expenses (paid largely by federal taxes) are strongly progressive while operating subsidies (dominated by local and state funds) are mildly regressive and total subsidies moderately progressive. However, the declining role of the federal government in providing operating subsidies will undoubtedly make the distribution of operating revenues less progressive.

A more extreme local example is pointed out by Webber (1976) in his analysis of the tax incidence for the Bay Area Rapid Transit system (BART). Webber found that the progressive nature of the sales tax revenue used to support BART stands in sharp contrast to its user profile (Figure 13-2). Given the goal of BART, to connect downtown San Francisco with its suburbs, it is no surprise that its ridership is dominated by middle- to upper-income categories, though the tax burden falls disproportionately on lower-income households.

The Distribution of Expenditures

Given the large subsidies supporting transit operation, it is natural that a major portion of the research investigating transit equity should focus on the distribution of transit expenditures, typically measured in terms of the net cost per transit trip. As indicated earlier, transit performance is related to relative location within a metropolitan area and hence to the underlying geographic patterns of employment and social groups. It is important to bear in mind that virtually all transit systems are oriented toward the city center. Therefore, cross-town travel and cross-suburb travel are difficult, since the historic patterns of transit facilities, especially rail systems, radiate out from the city center and have served principally to bridge the spatial gap between suburban homes and CBD jobs.

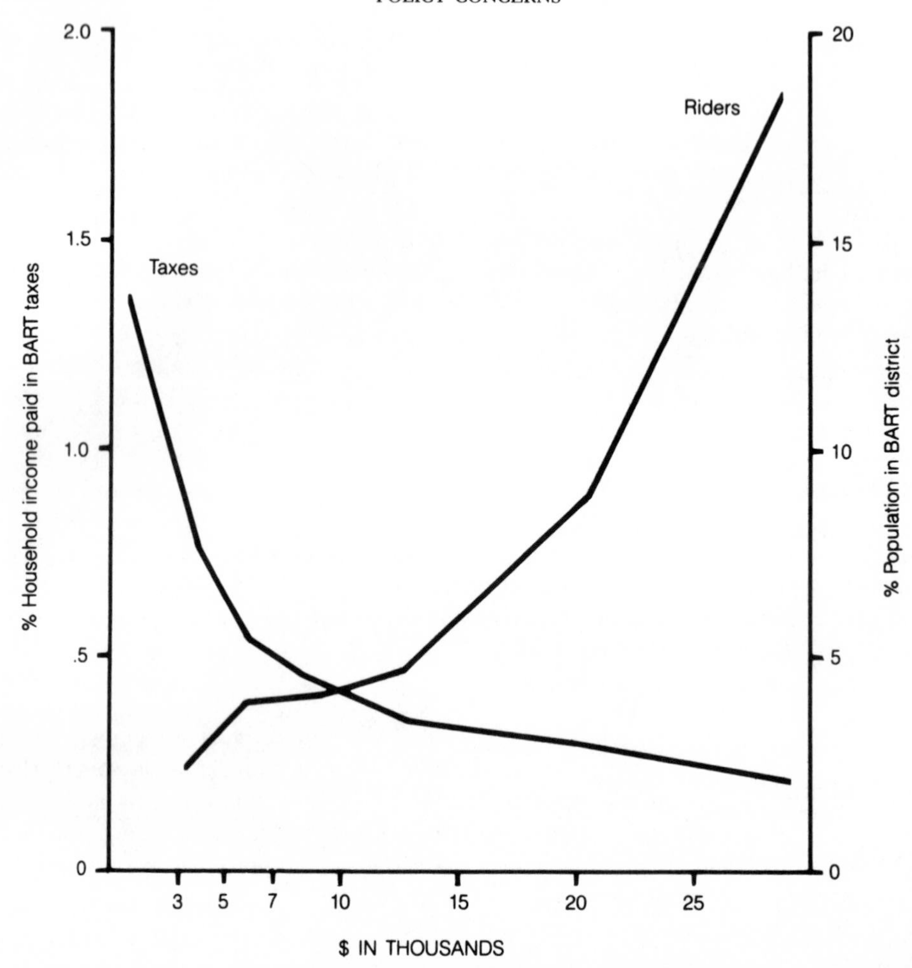

Figure 13-2. Income comparison of BART taxpayers and BART riders. From *The BART Experience: What Have We Learned?* by M. Webber, 1976, Monograph No. 26, Berkeley: University of California, Institute for Transportation Studies. Copyright 1976 by the University of California. Reprinted by permission.

As a result there is a fairly regular and predictable outcome of transit decisions regarding the choice of transit type, its location, and temporal pattern of service which affects the distribution of expenditures by social and spatial group.

Intermode subsidies occur as a result of differences in subsidies by mode. Commuter rail, for example, has an average subsidy of $1.89 compared to a bus average of $.63 per passenger trip and a heavy rail average of $.32. While these statistics appear to favor commuter rail by a wide margin, it is interesting to note that using either subsidies as a percentage of costs or subsidy per passenger mile as measures changes the relative standing of these transit modes, with bus subsidy costs the highest.

Interclass subsidies arise from differences in ridership patterns among the different transit modes for different income classes. Buses are used disproportionately by lower-income riders, subway by lower middle-income riders, and commuter rail by higher-income riders. We would expect this pattern, given the nature of the underlying geographic pattern of income groups and the service areas of the different transit types. It is borne out by differences in length of the average trip as well. A comparison of whites and nonwhites reveals a similar pattern. Although blacks account for only 10.4% of all urban trips, they account for 30.3% of bus and streetcar trips, 16.9% of subway trips, and only 6.3% of commuter rail trips (Pucher, Hendrickson, & McNeil, 1981). Hispanics make up 5.0% of all urban trips but are concentrated in bus travel (10% of all bus trips) and especially subway travel (an astounding 25% of all subway trips) as a result of their concentration in New York City. It is important to note, however, that all income and racial groups participate in all types of transit and, even more important, that the total distribution of operating subsidies by transit type favors buses (68.8% of operating subsidies) rather than commuter rail (12.3%). Even though commuter rail receives a disproportionate share of subsidies (as measured by subsidies per passenger), the total amount of funds redistributed between modes is relatively small.

Capital expenses, on the other hand, favor subway and other heavy rail systems. Pucher (1981) estimates that between 1965 and 1979, 64% of all capital expenditures on transit went to rail transit, 12% to commuter rail, and only 24% to buses. However, with the exception of the lowest income category, lower-income classes and racial minorities stand to benefit from the investment in rail systems.

Intramode subsidies, that is subsidies between passengers using the same mode provided by the same agency, have been studied more because of the clearer responsibility/accountability for providing a fair distribution of services. Previous research has shown that fares, costs, and subsidies are related to fundamental geographic and temporal characteristics of systems. For example, the amount of subsidy depends on length of trips, population density of service areas, and time of day. These factors play important roles because most transit systems charge a flat rate to riders, that is the fare charged is the same regardless of the distance traveled or costs paid to provide the service. Although most agencies have very coarse zones (usually just an inner zone and a suburban zone) and impose a surcharge that partially accounts for the typically longer trip of riders crossing zones, the match between costs and fares is generally poor and thus gives rise to variations in subsidies.

In one of the most detailed assessments to date, Cervero, Wachs, Berlin, and Gephart (1980) examined these relationships in case studies of transit systems in San Diego, Los Angeles, and Oakland. Sample routes reflecting a range of transit situations were selected in each city, and the costs per mile of service on each route by time of day were computed. User data, including fares, socioeconomic status, and distance traveled were obtained through on-board surveys of more than 10,000 users in each of the cities. The first finding is that subsidies are directly related to the length of the transit trip. Trips of less than one mile (between 5% and 15% of all trips) generated a profit in all three cases while trips of less than 4 miles (between 57% and 71% of all trips) on average paid a larger proportion of total fares than they consumed in total costs. In short, trips of under 4 miles, and especially trips of under 1 mile, are subsidizing the longer trips.

A second factor contributing to inequities in the distribution of subsidies within a transit service area involves the geographic location of the service, especially the pattern of residential density. Frankena (1973), for example, cites examples from Regina, Saskatchewan, and London, Ontario, in which service to outlying low-density areas was heavily subsidized by inner-city routes. In some of the outlying areas of Regina fares covered only 20% of the real costs which required subsidies of up to $1.00 per trip, or $500 per year for a regular commuter. Frankena goes on to note that because of the tendency of average income to increase with distance from the CBD and to be higher in corresponding low-density neighborhoods, the net effect of transit operations in these cities was a redistribution from the lower-income central city to the higher-income suburbs. The same pattern of population distribution by income group as well as contrasts by race occur throughout the United States, suggesting the potential for a similar outcome. In part higher fares compensate for the differences between cost of trips from more distant locations, but the increment, if it exists, doesn't usually account for enough of the cost differential. Frankena found that elimination of the $.15 zone charge resulted in a net additional transfer of $95.00 per year to suburban residents commuting to the central city. Other examples throughout the United States confirm this pattern. Pucher (1981) found that subsidies in central cities in northern New Jersey are half those of bus routes in other areas. The results are even more dramatic for newer rail rapid transit systems which are designed to bring suburban commuters into the CBD. A study of the Washington, D.C., METRO found that, when completed, the 101-mile system will have operating deficits in outer areas ten times larger than those in the inner city (Faucett Associates, 1976).

A third major factor found by Cervero et al. (1980), as well as others, is that off-peak users subsidize peak users. In spite of higher load factors, the costs of capital purchases of equipment that is underused, similar problems with scheduling labor efficiently, and deadheading (running buses empty one way) are not covered by fares or even surcharges that some cities have instituted. In the case of the three California cities, off-peak fares accounted for 42 to 55% of costs while fares accounted for only 32 to 38% of peak-hour expenses. Other studies (Morgan, 1976; Wabe & Coles, 1975) also found that peak-hour expenses are more than double those in off-peak hours.

Significantly, the geographic factors affecting transit subsidies correspond to the spatial distribution of social groups and thus also create unequal patterns of subsidy among these groups. In the study of California cities, Cervero et al. (1980) found that in two out of the three cities lower-income areas were subsidizing higher-income ones, and minority populations were subsidizing white populations. Although in these instances the cross-subsidies are not dramatic, the correspondence between transit circumstances and social status holds. Pucher, Hendrikson, and McNeil (1981), for example, found that average trip length varies directly with income. Lower-income passengers travel shorter distances and receive lower subsidies. Similarly lower-income households are less represented in peak-hour transit travel and more represented in off-peak travel as are racial minorities. As noted earlier, Webber (1976) found that BART ridership was composed overwhelmingly of higher-income households as we would expect given the geographic connections embodied in the system.

As is reflected by these studies, the measurement of subsidies by social groups and spatial areas is a complex task that requires large quantities of data and a sophisticated methodology in order to go beyond the initial observations of subsidies (and social groups) being related to distance traveled, geographic location, fare structure, and time of day. The measurement task is made especially difficult by the array of spatial interdependencies within the transit service area. A transit system typically has a flat fare for passengers from multiple origins to multiple destinations (and therefore traveling different distances) at different times of the day on routes with different load levels. The work of Cervero et al. (1980) is one of the best research efforts to date in its calculations of distance traveled by passengers and an average cost per mile of the route traveled used to compute the ratio of cost per mile and fare per mile for each passenger. However, their methodology suffers from their lack of attention to the particular part of the route the passenger is traveling (in addition to the problems of transfer passengers). Given the radial nature of most transit systems, we might expect that riders nearer the center of the city would be traveling in buses with higher load factors than their counterparts at the periphery. In other words a 2-mile trip near the city center would have subsidy characteristics different from a 2-mile trip near the periphery where load factors and costs per passenger mile would be greater.

In an attempt to account for this complexity, Hodge (1981) created a model that explicitly measures transit performance of individual subsections of every route; he then used the model to assess transit performance and social and spatial subsidies in a simulated city. Although it is not appropriate here to discuss the methodology in great detail, a brief description of the model and the simulated analysis will clarify the manner in which fares, costs, and subsidies are distributed. The first step is to represent the transit system as a network of nodes connecting individual route segments. In this way a route can be defined as a sequence of segments, as can an individual transit trip. Costs are assigned to each segment according to the length of the segment and time required to travel the segment (which partially accounts for the congestion factor nearer the city center). The key to the model, however, is the method by which flat-rate fares are converted into variable-rate fares, that is, the assignment of the appropriate proportion of the fare paid to each of the segments traveled during a transit trip. Unlike Cervero et al., who looked at fares as being equally distributed in equal proportions as a simple per mile basis, this model allocates fares in proportion to the cost per passenger determined for each segment traveled. Repeating this procedure for all trips yields the total fares generated for each route segment. The amount of subsidy for each segment is determined by subtracting total fares allocated to each segment from total costs which can then be expressed as subsidy per passenger, subsidy per mile, or subsidy per passenger mile.

To clarify this, consider the simple example displayed in Figure 13-3. Transit performance and the amount of subsidy per passenger vary greatly from one segment to another. The amount of subsidy also varies directly with location (and length of trip). Passengers starting at *A* have an average subsidy of $1.43 compared to a subsidy of −$.07 for those starting at *C* and traveling only one segment.

The effect of this variation on an urban system can be observed in the simulated city whose distributions of population, employment, and racial minorities are portrayed in Figure 13-4. Using a typical trip assignment model to allocate trips to segments, the performance of each transit segment is determined; the result is a fairly clear pattern of lower operating ratios (fares/costs) in peripheral locations (Figure 13-5a) where load factors are smaller. The average subsidy per passenger is highest in the same general peripheral locations (Figure 13-5b). As demonstrated in Figure 13-3, these passengers generally travel longer distances over less crowded routes, so that even the higher suburban fare ($.60 compared to $.40) fails to cover the additional costs. Given the typical pattern in U.S. cities of minority concentration near the city center, it is also not surprising that the average subsidy per white passenger is more than ten times the average subsidy per black passenger, who typically rides a shorter distance on portions of routes with higher load factors.

Distribution of Transit Outputs

Nowhere is the issue of the equity implications of transportation decisions more explicitly considered than in the reporting requirements established by UMTA to show compliance with Title VI of the 1964 Civil Rights Act, Section 601, which states:

> No person in the United States shall, on the ground of race, color or national origin, be excluded from participation in, be denied the benefits of, or be subjected to discrimination under any program or activity receiving Federal financial assistance.

The guidelines for providing compliance were the result of a study commissioned by UMTA in 1977, which established methods and demonstrated that these methods could realistically be expected of all recipients of UMTA operating on capital subsidies (Miller, 1977). The guidelines require local operators to show that minority populations (defined as blacks, hispanics, Asian or Pacific Islanders, and American Indians or Alaska natives) receive equal quality of service. Most of the guidelines are devoted to documentation of equity

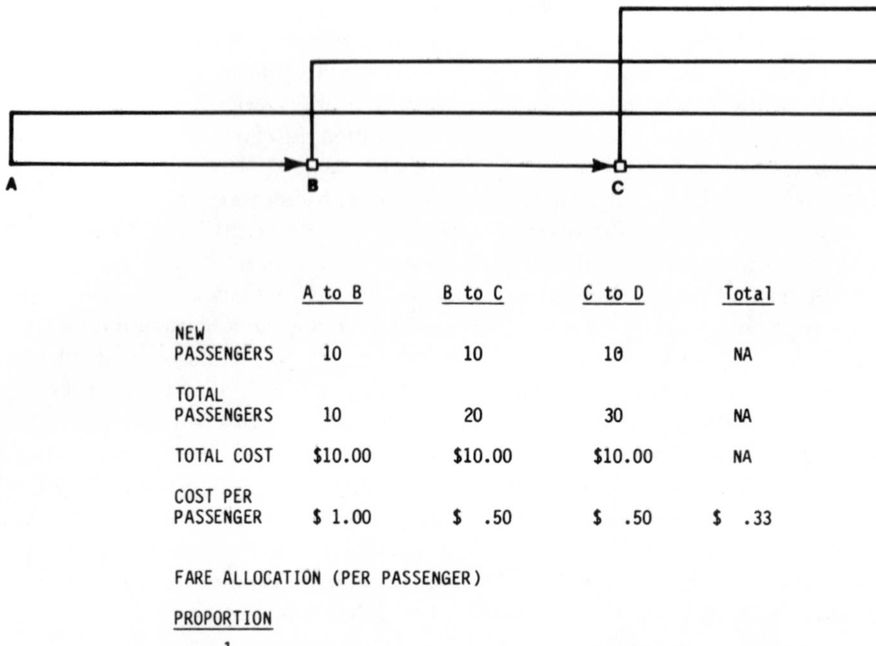

	A to B	B to C	C to D	Total
NEW PASSENGERS	10	10	10	NA
TOTAL PASSENGERS	10	20	30	NA
TOTAL COST	$10.00	$10.00	$10.00	NA
COST PER PASSENGER	$ 1.00	$ .50	$ .50	$ .33
FARE ALLOCATION (PER PASSENGER)				
PROPORTION				
(Fare[1])				
From A	1.00/1.83 (.22)	.50/1.83 (.11)	.33/1.83 (.07)	$ 1.00 (.40)
From B	--	.50/.83 (.24)	.33/.83 (.16)	1.00 (.40)
From C	--	--	.33/.33 (.40)	1.00 (.40)
NET SUBSIDY (PER PASSENGER)[2]				
From A	.78	.39	.26	1.43
From B	--	.26	.17	.43
From C	--	--	-.07	-.07

[1]Assumes $.40 flat fare

[2]Cost per passenger minus fare

Figure 13-3. An example fare allocation.

in transit outputs measured as (1) the quality of vehicles assigned to minority neighborhoods, (2) load factors for routes running through minority areas, and (3) the level of accessibility to various urban activities. In addition, operators are required to document the impact of new fixed facilities on minority neighborhoods and businesses as well as to describe the procedures for community involvement in decision making. It is useful to review the procedures related to these outputs to understand the ways in which such decisions affect social groups differently.

It has been more than two decades since Jim Crow laws, creating separate (and usually unequal) transit systems, were ruled illegal. But as in all too many cases, the change in law has taken much longer to translate into real changes in the way people behave. One of the most persistent charges against local operators has been the unevenness of service between minority and low-income neighbor-

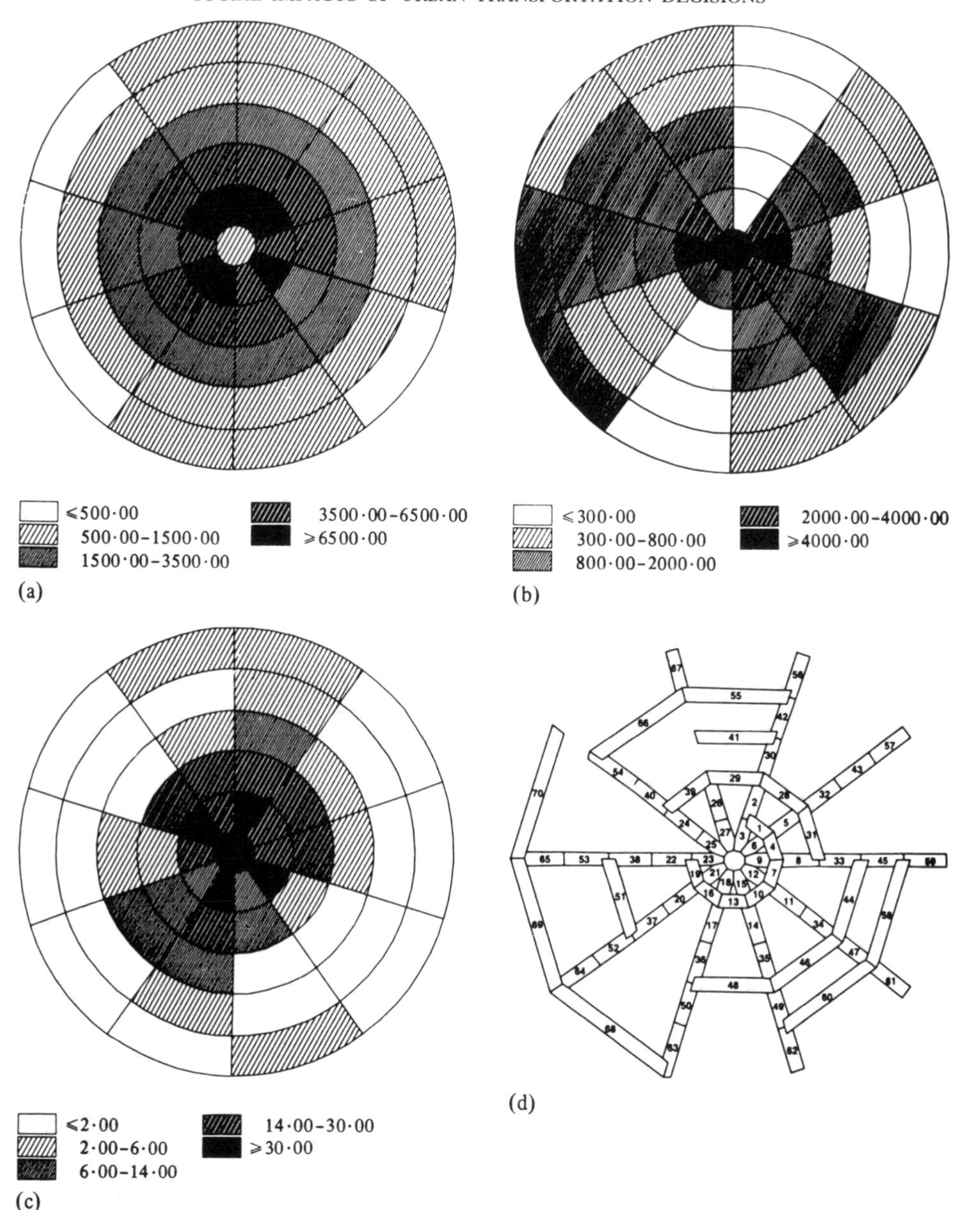

Figure 13-4. Spatial structure of a simulated city. a) Population density (per square mile), b) Employment density (jobs per square mile), c) Percentage black, d) Transit network structure. From Modelling the geographic component of mass transit subsidies by David Hodge, 1981, in *Environment and Planning A; 13;* 581–599.

hoods and their wealthier and whiter counterparts. This discrimination has been most obvious in terms of differences in the quality of equipment. Poorer, minority neighborhoods were often allocated old, unreliable buses without any amenities like air conditioning. The new guidelines require an operator to identify the average age of equipment on all lines during both peak and off-peak periods and to identify all routes that travel through minority neighborhoods. The operator is then required to justify any allo-

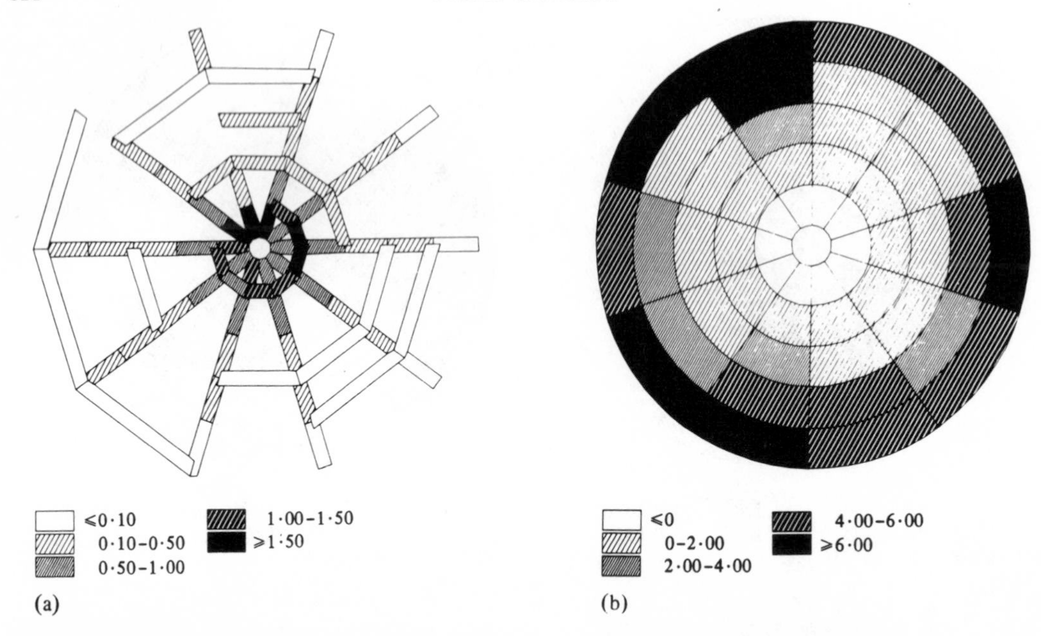

Figure 13-5. Simulated transit system performance. (a) Operating ratio (fares/costs), (b) Per passenger subsidy by origin. From Modelling the geographic component of mass transit subsidies by David Hodge, 1981, in *Environment and Planning A; 13;* 581–594.

cation of equipment that averages two years older than the system average age. The same logic applies to the evaluation of load factors. Again the operator is required to identify routes servicing minority neighborhoods and to justify any situation where the load factor at the maximum load point in peak hour in the peak direction exceeds the average by more than 10%.

The final measure of system performance (output), accessibility, is potentially one of the most interesting and important. After all, the primary function of a transit system is to provide the connections between activities, with comfort as a secondary concern. The guidelines are concerned with access to six major classes of activities, some of them subdivided into a total of 10 different accessibility measures. Although the definitions of accessibility vary a little, in general they all measure the percentage of population within a standard travel time of certain locations; the maximum time permitted in the guidelines varies according to the activity and the size of the metropolitan area (Table 13-2). The method for computing these measures involves estimating the travel time between sample zones (including a waiting factor related to the length of headway) and all zones that contain activities relevant to the particular measure. Sample zones are divided into a specified number of minority areas and corresponding areas (in terms of their distance from the CBD and other characteristics) in nonminority neighborhoods.

THE SEATTLE METRO: A CASE STUDY

In order to illustrate the concepts and issues regarding the social impact of urban transportation decisions, especially the question of equity, we will investigate the Seattle, Washington, METRO. METRO (the Municipality of Metropolitan Seattle) was formed in 1958 as an interjurisdictional agency to clean up heavily polluted Lake Washington. In response to its success, METRO's responsibilities were expanded in 1972 when it was given authority to acquire and manage public transit operations in King County. With a 1977 population of about 1.2 million, King

Table 13-2. Mobility/accessibility comparisons for equity determination

Mobility/access category	Time of day to be analyzed	Measure	Standard travel time by area population (minutes): Up to 200,000	200,000–1,000,000	1,000,000 and over
Employment	A.M., P.M. peaks	Percent of total employment with "standard" travel time by transit.	40	45	60
Shopping	Midday	a. Percent of population within "standard" travel time by transit of a regional shopping center (500,000 sq. ft. of floor area).	45	45	45
		b. Percent of population within "standard" travel time by transit of community shopping center or area.	30	30	30
Medical	All day	a. Percent of population within "standard" transit travel time of county or other major public hospital.	40	45	60
		b. Percent of population within "standard" transit travel time of outpatient clinics located within two miles of residence zone or tract.	30	30	30
Social services	Midday	a. Percent of population within "standard" transit travel time of nearest food stamp distribution center.	30	30	30
		b. Percent of population within "standard" transit travel time of social service (welfare) office serving particular residence zone being analyzed.	30	45	45
		c. Percent of population within "standard" transit travel time of nearest police precinct house.	40	45	45
Education	All day	a. Percent of population within "standard" transit travel time of nearest community college (if no community college, then nearest college generally open to public enrollment).	40	45	45
	Night	b. Percent of population within "standard" transit travel time of nearest night school.	40	45	45
Downtown (CBD)	All day	Percent of population within "standard" transit travel time of downtown.	40	45	45

Note. From *Equity of Travel Service* (p. 16) by D. R. Miller, 1977, No. DOT-UT-50029, Washington, D.C.: Urban Mass Transportation Administration.

County includes the city of Seattle, a number of incorporated suburbs, and a substantial amount of unincorporated territory. King County is sandwiched between Snohomish County to the north (the site of Everett and the rest of the official SMSA) and Pierce County to the south, which is the home of Tacoma and its SMSA of some 400,000 people. Both Snohomish and Pierce counties maintain their own transit systems, which partially connect to destinations within King County.

The eastern half of King County is rural and includes large forested areas of the Cascades; Puget Sound provides an abrupt edge to the west. White-collar employment is concentrated in the Seattle CBD but is growing rapidly in Bellevue and other suburbs east of Lake Washington. Blue-collar employment is concentrated in a sector running south of the Seattle CBD in the Duwamish Valley and out along the Green River valley (Figure 13-6). Although there is a great deal of heterogeneity in the distribution of social

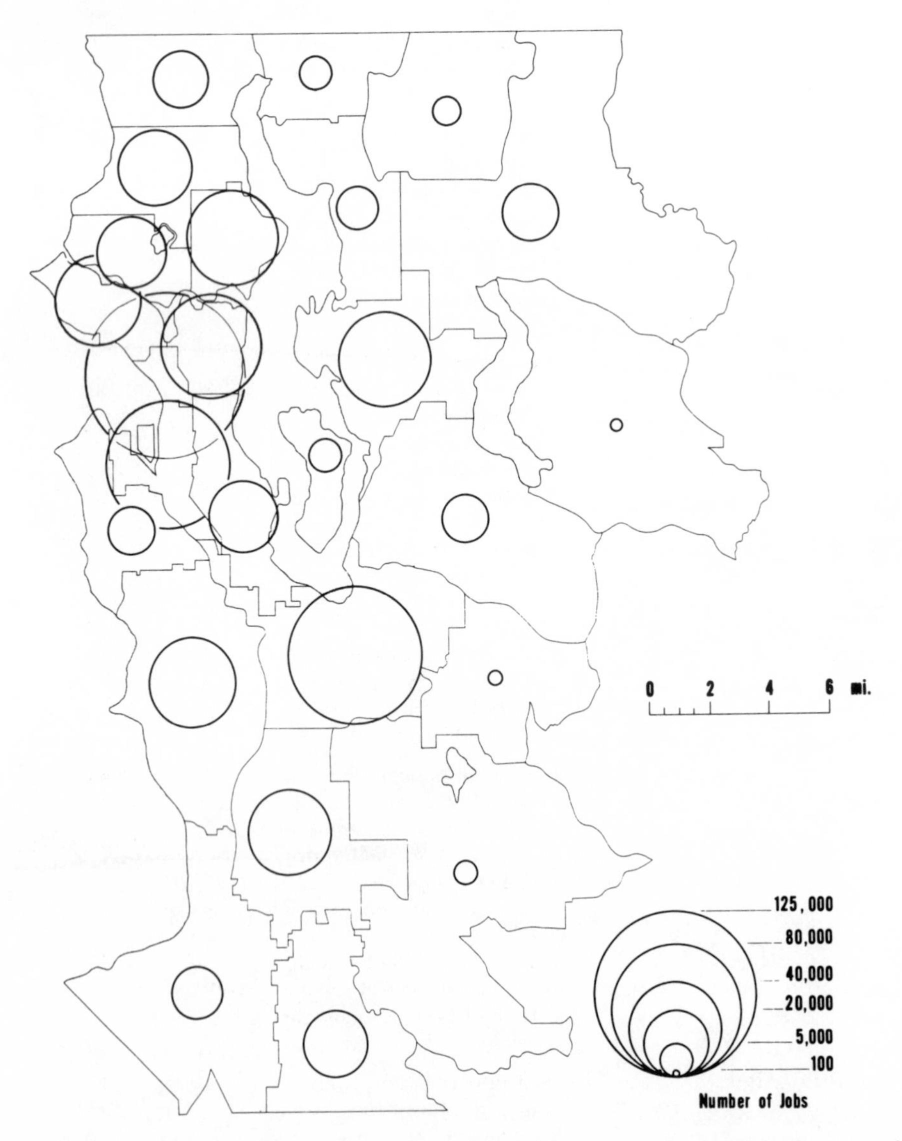

Figure 13-6. King County total employment, 1980. From *Population and employment forecast*, 1984, Puget Sound Council of Governments.

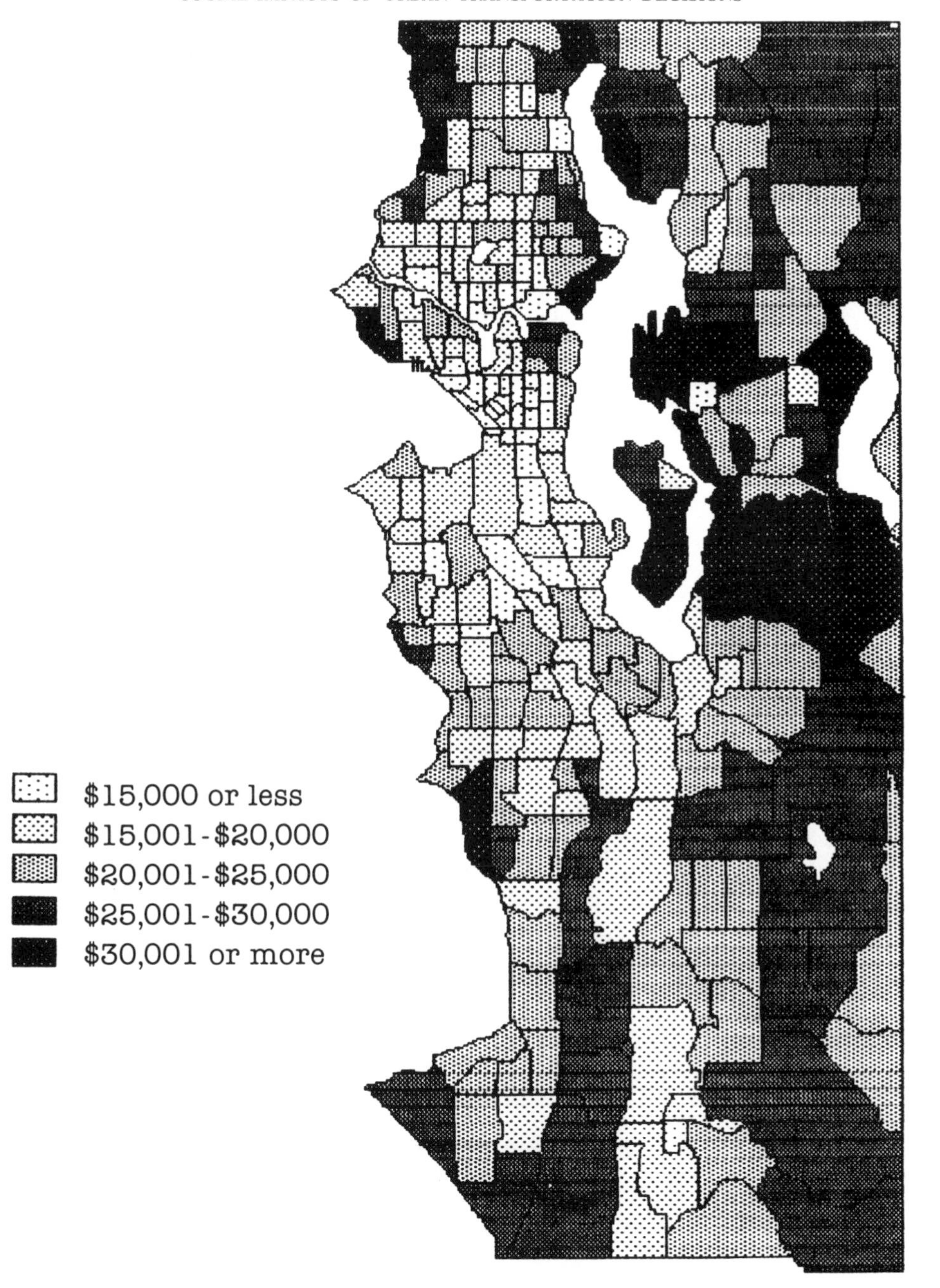

Figure 13-7. Mean household income, 1970 (dollars), Seattle.

groups in Seattle because of the view-laden terrain, many of the spatial patterns of social groups common to U.S. cities can be detected in the distribution of median income (Figure 13-7). Lower-income households are more concentrated in the inner city and wealthier households in the suburbs. However, it is important to bear in mind that there are significant numbers of higher-income persons in the central city as well as low-income households in the suburbs.

The north-south and east-west contrasts are noticeable as well in transit patronage (Figure 13-8). Our analysis of transit use and

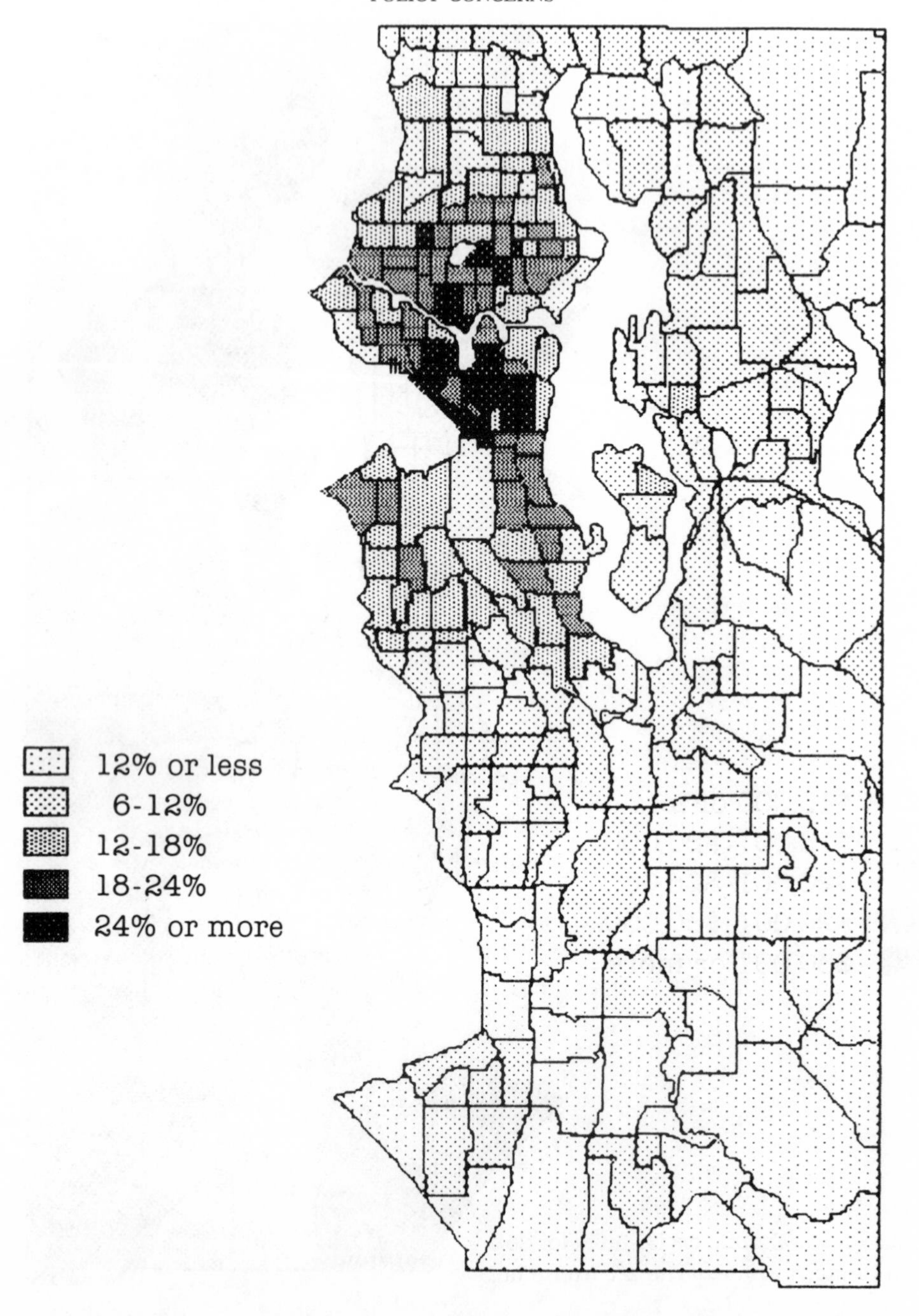

Figure 13-8. Percentage of workers using transit, 1980.

the distribution of equity in METRO is focused on 1976–1977 when the latest origin–destination survey was completed. In that year METRO operated more than 600 buses on a network of about 100 routes. There is no fixed rail service in the region.

The Distribution of Revenues

The total operating budget for METRO in 1976 was about $48 million, with riders contributing $10.3 million (21%), sales tax $18.5 million (38%), motor vehicle excise taxes $12 million (25%), federal operating assistance $6 million (12%), and other revenues adding $1.2 million (3%). The task is to determine who, located where, contributes how much to this operating budget. Our methodology is based on determining the average contribution for each income group to each of the first four classes of revenue, multiplying the appropriate average revenue contributions by the number of persons in each income group in each census tract, and then summing over all income groups to determine a tract total. These totals are in turn divided by the number of households in each tract and compared to the socioeconomic status of residents to reveal any possible social bias.

The sales tax and motor vehicle revenues depend upon the amount of taxable expenditures for each income group. These were estimated using the Bureau of Labor Statistics (BLS) consumer expenditure survey conducted in 1973 inflated to match 1977 incomes. As noted earlier, the sales tax is generally considered a regressive tax, as lower-income groups have more of their income subject to sales tax. Data from the BLS, for example, reveal that the $3,000 to $4,000 income category had 62% of its income subject to sales tax. This figure drops steadily with increasing income with households with an income greater than $25,000 spending only 27% of their income on items subject to sales tax. It should also be noted that a significant portion of the total sales tax collected comes from people from outside King County who purchase goods and services within the county.

The Seattle METRO is unusual in its dependence on automobile excise taxes for 25% of its operating revenues. The 2% excise tax, levied yearly on the entire fleet of privately owned automobiles, is collected by the Washington State Department of Revenue. For vehicles owned by residents of King County half of the tax is returned to METRO. For all other counties in Washington state, the entire tax flows directly to the state. This raises an interesting question of equity at a larger geographic scale, as all counties other than King County lose direct local control of this revenue. The distribution of excise taxes paid by different income groups within King County was determined in two parts. Estimates of average household expenditures on new autos and newly purchased used autos by each income group is estimated from BLS statistics. Given the average number of new cars and newly purchased used cars, the total number of used cars being taxed could be determined from the total number of autos in each tract as recorded in the U.S. census. Federal taxes supporting METRO operations accounted for only 12% of METRO's budget but, because of their progressive nature, help provide some contribution to equity.

The generally regressive funding of METRO operations is reflected in the spatial patterns of the incidence of tax burdens (Figure 13-9). In direct contrast to the pattern of median income (Figure 13-7), central census tracts, which are characteristically of lower income, show higher levels of tax contribution as a percentage of median tract income. It is important to note, however, that this measure reflects tax *burden* (that is, the proportion of income paid to support METRO) and not *absolute* taxes. The distribution of absolute levels of taxes per household would correspond more directly with the level of income. The regressive nature of the tax burden is further supported by correlations with median family income ($r = -.43$), median housing value ($r = -.31$), and the percent black ($r = .19$) for census tracts (using 1970 census data).

The Distribution of Expenditures (Subsidies)

Analysis of the distribution of transit subsidies in King County follows the model of Hodge (1981), described earlier, making extensive use of a 1977 origin–destination survey, which provides the necessary quantity and detail of information to make the assessment. The METRO system was abstracted into 517 nodes, which defined segments that composed 99 routes. For each of some 37,000

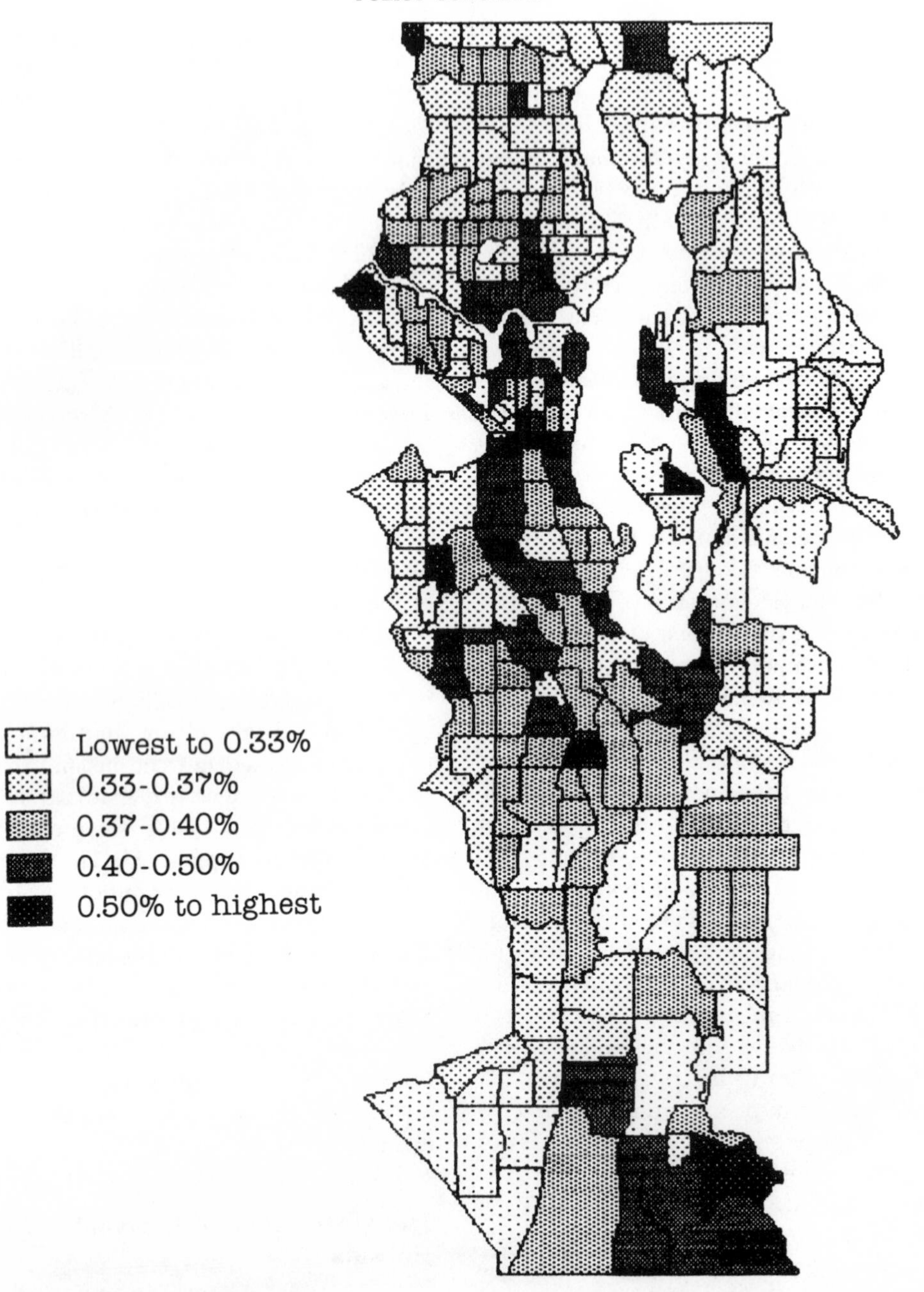

Figure 13-9. METRO tax revenue as a percentage of 1970 median tract income, Seattle. From *Spatial equality of mass transit service: The Seattle Metro*, by John MacGregor, 1981, unpublished M.A. thesis, University of Washington.

individual trips identified in an on-board survey, the closest node was assigned as the trip beginning and the corresponding destination node as the trip end with all segments in between identified as the trip itself. In this way the total number of passengers traveling over each link in the system is determined. Dividing the cost of service per link yielded the cost per passenger and cost per passenger mile.

Results of the analysis of routes in Seattle, while demonstrating the same general trend as that displayed in the example, also reveal more of the intricacies of a real system. In general the pattern of performance, measured by the operating ratio (fares/costs) varies with distance from the city center, but the geography of total system performance shows more variability than was expected. The overall correlation of the operating ratio with log of distance from the CBD is only a modest −.41. To map system performance by individual segments, it was necessary to generalize at a given location in order to handle multiple segments (either different times of the same route or different routes in the same place) and to "smooth out" small idiosyncracies. To accomplish this, a grid of 118 points was laid over the transit system, and all segments in the immediate neighborhood of any point were averaged into a single value. The result is a map of operating ratios generally occurring in different parts of the service area (Figure 13-10). Interpretation of the map requires some understanding of the unique geographic conditions of Seattle, most notably the effect of the separation of Seattle and a large part of its suburbs by Lake Washington. As a result there are distinct corridors of movement and activity. In addition it is important to recall that industrial development dominates south Seattle and south King County with white-collar employment concentrated in the Seattle CBD and more connected to residential areas to the north and east of the CBD. As a result, performance is high in the corridor running to the north of the CBD and the two corridors to the east, corresponding roughly to the two major transportation corridors crossing Lake Washington, and low to the south.

These patterns of system performance give us the first indication of the incidence of transit subsidy by the location of the home residence of transit users. Again the pattern to the north of the CBD and the corridors to the east stand out in terms of their lower per passenger transit subsidies (Figure 13-11). South King County followed the system performance as well, with high subsidy levels

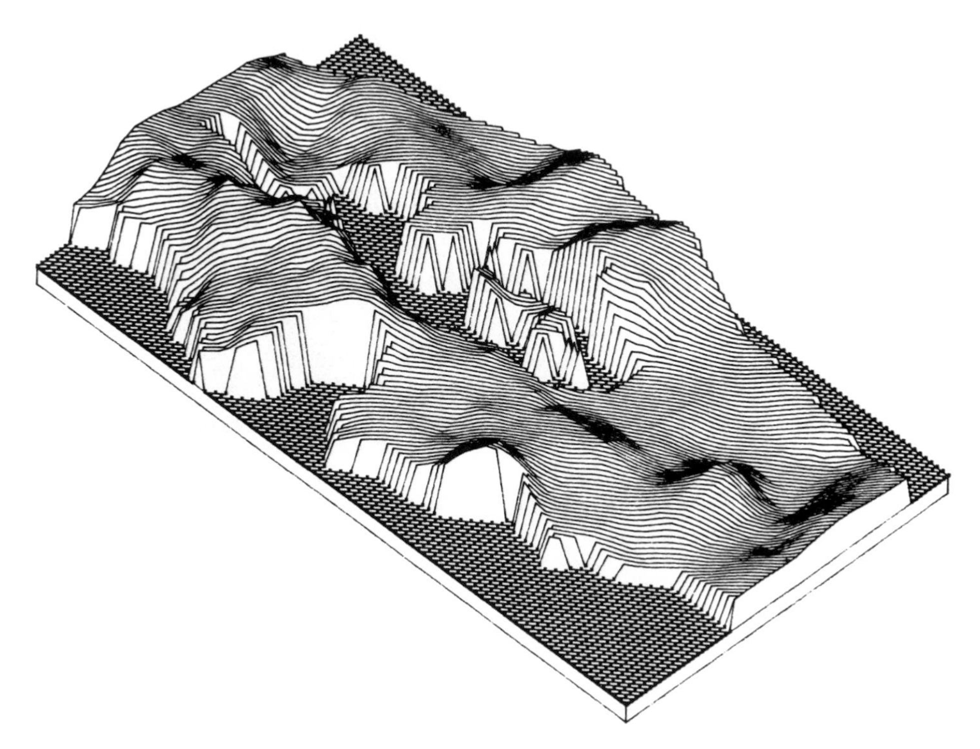

Figure 13-10. METRO operating ratios (ratio of fares to costs), Seattle.

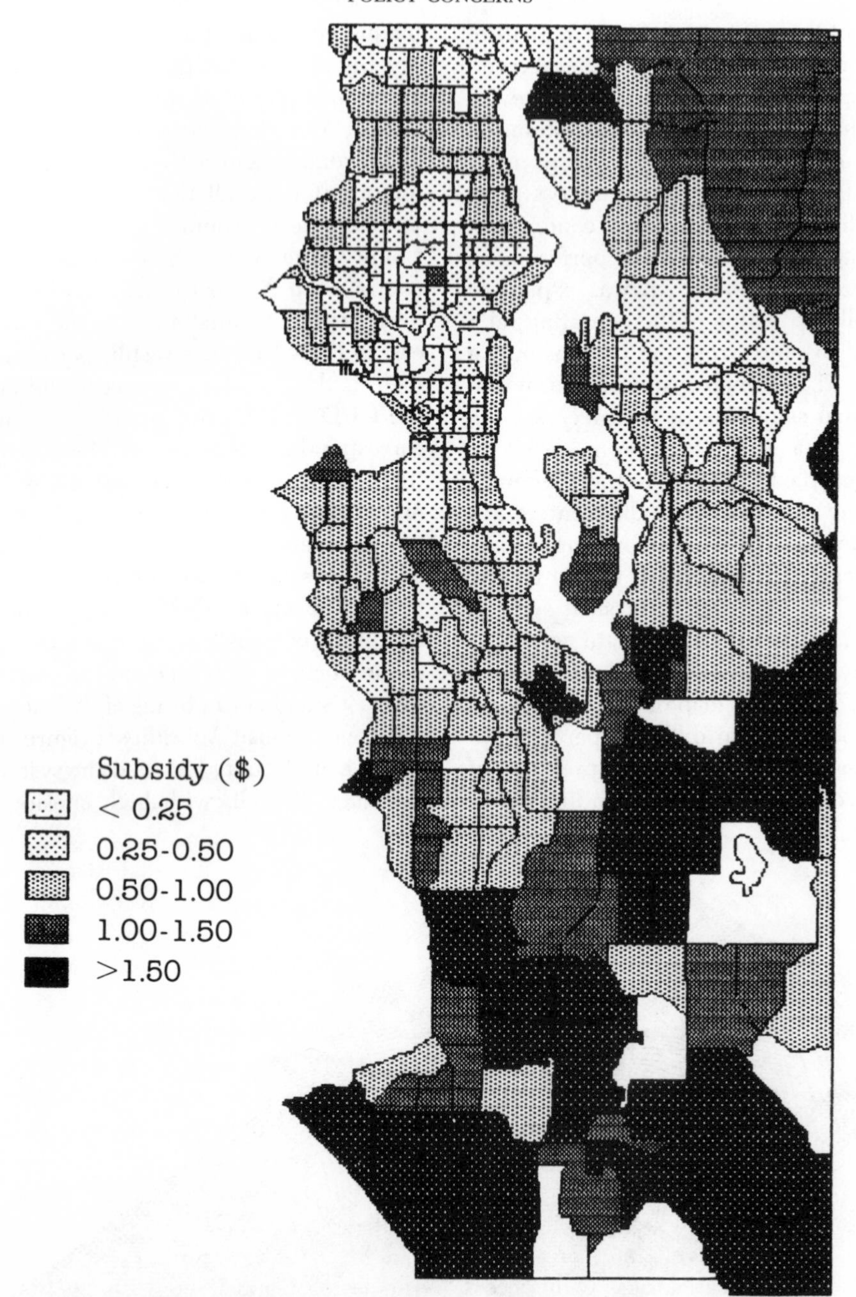

Figure 13-11. Per passenger subsidy by home location, Seattle.

per passenger. There are some surprises, however. Residents of the area to the far north of the CBD, for example, received lower subsidies than areas closer in. This outcome is most likely a result of the fare and route boundary corresponding roughly to the Seattle city limits. Riders in this area are at the far edge of the inner zone and hence likely to receive larger subsidies whereas residents farther north pay higher fares and rely more heavily on express buses which operate at relatively high load factors.

The biggest surprise of this analysis, given the general trends identified elsewhere, is the lack of a pattern in the incidence of subsidies by social group (Table 13-3). In the case of Seattle, at least, the longer trips of higher-income groups are offset by higher fares imposed on suburban riders into the central city. Lower-income groups do well as do elderly whose fares cover only 20% of their trip costs. It is important to note also that in many ways Seattle is an atypical American city. While it does have racial segregation on a limited scale (with a small minority population), there is substantial integration by income group, particularly at the census tract level, which often combines wealthy "view" neighborhoods with lower-income valley neighborhoods even though the overall pattern of income levels follows the general pattern (Figure 13-7). In any event the operations of the Seattle METRO appear to provide a remarkably equitable distribution of operating subsidies. Unlike other studies, distinct patterns of spatial bias in the distribution of subsidies do not translate into patterns of social bias.

Table 13-3. Transit subsidies by social group

Group	Persons	Average subsidy	Average fare/cost
Age			
Under 18	8,134	$.77	29%
19–34	33,570	.53	38
35–64	21,132	.52	39
65+	7,705	.61	20
Income			
$ 6,000	16,137	.60	30%
6–10,000	12,078	.55	34
10–15,000	11,664	.50	38
15–20,000	8,803	.57	36
20–25,000	6,382	.52	39
25,000+	7,599	.59	36
Auto ownership			
None	20,054	.59	31%
1	28,601	.52	37
2 or more	22,416	.61	36

The Distribution of Service Outputs

Evaluation of the quality and quantity of transit service provisions by METRO shows that Seattle is very much in compliance with UMTA equity guidelines (METRO, 1984). Although some routes serving minority areas have older equipment, the discrepancy is not great and is being reduced. Measures of accessibility show that minority areas in fact have a greater proportion of their population within the standard travel times prescribed by UMTA for most categories. Only access to medical outpatient clinics is below the nonminority comparison areas (71% compared to 93%) and is the only measure with less than 96% of the minority population within the standard distances.

Such results are not surprising given the looseness of access standards (see Table 10-2) and the central location of minority populations. As McLafferty (1982) has pointed out, the relatively high scores for minority accessibility are virtually a geometric imperative, since minority populations are located near the center of metropolitan areas. A more sensitive measure of the quality of service outputs was developed by MacGregor (1981), who reasoned that the quality of service cannot be measured on any constant absolute standard. It makes no sense, for example, to use headways as the measure of service output when comparing low-demand peripheral areas with higher-demand inner-city routes. Therefore, MacGregor suggested using a ratio of the time of transit trips to the length of time that would have been required if the trip had been made by automobile. One could think of the auto trip time as representing an "ideal" transit trip—one that implies door-to-door service upon demand, along the fastest route, and with no transfers or intermediate stops for others. How closely this variable ideal is met and shared by transit users in different areas and social groups can serve as an important indicator of equity in service outputs. MacGregor (1981) applied this concept to the Seattle METRO.

Without going deeply into the methodology, it should be noted that the total travel time calculated for transit takes into account access time to the bus, waiting between buses, and the amount of time required from deboarding to reach the destination all weighted more heavily than actual travel time. Similarly, calculated auto travel times include values for time spent parking and so on, which are also weighted more heavily than time spent actually traveling. Data for transit trips in King County come from the METRO 1977 origin–destination survey, and data for comparable auto trips came from the Puget Sound Council of Government's (PSCOG) Activity Allocation Model.

The geographic pattern of service levels again reflects the strong north-south contrasts with the southern part of King County suffering from inferior service (Figure 13-12). The service levels of south King County, however, are part of a "chicken-and-egg" scenario. Service levels are low because demand is low, and demand is low because service levels are low. In many ways the situation is a function of the dispersed employment centers in the area and the lower connections to the CBD. The dominant cross-suburban travel of south King County is simply ill-suited to transit service. The correlation between service level (note: the lower the value the better) and percent black ($r = -.13$) indicates a small bias in favor of minority neighborhoods. The correlation with median income ($r = .34$) and percent of households below the poverty level ($r = -.34$) reflect a similar pattern favoring poorer areas.

In general, service levels are inversely related to non-farebox revenues when revenue is expressed in absolute dollars per household ($r = .29$) but is unrelated to revenues expressed as a percentage of tract median income. Ironically, those who pay most (in absolute terms) get the worst service! The spatial pattern of the correspondence between service levels and tax burden (Figure 13-13) shows little in the way of systematic variation. However, there is a strong correlation between service levels and fares, either in absolute terms ($r = -.40$) or as a burden measure ($r = -.54$). Those who pay the most do have the best service. A final word of caution is in order, however. This analysis is based on a comparison of expressed demand only, that is, transit trips people take and not the ones they might like to take. It is not unreasonable to assume that those who have some mode choice will opt for transit if and generally only if it provides competitive service while those who are transit dependent must use the service regardless of whether or not it is an inferior good.

CONCLUSIONS

There is little doubt that the modern city offers unparalleled mobility for most of its residents. There is little doubt also that, in spite of a number of negative externalities associated with the automobile, most urban residents are far better off because of the transportation system developed in cities. But not all residents share equally in either the benefits or the costs associated with urban transportation. In many instances individuals are excluded from certain benefits because they cannot afford the cost of entry into the auto age. In other instances individuals face obstacles imposed by design and operating decisions that interact with underlying spatial patterns of residential and other land use activities. Low-income households face exposure to higher-than-average levels of air pollution because of their location near city centers where auto traffic and pollution are most concentrated. Historically, low-income and minority populations have suffered from inferior transit service as a result of discrimination by local operators. Although, at least in the case of racial minorities, such discrimination is now outlawed, the inferiority of transit service to the auto and the historical inertia of patterns of residential location vis-à-vis employment opportunities continue to work to the disadvantage of these groups.

The question of equity needs to be raised whenever it is clear that decisions will create different levels of social impact and especially where such impacts potentially discriminate systematically against a protected subgroup like racial minorities. However, providing an

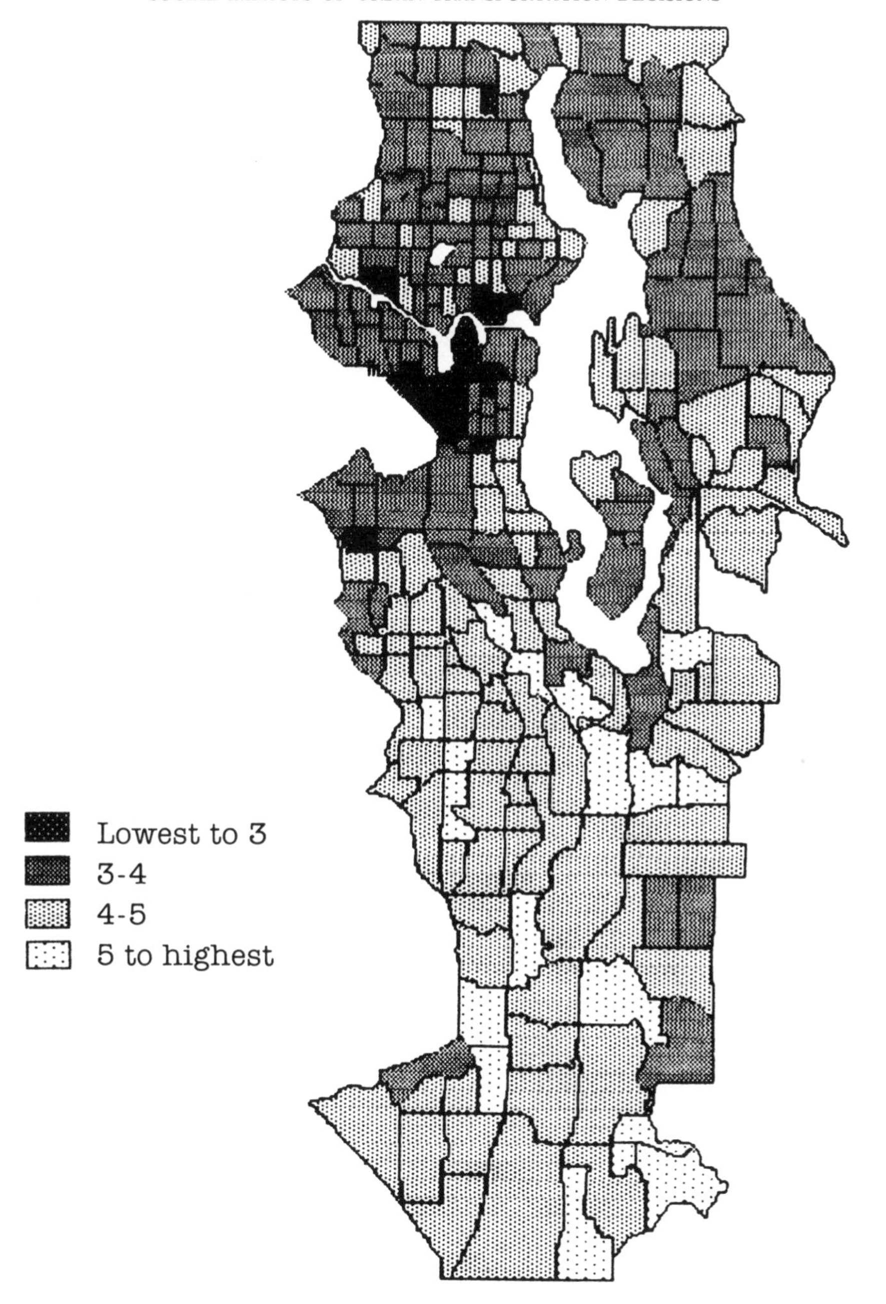

Figure 13-12. METRO service levels, 1977 (ratio of observed transit trip times to similar trips by automobile), Seattle. From *Spatial equity of mass transit service: The Seattle Metro* by John MacGregor, 1981, unpublished M.A. thesis, University of Washington.

absolute answer to the question of equity is impossible, for such answers depend upon the goals of a system, which features are to be evaluated, and eventually the rules used to evaluate distributions. Is a uniform distribution of costs or benefits any more or less defensible than a progressive or regressive distribution? We have seen how it is possi-

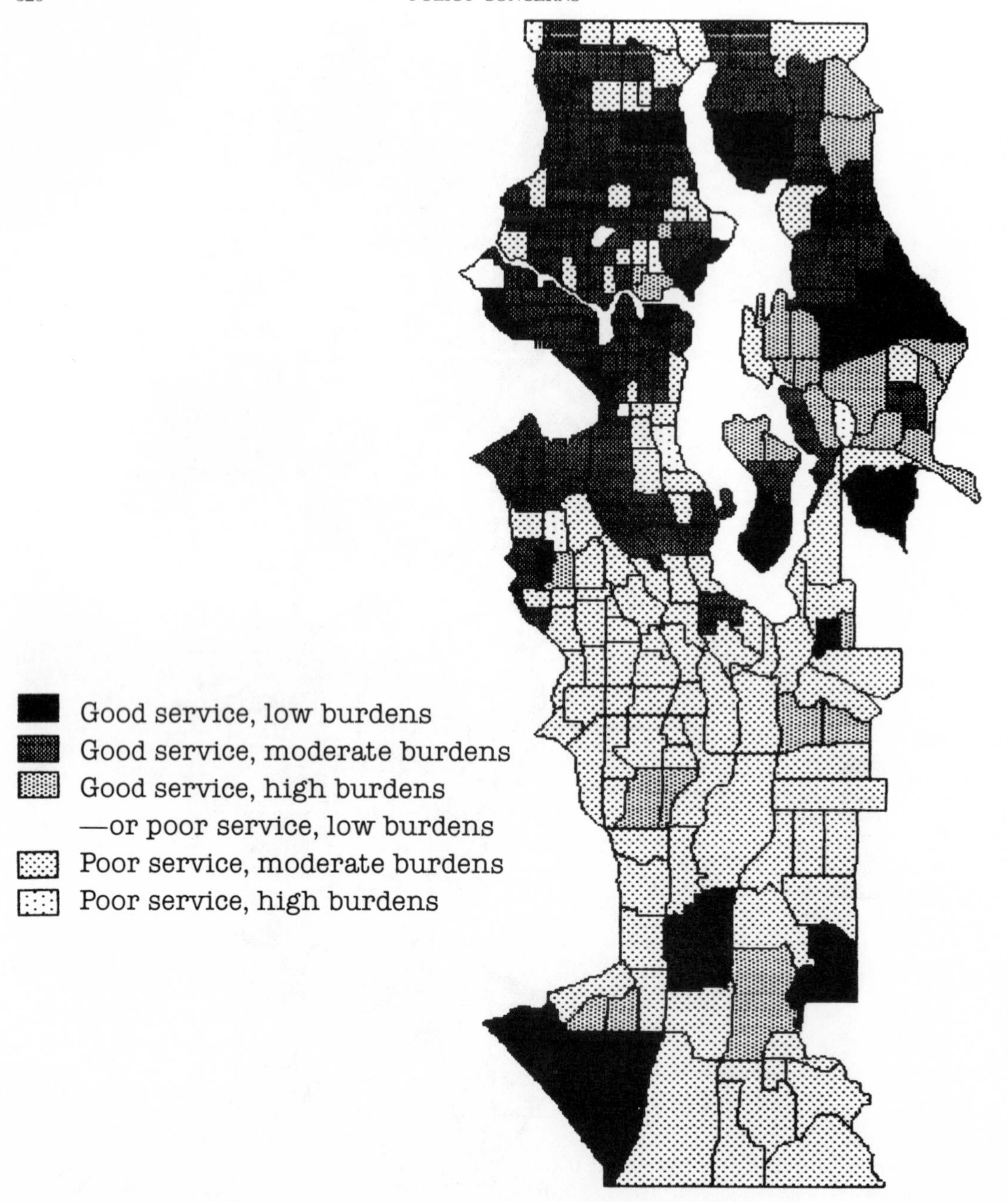

Figure 13-13. METRO service delivery and tax burdens, Seattle, 1976–1977. From *Spatial equity of mass transit service: The Seattle Metro* by John MacGregor, 1981, unpublished M.A. thesis, University of Washington.

ble to measure distributions of benefits and costs from a wide variety of perspectives with considerable disagreement over methods.

Still, while it may not be possible to provide a definitive answer to this fundamental social question, raising the question and measuring social impacts as best we can serves the decision process well. Whether it be a decision regarding something as large as routing a new urban freeway or as small as assigning a specific bus to an existing route, identifying the full range of impacts from a disaggregated perspective will ensure that all will be represented. The geographic perspective is vital to this process because of the close correspondence in this society between social status and geographic location. Even routine urban transportation decisions have clear spatial implications that often result in systematic patterns of social impacts. Identifying these impacts is fundamental to an open and equitable decision process.

The author would especially like to acknowledge the contribution of John MacGregor who is largely responsible for the research on tax burdens and service levels for METRO and the National Science Foundation who supported some of the research reported here.

References

Building the American city. (1969). Report of the National Commission on Urban Problems. New York: Praeger.

Cervero, R. B., Wachs, M., Berlin, R., & Gephart, R. (1980). *Efficiency and equity implications of alternative transit fare policies.* Washington, D.C.: Urban Mass Transportation Administration.

Faucett Associates. (1976). *Washington area Metrorail system: A current perspective and a preliminary appraisal of alternatives.* Washington, D.C.: Congressional Research Service of the Library of Congress.

Frankena, M. (1973). Income distributional effects of urban transit subsidies. *Journal of Transport Economics and Policy, 7,* 215–230.

Greenland, D., & Yorty, R. (1985). The spatial distribution of particulate concentrations in the Denver metropolitan area. *Annals of the Association of American Geographers, 75,* 69–82.

Hodge, D. (1981). Modelling the geographic component of mass transit subsidies. *Environment and Planning A, 13,* 581–599.

Lakshmanan, T. R., & Chatterjee, L. R. (1977). *Urbanization and environmental quality.* Resource Paper 77-1. Washington, D.C.: Association of American Geographers.

MacGregor, J. (1981). *Spatial equity of mass transit service: The Seattle Metro.* Unpublished M.A. thesis, University of Washington.

McLafferty, S. (1981). Urban structure and geographical access to public services. *Annals of the Association of American Geographers, 72,* 347–354.

METRO. (1984). *Equity of transit service study.* Seattle: Municipality of Metropolitan Seattle.

Miller, D. R. (1977). *Equity of transit service.* No. DOT-UT-50029. Washington, D.C.: Urban Mass Transportation Administration.

Morgan, R. T. (1976). *Buses in Bradford: final report.* Bradford, England: West Yorkshire Metropolitan County Council.

Pucher, J. (1981). Equity in transit finance: Distribution of transit subsidy benefit and costs among income classes. *American Planning Association Journal, 47,* 387–407.

Pucher, J., Hendrickson, C., & McNeil, S. (1981). Socioeconomic characteristics of transit riders: Some recent evidence. *Traffic Quarterly, 35,* 461–483.

Pucher, J., & Hirschman, I. (1981). Distribution of the tax burden of transit subsidies in the United States. *Public Policy, 29,* 341–368.

Schou, K. (1979). Energy and transport: Strategies for energy conservation in urban passenger transport. *Environment and Planning A, 11,* 767–780.

Seley, J. (1970). Spatial bias: The kink in Nashville's I-40. *Research on conflict in locational decisions.* Philadelphia: Regional Science Department, University of Pennsylvania.

Transit fact book. (1981). Washington, D.C.: American Public Transit Association.

Wabe, J. S., & Coles, O. B. (1975). The long- and short-run costs of bus transport in urban areas. *Journal of Transport Economies and Policy, 9,* 127–140.

Webber, M. (1976). *The BART experience: What have we learned?* Monograph Number 26. Berkeley: University of California, Institute for Transportation Studies.

14 ENVIRONMENTAL IMPACTS

FREDERICK P. STUTZ
San Diego State University

Changing priorities in the national consciousness in the United States have resulted in greater concern for the environmental impacts of transportation development, rather than concern mainly for pure efficiency of movement.

The impacts of transportation projects can sometimes be anticipated before any studies are begun. Newspapers abound with examples of transportation improvements or proposed additions that have created a great amount of controversy. Citizens protest a freeway in New Orleans because they fear that noise levels will dull the jazz music pouring from the city's famous Latin Quarter. A freeway is halted in midconstruction in San Francisco because it will interfere with the visual character of the city. In Philadelphia, the location selected for I-287 is contested because it will cause carbon monoxide levels to double in certain neighborhoods. These are examples of impacts that we might anticipate; but how can we predict what the actual effects of transportation projects might be?

The task of demonstrating the spatial effect of these impacts is frequently the job of the geographer. Although certain transportation-related impacts are subjective and difficult to qualify (such as the effect of a highway on an established neighborhood or on the aesthetics of an area), it is relatively easy to evaluate the air and noise pollution that will result from a transportation project. This chapter focuses on what are sometimes called the "natural systems impacts," "physical impacts," or "environmental impact" of transportation improvements. The discussion will emphasize noise pollution and air pollution because these impacts have caused the most public concern and because models to assess them are more highly developed than are models of other physical impacts (such as water quality, changes in vegetation, soil-loss problems). Visual impacts and the alteration in urban climates resulting from the warming of the built environment are additional physical impacts that are only beginning to be considered by agencies responsible for urban transportation improvements. Toward the end of the chapter special treatment is given to the environmental review process and to evaluating environmental impacts in a benefit/cost framework.

NOISE POLLUTION

Noise pollution is best defined as sound that is unwelcome or sound that is detrimental to the quality of life. Noise is commonly measured in decibels (dBA), a unit of sound pressure level. (The decibel is approximately 20 times the logarithm of the ratio of the measured sound pressure to zero dBA, or 20 N/m^2.) In urban areas noise usually comes from three sources: (1) transportation systems, (2) occupational or industrially related sources, and (3) community background noise, with transportation noise having by far the greatest impact on urban residents. Each community has its own "noise signature," or background noise, which is affected by its type and level of activities and its proximity to transportation routes. The overall size of a city has little effect on noise signature, and surprisingly, neither does traffic density (Federal Highway Administration, 1978).

Typical noise levels emitted from different transportation sources are shown in Figure 14-1. Though a few communities suffer from significant levels of aircraft or industrially produced noise, highway traffic is the major contributor of noise in the United States as a whole. With the growth of freeways and surface streets, the increasing use of the automobile, and the plans for numerous new rapid transit systems springing up across the nation, the need to assess the effects of noise on the physical condition of human beings and other noise "receptors" has become imperative.

Four major effects of traffic noise have been identified: sleep interference, speech interference, annoyance, and the impairment of hearing that comes from long exposure to high noise levels. Anyone who has been awakened night after night by a barking dog has some idea of the effect that *sleep disturbance* can have on a person's well-being. Exposure to noise can interfere with sleep in two ways. It can delay the onset of sleep, or it can cause sleep stages to shift. Studies of the effects of noise on sleep have not yielded a set of consistent results that would enable us to predict the sleep disturbance resulting from traffic noise. *Speech interference* from highway noise refers to the masking of speech by noise from the flow of traffic, whether from a single vehicle or from

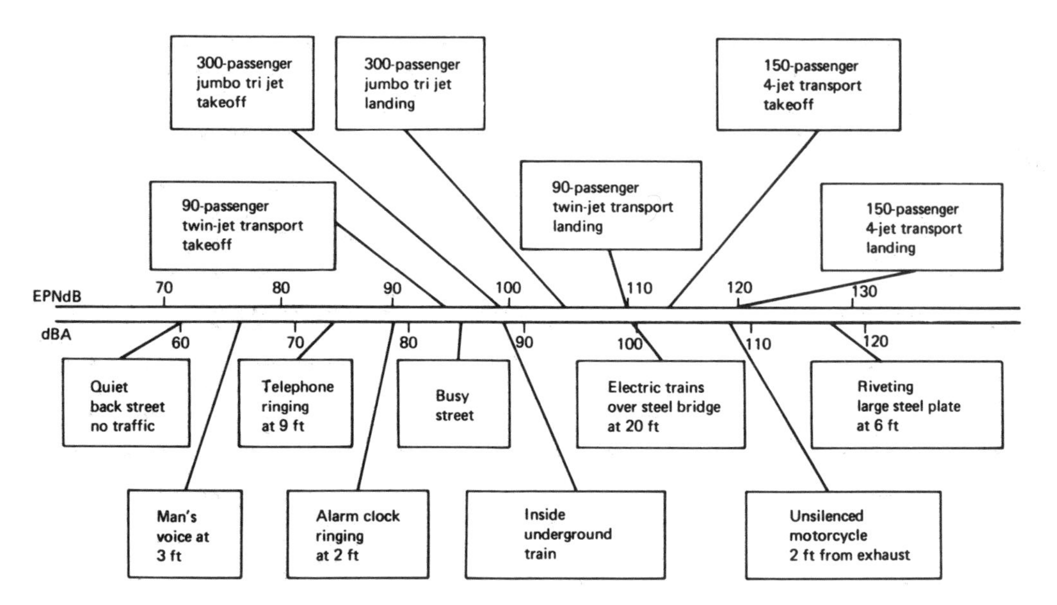

Figure 14-1. Typical noise levels in decibels (dBA) and (EPNdB) of some common noises. Approximate relation between EPNdB scale and dBA scale is shown in this comparison of various sources. From *Transportation Noise and Its Control* (p. 3) by U.S. Department of Transportation, 1972, Washington, D.C.: U.S. Government Printing Office.

high levels of traffic. Research on this aspect has been conclusive enough to permit reasonably accurate quantification of the effects on the basis of physical measurements of traffic noise. In addition, it is reasonable to conclude that noise levels that minimize speech interference are also low enough to minimize sleep interference.

Annoyance is a complex phenomenon that is difficult to define in terms of noise pollution. Although there is little doubt that high noise levels will annoy large numbers of people because of sleep or speech interference, there is no guarantee that low noise levels will not also cause significant annoyance. Most of the data on this question come from surveys of people's reactions to aircraft-related noise; these surveys show that as noise levels increase, so does the percentage of highly annoyed people. One recent study showed that whereas only 20 percent of the individuals sampled were highly annoyed when noise levels averaged 55 decibels, approximately 50 percent were highly annoyed when noise levels averaged 70 decibels (U.S. Environmental Protection Agency, 1974). It is known there are lower levels of school performance for students in schools situated in noisy zones (U.S. Department of Transportation, 1972).

Vibration produced by urban transportation is not likely to produce a significant impact on a whole community, but vibration (as well as noise, odor, dust, and lights) can be a problem for the three rows of houses closest to an urban highway (Figure 14-2). Objection to the Tokaido train in Tokyo and its proposed counterpart, the San Diego–Los Angeles 160-mile-per-hour bullet train, centers on vibrations in surrounding residential areas. Partly because of such environmental impacts, plans for the U.S. bullet train have recently been canceled. In U.S.

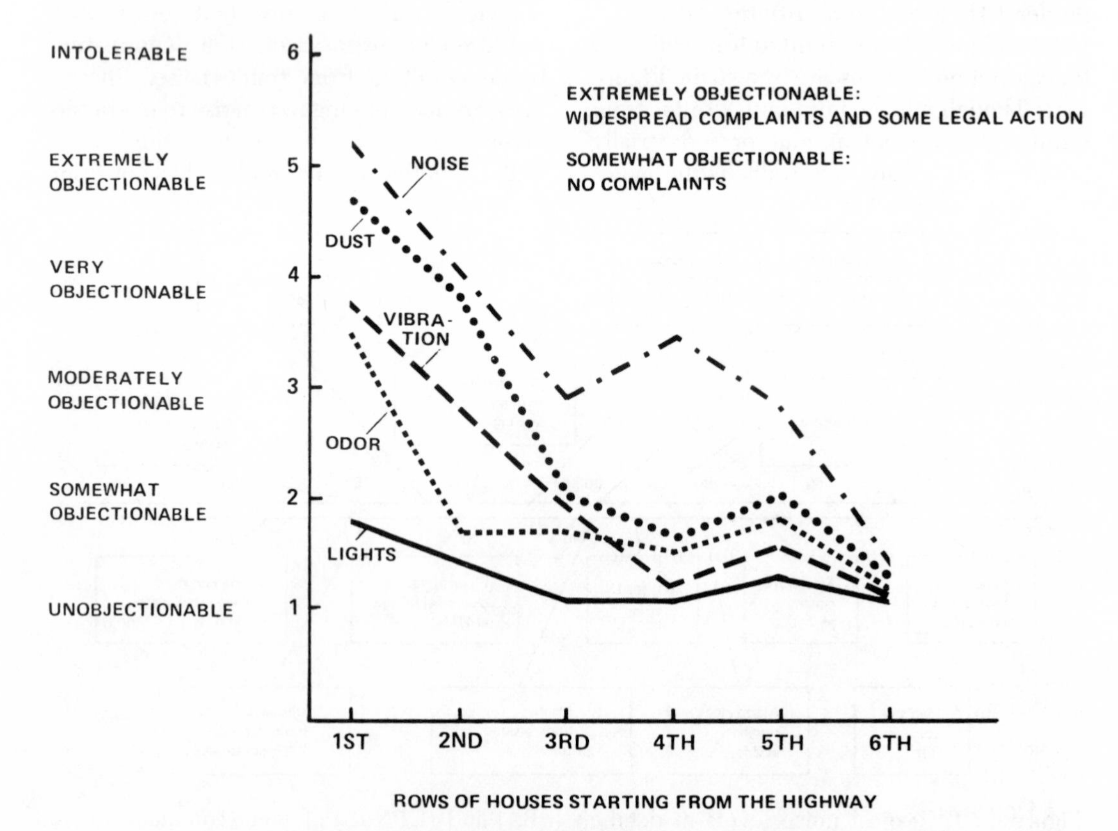

Figure 14-2. Noise and other negative highway impacts. From "Valuation and Compensability of Noise Pollution" by Jack Faucett Associates, *Quarter Progress Report,* December 1974. Copyright 1974 by CALTRANS. Reprinted by permission.

cities rail traffic is so rare that it does not usually create a severe noise problem. In Europe and Japan, however, where this mode of urban travel is important, considerable attention has been given to the design of equipment and railbed.

Noise Standards

Various communities and regions have set noise level standards, criteria, guidelines, and regulations. It is important to have an understanding of what these terms mean. *Standards* are a widely agreed upon and formally defined means of measuring noise levels. A *criterion* is a statement of the effects of noise on people. Neither a law nor a recommendation, it is simply a relationship based on scientific knowledge. *Guidelines* are usually an interpretation of criteria, not a legal statute, but an indication of an official position taken by a regulatory agency. A *regulation* is a legal statement of permissible noise levels and is based upon criteria.

Highways built with federal funds should not exceed maximum noise-level tolerance criteria as established by the Federal Highway Administration (FHWA) (U.S. Department of Transportation, 1976). There are no federal regulations forbidding construction of highways that exceed these levels; however the Federal Highway Program Manual does state that if a significant noise impact is identified, abatement measures must be considered. If noise-abatement procedures are studied and found to be physically infeasible or economically unreasonable, the project can be approved anyway. State regulations vary widely. In New York, California, and some other states traffic noise near an existing school is regulated. Of the numerous guidelines, criteria, and recommendations, the two most important are Environmental Protection Agency (EPA) guidelines for noise levels to protect the public health and welfare (within an adequate margin of safety) and Department of Housing and Urban Development (HUD) criteria for assessing the acceptability of noise levels at HUD-involved new construction or redevelopment projects. The various guidelines are discussed below.

The FHWA guidelines place the affected land into one of five land use categories and assign an acceptable noise level for each, which cannot be exceeded more than 10% of the time. The categories cover both developed and undeveloped land. Exterior noise levels are set at 60 to 75 dBA, and interior noise levels are set at 55 dBA. These levels are much lower than those in the California guidleines, and the federal requirements will probably become more stringent through time.[1]

EPA noise-level recommendations (45 to 70 dBA) are generally more stringent than those set forth by the FHWA. These recommendations may never have the force of law but should probably be regarded as long-range goals.

The HUD criterion is a more sophisticated approach to noise control. Noise levels are divided into four categories which range from "clearly acceptable" to "clearly unacceptable" (U.S. Department of Housing and Urban Development, 1971). The percentage of time that a noise level as measured on the dBA scale may be exceeded and still remain in any one category has been specifically defined. This relationship is presented in Figure 14-3.

Noise Assessment

Regardless of the set of standards or criteria applied, an assessment procedure must be followed in order to evaluate the impacts properly. This is usually a two-part process consisting of an initial evaluation and a refined evaluation.

An *initial evaluation* determines whether severe noise impacts will result from any of the proposed alternatives. This evaluation is intended to be quick and does not consider any mitigating effects of adjacent topography or structures. Therefore, noise impacts are

[1]Section 216 of the California Streets and Highways Code addresses noise level limits for schools only. Section 216 was amended in 1984, and as a result the state noise-level limit for schools is now the same as the federal criterion.

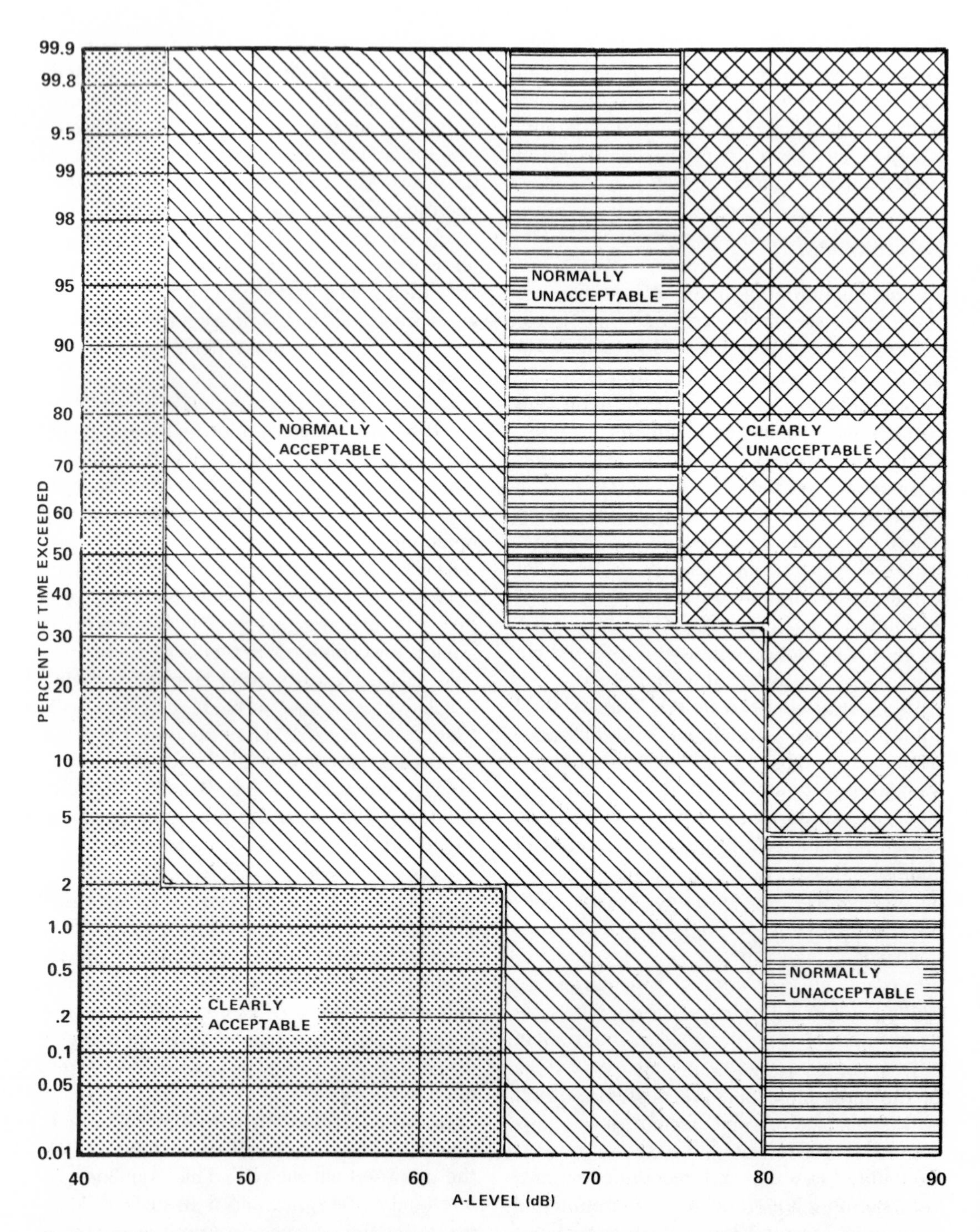

Figure 14-3. Current HUD criteria for nonaircraft noise. From *HUD Noise Assessment Guidelines, Technical Background* by U.S. Department of Housing and Urban Development, 1971, Washington, D.C.: U.S. Government Printing Office.

usually overstated in the initial evaluation. At this stage available data are used to calculate existing noise levels and predict future noise levels. This enables researchers to identify the noise zone of influence and to focus on noise-sensitive areas and facilities within this zone of influence. Finally, an estimation is made of the type of mitigating measure needed to reduce the predicted noise levels to an acceptable level.

A *refined evaluation* is done after the initial evaluation is complete. It is more comprehensive and includes comparison of future and existing noise levels, and an impact–perception–response evaluation (California Department of Transportation, 1979). The main difference between the initial and refined evaluations is the nature of data used. The first step in the refined evaluation is to gather new data on existing noise levels. Computer techniques are then used to predict future noise levels. A comparison is made of existing and future noise levels, and the findings are mapped along with areas and facilities that would be noise sensitive. Ways to mitigate the predicted effects are then assessed in terms of altering the findings. Finally, the significance of the impacts upon the various groups and communities is assessed on the basis of documented environmental noise research findings (California Department of Transportation, 1979).

Noise Models

There are a number of models available to help in the assessment of noise impacts. Some of the most important are described below in the context of some actual applications. Figure 14-4 shows the effects that additional cars added to a steel wheel/steel rail transit train have on noise levels. Although noise increases with train length, noise levels decline as distance from the tracks increases. At 50 feet, all trains produce unacceptable noise levels. At 100 feet, the one- and two-car trains have dropped into a range that is generally considered at the highest level of acceptance (normally 75 dBA). At 500 feet, the very long train has dropped into this category, and the one- and two-car trains have dropped into the category that HUD defines as normally acceptable 100% of the time (see Figure 14-1). However, at 1,000 feet, the very long train is still emitting noise at a level that HUD finds normally unacceptable if the level is maintained more than 32% of the time (which would be quite conceivable if trains are scheduled to run at frequent intervals). Noise is related not only to the length of a train, but also to its speed. The Urban Mass Transportation Administration (UMTA) has produced a formula for estimating noise as a function of a train's velocity *(V)* (Figure 14-5).

A 1975 draft of an environmental impact statement on San Francisco's Bay Area Rapid Transit (BART) revealed a number of factors related to noise and vibration (Gruen Associates, 1975). Residences 250 feet from the centerline of the track and in an unbroken line of sight were found to be the greatest recipients of noise from the system. A system that consisted of 10-car trains traveling at 70 miles per hour every 6 minutes would generate a noise level of about 65 dBA at a distance of 50 feet from the track if it had the tie-and-rail at-grade type of tracks. The same train traveling on elevated tracks would generate about 70 dBA. Overall, at-grade-track trains produced about 5 dBA less than a train traveling at the same speed on an elevated track. At-grade-track trains also produced less vibration. If headways were reduced from 6 minutes to 2 minutes, noise levels would increase by about 5 dBA. There will be a difference of about 8 dBA between trains traveling at 36 miles per hour and those traveling at 70 miles per hour (which represents the average low and high operating speeds). The row of houses closest to the tracks acts as a shield for the houses in the rows behind. The first row, in fact, absorbs most of the impact. Noise from nighttime BART operations is more noticeable than the daytime noise because community background noise drops by 8 to 10 dBA at night.

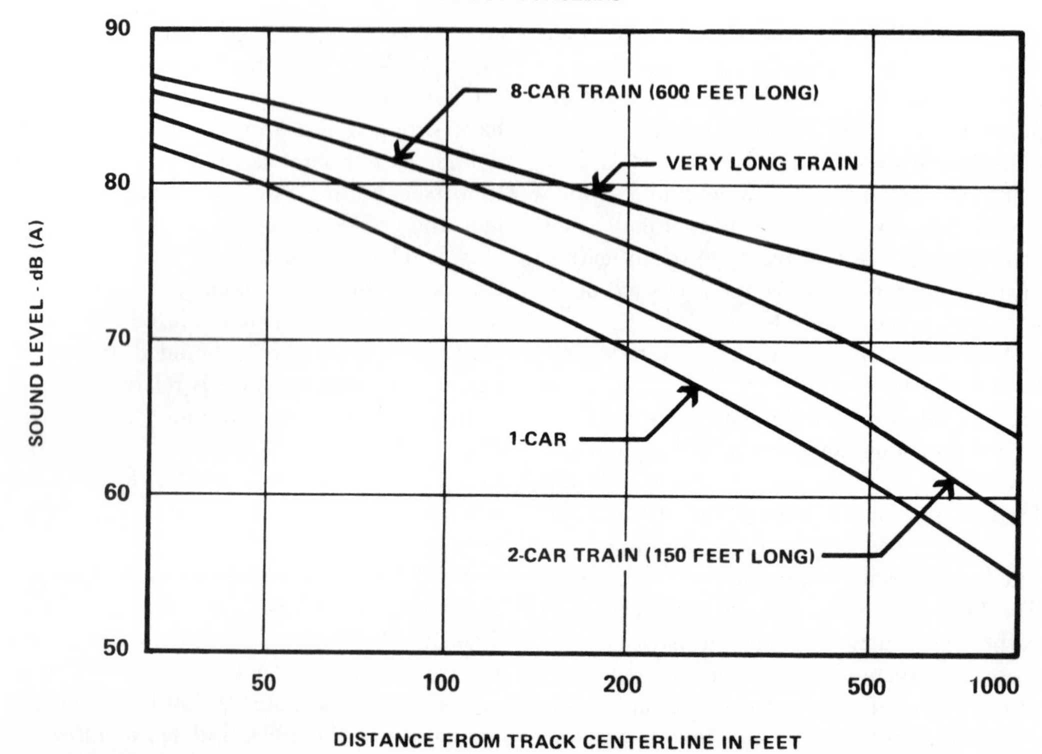

Figure 14-4. Wayside noise level for transit trains of various lengths (at 40 mph). Note: Assumes welded rail and appropriate rail and wheel maintenance to maintain true rolling surfaces. From *Transportation Noise and Its Control,* p. 17, by U.S. Department of Transportation, Washington, D.C.: U.S. Government Printing Office.

Design for Noise Mitigation

Many measures may be taken to control noise pollution. Sound is not greatly reduced by vegetation, but barriers such as earth berms, walls, hills, or other buildings do absorb noise. Focusing on the source of sound (such as improved engine and exhaust design or rubberizing rail transit wheels) may also reduce noise a great deal (U.S. Department of Transportation, Urban Mass Transportation Administration, 1974).

Land use control is perhaps the most sensible method of reducing noise impacts. If residential areas were not located near freeways, if open space, commercial, or utility areas separated highways from residential areas, then traffic noise would not pose a problem. Ownership of land facilitates controls on the part of local government. Building and health codes are an obvious form of control; regulating building materials and design, such as the orientation of housing away from noise sources, is effective. Control over lot subdivision, zoning, and the use of tax incentives comprise useful tools in noise control (U.S. Department of Transportation, Federal Highway Administration, 1978).

Airports pose great noise problems, mainly because housing developments are now located too close to them (Los Angeles International, O'Hare in Chicago, New York's J. F. Kennedy, and Logan in Boston are examples). Land use planning should preclude residential development adjacent to airports and encourage compatible uses such as industrial and commercial activities which are airport-related.

AIR QUALITY

Continued population increase and the growth in consumer demand is directly related to

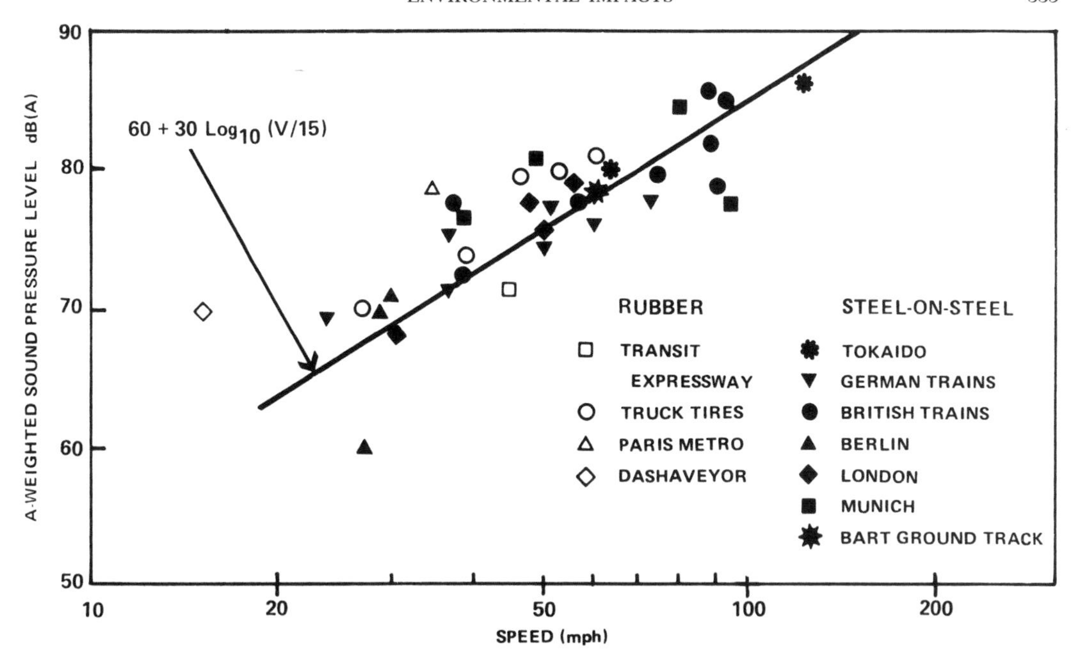

Figure 14-5. Noise from welded track and rubber tire systems normalized to single car at 50 feet. From *Transportation Facility Proximity Impact Statements* by the California Department of Transportation, 1979, Sacramento: State of California.

increased air pollution. Pollution is "an undesirable change in the physical, chemical, or biological characteristics of our air, land, and water that may or will harmfully affect human life or that of any other desirable species, or industrial process, living conditions, or cultural assets; or that may or will waste or deteriorate our raw material resources."

Although increases in local air pollution concentrations are not always readily perceived by the affected population, it is certain that adverse health effects can occur when air quality is low. Sulphur oxides exacerbate respiratory conditions in humans, and carbon monoxide can contribute to psychomotor dysfunction and have a serious effect on the cardiovascular and respiratory systems. The effects of air pollution on receptors is thus considerable (Table 14-1). Transportation is the major source of air pollution, putting over 100 million tons per year of carbon monoxide (CO), hydrocarbons (HO), and oxides of nitrogen (NO) into the air.

Great amounts of pollution are added to the air in the United States each year, including some 90 million tons of carbon monoxide, 70% of which comes from highway vehicles. CO is not nearly as harmful as SO_x, however, and vehicles contribute very little of the latter. Vehicles do add significant proportions of nitrogen oxides and hydrocarbons (33%) and 6% of the particulates. In total, 60% of the pollutants in U.S. cities is transportation-related, and almost all of this amount comes from the private automobile (U.S. Environmental Protection Agency, 1978). The various kinds of pollutants have different harmful effects, but the interaction effects of combined pollutants can be far more harmful than each pollutant individually (for example, photochemical smog).

Transportation sources also account for significant amounts of air pollutants such as sulfur, dust, rubber, asbestos, and lead. Not only is carbon monoxide the most abundant pollutant by weight, but it is also the most lethal of the gases produced by urban transportation sources because it attacks and kills living cells. As it is odorless, tasteless, and

Table 14-1. Effects of some major air pollutants.

	Effects on humans	Effects on vegetation	Effects on the atmosphere
Sulphur oxides and suspended particulates (dust)	Exacerbation of bronchitis and other respiratory conditions. Excess of mortality among those with previous respiratory or cardiac conditions.	Plants most sensitive to sulphur pollution are those with leaves having high physiological activity—alfalfa, grains, squash, cotton, grapes, white pine, apple, and endive.	Dust may reduce radiation and lead to reduction in earth surface temperature. Sulphur dioxide combines with atmospheric water to produce a corrosive and visibility reducing mixture of dilute sulphuric acid droplets. The acid rain falling in Scandinavia is due to sulphur derived from fossil fuel combustion in western Europe. SO_x forms acid rain.
Carbon monoxide	Effects on psychomotor functions. Effects on cardiovascular system. Potent inhibitor of hemoglobin causing asphyxia. Increases heart rate.	Effects respiration in plants.	
Photochemical oxidants (includes Ozone and PAN: peroxyacyl nitrates)	Respiratory system. Eye, nose and throat irritation. O_3 in acute form causes lung edema and irreversible emphysema.	Cause lesions on leaves of plants. Decline in agricultural crop yields. Ozone affects field crops and trees. PAN toxic to many vegetables and ornamental plants. O_3 is main cause of chlorosis and early needle drop in pine.	Ozone in the stratosphere absorbs solar ultraviolet radiation and is essential for life on earth. O_3 and PAN are photochemical products of NO_x and unburned hydrocarbons. They are "smog."
Nitrogen oxides	Respiratory tract and lung infections. Eye and nose irritation.	Restrict growth and cause injuries similar to those caused by SO_2. Acid rain kills trees.	NO reacts with oxygen to form NO_2. NO_2 being brown in color reduces visibility. NO_x forms acid rain, acidifies lakes causing fish kills. NO_x in upper atmosphere lowers ozone shield raising ultraviolet light levels.
Lead	Effects on metabolism, blood and kidneys. Lead accumulates in the body more rapidly than it is excreted, causing anemia and mental retardation if injested by children. 90% of our lead comes from air; only 5% from food.	Accumulates on plants close to sources of emission, e.g. in gardens along major motorways and main roads.	Lead can be carried long distances like SO_2 and fly ash. The history of airborne lead can be "read" in arctic snowfall cores.
Fluorides	Eye irritations, nose bleeds, inflammations of the respiratory tract and severe difficulties in breathing.	Cause withering of shrubs and damage to flowers and citrus groves. Low concentrations of fluoride damage gladioli, pine trees, stone fruit, and azaleas.	Can be carried long distances. Fluorocarbons, especially CCl_2F_2, destroy O_3 in stratosphere.

Note. Adapted from Douglas, 1983, and expanded by author.

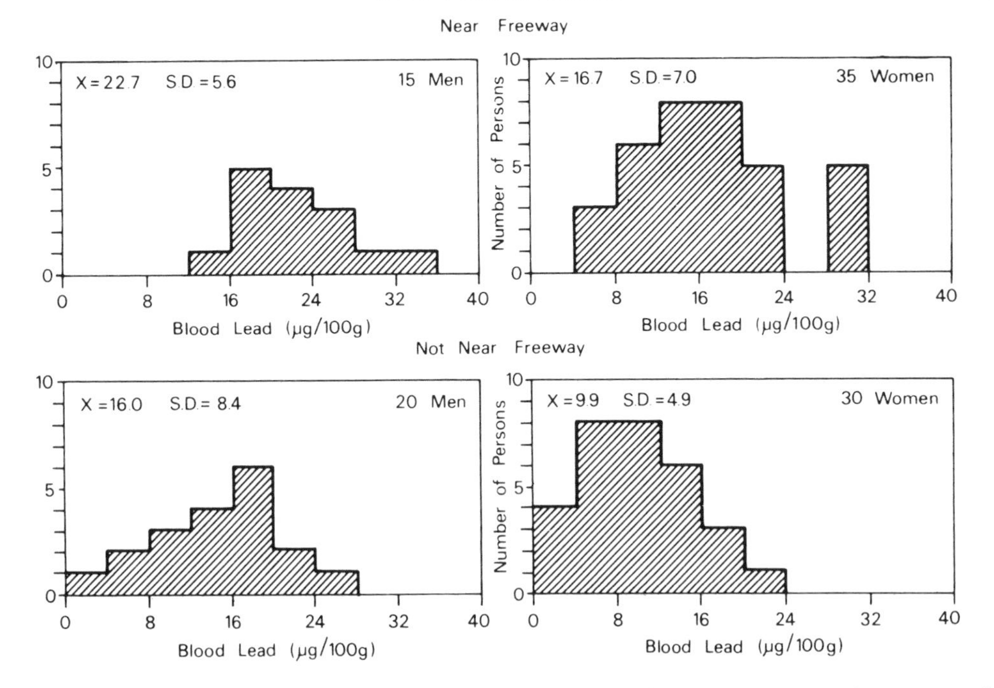

Figure 14-6. Lead in blood near freeways. From *The Urban Environment* by I. Douglas, 1984, Baltimore: Edward Arnold. Copyright 1984 by Edward Arnold. Reprinted by permission.

colorless, it is difficult for humans to perceive or to measure. Lead is the most troublesome of the solids because it can poison the circulatory system of living organisms. The increased levels of lead in the bloodstreams of people within 1 mile of freeways is shown in Figure 14-6. Lead levels in "clean air" average .001 parts per billion (ppb), average U.S. urban air 1 to 4 ppb, air at 45th Street in New York City 8 ppb, plants within central cities 60,000 ppb, soil within 100 feet of freeways 100,000 ppb, and earthworms within 100 feet of freeways 300,000 ppb (Ehrlich, Ehrlich, & Holdren, 1977).

The Clean Air Amendments of 1970 set air quality standards on a national level (Table 14-2). Primary standards are those a state must meet within 3 years after its implementation plan is approved by the EPA. Secondary standards must be attained by the state within a "reasonable time" after the implementation plan has been approved by the EPA. In addition, each state has air quality standards of its own. Table 14-2 gives a comparison of federal and California standards. The standards are listed as maximum parts per million (ppm) acceptable within a corresponding amount of time.

Air quality standards set quantitative limits on the various pollutants. For example, that of CO set by the EPA is 9 ppm maximum 8-hour concentration, and 35 ppm maximum 1-hour concentration may not be exceeded more than once per year.

The Clean Air Act Amendment (1977) set more precise air pollution standards. Every state is now required to have a state implementation plan, and the EPA provides funding in programs for "nonattainment" areas, areas in which air standards are already violated. Local and state governments are required to take the necessary steps in funding and compliance enforcement, so that standards are met. Areas at present not violating standards, especially in natural conservation areas, may be designated as zones of "prevention of significant deterioration." Air pollution concentrations have been reduced in some U.S. cities, and this is largely due to controls over vehicle emission.

Table 14-2. Ambient air quality standards applicable in California

Pollutant	Averaging time	California standards	Federal standards[c]	
		Concentration	Primary[a]	Secondary[b]
Photochemical oxidants (corrected for NO_2)	1 hour	0.10 ppm	0.08 ppm	Same as primary standards
Carbon monoxide	12 hours	10.00 ppm	—	Same as primary standards
	8 hours	—	9 ppm	
	1 hour	40.00 ppm	35 ppm	
Nitrogen dioxide	Annual average	—	0.05 ppm	Same as primary standards
	1 hour	0.25 ppm	—	
Sulfur dioxide	Annual average	—	0.03 ppm	0.02 ppm
	24 hours	0.04 ppm	0.14 ppm	0.10 ppm
	3 hours	—	—	0.5C ppm
	1 hour	0.50 ppm	—	—
Suspended particulate matter	Annual geometric mean	60 $\mu g/m^3$	75 $\mu g/m^3$	60 $\mu g/m^3$
	24 hours	100 $\mu g/m^3$	260 $\mu g/m^3$	150 $\mu g/m^3$
Lead (particulate)	30-day average	1/5 $\mu g/m^3$	—	—
Hydrogen sulfide	1 hour	0.03 ppm	—	—
Hydrocarbons corrected for methone	3 hours (6–9 a.m.)	—	0.24 ppm	Same as primary standards
Visibility-reducing particules	1 observation	Visibility to 10 miles when the relative humidity is less than 70%	—	—

[a]National primary standards: The levels of air quality necessary, with an adequate margin of safety, to protect the public health. Each state must attain the primary standards no later than three years after that state's implementation plan is approved by the Environmental Protection Agency (EPA).

[b]National secondary standards. The levels of air quality necessary to protect the public welfare from any known or anticipated adverse effects of a pollutant. Each state must attain the secondary standards within a "reasonable time" after implementation plan is approved by the EPA.

[c]Federal standards, other than those based on annual averages or annual geometric means, are not to be exceeded more than one per year.

Note. From *Transportation Facility Proximity Impact Statements* by the California Department of Transportation, 1979, Sacramento: State of California.

Air Quality Assessment

As with noise control there is an accepted method of assessment for air quality impacts, using an initial evaluation followed by a refined evaluation. In the initial evaluation process the first step is to estimate the influence zone and define it in terms of distance from a transportation route or facility. Areas and facilities within the zone of influence that are likely to be air sensitive, either directly or indirectly, are identified and mapped. Next, the transportation alternatives are compared and rank-ordered according to the potential air quality impact findings in the previous step. Because this initial evaluation is intended to identify possible major problems, any spatial clustering of impacts identified by the mapping will be particularly significant (California Department of Transportation, 1979).

Once the comparisons have been made and route alternatives defined, the refined evaluation is made. A computerized microscale analysis is used to calculate localized concentrations of various pollutant levels. (Current technology allows accurate measurement of CO levels only. With the advent of emission control devices on automobiles, this assessment is rapidly decreasing in importance while ways to measure the levels of other gases and particulates are increasing in importance.) The numbers and characteristics of the population living within the zone of influence are then calculated.

The final step is to identify those groups that would be subjected to excessive pollution levels or levels that are significantly greater than they are at present. This step will benefit greatly from improvements in the quantification of factors related to air pollution.

Although air pollution is probably more of a health hazard than noise pollution, the former is more difficult to perceive. This is apparent in Figure 14-2, which shows that complaints about odor are not as intense as are complaints about noise. The residents of the first row of houses next to an air polluting facility evidently notice the pollution much more than those living in the next row of houses. Moreover, the decline is more abrupt than it is with noise pollution. A method of predicting impacts that can show residents what specific levels they can expect and what hazards these levels are capable of producing will improve the process (U.S. Environmental Protection Agency, 1979).

Air Quality Models

As mentioned previously, the only pollutant that can be quantified easily is carbon monoxide (CO). There are microscale models that can simplify the process of estimating CO concentrations. A microscale model is one that determines the influence of emissions within about 0.3 kilometers from the source. Beyond that area, the emissions from a single transportation project would probably be negligible. One of the models is the CALINE model, which uses a simple graph to make a quick, rough estimate of the CO concentrations (Benson, 1981). A number of factors are assumed in this process and can be thought of as constants. It is assumed that the highway is 30 m wide and an at-grade road that is straight and level with four paved lanes and a median that is less than 10 meters wide. An atmosphere stability (wind speed) class and wind angle must be known and then the appropriate graph selected (Figure 14-7).

The CO concentration for a receptor distance is determined by reading Figure 14-7 horizontally across from the emission factor to that distance and vertically down to the CO concentration. For example, CO has a G/mile of 50 at a wind speed of 4m/sec. This raw value of 16 on the graph is then adjusted up with increasing traffic volumes, or down with increasing wind speed. When receptor distances are not known, linear interpolation may be used. Such graphs as this are used for other air pollutants as well. Each pollutant has its own emissions factor (G).

Monitoring air quality reveals that concentrations are greatest (1) along transportation arterials (for example, CO was above 20 ppm during peak hours on highways, far exceeding the standards), and (2) during winter. There is also a great fluctuation in concentration according to topography, atmospheric conditions, and the time of day (peak hour).

Emissions are affected by traffic speed and by the amount of stopping and starting; improved traffic flow greatly reduces emissions. Nitrogen oxides, however, increase as speed increases. Automobile maintenance has, contrary to most opinions, little influence on hydrocarbon emissions (U.S. Environmental Protection Agency, 1978).

Air Pollution Mitigation

Once the evaluation process is complete, mitigating measures can be examined and the appropriate ones chosen and implemented. Air pollution concentrations are affected not only by vehicle miles traveled (VMT) but also by factors such as atmospheric conditions, the location and structure of freeways, topography, and land use development. Although VMT may have a little influence if other local factors such as wind speed are relatively more important, because VMT can be controlled most easily, many of the control measures are directed at reducing VMT. These measures include increasing fuel prices; improving parking control (and pricing), idling controls, vehicle retrofits; and

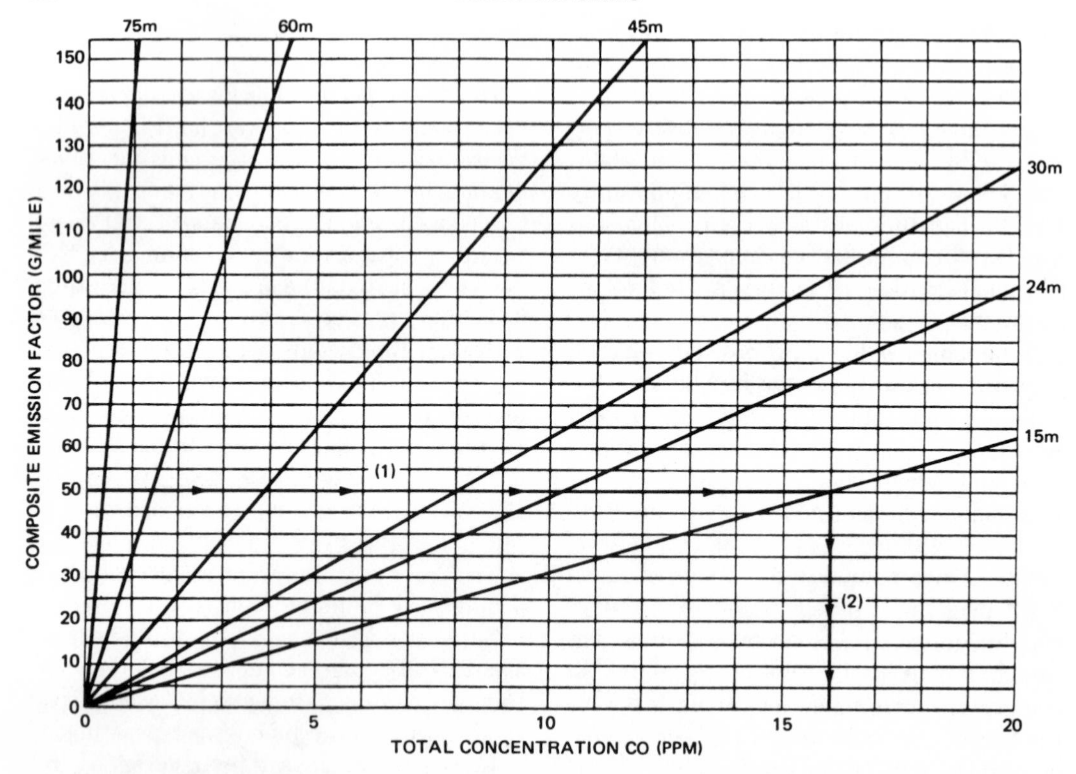

Figure 14-7. Nomograph for determining unadjusted CO concentration at various receptor distances (m) (stability class F; wind angle at 10°). From *Caline 3—A Graphical Solution Procedure for Estimating Carbon Monoxide (CO) Near Roadways* by P. E. Benson, FHWA Technical Advisory #T-6640.6, Washington, D.C.: U.S. Department of Transportation, Federal Highway Administration.

promoting carpooling, periodic bans on driving, inspection and maintenance, and staggered work hours. Hydrocarbon emission would particularly be reduced by improved inspection and maintenance, whereas increased transit and carpooling (freeway bus and HOV lanes) would reduce CO emissions the most (U.S. Department of Transportation, Federal Highway Administration, 1980).

Land use planning and transportation planning should be coordinated, so that (1) residential areas experience the lowest levels of pollution, (2) VMT are minimized via efficient transportation networks, and (3) transit is encouraged. Consideration of freeway location and design offers the best chance to improve air quality without changing the structure of American society. Proposed highways can be shifted to alternate sites if the original site will affect sensitive groups or facilities. Freeway design can mitigate the impact of air pollution by using, for example, earthen berms, dense landscaping, and elevated or sunken highways.

OTHER IMPACTS

Heat Island Effect

Closely related to the air pollution problem discussed above is thermal pollution of the atmosphere by urban transportation. The urban heat island effect is the phenomenon of higher temperatures within the city center compared to the surrounding, less densely developed countryside. Transportation systems are great contributors to this phenomenon. The widespread use of asphalt and concrete paving has provided a sizable increase in available heat storage media. When winds are weak, there is a flow of air from the country to the city; this flow is strongest in the morning and early evening hours. Temperatures within the city rise faster because of increased available heat storage from paved transportation surfaces and buildings. This heated air rises, drawing in cooler air from the surrounding countryside. The "urban winds" create a "dust dome" which extends beyond the urban fringe. Air is warmed, polluted, and reduced in humidity and in circulation (Landsberg, 1981).

Figure 14-8 shows this effect of land use on surface temperature. Three aerial infrared photographs have been taken of downtown Baltimore, Maryland. The light areas indicate higher temperatures and the dark areas indicate lower temperatures. As early as 6:00 A.M., the streets of the downtown area are clearly delineated and are warmer than surrounding areas. The streets appear to be comparable in temperature to the water in the harbor (the light area slightly to the left of center). The temperature of the harbor water should remain nearly constant throughout the day and night, thus providing a point of reference. By 10:30 A.M., the areas between the streets are warmer than both the streets and the harbor. By 1:45 P.M., the only remaining cool areas are the harbor and occasional open spaces within the city.

Effects on Water Resources

The development of urban transportation systems has had considerable impact on water resources. A high percentage of paved surfaces, especially in heavily automobile-dependent areas, has caused significant disruptions to the natural hydrologic cycle. Paved streets, freeways, parking lots, and airports cover much of urban space—from 20% (in suburban areas) to 70% in CBDs such as Los Angeles.

The widespread use of pavement has caused great reductions in the amount of water that can penetrate the ground surface and enter groundwater storage areas or aquifers. This reduction of penetration and concomitant increased use of paved storm drains and stream channels has meant that the aquifers are no longer being replenished as quickly or as regularly as they once were. The loss of infiltrating water has led to depletion of groundwater supplies, which can create problems such as contaminated fresh-water aquifers in coastal areas where excessive pumping has allowed penetration by brackish marine water (as in Galveston, Texas, or on Cape Cod, Massachusetts). Land subsidence is especially prevalent in areas where there has been active pumping of groundwater. In Tokyo, over 60 km of land lie below sea level because of extraction of groundwater and subsequent soil compaction. Areas of Bangkok, Thailand, are sinking at rates of 4 to 14 cm per year. Some areas of the city dropped from above sea level to below it in only four years (Douglas, 1984).

In areas with a high percentage of impervious ground surface, there is also a higher-than-natural rate of evaporation. Water that would normally soak into the ground is held above ground and exposed to sun and air, making less water available downstream from the developed region.

Increased use of paving also has serious effects on the reaction of a stream basin to increased water flow; the flood surge is greater as less water is able to penetrate the ground. The peak of the floodstage is reached sooner, and the flow rate drops off faster with increased paving coverage. This more rapid return to base flow means there is less time available for flood waters to be absorbed into water reserves.

In addition to the quantity of water in a stream channel, the quality of that water may also be seriously affected by the development of transportation routes, especially during the construction phase and later by the application of road salts. When large areas of land are stripped of vegetation, the rate of soil erosion increases sharply. Eroded soil enters the stream flow and is responsible for poor water quality and increased siltation. Soil erosion is, therefore, also a nontrivial by-product of transportation projects and its effects can be predicted. Soil erodibility is a function mainly of size of soil particles, com-

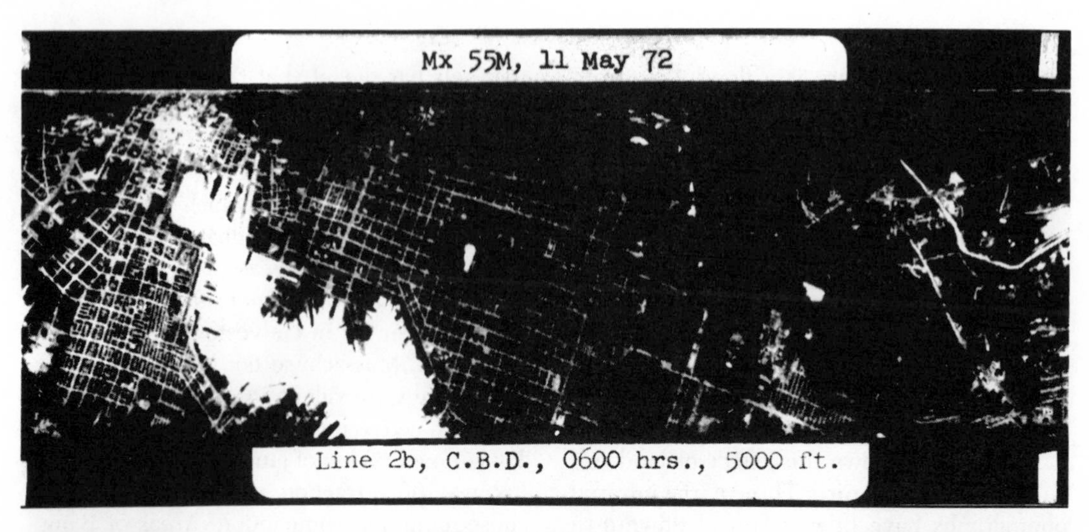

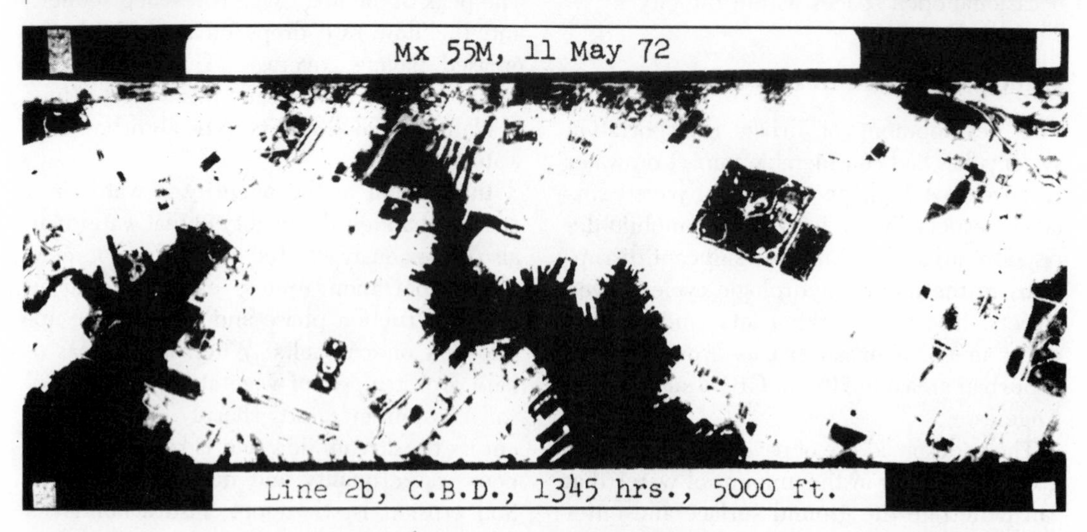

Figure 14-8. Thermal infrared images of downtown Baltimore show warm areas (bright shading) and cooler areas (dark shading). Contrasts exist by time of day and by built up areas versus urban green areas. Photos by John Lewis and Robert Pease for USGS.

paction, the amount of organic matter available in the soil and moisture content.

Effects on Plant and Animal Life

Establishment of transportation corridors also has noticeable effects on local plant and animal communities. Cut-and-fill grading may affect the quality of soil and therefore the nutrients available to plants. Removal of long strips of native plants along a right-of-way leaves a disturbed environment that is well suited to invasion by weedy species. Roadside landscaping is often intentionally undertaken with introduced species as well, thus leaving little opportunity for slower-growing native vegetation to reestablish itself. This disruption of the natural ecological succession in the plant community is equally disruptive to the animal community. The animals found in a well-established stand of native vegetation are not able to subsist on the weedy species of plants that will take over. These disruptions may reduce the number of both plant and animal species present. However, at least two counterintuitive results suggest that freeway construction may actually increase plant and animal proliferation. Plant seeds are known to be pulled along traffic corridors by the vacuum of moving vehicles, and certain animals, such as the white-tailed kite (a bird), once near extinction, have spread via the freeway right-of-way, which offered fencing against predators but not against rodents serving as food.

Visual Impacts of Urban Transportation

Transportation developments often have negative visual impacts. The pleasure experienced from a view may be reduced by unsightly land uses, overhead utility lines and poles, and excessive signs. Sign control programs are common nowadays; for example, the FHWA subsidizes state departments that control billboards along the interstate highway system. Sign control is also necessary to reducing accidents, as too many signs contribute to confusion. More efficient communication with drivers is required.

Freeways reduce visual pleasure by their actual physical structures which reduce natural harmony, and also indirectly by the land use developments they stimulate. The visual impact of freeways, tracks, and large parking lots may be mitigated through the use of appropriate landscaping and vegetation. The division of large lots into smaller areas separated by greenery and the use of islands improve the visual effect and concomitantly driver orientation and traffic flow (U.S. Department of Transportation, Urban Mass Transportation Administration, 1983).

Planning and design of freeways for pleasurable visual qualities is important. Some principles in this regard are (Simonds, 1970):

1. Freeways entering the city should expose the most interesting and distinctive features of the city.

2. Freeways should provide a pleasant driving experience to reduce stress and improve wayfinding.

3. Beauty is the result of careful consideration of the cumulative effect of details such as scale, location, structural simplicity, alignment, landscaping, and harmony with both the natural and architectural environment.

4. High-speed travel means that views directly ahead must be emphasized. A narrow angle of vision and a focus that is relatively far ahead must be considered; views that are not fragmented on gently curvilinear routes are thus best.

5. The view from the freeway is as important as the view of it from the city or adjacent communities. The freeway should blend with the community and the cityscape; attention to topography, location along existing boundaries between communities or land uses (for example, in canyons), the street pattern, vegetative screening, and landscaping is essential.

MEASUREMENT AND VALUATION OF ENVIRONMENTAL IMPACT OF TRANSPORTATION DEVELOPMENT

The assessment of environmental impacts of any transportation project is usually required by law in order to promote optimal development planning, that is, planning that maximizes benefit but minimizes adverse effects on the environment. But no best method for such assessment has been developed, and methodologies vary from agency to agency. This lack of prescription in method has encouraged research into different ways of assessing and evaluating impacts. Measurement and valuation would enable the overall effects, both negative and positive, to be assessed rationally in the process of decision making. Inadequate attention has been paid to the consistent valuation of environmental quality variables; most assessments provide inadequate information upon which to base planning decisions and as a result planning is highly subjective. There is an urgent need for valuation and translation of impacts into monetary terms—terms that are easily interpretable, that are additive in that different impacts may be added together to find total costs and benefits of a project, and that facilitate comparison of alternative projects.

Many environmental impacts are not easily quantified, but given the shortcomings of qualitative assessment, it is essential that research proceed to find more credible and useful methods. Figure 14-9 describes the process of assessment and the role of measurement and valuation in improving the process, ultimately aiming for improved planning decisions and optimal resource use.

Environmental Impacts of Transportation Developments

There are three methods of environmental impact assessment. The first is the expert committee using a Delphi technique (general agreement), as well as checklists and rating matrices. Though the least scientific, it is still the most commonly used approach. The second method is suitability mapping or the overlay technique. It uses information gathered on each geographic cell of the study area and creates a geographic information system. Overlays are weighted by cell according to land attributes and location and may be produced manually (but more frequently by computer). Maps show the highest and lowest impact areas (Figure 14-10). Because of the massive data inputs and computer time required by this technique (which was developed for major government projects and private utility corridor planning), it is just starting to be used at the state and local levels (Stutz, 1983; NewKirk, 1982). The third technique is benefit/cost analysis, an effort to add valuation in a monetary framework to each impact.

Because this discussion is concerned with valuation, it focuses on benefit/cost analysis. Different methods are suitable for different applications, but what is most important in assessing complex environmental interactions is that a consistent, systematic method be used to facilitate decision making. Proper analysis of impacts requires detailed understanding of how the transportation project will affect certain "receptors." A natural system such as an urban park may be the receptor of transport impacts if construction activity is planned there. A list of receptors is given in Table 14-3. The changes in the quality of the system, the extent of the exposure to the changes, as well as the spatial and temporal distributions of the changes, should be specified. Only then can the precise physical effects on the receptor be estimated. Often difficulties in estimation are experienced. Spatial and temporal patterns may fluctuate, for example, air pollution concentration from a line source such as a highway. Effects on people vary with perceptions of the positive or negative values associated with a project. Despite difficulties, accurate estimation is possible with a number of approaches.

Statistical Models

Models simplify the real processes and changes in the environment by focusing on a few of the many interrelated variables. Selected

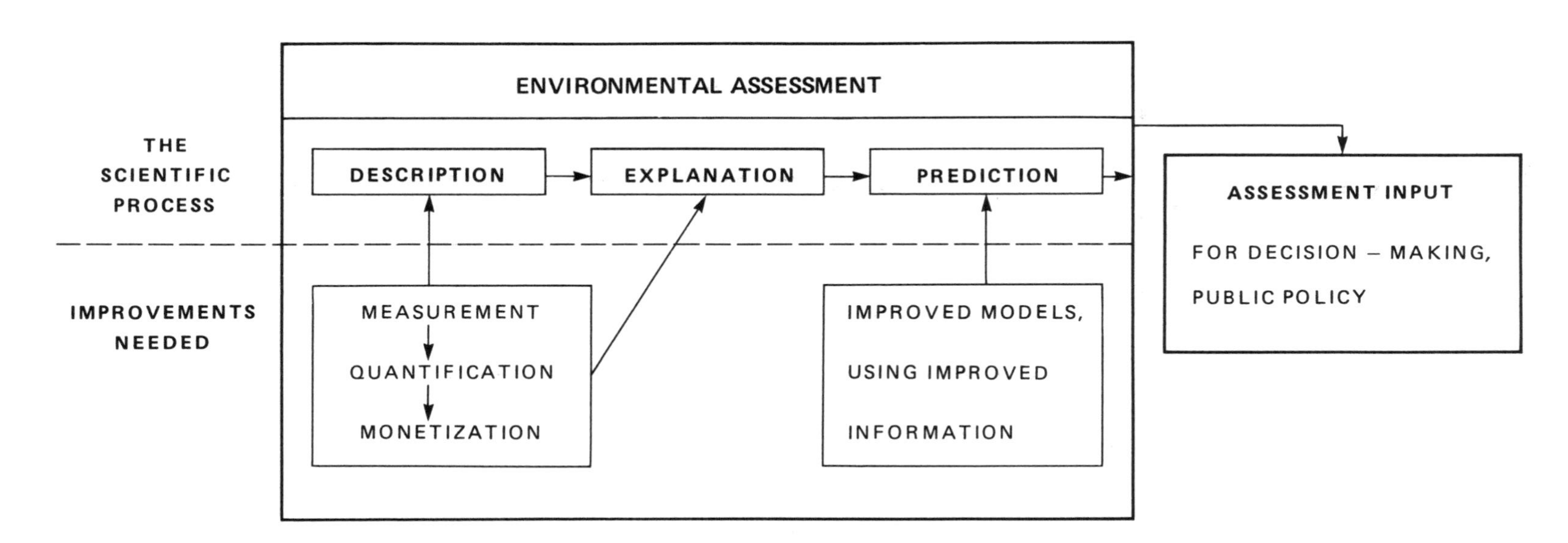

Figure 14-9. The environmental assessment process is sorely in need of improved data, models, and measurement techniques.

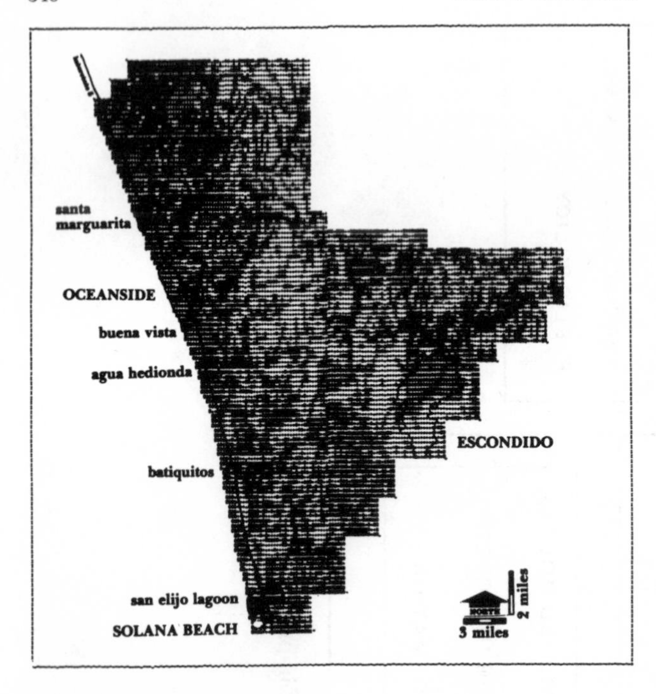

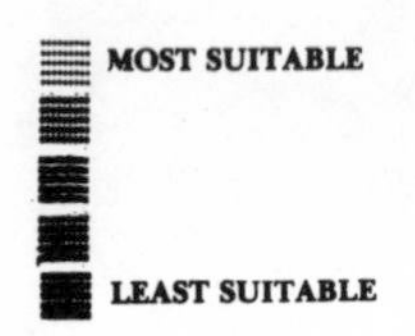

Figure 14-10. An urban suitability model on the regional scale of San Diego, California. From "Land Use Suitability Modelling and Mapping" by F. P. Stutz, 1983, *Cartographic Journal, 20*, p. 39–49. Copyright 1983 by Cartographic Journal. Reprinted by permission.

Table 14-3. Typology of receptors

Receptors	Stationary	Mobile within fixed habitat	Mobile
Resident fish species		X	
Migratory fish species			X
Resident animal species		X	
Migratory animal species			X
Natural vegetation	X		
Materials in structures	X		
Materials in vehicles			X
Agricultural and forestry activities	X		
Industrial and commercial activities	X		
All other activities	X		
Residences	X		
Humans			X

Note. From *Environment Natural Systems and Development: An Economic Valuation Guide* by M. M. Hufschmidt, 1983, Baltimore, Md.: Johns Hopkins University Press. Copyright 1983 by Johns Hopkins University Press. Reprinted by permission.

variables can be mathematically related and used to predict a particular change. Models range from simple qualitative cause–effect relationships to quantitative correlation and multiple regression models. In regression models of environmental assessment, the dependent variable *Y* would be some measure of environmental quality, such as air or noise pollution levels, while the independent variables would be measures believed to affect *Y*, such as traffic volume or distance from a highway. Regression has been used to estimate human health responses to "doses" of air pollution and chemical residuals (Hufschmidt, 1983).

Selection of Approach or Model for Impact Assessment

Choice of methodology depends upon the constraints within which the researcher is working such as time, level of expertise, finances, and data availability. Hufschmidt (1983) suggests a systematic way of selecting an appropriate method, using four criteria: (1) the level of accuracy required, (2) the ability to measure receptor response to different project impacts, (3) the extent of commonality a method has with previous studies in the area, and (4) future usefulness of the method in the area. The complexity of the model that is used will affect its accuracy, cost, the amount of data required, and the extent to which the data requirements can be fulfilled to calibrate and verify the model. For example, complex models may require more detailed data than are available (Hufschmidt, 1983). These relationships are shown in Figure 14-11. If a model is unnecessarily complex, it will be difficult to interpret.

Valuation of Impacts in Monetary Cost and Benefit Terms

Rational decisions involving project-alternative selection are based on economic efficiency. Neoclassical welfare economics, upon which the valuation techniques discussed here rest, is based on the "economic man" assumption—that people want to maximize the profit to be derived from any project, that is, to maximize the "surplus of benefit over costs" (Barlowe, 1978). The project alternative with the greatest surplus should be selected if economic efficiency in resource use is to be followed. Benefits may be primary (the value of the immediate project effects) or secondary (value derived from effects of a project). Total project costs include costs of construction labor, land and materials, op-

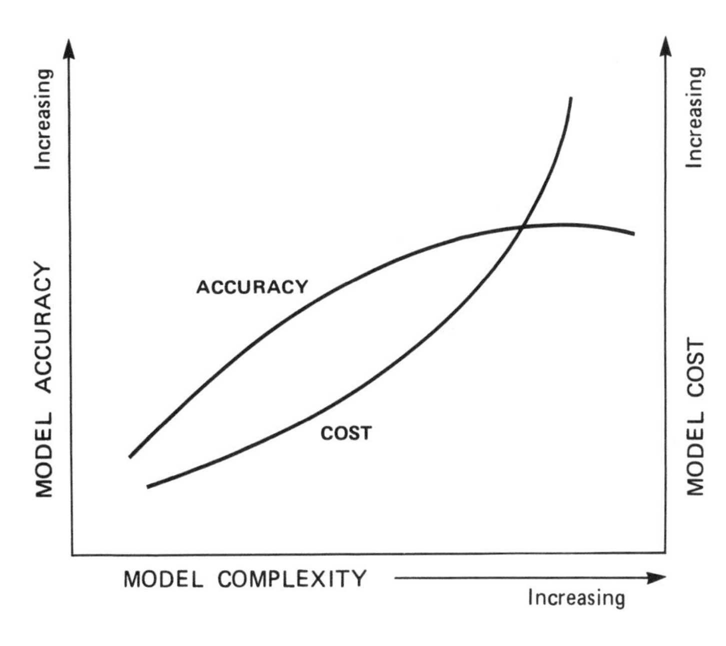

Figure 14-11. Relationships among model complexity, model accuracy, and model cost. From *Environment Natural Systems and Development: An Economic Valuation Guide* by H. M. Hufschmidt, 1983, Baltimore: Johns Hopkins University Press. Copyright 1983 by Johns Hopkins University Press. Reprinted by permission.

erating and maintenance costs, and costs of mitigating adverse impacts.

Market prices are used whenever possible as an indication of value. For example, present prices of the land to be used by a project indicate a value that can be added to the benefit/cost analysis. Future benefits and costs require projections to obtain future price levels. Discounting methods adapted from investment economics enable the present value of a stream of benefits and costs over time to be calculated. Intangible social values, such as aesthetic value or recreation opportunities, provide problems of monetary assessment because their relationship with market prices is not easily defined, but recent applications of welfare economics and microeconomics theory has yielded some reliable methods.

The Benefit/Cost Model

The concept of Pareto efficiency is a crucial but simple criterion to use in benefit/cost analysis. A Pareto-efficient use of resources or project development maximizes social welfare in that any further increases to user benefits for one part of the system would mean cost increases that outweigh the benefit. The comparability of value over time is important in resource allocation decisions, and inter-temporal Pareto efficiency extends the basic concept. A resource development is Pareto efficient over time "if and only if it is not possible to increase the utility of the affected parties at any point in time without decreasing utilities at other points in time" (Howe, 1979).

"Utility" reflects economic welfare and may be measured by willingness to pay prices for having commodities, in this case environmental quality services. Market prices paid indicate willingness to pay, and where individuals do not pay, (for example, for communal commodities like most environmental quality services), prices may be imputed from observing human behavior. Investment economics may be applied to translate streams of future benefits and costs into a present equivalent. Inter-temporal Pareto efficiency is adhered to via the net present value *(NPV)* criterion; *NPV* must be maximized and can be calculated as follows:

$$NPV = \sum_{t=0}^{n} \frac{NB_t}{(1+r)^t} \qquad (1)$$

where *NB* is net benefits (benefits minus costs), at time t (a particular year), r is the discount rate, and n is the time horizon used in the planning of a project. Thus, the sum of the stream is:

$$NB_0 + \frac{NB_1}{(1+r)^1} + \frac{NB_2}{(1+r)^2} + \; , \; . \; . \; . \; , \; + \frac{NB_n}{(1+r)^n} \qquad (2)$$

The discount rate embodies opportunity cost, the innate desire of people to have the benefits in the present rather than to wait. Thus, for each time interval in the future, the value of *NB* decreases as the denominator increases. Alternative projects may easily be compared given their particular *NB* values and time horizons.

The remaining step in the process is economic valuation of impact. Particular attention is focused on the problematic area of intangible environmental values. Valuation may be approached from the perspective of either benefits or costs.

A variety of approaches, each with its own techniques, has been applied to valuation of benefits. They are, in order of discussion:

1. Market value
2. Opportunity cost
3. Human income loss
4. Surrogate market values
5. Attitudinal survey

1. Market Value Approaches

This basic approach is used in most benefit/cost analyses. Changes in environmental quality produced by transportation systems affect the productivity of natural systems causing changes in quantities produced. These quantities can be measured. Economic valuation is possible using the prices of products. For example, a highway may increase air pollution levels, thus lowering productivity of truck farms or nurseries. The reduction in output, valued in loss of goods at their price level, is an indication of the benefit to be derived from pollution abatement measures. Polluted highway runoff may be dealt with in a similar manner if it is thought sufficient to damage groundwater and plant growth.[2]

We have examined the general pattern of air and noise pollution levels around highways. They are significantly higher close to the highway but diminish rapidly with distance until the average levels of an area are reduced. Productivity of natural systems is thus only significantly impaired adjacent to the highway. Reduced output could cause a reduction in land value, a price change that may also be used in valuation. The market price approach relies on measuring productivity of natural (or human) systems. Output is viewed as a benefit, but potential loss of this benefit because of adverse impacts may also be viewed as cost. Freeman (1979) discusses the reduction and measurement of land values in zones near urban transportation facilities.

2. *Opportunity Costs*

This approach is often used when alternative land uses are possible for an area chosen to be used for a transportation development. The value of the land, if it were allocated to other uses such as housing or parks, would be indicated by the land prices or income gained from output. These opportunities for other benefits would be lost; they are thus viewed as costs and should be considered in any land use plan. This approach is very useful when the real social benefits cannot be easily estimated. Opportunity costs represent the largest social benefit forgone, and therefore their measurement is an important part of benefit/cost analysis.

3. *Human Income Loss*

This approach views people and their labor as capital in the production process, and their income lost due to decreased environmental quality is measurable. Pollution and damage to natural systems may cause increased illness, medical expenses, absenteeism from work, and even decreased life expectancy. Disposable income is lost by all these impacts and comprises a forgone value, as an indirect effect of a transportation project.

Increased costs of illnesses has been related to increased levels of ambient air pollution. Schwing et al. (1980) measured health damage function in a benefit/cost analysis of reductions in auto emissions. Air pollution control in urban transportation planning has health benefits which are thus measurable (Baumol & Oats, 1979).

4. *Surrogate Market Values*

Surrogate or indirect values are often used to assess environmental quality benefits where direct market values are difficult to obtain.

Travel cost as a surrogate for unpriced environmental benefits/services is often used in analyses of recreation services. The value of the recreation experience derived from an urban park threatened by a transport project, for example, has no market price, but the total cost the user is willing to pay for the benefit, including travel costs, is a measure of the value. Using survey data, one can calculate a demand curve, showing willingness to pay varying travel costs in order to obtain the experience. The area under the demand curve represents the value of the park. Travel time may be added to travel costs for a more realistic valuation. A similar procedure may be followed to assess the value to users of a transportation facility such as an airport or commuter train station. The

[2] In a recent lawsuit, Caltrans (the California Department of Transportation) was found liable and ordered to make restitution for a drop in property values related to the freeway's (SR-125) proximity to a residence (in particular, a noise increase was evident although not over federal criteria). This is apparently the first time Caltrans has paid damages on a property that was not directly affected by a project. In this case Caltrans has decided to construct a noise barrier.

distance traveled to the facility is the basis for time and energy travel costs; this distance would provide a useful surrogate measure of value, and the distance users are willing to travel to a freeway could establish its value and service area.

The property value, or market prices of land, may be used as a surrogate measure of willingness to pay for environmental services. The present net value of a property may be reduced by diminished environmental quality. The drop in property prices due to air pollution, for example, indicates the value people place upon clean air or their willingness to pay for improved air quality. Precise calibration for the areas affected is necessary to determine the extent of the relationship and the exact demand curve (willingness to pay) for air quality. Not all transportation developments impair overall air quality; many redistribute the pattern or even reduce pollution. A highway may remove congestion from local streets, thus improving local air quality, and hence, property prices.

Regression equations may be constructed showing relationships between housing prices and any number of relevant independent variables including environmental quality variables. The extent to which a particular quality effects price, keeping other variables constant, may thus be found (Freeman, 1979). It is possible to value any change in quality, be it any form of pollution or change in amenity such as accessibility or congestion.

Transportation projects often convert land from open space to other uses. The value of such land, whether recreational, ecological, aesthetic, cultural, wildlife values, is often difficult to determine because people who benefit from present land use are distributed over space and time. Society as a whole, now and in the future, benefits. Land prices paid for such land, or the costs of replacing the lost value (that is, buying other land of equal value for preservation), may prove useful surrogate values.

5. Attitudinal Surveys

In the absence of reliable market prices for valuation, surveys to elicit the extent of people's willingness to pay for environmental qualities are popular. Human preferences and the value they place on any quality are obtained via a variety of questioning techniques. Many theorists regard surveys as the only useful approach in the valuation of intangibles such as historical or aesthetic value, or the value of preserving a rare plant for posterity. The value of a quality may be estimated as the amount people are willing to pay for it, or the amount they are prepared to receive as compensation for its loss. Survey-based valuation methods range from bidding and trade-off games to the simulation of quantity choices of environmental qualities.

Cost Valuation

The financial costs of resources used in a transportation project, costs of mitigating environmental impacts, and the opportunity costs of directing these resources toward other uses are measurable from the cost perspective in benefit/cost analysis. Various methods are used for determining the lower level of environmental protection and for valuing these costs to compare project alternatives. This discussion describes the use of replacement costs and preventive expenditures for valuation.

Replacement-Cost

Replacement-cost techniques measure the cost of replacing lost quality. Such costs may be viewed as the minimum level of demand or benefit. A recent example is provided by a highway development scheme in San Diego which will destroy rare vernal pool habitats. A decision to replace these pools by buying and preserving other pools nearby involves substantial costs, given the high land values in California. The Lavine-Meyberg (1979) energy method also fits into this approach.

Preventive Expenditures

This approach focuses on the willingness of people to pay the costs of reducing adverse environmental quality levels. Lower air or noise quality adjacent to a highway, for example, may be monetized by valuation of people's willingness to move elsewhere (movement costs) or to pay for abatement measures. Hufschmidt (1983) describes a model for decisions to accept movement cost, prompted by noise disutility (Δ), which is the depreciation in the value of property as a result of noise, and by the costs of moving (R), such as searching and hauling. Movement will occur if:

$$N > S + \Delta + R \tag{3}$$

where N is a household's subjective assessment of the disutility of noise and S is the consumer surplus experienced by the household (the household's estimate of the price, for which they would sell the accommodation over and above its market price, for example, rent). Additional costs of abatement, such as noise insulation measures or installation of air conditioning in the case of valuing air quality costs, may be added to the decision-making model. Consider the demand curve for theoretical willingness to pay for abatement of either noise or air pollution. Stress caused by microscale pollution, such as by traffic or automotive emission control, will shift the curve to the left, shifts which can be monetized as changed value of noise or air quality.

Conclusion

A variety of tools and approaches is used in monetization of environmental impacts, as this review has shown. Most are summarized in Figure 14-12. Many of these are imperfect and subject to criticism. This new and expanding field has attracted the attention of many researchers in land, resource, and environmental economics. Methodological improvements in this area may be expected because of the urgency of facilitating decisions about the use of scarce resources. Decision making at present is difficult and in many instances suboptimal and wasteful, owing to poor benefit and cost information and diverse impacts that are not additive. Subjective decisions, easily swayed by individuals and small-scale politics at the planning level are therefore common, rendering many of the impact assessment and benefit/cost analyses useless. A more objective valuation method is vital to provide transportation planners with valid and credible information that will establish common and acceptable ground for decision making. In the absence of monetized value, decisions must still be made. The adherence to minimum standards of environmental quality is always important; in the absence of monetized value, decisions are made in a relative vacuum and thus rely more on such standards as guidelines.

DRAFTING THE ENVIRONMENTAL IMPACT STATEMENT

The environmental impact era for urban transportation began with the National Environmental Policy Act (NEPA) of 1969. NEPA required a written environmental impact statement (EIS) for all major federal development projects. This action was followed by legislation in many states requiring EIS's to be written for major private projects as well. The overall effect on urban transportation projects created by this movement has been to (1) slow the development of new projects, (2) to involve public policy groups in the decision-making process to a greater degree, (3) to increase public awareness and preservation of human and natural communities and environments, and (4) to utilize present transportation infrastructures to a more efficient level (Cohn & McVay, 1982). The drafting of the EIS is now an important early step in most significant urban transportation projects funded by local and state agencies and is required by federal agencies such as DOT, and FHWA and UMTA, which are part of DOT.

Purpose and Need

The first step is to state the purpose and need for transportation improvements (Table 14-4). This step includes methods to identify and describe the transportation problems that the proposed action is meant to improve.

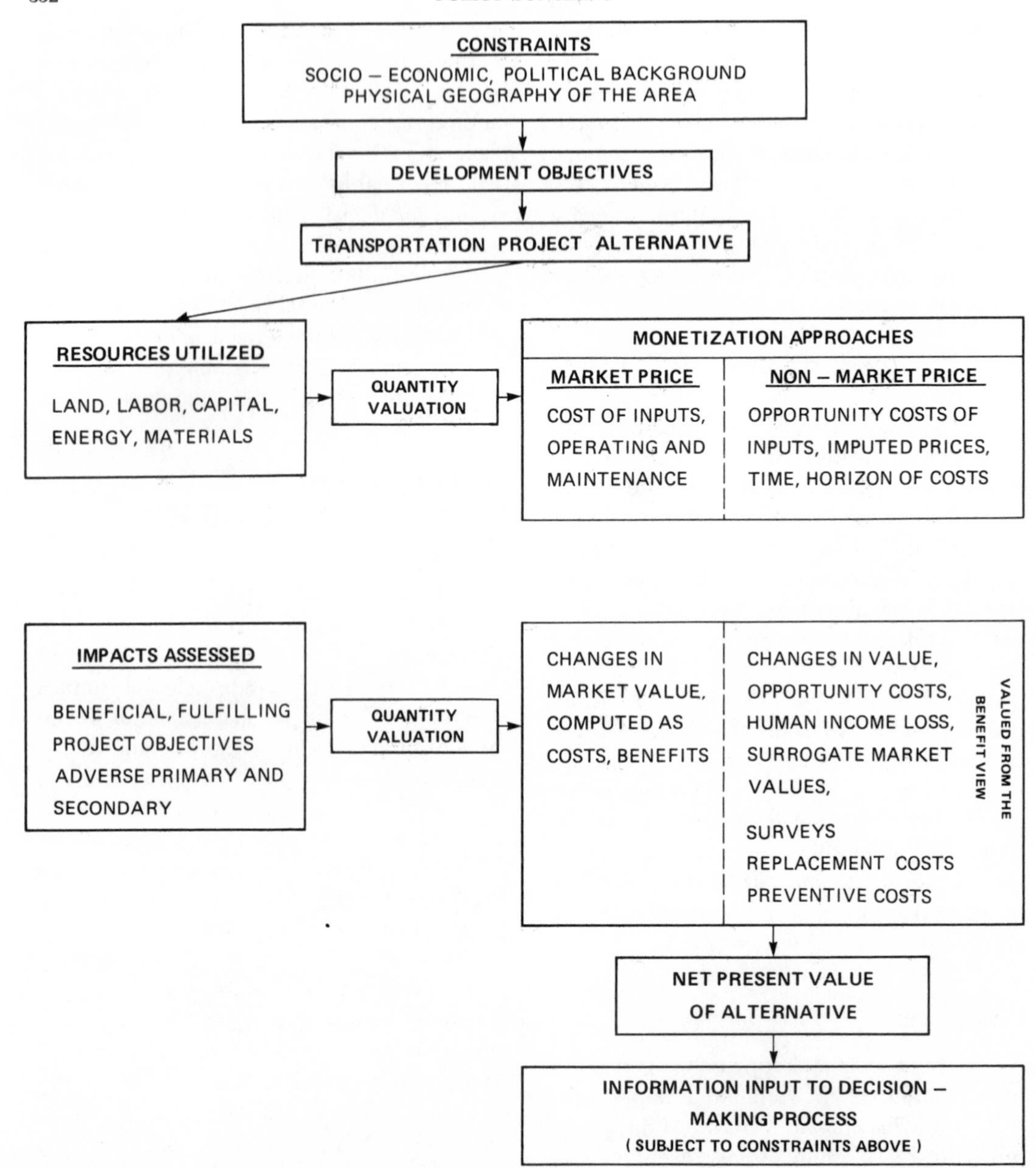

Figure 14-12. Schematic diagram of benefit/cost assessment process for a transportation project.

Following are some broad guidelines that can be used to explain the need for a proposed subject (San Francisco, 1983).

System Linkage. How does the proposed project fit into the overall transportation system? Will it be a connecting link or will it close an essential gap in the present urban system?

Capacity. What capacity and level of service are needed for the present and future traffic levels?

Transportation Demand. What is the current demand and what is its relationship to regional, statewide, or previously adopted urban transportation plans?

Legislative Authority. What is the federal, state, or local government authority (legislation) directing the action?

Table 14-4. Drafting the environmental impact statement

I. Purpose and need for the proposed project
  A. How is project linked to total system?
  B. What future capacities are needed in system?
  C. What are the current and future demands?
  D. What legislative authority will direct project?
  E. What are the economic and social effects?
  F. What are the intermodal relationships of the project?
  G. Will there be overall safety improvements?
II. Alternative projects proposed
  A. No-build, status quo option
  B. Transportation systems management option (TSM)
  C. Small-scale improvement alternative
  D. Medium-scale improvement alternative
  E. Large-scale improvement alternative
III. Affected environment and consequences
  A. Existing social, economic, and physical setting
  B. Expected social and economic impacts and mitigation
  C. Expected relocation impacts and mitigation
  D. Expected land use impacts and mitigation
  E. Expected noise and air quality impacts and mitigation
  F. Expected energy and conservation impacts and mitigation
  G. Stream, river, and floodplain impacts and mitigation
  H. Wetlands and water quality impacts and mitigation
  I. Threatened or endangered species impacts and mitigation
  J. Prime and unique agricultural lands impacts and mitigation
  K. Expected visual impact

Economic and Social Demands. What projected economic development and use changes indicate the need to improve or add to highway capacity, such as new employment, schools, land use plans, recreation areas, residential communities, and the like?

Intermodal Relationships. To what degree will the proposed facility complement and join existing rail lines, port facilities, airports, mass transit services, highways, and streets?

Safety Improvements. Will the proposed project correct existing or potential safety hazards? Will it create new hazards?

Alternatives

The reasonable alternatives are set forth in this section of the EIS, and a concise discussion of how they were selected is presented. It describes other alternatives that were eliminated and why. The types of alternatives that are normally included in this section include the following:

No-Action Alternative. This includes only those short-term minor reconstruction activities that are a part of an ongoing plan for continuing operation of the existing urban transportation system.

Transportation System Management (TSM) Alternative. TSM includes activities designed to maximize the utilization and energy efficiency of the present system. Options such as high-occupancy-vehicle (HOV) lanes on existing roads, vanpools and ridesharing, fringe parking, and traffic signal optimization are possibilities of TSM. Evidence that the project is consistent with the National Energy Policy Act of 1978 promoting bicycle and pedestrian movement is important.

The Proposed Actions. The discussion of several alternatives (Table 14-4) in this section usually includes a brief written description of each alternative, supplemented with maps, photographs, drawings, or sketches, which assist in understanding the various alternatives, impacts, and mitigation measures. It should include a clear understanding of each alternative's termini, location, costs, and major design features and the effect each alternative will have on the natural surroundings or the community. Each alternative is listed in order of increasing cost or scale. The final EIS must identify the preferred alternative and discuss the basis for the selection.

Affected Environment and Consequences

Next a concise description of the existing social, economic, and natural environmental setting for the areas affected by all of the alternative proposals must be provided. All environmentally sensitive locations or features need to be identified. The statement should describe other related activities in the area, their interrelationships, and any significant cumulative environmental impacts.

The most important section is the Environmental Consequences section. This section discusses the probable social, economic, and environmental effects of the alternatives and the measures needed to mitigate adverse impacts. A discussion of impacts in both the draft and final EIS usually includes a summary of the studies undertaken and the major assumptions made. Enough data or cross-referencing is included to allow the reviewer to determine the validity of the methodology. Sufficient information is included to establish whether or not the conclusions on impacts are reasonable. The mitigating measures are described. These measures are usually investigated in detail before the completion of the final EIS. Charts, tables, maps, and sketches illustrating the differences among the alternatives are usually included (U.S. Department of Transportation, 1980). Examples of the potentially significant impacts of urban transportation projects are described below.

The statement should discuss how the proposed project would change neighborhoods or community cohesion for various groups. Changes in travel patterns and accessibility should be discussed, and impacts on school districts, recreation areas, churches, businesses, retail sales, police, and fire services must be analyzed. In particular, the effects on groups such as the elderly, the poor, nondrivers, transit dependents, and minorities should be discussed to the extent that impacts can be reasonably predicted. Relocation information for both households and businesses, which is necessary for the draft EIS, must be summarized in enough detail to explain the relocation situation along with a resolution of anticipated or known problems.

A description of current development trends and state and/or local government plans and policies on land use must be included. How the proposed project alternatives fit into the area's comprehensive development plan should be analyzed. Secondary social, economic, and environmental impacts of significant induced development should be presented, including historical and archaeologic aspects. Any proposed alternatives that will stimulate low-density, energy-intensive development in outlying areas and that will have significant adverse effects on existing communities should be noted.

The EIS must contain a brief discussion of air quality effects and a summary of the CO analysis if one is performed. The noise analysis in the EIS should include a brief description of noise sensitive areas, including information on the numbers and types of activities that may be affected, the extent of the impact, mitigating measures that can be taken to solve noise problems, and problems for which no viable solution exists.

Both the draft and final EIS should include a general discussion on the energy requirements and conservation potential of various alternatives being considered. The final EIS should identify any energy conservation measures that will be implemented for each alternative (U.S. Department of Transportation, 1980).

The draft EIS should contain a summary of the location and hydraulic design of encroachments on rivers and floodplains. It should also discuss any significant impacts on fish and wildlife resources, including direct impact to fish and wildlife, loss or modification of habitat, and degradation of water quality. Projects within wetlands now protected in the United States must include sufficient information in the draft EIS to identify the extent of the wetlands, describe the impacts of construction, evaluation alternatives that would avoid these wetlands, and identify practicable mitigating measures. The EIS should discuss any locations where road or construction runoff may have a significant effect on downstream water uses, including existing wells. Procedures for estimating pollutant loading from highway runoff must be spelled out. Proposed mitigation measures should include evidence of coordination with the Corps of Engineers. If a selected alternative affects an aquifer, a design must be developed to assure, to the satisfaction of EPA, that it will not contaminate the aquifer.

The transportation authority must request information on whether any species listed or proposed as endangered or threatened may be present in the proposed construction area. If it is determined that an endangered species is present, the distribution of the species, its habitat needs, and other biological requirements, analysis of possible impacts to the species, and an analysis of possible mitigating measures must be made. The EIS should contain a map showing the location of prime and unique agricultural lands in relation to the project alternatives, a summary of results of consultations with the USDA, and copies of correspondence with USDA regarding the project. Measures to avoid or reduce direct and indirect effects on these lands should be identified.

Finally, as assessment of the visual impacts of the proposed project should be included. The draft EIS should be circulated to state and local arts councils that have been officially designated to review impacts of such projects on the aesthetic characteristics of the community.

SUMMARY

Environmental impacts are physical effects from the highway or urban transit improvement that are felt within a short distance. Various types of impacts produce different proximity-sensitive zones. For example, noise effects from a highway tend to dissipate more quickly with distance than air pollution impacts. Environmental impact analysis is the orderly and logical process by which the potential effect of a proposed development project on its immediate and more distant environment is analyzed. Types of analyses may range from impact on animal and plant life to impact on urban health, depending on the nature and location of the development project.

Environmental impact assessment begins after the initial formulation of alternatives (including the "no-build" option), though results of the subsequent assessment may lead to a significant refinement of alternatives or may alter the trade-offs (cost versus service, versus environmental impacts, and so on) between alternatives and thereby alter the final selection. Environmental considerations are important inputs to the development of alternatives, either directly from the goals and objectives, or as a result of the environmental awareness of the project team. It is important to identify environmentally sensitive areas early in the transportation planning process.

Environmental impacts are part of the overall environmental evaluation effort, along with consideration of regional-scale impacts (such as urban sprawl) and other effects (such as displacement and relocation) which fall outside the scope of environmental impacts. Environmental impact assessment forms a vital input to the total assessment process, along with the necessary regional-scale assessments. The ultimate outputs of the environmental analysis should be cycled back to the land use planning elements of the process, since the environmental findings may indicate that significant changes to the underlying land use and transportation assumptions are needed.

References

Barlowe, R. (1978). *Land resource economics.* Englewood Cliffs, N.J.: Prentice-Hall.

Baumol, W. J., & Oates, W. E. (1979). *Economics, environmental policy and the quality of life.* Englewood Cliffs, N.J.: Prentice-Hall.

Benson, P. E. (1981). *Caline 3—A Graphical Solution Procedure for Estimating Carbon Monoxide (CO) Concentration Near Roadways,* FHWA Technical Advisory #T 6640.6. Washington, D.C.: U.S. Department of Transportation, Federal Highway Administration.

California Department of Transportation. (1979). *Transportation facility proximity impact assessment.* Sacramento: State of California.

Cohn, L. F., & McVoy, G. R. (1982). *Environmental analysis of transportation systems.* New York: John Wiley.

Douglas, I. (1984). *The urban environment.* Baltimore: Edward Arnold.

Ehrlich, P. R., Ehrlich, A. H., & Holdren, J. P.

(1977). *Ecoscience: Population, resources, environment,* 3d ed. San Francisco: W. H. Freeman.

Faucett, J. (1974). Valuation and compensability of noise pollution. *Quarterly Progress Report.*

Federal Highway Administration. (1978). *Noise standards and procedures, planning and procedure manual 90-2,* Pt. 1. Washington, D.C.: U.S. Government Printing Office.

Freeman, A. M. (1979). *The benefits of environmental improvement: Theory and practice.* Baltimore: Johns Hopkins University Press.

Gruen Associates. (1975). *Phase I draft report of BART impacts on environment.* San Francisco: Bay Area Council of Governments.

Howe, C. W. (1979). *Natural resource economics.* New York: John Wiley.

Hufschmidt, M. M. (1983). *Environment natural systems and development: An economic valuation guide.* Baltimore: Johns Hopkins University Press.

Landsberg, H. E. (1981). *The urban climate.* New York: Academic Press.

Lavine, M. J., & Meyberg, A. H. (1979). Use of energy analysis for assessing environmental impacts due to transportation. *Transportation Research, 16,* 35–42.

NewKirk, R. T. (1982). *Environmental planning for utility corridors.* Ann Arbor, Mich.: Ann Arbor Science.

San Francisco. (1983). *Guidelines for environmental review: Transportation impacts.* San Francisco: Department of City Planning.

Schwing, R. (1980). "Benefit Cost Analysis of 1980 Auto Emission Reduction." Journal of Environmental Economics and Management, 7, 44–64.

Simonds, J. O. (Ed.). (1970). *The freeway and the city: Principles of planning and design.* Washington, D.C.: U.S. Government Printing Office.

Stutz, F. P. (1983). Land use suitability modelling and mapping. *Cartographic Journal, 20,* 39–49.

U.S. Department of Housing and Urban Development. (1971). *Noise assessment guidelines, technical background.* Washington, D.C.: U.S. Government Printing Office.

U.S. Department of Transportation. (1972). *Transportation noise and its control.* Washington, D.C.: U.S. Government Printing Office.

U.S. Department of Transportation, Federal Highway Administration. (1977). *Federal Highway Program Manual* (FHPM), 7-7-3. Washington, D.C.: U.S. Government Printing Office.

U.S. Department of Transportation. (1980). Environmental impact and related procedures: Final rule and revised policy on major urban mass transportation investments and policy toward rail transit. *Federal Register, 45* (212), October 30.

U.S. Department of Transportation, Federal Highway Administration. (1978). *Noise standards and procedures, planning and procedure manual 90-2,* Pt. 1. Washington, D.C.: U.S. Government Printing Office.

U.S. Department of Transportation, Federal Highway Administration. (1980). *Procedures for estimating highway user costs, fuel consumption, and air pollution.* Washington, D.C.: U.S. Government Printing Office.

U.S. Department of Transportation, Urban Mass Transit Administration. (1974). *Wheel/rail noise and vibration control.* Washington, D.C.: U.S. Government Printing Office.

U.S. Department of Transportation, Urban Mass Transportation Administration. (1979). *BART in the Bay Area: The final report of the BART impact program.* Springfield, Va.: NTIS.

U.S. Department of Transportation, Urban Mass Transit Administration. (1983). *Visual aesthetic impacts of AGT guideways.* U.S. Department of Transportation. Morgantown: West Virginia University.

U.S. Environmental Protection Agency. (1974). Noise abatement: Interstate motor carrier noise emission standards, 40 CFR Part 202, *Federal Register, 38.* Washington, D.C.: U.S. Government Printing Office.

U.S. Environmental Protection Agency. (1978). *Mobile source emission factors.* Final Document, EPA 400/9-78-006. Washington, D.C.: U.S. Government Printing Office.

15 METROPOLITAN EXPANSION AND THE COMMUNICATIONS–TRANSPORTATION TRADE-OFF

DONALD G. JANELLE
University of Western Ontario

Picture yourself, a successful business consultant, sitting in the rustic comfort of a mountain retreat 200 kilometers from the downtown office. A new radio tower to serve the area's expanding number of cellular telephone users has made it possible to work from the cottage. Along with a trusted microcomputer, the cellular phone gives instant retrieval of data from office files, allows for the transfer of instructions and information with business associates, and offers access to a specialized daily news service. So far, this combination, along with a facsimile machine for transmitting and receiving documents has proven adequate. Recently, it has been necessary to commute to the city only two days per week, to conduct more complicated negotiations and to meet prospective clients—the kinds of activities where physical presence is needed to size up the situation and to add an essential personal touch to the relationship. Even on the trip to the office and back, the cellular phone permits communications that are important to the productive use of time. But, aside from business considerations, you and the family have discussed the possibility of selling the home in the city and on improving the cottage as a permanent residence. The wholesome rural ambience, fresh air, and interesting neighbors would, in your opinion, be good for the family. The new communication technologies and the limited-access highway to the city have given you the freedom to enjoy the best of both worlds—the cultural amenities and business opportunities of the city, and the rural and natural amenities of the countryside.

This image of "telecommuting" to work from one's "electronic cottage" is a reality today for a few people in certain professions and, no doubt, many people entertain thoughts of similar life-styles. As attractive as all this seems, there are serious constraints to the widespread substitution of communications for transportation. In addition, the consequences of their rapid adoption bear careful consideration. This chapter concerns the role of new technologies in communications, their relationships with transportation, and their likely impacts on the development of urban areas.

Many of the questions raised in previous chapters about urban transportation are equally appropriate for urban communication. How will new and more widespread technologies influence land use patterns in metropolitan regions and in their exurban hinterlands? Will they encourage new functional and spatial relationships between the home and the workplace that will alter the journey to work, the frequency of trips, or the choice of travel mode? In general, these questions are not easily accommodated by standard predictive models. In most instances there is simply too little known about these changes and the responses to them to permit even simple approximations by regression techniques. The answers lie in understanding the complex and revolutionary changes that are affecting the very nature of interdependence between individuals and their socioeconomic structures.

This chapter will focus on the implications of communications technologies in the context of postindustrial societies, using primarily European and North American examples. It will describe (1) how structural and functional changes in the metropolitan economy have placed greater reliance on the quality and accessibility of information needed by corporate and public decision makers, (2) how technological changes have altered the ways that information needs are met, (3) how firms have changed their organizational structures and locational patterns, (4) how new locational and functional relationships between the home and the workplace have developed, (5) how the nature of the work task controls the extent to which communications may be substituted for transportation, and (6) how the changing transportation–communication relationship poses general planning and policy issues for cities and their metropolitan regions. However, before considering these matters, it is useful to develop a conceptual framework that relates the structure and function of urban regions to changes in the significance of distance.

URBAN STRUCTURE AND DISTANCE

Urbanization is the very embodiment of communications—by concentrating a wide variety of creative specialists in a region of limited extent and of high connectivity, the need for costly movement of goods and people is minimized. This is most evident in the central business areas of large cities where the chief executives of major firms and public agencies have unexcelled opportunities for face-to-face exchanges. However, these urbanization benefits are achieved at the price of crowding, congestion, excess demands on the natural environment, and other social and economic costs that may propel a dispersion of activities and people to outlying areas. The root of this dilemma (the need for specialized interaction and information, and the desire for "elbow room" and amenity) is, in part, the product of temporal and spatial constraints on human behavior.

In their elemental form, these constraints are internal, defined by one's biological clock (the need for sleep, food, and so on) and life span, and by the inability to be in more than one place at a time. The importance of these constraints, however, is as much due to social and technological factors as it is to biological traits. Whereas the biological traits are fixed by a genetic code, the social and technological factors are subject to modification through human innovation.

Social and Technical Constraints on Distance

Social practices prescribe the precise timing and locations of events (such as work and school) that confine human movement to set rhythms and that result in general maximum distance limits between activity sites (such as home and work). Technology provides potential extension of these distance limits by altering the amount of time required for given tasks and by lessening the time required for movement between places—time that might be applied to other activities or that could be used to allow for even greater distances between one's home and work locations or between one's business associates. These interrelationships are listed in Table 15-1. It is contended that any critical discussion on the transportation–communication trade-off must be related to the elemental spatial and tem-

Table 15-1. Constraints on individual behavior

	Temporal	Spatial
Biological	Prescribes essential life-support events (eating, sleeping, hygiene) according to distinct patterns of time	Sets limits on the geographical range of movement
Social	Prescribes events in time (church attendance, work hours, school, holiday periods)	Allocates activities in space and the distances between them
Technological (transportation, communications, production)	Frees time for other activities, allows faster response times for meeting needs, permits storage of information for use at any time	Alters the distance that can be covered per unit of time, allows linkages between distant places, permits different levels of separation between production units

poral constraints and their associations with biological, social, and technological factors.

Geographers Joseph Spencer and William Thomas (1969) referred to transportation and communication innovations as "space-adjusting technologies" that change the significance of distance and allow for higher degrees of accessibility to specific locations. However, it may also be appropriate to call them "time-adjusting technologies" in that time is freed for alternative use in a variety of activities, not just in increasing the number of kilometers logged per day.

Increasing Space Standards and Distance

While much of the time freed from work and travel has gone to other creative efforts and leisure, it is also true that some of it has been used to achieve more space per person, or higher space standards (such as larger house lots, extra space for potential factory and store expansion, space for parking more and more cars, and the like). In addition, new technologies (like single-floor industrial plants) have increased the average distances that separate activities in urban regions. These increases have not been systematically documented, but illustrative speculations include the claims by the famous Greek planner, Constantinos Doxiadis (1968), that, in the 40 years before 1968, average densities in several major cities dropped by two-thirds, implying that each individual was using three times the space used by people in the 1920s. Ivan Illich (1974), the noted social critic, has argued that increased speeds create distance by allowing for increased separations between related activities, and a highly readable book by Schaeffer and Sclar (1975) on transportation in Boston laments the increasing proportion of trips that are dependent on the automobile.

The Distance Between Home and Work

One of the most basic distance constraints on human activity is the separation between the home and workplace. This distance is crucial to interpreting urban transportation patterns and to gauging the possible transfer of many work-related activities to the home, via communications technologies. The home is the base for satisfying most of one's daily and lifetime biological needs and for meeting most of one's critical social obligations. Increasing the home's distance from the workplace comes at the expense of time that could go for household and other activities. Although the documentation is sparse, evidence points to a long-term trend of increasing distances. *The 1969–1970 Nationwide Personal Transportation Study* (U.S. Department of Transportation, 1973) indicated a mean home–work trip length of 9.2 miles. *The Journey-to-Work Supplement to the 1979*

Annual Housing Survey (U.S. Bureau of the Census, 1982) indicates that, irrespective of concerns for energy supplies and the price of gasoline, the average distance from home to work for 54 million employed householders was 11.1 miles in 1975 and 11.9 miles in 1979.

Such trends suggest interesting questions about human behavior and motivations. What are the social and technological factors that permit greater separation between residences and workplaces? Why do people spend their time-resource and travel-speed gains for home locations that require longer journeys to work? What are the principal constraints that set the outer limits of this distance relationship? And how might achievements in communications alter these limits? Two factors that offer partial interpretation of these trends are changes in the terms of employment and time–space convergence.

Terms of Employment

It was observed earlier that social factors and technologies for overcoming distance play important roles in determining the temporal and spatial constraints on human activities. Jean Gottmann (1977a), the geographer who coined the term "megalopolis" to describe the massive coalescence of urban development along the East Coast of the United States, calls attention to the terms of employment as probably the most significant social determinant of urban spatial patterns. They represent contractual agreements between the employee and the employer concerning such matters as compensation, hours of work, time for holidays, work location, and so forth. Historically, the trends to shorter work weeks, longer paid vacations, flexible work hours, and even job sharing and early retirement, have given employees greater control over how they spend their discretionary time. In the early period of the Industrial Revolution, fourteen-hour workdays and six-day workweeks constrained workers and their families to live within short distances of the factories. By the early twentieth century, concessions granted to labor provided some release from these temporal constraints and permitted employees to use some of their freed time to commute from greater distances. Gottmann sees this as contributing to a broader range of living environments and to the potential for massive suburbanization. Some of the free time could be channeled into creative and pleasurable activities (such as sports, hiking, arts) but, in turn, these two generate needs for space—for parks, stadia, museums, and other cultural and recreational facilities. Two-day weekends permitted day trips into the countryside and to weekend cottages (secondary residences), and paid vacations helped to promote seasonal tourism, the growth of specialized resort centers, and urbanization in the coastal and mountain zones of high amenity.

Concurrent with these social changes, the advent of the automobile and the paving of roads ushered in new possibilities for expanding the commuter fields of cities, the residential options for households, and the geographical range for leisure activities. The space-adjusting consequences of transportation innovation are assessed directly by the concept of time–space convergence (Janelle, 1968, 1969).

Time–Space Convergence

If it required 74 hours to travel by stage coach from Boston to New York in 1800 and only 5 hours by automobile in 1980, then these two places have been approaching each other in time–space at the average rate of 23 minutes per year. Such measures provide a convenient way of characterizing the degree of transportation innovation between any pair of places and for determining relative changes in their overall accessibility with other places in the urban system. Needless to say, time–space divergence is also a possibility, reflecting the deterioration of transportation service between locations. These measures of convergence and divergence objectively express what has long been characterized as a "shrinking world" at regional and global scales and as "congestion" at local scales.

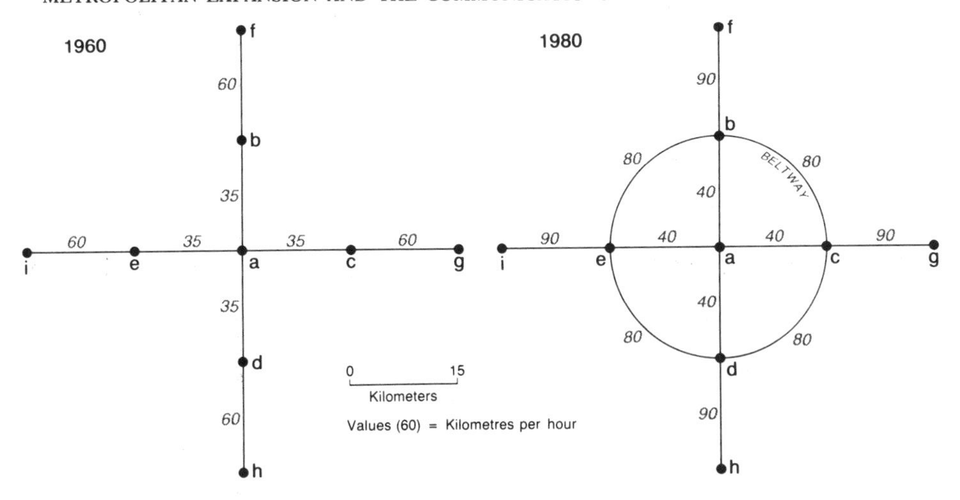

Figure 15-1. Hypothetical urban highway network and travel speeds used in calculating rates of time-space convergence (Table 15-2).

Convergence rates vary depending on the degree and timing of innovations and on the degrees of route deterioration and congestion. However, even under conditions of uniform changes in transportation quality, the convergence rates between more distant places are always greater than for closer places. An example will illustrate this point.

In Figure 15-1, a simple hypothetical urban highway network is focused on a single city. The travel-time distances for 1960 and 1980 and the resulting time-space convergence rates for each of nine locations within the metropolitan area are shown in Table 15-2. Whereas the 1960 pattern in Figure 15-1 shows a simple crossroads situation, the 1980 pattern represents a typical limited-access beltway with speeds of 80 km/hr. The outer fingers of the network are improved to 90 km/hr by 1980 and, because of the diversion of traffic around the city, speeds within the city increase from 35 to 40 km/hr. For each place, the average distance and the average convergence to all other places are shown in Table 15-2. The greatest accessibility gains have been recorded for the more distant

Table 15-2. Convergence rates and distance in a hypothetical metropolitan setting

	Average kilometers (each place to all eight other places)			Average travel times (min.) (each place to all eight other places)			Convergence rates 1960–1980 Average minutes per year (each place to all eight other places)	
Places (letters refer to the places shown in Figure 15-1)	1960 Network	1980 Beltway	Straight Line	1960	1980 (using shortest distance paths)	1980 (using shortest time paths)	Using shortest distance paths	Using shortest time paths
a (city center)	22.50	22.50	22.50	33.24	27.54	27.54	0.29	0.29
b, c, d, e	31.80	32.98	28.68	49.26	27.90	25.50	1.07	1.19
f, g, h, i	45.0	46.1	41.80	62.40	36.66	34.26	1.28	1.40

journeys, demonstrating the distinctive advantages that peripheral locations have over central locations.

Two critical conclusions may be drawn from this simple example. (1) The center of the city, because it is central, has the shortest average distance to all other points in the city. Hence, it stands to make the least absolute gains from overall urban transportation improvements. (2) In contrast to the central city, any overall increase in speed and flexibility of transportation, such as that facilitated by the automobile, means that the urban edge will converge, on average, more rapidly with other places in the city than either the center or any points intermediate between the center and the edge. Since space limitations are not as critical in the outskirts as in the central city, the urban periphery has taken full advantage of automobile benefits.

Most urban transportation improvements over the last quarter century have been automobile oriented, emphasizing a combination of limited-access radial and circumferential highways. These innovations have been of primary advantage to the urban edge, allowing easier cross-town movements by private vehicles and easier service of both central and regional hinterland positions. In response to its improved accessibility, the emergent dominance of the urban periphery has been reinforced by the inflow of capital investment (chapter 2) and by the in-migration of talented residents. The locations of airports and other transshipment depots augment still further the accessibility advantages of the urban edge.

Changing patterns of accessibility within urban regions and changes in the terms of employment as described by Gottmann (1977a) are seen as very significant facilitators for greater separation between home and work and for establishing functional linkages over greater and greater distances. However, other factors also contribute to such trends. The extended open hours of stores and shopping centers permit people to travel longer distances, since with less scheduling constraint they can more easily budget their trips according to their personal needs and availability of time. In addition, rising incomes and trends toward more than one income per household have lessened the fiscal restraints on the use of household income. Ginzberg and Vojta (1981) note that, with rising incomes, an increasing proportion goes for products and activities related to self-identity and life-style, and less to basic necessities, comfort, and convenience. One manifestation of this is the dispersal of exurbanites to the peripheral regions of the urban field—in part, a trend related to the quest for amenities and a trend that, as noted in chapter 11, contradicts standard expectations from location theory. Nonetheless, there appear to be basic regularities in this process that should not be ignored. The "law of constant trip rates and constant travel time" is one such regularity that suggests outer limits to the time and resources that people will devote to travel.

Transportation Expansion and the Law of Constant Travel Time

The increase in the personal consumption of distance has been documented in this chapter and in chapter 1. Yet, in general, people neither spend more time traveling nor make more trips today than in previous decades. Time, a finite resource in people's lives, is by definition a scarce resource that must be allocated according to subjective notions of utility, whether measured in dollars per hour or pleasure per hour. Dutch transport analyst Geurt Hupkes (1982) does not deny the expanding mobility of people in affluent, auto-owning societies, but his investigations indicate that the number of daily trips and the amount of time devoted to travel remains constant over time. He cites documentation for these tendencies in many different countries and provides an aggregation of results from time-budget surveys in the Netherlands for 1962 and 1972. Table 15-3 presents these data. The numbers of trips, travel hours, and passenger kilometers per person per annum are presented for several different transport modes. Although substantial shifts occur

Table 15-3. The three dimensions of mobility per person, the Netherlands

	Trips		Travel time (hours)		Passenger kms	
	1962	1972	1962	1972	1962	1972
Motorcar	110	439	39	145	1,867	7,237
Motor bicycle	0	3	3	1	124	42
Bicycle	629	326	151	81	1,886	848
Moped	130	164	29	34	648	720
Walking	594	540	159	121	653	486
Railway	19	16	13	11	781	712
Local public transport	55	44	10	9	171	133
Regional public transport	35	23	17	12	426	295
Other collective transport	7	8	12	13	448	483
Taxi	5	5	1	1	31	31
Airplane	0.08	0.3	0.3	1	121	491
Total	1,584	1,568	434	429	7,156	11,478

Note: From "The Law of Constant Travel Time and Trip-rates" by Geurt Hupkes, 1982, *Futures, 14,* p. 39. Copyright 1982 by *Futures.* Reprinted by permission.

among the modes, the remarkable stability in the overall number of trips and in the total time devoted to travel provides compelling evidence for Hupkes's claim.

In Table 15-4, I have used the data of Table 15-3 (excluding airplane travel) to calculate average travel speeds, trip distances, and convergence rates. It is clear that substantial gains have been achieved for average speeds by motorcar and, although the convergence rates by rail are double those of the automobile, the shift of trips from rail and from bicycle to the automobile indicates that the auto's combined attributes of speed and freedom from externally set schedules and fixed routeways offer greater access to a wider range of residential and life-style options. On average, people have used the advantages of

Table 15-4. Average trip characteristics, the Netherlands

	Average speed (km per hour)		Average trip length (km)		Average convergence rate (km gained or lost per hour of travel)
	1962	1972	1962	1972	1962–1972
Motorcar	47.87	49.91	16.97	16.49	+2.04
Motor bicycle	41.33	42.00	—	14.00	+0.67
Bicycle	12.49	10.47	3.00	2.60	−2.02
Moped	22.34	21.18	4.98	4.39	−1.16
Walking	4.10	4.02	1.10	0.90	−0.08
Railway	60.08	64.72	41.12	44.50	+4.68
Local public transport	17.10	14.78	3.10	3.02	−2.32
Regional public transport	25.06	24.58	12.10	12.82	−0.48
Other collective transport	27.33	37.15	64.00	60.38	+9.82
Taxi	31.00	31.00	6.20	6.20	0.00
Averages for all trips (weighted by number of trips per mode in 1972)	16.22	25.67	4.44	7.01	+9.45

extra speed to broaden the geographical range of their activities rather than to increase the number of trips. In addition, the average number of minutes devoted to travel per day has not changed appreciably. Calculations from Table 15-3 give an average per capita daily expenditure of 71.3 minutes in 1962 and 70.5 minutes in 1972.

Of course, there is no law that dictates the maximum amount of time that individuals allocate to travel. Within any society there are variations in what is acceptable to different individuals. Hupkes suggests that, beyond a certain point, extra travel might generate biopsychological stress, which could dictate a cap to the time judged as generally acceptable. His most interesting explanation, however, is based on a utility-optimizing approach that may shed light on the debate over the transportation–communications trade-off.

Figure 15-2a illustrates the *intrinsic utility* that travelers derive from the journey itself (for example, the sense of speed, the opportunity for fresh air or exercise, or other benefits). But, as trips continue and boredom or fatigue begin to dominate, this utility may decline, and additional increments of distance may even yield negative utility. *Derived utility* is related to the value obtained

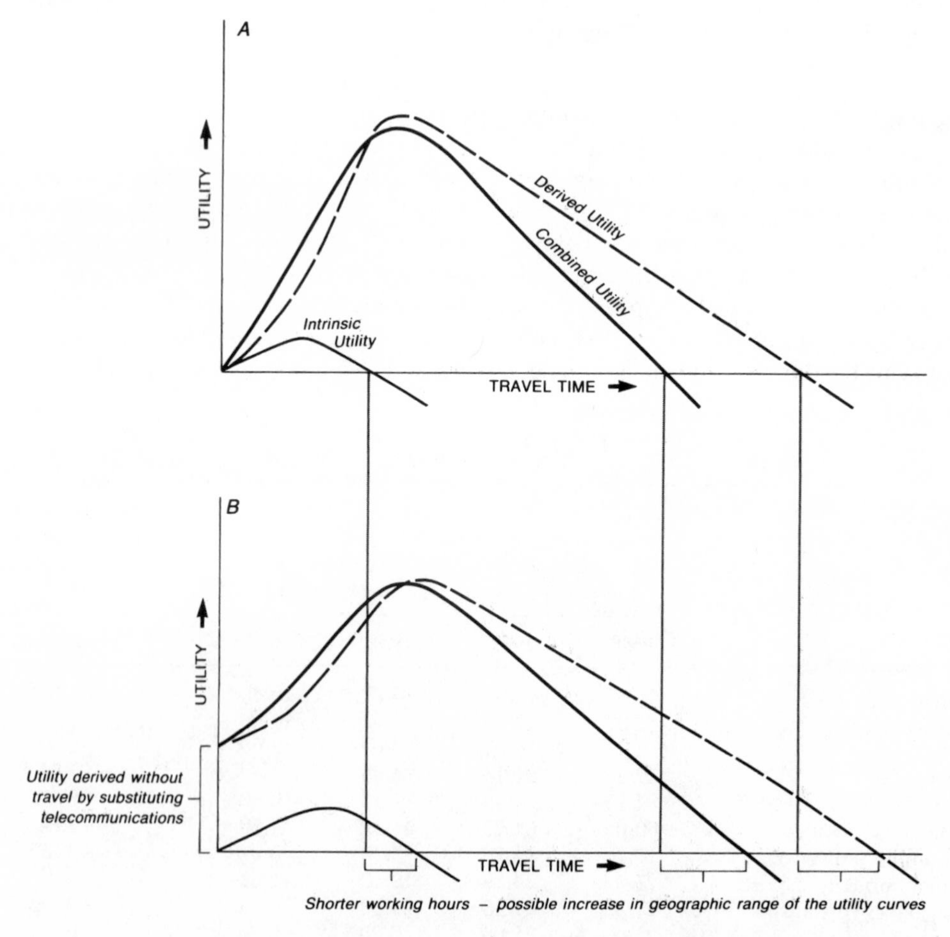

Figure 15-2. (a) The utility of travel time; (b) Changes in the utility of travel time following the introduction of telecommunication systems and reduced work hours. Adapted by the author from "The Law of Constant Travel Time and Trip Rates" by Geurt Hupkes, 1982, *Futures*, volume *14*, pp. 42, 45. Copyright 1982 by *Futures*.

from activities that traveling makes possible, but this too will peak at some point beyond which extra travel cuts into the time supply for these activities or is judged as too high a price to pay. These intrinsic and derived utility curves will differ for different individuals and may even shift as one progresses through the life cycle. Furthermore, technological changes may result in significant variations.

In Figure 15-2b, advanced telecommunications (picture phones, computers, and the like) are assumed to be cheap and easy to use and to have achieved a wide distribution among businesses and households, allowing for some substitution for working, shopping, and personal business trips. In this case, the upward shift in the utility curve at the origin reflects the utility derived without trips. In addition, Figure 2b illustrates the effects of a reduction in work hours. Following Hupkes's assumption, if travel outside of working hours is evaluated for 20% of leisure time, then a reduction from 8 to 7 working hours per day would permit an extra 12 minutes for travel, some of which might be allocated to broadening one's geographical base of operation; some might go for more pleasurable but slower trips (such as walking).

The merits of Hupkes's approach require empirical evaluation. There is no definitive proof that an optimal amount of time to devote to travel exists. Although the average value is relatively stable, it is subject to change as a consequence of social and technological forces, and it would be sensitive to policies on such matters as speed limits, transport investments in different modes, the availability of fuel, and support for technological substitutes for transportation.

The possible uses of time, freed as a consequence of technical change or of changes in the terms of employment, depend in part on the choices that people make about where to live in relation to jobs and other activity sites. In general, an outcome of these decisions has been the outward migration of the city. It is important, therefore, to view the substitution of communications for transportation in terms of what motivates people to spread out at lower overall population densities.

Motivations and Opportunities for Exurban Life-Styles

Many factors motivate the migration of populations to regions beyond the edges of outwardly expanding North American cities. The symbolic "newness" of suburbs and the opportunity to start with a clean slate, where land is unencumbered by existing structures, may imbue a certain existential character to the process that is hard to label. Cheaper land (the possibility of having more of it per capita) and lower taxes have been important inducements for the centrifugal spread of city populations. But increasingly the traditional economic argument of a trade-off between more housing and land, on the one hand, and less transportation cost, on the other, may be overshadowed by preferences for "life-style locations." Geographer Wilbur Zelinsky calls these "voluntary regions" (1975), where people with like concerns stake out and create territories or places that cater to their interests. Residential communities built around golf courses, within the commuting range of large cities, represent an obvious example of such places. The post-1950 development of "dispersed cities" (Burton, 1963) and "counter-urbanization" (Berry, 1980) has resulted in a great variety of specialized communities, a trend that is not unrelated to the relaxation of social and technological constraints on where one chooses to live. A simple example illustrates one dimension of this relaxation of locational constraints.

If one assumes an increase in average commuting speeds from 60 to 70 kilometers per hour, then the distance that can be reached within a 30-minute trip from a central point (for example, the workplace) increases by 5 kilometers. Yet, this 17% gain in distance will increase the accessible land resource from 2,827 to 3,848 square kilometers, a 36% gain. Thus, a small increment in travel speed produces a considerable expansion of land resources, one that may increase the potential variety of living environments several fold (for example, communities of different sizes and social characteristics, and sites associated with different physical geographical features).

Economist Wilbur Thompson (1968, p. 356) poses a related example. If the population of a city were to double and its population density remain constant, the distance from the edge of the city to its center would increase by only 40%. But, if there were a 50% drop in average population density, the size of the urban area would quadruple, and one would have to double the transport speed in order to maintain the same journey times from the edge to the center. He suggests the possibility that transportation developments in North America may have maintained this constant relationship over the past century. But even in this situation, the areal extent of cities would be expanding faster than the average distances of commuting trips. Thus, as the supply of land within a tolerable commuter range increases, the choice of residential site can be matched more closely with individual life-style goals. Such choices become less constrained by the locational selections made by others, and more attention can be given to weighing the amenities to be gained from specific physical and cultural environments.

Amenities and Urban Expansion

The combination of increasing discretionary money and increasing discretionary time gives a strong leisure orientation to many locational decisions. Edward Ullman (1954) equated these trends with the growing significance that amenity plays in shaping the development of urban systems. He noted the rapid growth of Sunbelt cities and of mountain and coastal resort centers at regional and national scales. He also saw amenity as a leading contributor to suburbanization, to the increase of scattered nonfarm residents in countrysides, and to the growth of towns and small cities located within the fields of influence of major metropolitan centers.

In one of the few systematic studies on the topic, Philip Coppack (1985) attempts to come to grips with the meaning of amenity and he adds significant evidence of its growing importance in one of the most urbanized parts of North America: the zone within a 2-hour driving range of Toronto. Coppack defines amenity not in terms of the facilities that meet human physical needs but in terms of the ability of places to satisfy human psychological needs. Whether real or imagined, the degree of amenity is seen as having a positive association with how people perceive the aesthetic quality of physical and human environments, with a place's representativeness of past (and treasured) values or rural ambiance, or with its ability to offer a wide potential of pleasant activities. More pragmatic (but still difficult to define) concerns for "a good place to raise children" and the opportunity to escape from the less desirable features of urban life (congestion, pollution, crime, and so on) are also embraced in this concept of amenity. However intangible such notions as scenic beauty and country atmosphere might be, Coppack describes amenity as a high-order commodity that people are willing to pay for and to travel long distances to obtain.

Coppack views the outward extension and transformation of urban fields (following Friedmann & Miller's concept, 1965) as direct products of amenity-seeking behavior. The resulting mix of permanent nonfarm residents, farm population, and temporary users of the urban periphery (weekend cottagers, day trippers, and weekend shoppers) contribute to satisfying the market needs of many lower-order commercial and service functions. In such a way, amenity (the motivation) is seen as the conceptual link between the expanding urban field and the changing functional offerings of central places, leading to the resurgence of places that once bore the brunt of population loss and decay stemming from an earlier era of cityward migration.

In principle, the supply of amenity owing to the land resource does not vary with increasing distance from a city. However, because of travel costs and bid rents for the use of land, the feasible supply of the amenity resource would develop a distinct concentric pattern around the city. Thus, Coppack describes an urban core of high cultural amenity; an inner field of low amenity supply, resulting from the competitive displacement by other land uses; a middle field that possesses a high amenity value owing to the presence of acceptable bid rents and travel costs; and an outer field, where the evaluation of the amenity is low because of the high travel costs, even though the bid rents are low.

Figure 15-3 expands on Coppack's model to suggest that, through time, the accessibility of amenity resources will shift outward. In part, this may be due to new opportunities provided by transportation and communication developments. However, in addition, the intensification of urban facilities and population growth in centers within the initial (time 1) area of high amenity use may convert it into the jumping-off point for a succeeding round (time 2) of outward expansion. This may be an attempt to regain the amenity that one sees slipping from one's grasp. Thus, through a cycle of amenity growth and decline, places at increasing distances from the major metropolis become the major foci of amenity attraction and the centers for the progressive outward expansion of urbanization influences.

If amenity is the driving force of expanding exurban settlement patterns, then the perceived outward shift of the zone of high amenity would likely result in one or a combination of the following responses: (1) demands for improved accessibility via transportation investments that permit longer distances without any appreciable increase in travel times—thus increasing the amenity supply, (2) the substitution of communication technologies for trips, or (3) the transfer of jobs to the exurban frontier. The first of these options has been the focus of discussion to this point and provides the benchmark for evaluating the remaining two options. However, this evaluation requires a shift in emphasis, from the individual and the household to a focus on changes in the employment structure and interaction requirements of corporate and public agents. The substitution of communications for transportation must

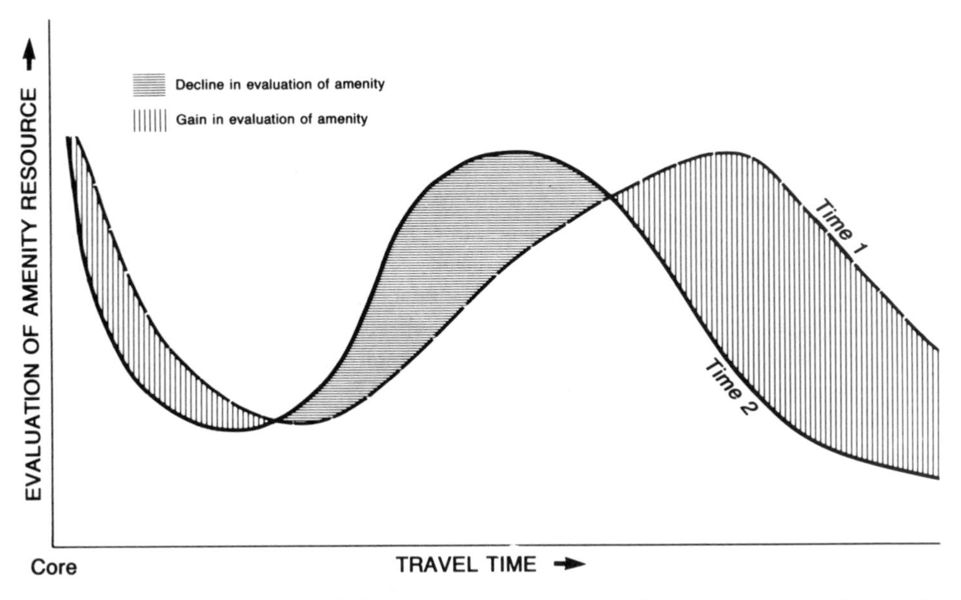

Figure 15-3. Perceived value of the amenity resource with increasing travel time from the core of a metropolitan region. See Philip Coppack (1985, pp. 72–77) for a discussion of the principles that underlay his original model, upon which this expanded model is based.

be considered both in terms of how such a shift would affect the ability of private and public establishments to carry out their objectives and in terms of the locational relationships between these establishments and the households of their employees.

STRUCTURAL AND FUNCTIONAL CHANGES IN THE METROPOLITAN ECONOMY

The formative period of urban growth in the United States was characterized by employment in primary (for example, agriculture and mining) and secondary (manufacturing and construction) activities. However, today the service or tertiary sector of the economy is the dominant provider of jobs. Between 1929 and 1978, its share of employment rose from 55 to 68% and represented the most significant contributor to economic growth (Ginzberg & Vojta, 1981, p. 48). According to calculations by Ginzberg and Vojta (1981, p. 50), 16 million of the 30 million new jobs created between 1959 and 1978 were in the service sector (professional, technical, managerial, administration, sales, and craft occupations). Gottmann (1977a) marks 1955 as the year when white-collar workers exceeded blue-collar workers in the United States, and through the 1970s, Armstrong (1979) affirms its lead in growth over other sectors. The significance of this trend is manifested in the growing importance attached to the flow of information within society.

The flow of information is particularly critical in the economic sector that Gottmann (1977b) identifies with quarternary occupations—primarily high-level managerial, research, and professional jobs that focus on processing, analyzing, and distributing information, and that rely heavily on consultation with specialists. Timely access to highly reliable information is vital to the success of many private and public establishments. Many of the larger corporate firms compete for information and for the best minds to process and interpret it, they invest in the most sophisticated equipment to store, retrieve, manipulate, and communicate it, and they pay vast sums of money to locate their corporate offices in the most information-rich areas of major metropolitan centers. Meier (1973) sees the continual improvement of communication linkages as a principal prerequisite for urban growth.

The spatial impact of information availability and use on the development of urban systems is related, in large measure, to how public and private agents have organized their operations to make effective use of information, labor, and other resources. Westaway (1974) observes that, in general, as businesses increase in scale and in complexity, the controlling decision-making functions of upper-level management have tended to centralize for maximum access to consultation and information, but productive and administrative activities, based more on routine operations, have become more widely dispersed.

In the United States, jobs have not dispersed as rapidly as population, but this varies with economic sector. Thus, Armstrong (1979) notes that, in the 1960s and 1970s, manufacturing jobs were fairly evenly distributed with population, whereas Standard Metropolitan Statistical Areas (SMSA) had disproportionate shares of white-collar workers. Her calculations indicate that 90% of jobs in the headquarters of major firms were located in SMSAs with populations in excess of 250,000, in areas where the density and quality of information is greatest. Törnqvist (1973) documents a similar pattern in Sweden, where more than half of the 250,000 "contact-intensive" employees were found in the three leading cities. These individuals, having advanced educations and incomes that are four times the national average, are seen as an attraction to related jobs in the business–service area.

Although the central business areas of large American cities have maintained their share of new office construction in recent years, there is considerable evidence of dispersal at the regional level. Looking to the future Armstrong (1979, p. 77) observes that

> the spread of development into continuous bands of low-density settlement, has significance for the future location of office activity. Most urban region corridors are comprised of several commuter sheds. Working-age resettlers tend to be youthful, educated, relatively affluent, and skilled in white-collar occupations. As smaller metropolitan areas grow in the shadow of major metropolitan centres and the interstices between SMSAs are settled by white-collar workers, the density of the available labour force at decentralized locations increases while the tolerance for lengthened work trips declines. In the New York Urban Region, which encompasses the New York–NE New Jersey SCA and nine other SMSAs midway on the Boston-to-Washington corridor, many more corporations relocated to the smaller regional metropolitan areas and outlying suburbs than moved to other parts of the nation.

Armstrong goes on to note, however, that most of the dispersed office activity in smaller SMSAs and in exurban centers has represented modest developments. In this sense, it does not appear that the moves to exurbia have had any substantial impact on the degree of concentration in leading commercial centers. However, the moves to outlying areas may represent locational shifts that are compatible with the information needs of different elements in a firm's organizational structure.

Information Needs, Corporate Organizational Structure, and Spatial Patterns

Hymer (1972) presents a four-stage process of business reorganization. It begins with a small, individually or family-owned, single-purpose firm. The second stage of development sees a consolidation of such firms into larger multipurpose corporations that serve several regions and that result in new administrative structures to coordinate their dispersed activities. Thus, specialized activities (finance, marketing, production, and so on) may find it useful to occupy different facilities, but a head office would be required to coordinate the corporation's planning function. In stage three, many successful corporations may seek greater diversity by decentralizing their operations into distinct divisions, each with its own product and head office, while maintaining a general office for planning the joint operations of the divisions. Finally, the emergence of multinational focus is seen as an attempt to gain control over resource supplies and to secure expanded markets.

At the second and successive stages, the specialized units of the corporation become dispersed in response to the availability of labor, markets, and raw materials. However, divisional head offices tend to concentrate in large cities that can meet their demands for white-collar workers, and the general offices of multidivision conglomerates are concentrated in those few very large cities that offer easy access to capital resources, media, related enterprises, and government. According to Knight (1982, p. 57), "as smaller firms are taken over through mergers and acquisitions, their high-level management and technical functions are brought under the corporate umbrella and shifted to the city where the parent company is headquartered." In essence, organizational changes in the corporate sector are resulting in a distinct sorting of the labor force according to the sizes and functional significance of cities, with the large cities having high proportions of upper-level administrators and lower proportions of operants (blue- and white-collar workers who are involved in the more routine manual or clerical activities) than small cities. Correspondingly, the significance of communication linkages can be expected to vary according to the labor composition of cities and to the presence of information-sensitive firms within them.

Communication Linkages and Contact Requirements

Communication linkages are needed to pass on or to receive information, to seek advice, to make assessments, to negotiate agreements, and so forth. Such task distinctions are critical to the evaluation of information needs and to determining the appropriateness of different means for transferring information. A classification of job-related tasks, used by Thorngren (1970) to analyze the communication needs of Swedish firms, distinguished among "orientation," "planning," and "programming" tasks, each directed to different kinds of objectives and requiring different kinds of contacts in order to be carried out successfully.

Orientation tasks may involve initiating new projects that require complicated discussions, difficult negotiations, and the assessment of prospects. In general, these activities entail considerable nonroutine decision making and coping with novel situations. The inherent uncertainty in new ventures makes face-to-face meetings particularly important —allowing for immediate feedback, flexibility in seeking on-the-spot clarification, and the use of various communication aids (demonstrations, exhibits, discussions, and the like). The complexity of many orientation tasks requires consultation with a variety of specialists. Thus, prearranged meetings, often of considerable duration and in many locations, make travel an essential component in the daily work life of many managerial, professional, and technocratic occupations.

Planning tasks represent efforts to implement decisions that have already been made; for example, organizing procedures for production, marketing, finance, and so on. Although each problem to be solved may be unique, there are established channels of communication used in resolving them. The contacts are generally less dispersed spatially and of less duration than for most orientation activities. Very often, indirect contacts, via telephone or mail, are sufficient to carry out planning activities.

Programmed tasks refer to routine (sometimes repetitive) exchanges of information and to decision making that is based on structured and often standardized procedures ("according to the book"). Many production, sales, and clerical tasks fall into this category. At this level, indirect contacts, particularly by telephone, are usually sufficient for most intra- and inter-firm communications.

In general, the transition from orientation to programmed activities is toward decisions that require less time and tasks that require less travel. Olander (1979) uses Hägerstrand's (1970) time–geography approach to depict the interrelations among these kinds of tasks, seeing them as integrated activity systems that result in distinct time–space patterns of movements (paths), linking individuals and places, both within and among corporate establishments (Figure 15-4). This focus on the daily activity patterns of a firm's employees calls attention to the spatial and temporal constraints that firms must consider in choosing the locations of their subsidiaries and corporate headquarters, and in matching their communications requirements with appropriate means of contact. Thus, issues concerning the relocation of a firm or parts of its operations (possibly to the amenity region outside the city), the separation between the home and work locations for different kinds of employees, and the possible substitution of new ways to exchange information, are seen in relation to the sensitivity of communication needs and to the daily activity systems of individuals.

Some work tasks are more contact intensive than others. For example, Törnqvist (1970) suggests that the following activities are very reliant on face-to-face exchanges: decision making, planning and negotiation, distribution of information, publicity and selling, control and intelligence activities, research, analysis and educational work, construction, and product development and designing. Fernie (1977) compared the frequency of face-to-face contacts by functional groups of firms for three medium-sized cities in the British Isles (Dublin, Edinburgh, and Leeds). Based on diary surveys of office linkages among firms in each of these cities, the

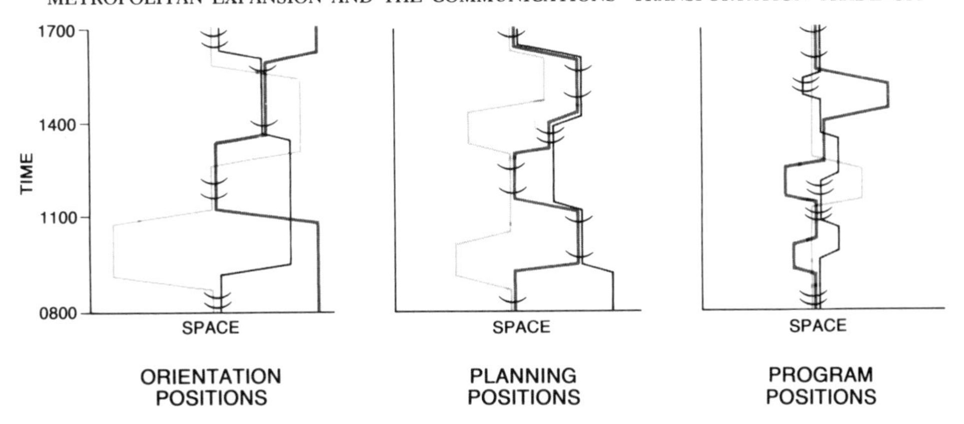

Figure 15-4. Activity paths of a firm's employees. Adapted by the author from "Office Activities as Activity System" (p. 16) by Lars-Olaf Olander, 1979, in P. W. Daniels (Ed.), *Spatial Patterns of Office Growth and Location*, Chichester, England: John Wiley and Sons. Copyright 1979 by John Wiley and Sons.

following functional groupings ranked highest: meetings with customers and clients, banking and other financial activities, contacts with lawyers, and meetings with competitors. Business with government ranked high in the cities of Dublin and Edinburgh; and, in Leeds, face-to-face meetings with firms in other cities, particularly London, were important. These findings accord well with Gad's (1979) observations. Based on diary surveys of employees from representative firms in Toronto, he concludes that law firms, investment dealers, banks, trust companies, customs brokers, shipping agencies, and the headquarters of mining firms require high levels of face-to-face communication within the core of the city's business area. In addition, Goddard's (1973) extensive analysis of face-to-face and telephone contacts among office establishments in London yields similar conclusions. All these studies confirm the importance of direct contact for tasks that are nonroutine in nature.

Evidence assembled by Abler (1977) for the United States suggests that occupations based on nonroutine activities have increased proportionately over routine-oriented jobs. The importance of face-to-face meetings would suggest that this occupational shift could augment still further the centralization of office functions within the central business areas of large cities. However, some (see, for example, Coates, 1982; Kalba, 1974) have suggested that advances in telecommunications will allow for the substitution of many direct contacts. Essentially, two related propositions have been advanced. First, telecommunications can substitute for transportation; and, second, these technologies will permit information-based office work to decentralize beyond the central business area and even beyond the city's built-up zone to outlying regions of high amenity. An evaluation of these propositions must focus on two questions. First, to what extent can electrically transmitted communications replace face-to-face meetings? And, second, how might telecommunication linkages alter the relationships between the spatial organization of firms and the locational patterns of employees' households?

SUBSTITUTING COMMUNICATIONS FOR TRANSPORTATION

Ronald Abler (1975, 1977) has carefully documented the improvement in telephone service in the United States. He notes that the time required for connecting two parties, one in New York and one in San Francisco, has declined from 14 minutes in 1920 to 30 seconds in 1970—an average time–space convergence rate of 16 seconds per year. Similarly, the rate of cost–space convergence between these locations has been remarkable—an average reduction in cost of 29.7 cents per year for a 3-minute, station-to-station call ($16.50 in 1919 and $1.35 in 1970). Based on changes in the consumer price index, this is equivalent to a convergence of 57.6 cents per year in 1970 dollars. According to Gottmann (1977a), the universal availability of the telephone has had a profound impact on the spatial form of the U.S. metropolitan system, on the organization of corporate and public agencies, and on the life-styles of individuals.

The availability of an inexpensive and reliable telephone service, made even cheaper by wide-area telephone services (WATS) and cable linkages, and made even more versatile by means of computer-related connections, should be a powerful incentive to substitute telecommunications contacts for the more costly and more time-consuming face-to-face linkages. To illustrate some of the potential benefits to be gained by such a substitution, consider the example illustrated in Figure 15-5. Figure 15-5a shows B traveling directly from home to A's work location for a 30-minute meeting. B has sacrificed alternative activities that could have been carried out in the time required for this trip. Of course,

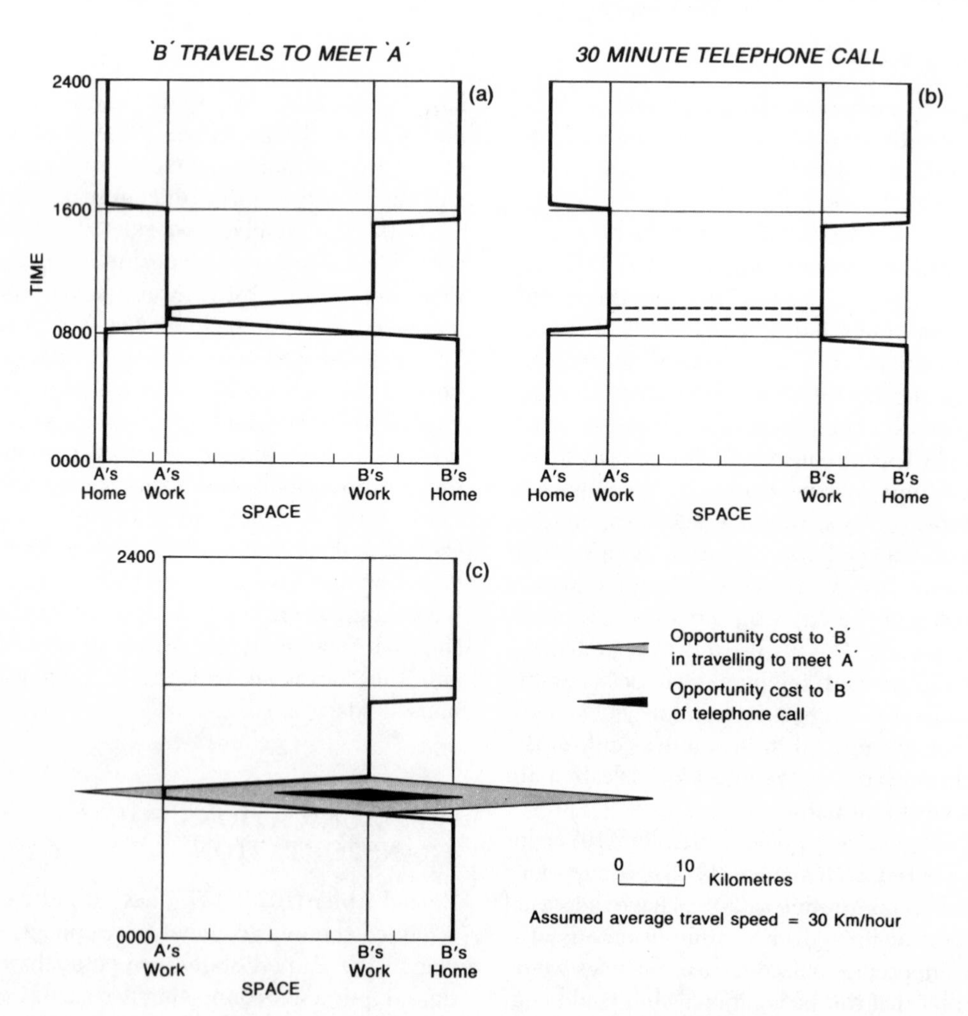

Figure 15-5. Comparative space-time opportunity cost of a telephone call and a business meeting.

Table 15-5. Communication systems in 1985

Audio
- Telephone (transmits sound—speech—over wire or radio)
- Conference calls (multiparty telephone connections)
- Cellular (mobile) telephone (access to all telephone services from a moving vehicle—currently only in large cities)
- Answering and recording machines (linked to telephone and activated by incoming calls)
- Broadcasting (one-way communication by radio or television)
- Citizen-band radio (interactive communication over short distances)
- Paging systems ("beepers," used by emergency personnel)

Visual
- Mail (requires transport)
- Electronic mail (transfer via computer connections)
- Printed publications (requires transport)
- Facsimile machines (image transfers of printed documents by radio or telephone transmission)
- Telegraph (electrical transmission of messages or signals)
- Telex (telegraphic transmission of printed signals to teleprinters)
- Computer conferencing (multiparty computer connections having either synchronous or asynchronous access; also called "networking")

Audiovisual
- Face-to-face meeting (requires transportation to meeting site)
- Picture phones (telephones that permit one to see the other party)
- Tele-conferencing (audio and visual transmission of interactive discussions by 3 or more people in 2 or more locations)
- Cable television (interactive broadband transmissions via coaxial cables)

this opportunity cost would have to be evaluated against the benefits of a face-to-face meeting over that of a telephone linkage. In Figure 15-5b a 30-minute telephone link between the work locations of A and B is represented by the horizontal broken lines. Although, in this example, it is not possible to assess the comparative quality of the communication for the two situations, it is possible to illustrate the general extent of the opportunity cost. Figure 15-5c represents the opportunity cost to B by means of space–time prisms (Hägerstrand, 1970). These define the potential space that is accessible to an individual from a specific location within a given span of time. In this case, the inner prism represents the space–time opportunity cost associated with the telephone call and the outer prism encloses the lost opportunities associated with the face-to-face meeting and the required journeys. It is clear that B's trip to meet A in person has preempted a greater range of alternative activities than would a similar exchange by telephone. The significance of this cost, however, will depend on the value that B attaches to time, on the differences in financial costs, and on the value that might be attached to the foregone options.

With technological advances, the telephone, radio, computer, and cable now provide incredible extensions in communication capabilities. Some of the available systems are listed in Table 15-5. Hiltz and Turoff (1978) compare several of these options, along with face-to-face meetings, with respect to speed of message transfer (talking rate or reading rate), system delays ("none" for face-to-face to "considerable" for telegram or telex messages), and memory (whether stored in the individual's brain or on magnetic media, as with computerized conferencing). However, a criterion that is most relevant to the substitution for transportation relates to the degree of spatial and temporal constraint on the use of the system. This is summarized in Table 15-6 by whether or not locational and/or temporal coincidence is required between communicants. Face-to-face contacts require that all participants be in the same place at the same time and, as noted earlier, they entail transportation expenses and the dis-

Table 15-6. Spatial and temporal constraints on communication systems

		Geographical coincidence of communicators required	
		Yes	No
Temporal coincidence of communicators required	Yes	Face-to-face meeting	Picture phones Telephone Teleconference (audio or audio-visual) CB radio
	No		Answering and recording machines Computer conferencing Mail Telegrams, telex Printed publications

placement or exclusion of activities that could have been completed during the travel period. In contrast, systems based on the permanent storage of messages (such as tape recordings and books) pose the lowest levels of constraint, since they can be accessed and responded to at any time that is convenient to the individual; the cost of such systems would depend on the seriousness of the delay.

The impact of communication technologies in the long term will depend largely on their accessibility to potential users. Radio and telephone are the most widely distributed (more than 450 million telephones in the world); but, increasingly, it is the ability to link the telephone with computers that offers entry to high-level telecommunications potentials. According to the McGraw-Hill Information Systems Company (MHIS, 1985), personal computers were available in 12.3% of all households in the United States in January 1985, up from 8.3% in the previous year. Even though they project this figure to reach 15.8% by the end of 1985, it is not likely that this powerful tool will be available to the majority of American households for some time. Even so, Coates (1982) sees a day when each household will invest annually in telematics (computers, related equipment, and software) the equivalent of what it now spends on automobiles. Interestingly, the move in this direction is more advanced in some parts of the country than in others. As of January 1985, the highest market penetration of home computers was in the New York metropolitan area, where MHIS reports more than 900,000 units. Not surprisingly, this urban region leads all others in the availability of banking-from-home and information-retrieval services, a factor that may account for the ability of its financial institutions to draw business from outside the state of New York.

In order to assess whether these technologies will bring disproportionate benefits to existing urban regions or allow for increasing dispersal of information-based jobs to remote amenity regions, more data on geographical variations in the market penetration of home computers and other technologies are required. Nonetheless, even in the absence of such information, speculations on the nature of an impending communications revolution abound. The work-from-home scenario figures prominently in many of these forecasts.

Working from Home—Telecommuting

The increasing accessibility of telecommunication–computer systems may have profound impacts on the spatial patterns of land use and traffic flow within urban regions and on the life-styles of many people. Martin (1978) and Toffler (1980) focus on one of the more interesting prospects—people working from home and shifting their residences and weekend retreats to more remote regions that have high levels of amenity. At the same time, they see this as a way of reducing urban traffic congestion and traffic-based pollution and a means of cutting the consumption of energy. Toffler maintains that the switch to information-based jobs makes the "electronic cottage" a distinct possibility for many people. He cites examples of firms that have made this option available to employees, with the homes and secondary residences of some top executives having computers, facsimile devices, and teleconferencing stations.

The electronic cottage may be more suited for some jobs than others. Suitable candidates are those who work in sales and who prearrange visits by phone; consultants and counselors, who have widely dispersed customers; and those who require long periods for reflective work—academic researchers, computer programmers, writers, artists, architects, and designers. In addition, some managerial and professional jobs combine periods of intense contact with others and periods to be alone, either to create a large block of uninterrupted time for work on a single-objective task (such as a report), or to withdraw temporarily from the pressures of the job and from "information overload." The selective access to public and private data banks, and to consultants via home-based telecommunication systems permit these and many other activities to be carried out from the distant amenity regions of exurbia.

Although the diffusion of the electronic cottage is a distinct possibility, there are obstacles to its widespread acceptance. Albertson (1977) cites evidence that the traditional workplace contributes to individual self-esteem, provides companionship, and offers the advantages of direct feedback and motivation. In addition, he raises the possibility that working at home could alter the social meaning of the home and place new burdens on the family unit. Even the journey to work, he notes, is seen by some as an opportunity to make the psychological transition from work to family roles. Salomon (1984) echoes these concerns. In a broad consideration of the nonmonetary impacts of work-at-home telecommuting options, he sees opportunities for those who are tied to the household by either social or physical constraints to enter the labor force. However, he cites possibilities for increased social stress resulting from the mix of family and work lives. Further, he notes how, for women with career aspirations, working from home might deny one the same opportunities enjoyed by those who are in closer personal contact with the job environment. These kinds of considerations may discourage the rapid and widespread adoption of telecommuting practices. They also illustrate how little is understood about the likely side effects of their use.

However important the concerns expressed by Albertson and Salomon are, it is evident that telecommunications do offer the possibility for altering the locational and transportation relationships between the home and workplace. But, aside from these relationships, it is important to consider the ability of different communications media to satisfy the business needs for carrying out orientation, planning, or programming tasks.

Assessing Substitution Potentials

The quality of information may be insensitive to the distance over which communication takes place but, in contrast, it is highly sensitive to the mode of transfer. Decisions on the use of specific means will depend on the suitability of a given mode for the task and on some assessment of the cost effectiveness. According to Collins (1980, p. 108), suitability would depend on behavioral and physical factors. Will the system allow the user to achieve the desired objectives (such as persuasion of the other individual or group)? Is it physically available to others? Can documents be transferred? Cost effectiveness would consider the price of terminals and related equipment, the tariff for transmission of information, and the time required to prepare for and to carry out the communication. Since some services, such as audiovisual teleconferencing, represent large investments, it is important that private firms and public agencies give careful thought to these concerns.

Some of the most exhaustive laboratory and field-trial evaluations of audio and audiovisual communication modes have been conducted by the Communications Studies Group (CSG) at University College, London (Short, Williams, & Christie, 1976). Summarizing their findings: (1) audio media in combination with document transmission facilities are as effective as audiovisual or face-to-face modes for collaborative problem solving between people who know each other; (2) face-to-face meetings are preferred for situations that require bargaining or negotiation; however, those with weak cases should consider the use of audio-only modes; and (3) audiovisual modes are preferred to audio-only modes in situations where it is important to get to know people.

In reviewing the CSG's investigations, Reid (1977) concludes that the absence of visual contact had no measurable effects on the outcome of conversations to exchange information or to solve problems. However, the communications media did influence the outcomes in situations where it was intended to form impressions about people or to resolve conflicts. Outside of the laboratory, Pye (1979) notes that participants in actual teleconferencing meetings regarded them as suitable for simple tasks (such as delegating work) but as less satisfactory for interpersonal interaction (for example, disciplinary interviews). Given the choice between teleconferencing and travel, the decisions depended on (1) the task of the meeting, (2) the familiarity among participants, and (3) the extent of travel that would be involved.

It is evident from the CSG studies that certain tasks lend themselves more than others for telecommunications use. Estimates on the possible levels of substitution for face-to-face meetings, based on surveys in the United Kingdom (Goddard, 1980; Goddard & Pye, 1977; Pye, 1979), indicate that from 53 to 70% of them could have been adequately carried out by audio and audio-graphic (with facsimile devices) systems. Among decentralized offices that have relocated outside of London, in only 22% of the meetings did the tasks absolutely require face-to-face contacts. Estimates for the United States (though questioned by Kalba, 1974, p. 39) suggest that 14 to 22% of urban vehicle trips could be replaced by telecommunications.

Aside from knowing what proportion of business meetings are candidates for substitution by telecommunications, it is important to know to what extent business people would actually use such systems beyond current levels. Investigations in New Zealand led Palm and Farrington (1976, p. 141) to conclude that "no matter what the potentialities for media substitution, we will expect that there will be some inertia which will maintain established forms of business contact." On the other hand, many of the estimates of possible potential substitutions were based on surveys in the early 1970s, prior to the entry of microcomputers and their current diffusion to households and schools. The emergence of a generation brought up with cathode ray tubes (CRTs), keyboards, computers, and modems makes speculation about future behavioral patterns risky. These computer-literate individuals may find telecommunications a more acceptable option for carrying on business, shopping, banking, and other kinds of transactions. Furthermore, external circumstances, such as energy shortages and rising transport costs, may also influence the level of substitution, adding still further to the uncertainty of predictive efforts in this area.

Equally difficult to assess is the likelihood that widespread telecommunications use will generate trips. For example, Coates (1982, p. 181) sees these technologies as "broadly socializing . . . , [and likely to] stimulate the desire for more and diverse flesh-to-flesh communication." Pierce (1977, pp. 164–165) echoes this view: "a telephone call sometimes substitutes for a trip, but more and faster communication tends to engender widespread associations and activities that result in trips." Based on communications' competitive yet complementary effects on trips, it is easy to agree with Cowan (1973) that "communication is . . . not a substitute for travel but a different medium of communication." Thus, rather than view it in isolation, it is necessary to consider the role of telecommunications in relationship to the following environmental circumstances: (1) the organizational environment of the firm or agency (does it have single or multiple units?), (2) the locational arrangements of the operational units within an organization (intra-firm) and between the organization and its essential contacts with other establishments (inter-firm), and (3) the journey-to-work relationships between the organization and its employees. One of the prominent areas of current research by geographers, studies on the location and relocation of office establishments, addresses all three of these considerations and relates clearly to other themes discussed earlier in this chapter.

OFFICE RELOCATION AND THE TRANSPORTATION–COMMUNICATION TRADE-OFF

There are many incentives for decentralizing the office work force. Many individuals prefer the living conditions of exurban amenity zones and would appreciate the opportunity to have their workplaces closer at hand. Certain kinds of corporate organizations would benefit from the lower land costs and taxes, and from the cheaper labor rates that are often associated with more peripheral regions. Also, public authorities may find office dispersal a suitable means for relieving population and land pressure, for curbing transportation congestion in the central parts of cities, and for reducing fuel consumption and derivative forms of pollution.

The rapid growth of the office sector makes it in many ways an ideal candidate for spreading employment opportunities to outlying regions. But in spite of the potential benefits, there remain doubts about the feasibility of increasing the levels of office dispersal and about the long-term desirability of such trends.

Limits to the possible degree of dispersal are imposed by the need for organizations to maintain essential communications. The costs of linkages from peripheral locations must, ideally, not exceed the gains to be derived from decentralization. This is particularly restrictive for those activities that rely most heavily on face-to-face exchanges for types of communication that are not easily transferred by telecommunication technologies and that may not be available either in peripheral office centers or near isolated office sites.

A theoretical understanding of the locational and functional needs of the growing quarternary sector is in a rudimentary state. Extensive empirical description of information flows within and among office establishments would be helpful for assessing the benefits and costs of relocating individual office establishments, both from the firm's and society's perspective. But, such work is scarce and in a developmental form. The influence of office moves on interaction patterns and on the quality of business decisions is largely one of conjecture and intuition. The known relationships between office relocation and the transportation–communication trade-off is sketchy, based on studies in only a few locations—most notably in Sweden by Thorngren (1970) and Törnqvist (1970), in the United Kingdom by Cowan (1973), Daniels (1973, 1979a, 1979b, 1981), Goddard (1973, 1980), and Pye (1979), and in North America by Gad (1979). Examples of research and primary findings for the United Kingdom and for North America are summarized briefly.

The United Kingdom

The problems of congestion in the London area and the need to spur employment expansion in other parts of the country prompted the official development of an office dispersal program. The Location Office Bureau (LOB) was established in 1963 to promote these objectives. By 1976 approximately 1,600 of the 3,500 firms who had consulted the LOB had actually moved. Of these 85% went to parts of southeastern England, most within 100 kilometers of London (Goddard & Pye, 1977). The decision on whether or not to move was often based on a firm's evaluation (usually subjective) of its communication needs. The rigorous identification of those firms or parts of firms that could leave the city's centralized office area without suffering increases in communication–transportation costs and damage to their operations posed a challenge for researchers. The most significant study on this problem in the United Kingdom was initiated by Goddard (1973).

Goddard analyzed the business-related communications of 705 executives from 72 commercial firms in central London. By means of 3-day contact diaries, the characteristics (for example, duration and mode) and objectives (to exchange information or to negotiate) of 5,266 telephone calls and 1,554 face-to-face meetings were recorded, each contact occurring outside the respondent's place of work. This extensive data set provided opportunities for evaluating the substitution potential of telephone-based contacts in relationship to the type of firm.

Goddard's research confirms that establishments involved with routine program-type contacts were the most locationally flexible and most suited for dispersal. Those concerned with planning and orientation activities needed the more highly centralized locations and tended to find telephone substitution less desirable. Manufacturing firms, particularly those that combine their office and production functions at single sites, were prominent among the movers, whereas multilocational business organizations, and tertiary- and quarternary-sector establishments that served national and international markets tended to retain central-city locations for their office and head-quarters functions. These nonmovers required the rich and timely information resources of central London, and because of the importance of face-to-face meetings, the minimization of travel time and related costs was of critical significance.

Studies of changes in the business contact patterns of executives, following their firms' moves from the city (Daniels, 1979b, 1981; Goddard & Pye, 1977), indicate that many of them retained linkages with the London-based establishments that they dealt with prior to moving. In some instances the substitution of contacts within the local area of a move could reduce the need for costly travel and communication to London; however, these contacts take time to develop and depend on the level of growth in the reception area's office component. Furthermore, because of what Goddard and Morris (1976) describe as London's shadow effect over a range of about 100 kilometers from the city's center, many firms found themselves in settlements that lacked the complement of business services that might be expected in more distant places of similar size. Indeed, they recommend that it would be more effective to move to the 100-to-150-kilometer zone, far enough from London to offer incentive for the local development of a greater mix of business services and, hence, where there would be more encouragement to use local services, thus breaking the costly communication and transportation links with the capital.

Aside from inducing changes in intra- and interfirm contacts, the relocation of office establishments can have considerable effect on the journey-to-work requirements of employees. Daniels (1973, 1979a) surveyed the employees of decentralized firms in 1969 and conducted a follow-up survey in several of the same firms in 1976. In some cases the dispersal of office jobs lengthened the average work journey, especially for firms that had to draw employees from several scattered settlements in outlying regions of low population density. However, in general, there was a decline in the average length of work trips following a move. In part, this is explained by the recruitment of clerical support staff from the areas of relocation. Not unexpectedly, there was a pronounced shift to private automobile use and, by 1976, an increasing degree of ridesharing among employees working in peripheral office establishments. In central London and in Greater London, a higher proportion of the sampled work force was using public transportation by 1976.

Notwithstanding the significance of the pioneering research carried out in the United Kingdom, there are several determinants of office location and, indirectly, of substitution possibilities and travel-to-work patterns that remain poorly understood. Among these, Daniels (1979b) cites the significance of the prestige factor in office location and the tendency of financial institutions to preempt the most favored locations, the role of large development companies who, in response to profit motivations, may "lead" demand by altering the pattern of office supply, the impact of public zoning and development controls, and the influence of corporate mergers and acquisitions. Even the tradition of the business luncheon may impose strong locational pulls that encourage the kinds of intense agglomeration found in places like central London.

Compounding the complexity of the research task, the rapidity of change in the business environment and in the physical development of cities calls for a degree of monitoring that is probably beyond the capabilities of interested researchers. For example, in the case of London, we would need to assess the impact of the Conservative government's recent abolition of the Location Office Bureau and of the need to obtain office development permits. Questions arise about the possible locational responses to such changes as the abolition of metropolitan planning departments, the completion of the M-25 orbital road (beltway) around London, the release of centrally located land for office and residential expansion in the abandoned dock lands, the construction of a new monorail between the dock-lands area and the financial district, and the significance of hypermarkets (large multipurpose shopping establishments) in providing services to outlying areas. All these changes exert influence on locational decisions that have bearing on the transportation–communications trade-off. Many of these kinds of changes are common to cities in other Western societies, including North America.

North America

In general, systematic empirical analysis of office location practices and of contact linkages among firms has been neglected by North American researchers. Daniels (1979b) attributes this to the absence of public policies on national and regional economic development and to the absence of government programs to encourage decentralization. Armstrong's (1979) work has added significantly to the understanding of broad national trends in office development, but it largely ignores the behavioral dimensions of office work and the nature of work-related travel within metropolitan regions.

Toronto

Research in Toronto by Gad (1979) provides the only significant North American study that is empirically and conceptually comparable with the work in the United Kingdom and Sweden. Based on 1,060 1-week diaries and 541 questionnaire surveys of employees in 30 establishments, he analyzed the external linkage patterns for face-to-face business meetings with individuals from other establishments requiring movement to other locations.

At the time of Gad's survey in 1971 Toronto was experiencing rapid growth in office activities. Although the metropolitan area contained only 11% of Canada's labor force, it accounted for 22% of the country's finance, insurance, and real estate establishments. Its stock market was second to New York's within North America. Between 1951 and 1971 the suburban boroughs had attracted a considerable proportion of manufacturing jobs. And while these boroughs accounted for only 10% of the office floor space in 1965, by 1976 this increased to 21%. Nonetheless, office space within the inner core of the city (approximately 4 square kilometers) increased from 8 million to 43 million square feet between 1953 and 1976, and office employment expanded by 65,000 workers. Seeing this expansion as a threat to the residential role of the central city and as a contributor to problems of traffic congestion, the City of Toronto attempted to contain this rapid growth. Through downzoning of land in the central core and through the use of restrictive height bylaws, the city has attempted to channel office expansion to suburban centers and thereby to enhance the residential role of the central city and to reduce the level of commuting to the city.

The rapid expansion of office jobs in Toronto and the city's attempts to establish controls over this process give Gad's work fundamental significance. In developing an index of communication damage, he illustrates how different categories of office industries vary in their decentralization potentials. The index is based on the intensity of contacts (such as links per communicant per week), the degree of contacts among categories of firms, and the spatial concentration of meeting partners. Using the number of links per communicant per week with meeting partners in the central corridor, he observes that law firms (11.3 links), public relations consultants (8.4), and executive search consultants (5.9) were the most dependent on central accessibility. The central corridor is an area "comprising only 2.6 per cent of Metropolitan Toronto's area . . . [that had] . . . about 77 per cent of the net rentable floor space in office buildings over 20,000 sq. ft. at the beginning of the 1970s" (Gad, 1979, p. 283). Good candidates for decentralization from this area would be life insurance firms (1.6), consumer and business finance firms (1.3), and manufacturing firms (1.1).

Expanding Gad's analysis, Code (1983) provides a systematic evaluation of Toronto's decentralization policy. He observes that, even with 1980 rental rates for office space in the downtown core 96% higher than in the suburbs, there was insufficient incentive for relocating from the center. Code claims that the resistance to outward migration of office activities is rooted in the contact requirements of firms, the strong preference for face-to-face meetings, and the need to "shop" among many possible information sources. He sees a "quality" element to the information opportunities in the core that is not present in the smaller suburban centers. Although he does not quantify the cost advantages of being accessible to quality information sources, he does demonstrate how the transportation and time costs of operating from a suburban center exceed the differential in rent between the core and the suburbs and, on this basis, illustrates the financial advantages to the firm in maintaining a central location.

Simulations of Dispersed Work Centers

Whereas the work on Toronto provides a strong empirical basis for evaluating patterns of office contacts and relocation potentials, alternative approaches have focused on scenario and simulation techniques to assess the probable impact and merits of hypothetical spatial patterns of office locations in metropolitan regions. The primary focus of work by Harkness (1973, 1977) and Nilles, Carlson, Gray, and Hanneman (1976) has been to identify changes in the journey-to-work patterns that would result from different spatial configurations of work centers and from greater use of communication linkages.

Nilles et al. (1976) estimated that 48% of the U.S. work force were candidates for telecommuting to work but observe that this would require significant reorganization of many businesses from the current dominant pattern of centralized single-site establishments and fragmented branch-bank-type structures or isolated parts of firms to dispersion of work forces to centers throughout urban regions. Workers would commute to the closest one. At higher levels of diffusion, they envisage neighborhood work centers that could serve several companies through telecommunication switching centers. These would be equipped with audiovideo conference facilities and computer terminals.

Through analysis of the regional office of a large insurance firm in Los Angeles, Nilles et al. (1976) demonstrate the utility of this approach. Much of this firm's work dealt with routine computer processing and updating of information. Calibrating a cost–benefit model with data on the firm's contact needs and the residential locations of its employees, they demonstrated how the introduction of 18 remote work centers in the Los Angeles basin could reduce the average commuting distance from 10.7 to 3.9 miles. Projecting such changes to the national level, it is evident that these schemes could have profound impacts on reducing fuel consumption and related pollutants. However, as the authors note, firms must see clear economic rewards for adopting the dispersed work station and telecommuting option.

In evaluating proposals for the dispersion and diffusion of work centers, Hiltz and Turoff (1978, p. 477) note that since commuting costs are borne by workers (private vehicles or public transportation) and government (transport facilities), and since the investment in work stations is at the firm's expense, there are no financial incentives for such developments. According to Salomon (1984), the assumption of the Nilles et al. (1976) research (and also that by Harkness) that up to 50% of the work force might adopt telecommuting work patterns is groundless and places in doubt the utility of current simulation approaches to evaluating and proposing shifts to dispersed work centers.

Evaluations of Office Location Research

As illustrated in the previous sections, analyses of office contact patterns give good indications of the types of firms that can successfully operate from noncentral locations, that can substitute communications for transportation, and that can locate in closer average proximity to their employees. Although the assumptions of some previous simulation analyses may be tenuous, these techniques have potential for evaluating some of the possible benefits and costs associated with various locational patterns of office sites and satellite office facilities. Nonetheless, office location patterns are controlled, in part, by a host of external factors that have not been carefully integrated into most predictive and cost–benefit models. Among those mentioned by Daniels (1979b) are the role of office development companies and the impacts of corporate mergers and acquisitions. In addition, one would have to consider the behavioral responses of corporate executives and commuters to changes in the transportation and communication systems.

A fundamental criticism of much of the research on the effects of office relocation has been advanced by Clapp (1983). Whereas most attempts to assess the costs and benefits of office relocation have focused on a single office establishment, Clapp argues that in order to maintain previous levels of face-to-face meetings with the relocated establishment, it is possible that the extra travel by employees of other firms will exceed the gains (such as lower wages and rentals) achieved by the relocated firm. Alternatively, abandoned contacts may result in damage to the quality of information that these firms must rely on.

Clapp's critique exposes once again the thin empirical and conceptual base that exists for evaluating the complexities of office relocation and their relationships to the transportation–communications trade-off. Any selective relocation of a given establishment must consider the impacts on client establishments as well. This system-wide perspective would require that existing diary survey methods for gathering research data on communication contacts be expanded to include those who have dealings with the relocated establishments. The importance of these contacts would, in principle, encourage centralization rather than dispersal of the office community, lending support for the arguments advanced by Code (1983). It appears, then, that Clapp may have detected a fundamental conceptual weakness in the benchmark studies on office contacts in London, Toronto, and Sweden, and in the extensive simulation efforts by scholars in the United States. At this stage, at least, there are sufficient uncertainties in our understanding about the role of office relocation to suggest that public policies that focus on office location as a basis for regional development and transportation planning should proceed with caution.

CONCLUSIONS

Although many office activities will continue to require central locations, it is evident that the technological options for more decentralized patterns of job locations will be enhanced in the future. Nonetheless, it is evident, too, that such development must recognize the specific information and contact requirements of individual establishments, and the needs of many information-sensitive activities to occur in metropolitan settings.

Characterizing the metropolis as a "giant feedback-mixer of society," Richard Meier (1973, p. 353) sees no limits to the importance of information exchange. In part, this is supported by Melvin Webber's (1973) concerns for related trends in the growth of "knowledge" industries (producing and distributing information and ideas) and the growing importance of communication technologies.

In an apparent contradiction to the centralizing forces of information and contact requirements, recent North American trends toward exurban dispersal of residences and jobs indicate the increasing importance attached to personal perceptions of amenity. While the contact requirements of the knowledge industries pose constraints on this process, time–space convergence represents a potential relaxation of such constraints. The penultimate geographic expressions of technological substitution for direct contacts are Webber's (1963, 1964) "community without propinquity" and "nonplace urban realms," but as of the mid-1980s, the prerequisite levels of mobility and access to advanced communications are less than he envisioned in the early 1960s.

This chapter has stressed the importance of behavioral constraints that set outer limits to the degree of acceptable physical separation from the centers of metropolitan life. Among those mentioned, the most significant are the need for face-to-face contacts and the law of constant travel time and trip frequency. In addition, the related biological (such as the need for rest) and social constraints (terms of employment that favor prescribed work hours and workdays) impose limits to the geographical range of daily movements. On the strength of these constraints I conclude that transportation technologies will remain more fundamental to the spatial structuring of urban regions than communication technologies. It seems axiomatic to the material basis of human existence that transportation provides the vital links to one's day-to-day system of life support and that the outer limits of transport feasibility set the range of one's social, economic, and other opportunities. Such a view portends gradual rather than revolutionary changes.

According to Brotchie (1984, p. 584) developments in computer, telecommunications, and transportation technologies will "induce further substantial changes in lifestyle, in patterns of urban activities, and in urban form." Among the changes of spatial relevance, he suggests reduced links between production centers and locations of final demand, reduced ties between residential and employment locations, changes in human activity patterns, altered space requirements for many activities and land uses, and changes in the geographical patterns of population density. However, evidence presented in this chapter indicates that such changes are manifested only indirectly through gradual modifications in accepted rules of behavior. An example that has bearing on the question of substitution may help to illustrate this point.

The day is regarded traditionally as the fundamental temporal unit for programming the choice, sequence, and duration of events in an individual's life. However, in pursuing desires for the amenities of an exurban environment, the week may become a more significant unit of program planning. Of many possible alternatives (if the terms of employment permit), one might decide to allocate only 3 workdays at the office for essential face-to-face meetings and to spend alternate weekdays at home engaged in work that might benefit from greater isolation.

In this example, one could consider relocating the home to an area of higher amenity at greater distance from the city office without increasing the weekly aggregate of time spent in traveling to work. Such a shift in behavior would expand the range of opportunities for exurban living. For instance, if one assumes a modest average travel speed of 80 kilometers per hour and a time allocation of 30 minutes per day (150 minutes per 5-day workweek) for the one-way journey to work, then the outer distance between home and office would be 40 kilometers. However, by substituting telecommunications from home for 2 days per week and allowing for the full 150 minutes per week for traveling to work, the range for three 50-minute work trips would be 67 kilometers. Thus, without increasing the weekly allotment of time in work-related travel and without altering the level of linkage to the city, the zone of potential exurbanite expansion increases by 180%, from 5,027 to 14,103 square kilometers.

Even under assumptions of very modest travel speeds and current standards of communications capability, the land supply for exurban residential expansion could increase dramatically. If one accepts scenarios about the development of automated control of automobiles on select highways by the year 2010, the availability of semiautomatic automobiles that would be safe at controlled speeds (even for the elderly and the handicapped to drive) by the year 2040 (see Svidén, 1983), and computerized improvements in traffic coordination, then the resulting relaxation of stress on drivers could result in an even much wider geographic choice of job–residence relationships.

As part of a research program sponsored by the Massachusetts Institute of Technology, Svidén has considered "automobile usage in a future information society" based on trends in Sweden. In general, his alternative scenarios indicate that with higher degrees of dispersed living, shifts toward an information society will be accompanied by higher—not lesser—levels of automobile use. He envisions the possibility of a high "info-mobility society" developing over the next two generations, a society where access to both automobiles and tele-data terminals is much more common than at present.

A high "info-mobility society" is an alluring prospect. But with it come uncertainties. Will expanded fields of household interaction go beyond the effective range of governance? Or will the new technologies allow for broader participation in the political process at local and regional levels? Will there be evidence of social bias in the selective accessibility to the new technologies and their associated life-styles? Will fears expressed by Webber (1973) and Abler (1975) that by allowing for more selective patterns of communication, new technologies could promote higher degrees of segregated social networks than currently exist come true? Appropriate forms of social innovation, structural changes in institutional organizations, even public subsidization for widely distributing the prerequisite equipment and education may be needed to adapt to the impending changes. No matter how society responds to these and other questions, the future will be as much a product of political choice as one of technological possibility. My best guess for the 1990s is that both mobility and telecommunications systems will expand in their synergistic relationships, that both decentralized exurban development and centralized urban patterns will continue to expand, offering the range of contact requirements essential for a complex postindustrial society, and that transportation planning must accommodate simultaneously the needs to support both patterns of development.

References

Abler, R. (1975). Effects of space-adjusting technologies on the human geography of the future. In R. Abler, D. Janelle, A. Philbrick, & J. Sommer (Eds.), *Human geography in a shrinking world* (pp. 35–56). North Scituate, Mass.: Duxbury Press.

Abler, R. (1977). The telephone and the evolution of the American metropolitan system. In I. de Sola Pool (Ed.), *The social impact of the telephone* (pp. 318–341). Cambridge: The MIT Press.

Albertson, L. A. (1977). Telecommunications as a travel substitute: Some psychological, organizational, and social aspects. *Journal of Communication*, *27*, 32–43.

Armstrong, R. B. (1979). National trends in office construction, employment and headquarter location in U.S. metropolitan areas. In P. W. Daniels (Ed.), *Spatial patterns of office growth and location* (pp. 61–93). Chichester, England: John Wiley.

Berry, B. J. L. (1980). Urbanization and counterurbanization in the United States. *Annals, American Academy of Political and Social Science, 451,* 13–20.

Brotchie, J. F. (1984). Technological change and urban form. *Environment and Planning A, 16,* 583–596.

Burton, I. (1963). A restatement of the dispersed city hypothesis. *Annals of the Association of American Geographers, 53,* 285–289.

Clapp, J. M. (1983). A model of public policy toward office relocation. *Environment and Planning A, 15,* 1299–1309.

Coates, J. F. (1982). New technologies and their urban impact. In G. Gappert & R. V. Knight (Eds.), *Cities in the 21st century* (pp. 177–195). Beverly Hills, Calif.: Sage Publications.

Code, W. R. (1983). The strength of the centre: Downtown offices and metropolitan decentralization policy in Toronto. *Environment and Planning A, 15,* 1361–1380.

Collins, H. (1980). Forecasting the use of innovative telecommunications services. *Futures, 12,* 106–112.

Coppack, P. M. (1985). *An exploration of amenity and its role in the development of the urban field.* Unpublished Ph.D. thesis, University of Waterloo, Waterloo, Ontario.

Cowan, P. (1973). Moving information instead of mass: Transportation versus communication. In G. Gerbner, L. P. Gross, & W. H. Melody (Eds.), *Communications technology and social policy* (pp. 339–352). New York: John Wiley.

Daniels, P. W. (1973). Some changes in the journey to work of decentralized workers. *Town Planning Review, 44,* 167–188.

Daniels, P. W. (1979a). Office dispersal and the journey to work in greater London: A follow-up study. In P. W. Daniels (Ed.), *Spatial patterns of office growth and location* (pp. 373–400). Chichester, England: John Wiley.

Daniels, P. W. (1979b). Perspectives on office location research. In P. W. Daniels (Ed.), *Spatial patterns of office growth and location* (pp. 1–28). Chichester, England: John Wiley.

Daniels, P. W. (1981). Transport changes generated by decentralized offices: A second survey. *Regional Studies, 15,* 507–520.

Doxiadis, C. A. (1968). Man's movement and his city. *Science, 162* (18 October), 326–334.

Fernie, J. (1977). Office linkages and location: An evaluation in three cities. *Town Planning Review, 48,* 78–89.

Friedmann, J., & Miller, J. (1965). The urban field. *Journal of the American Institute of Planners, 31,* 312–320.

Gad, G. H. K. (1979). Face-to-face linkages and office decentralization potentials: A study of Toronto. In P. W. Daniels (Ed.), *Spatial patterns of office growth and location* (pp. 277–323). Chichester, England: John Wiley.

Ginzberg, E., & Vojta, G. J. (1981). The service sector of the U.S. economy. *Scientific American, 244,* 48–55.

Goddard, J. B. (1973). Office linkages and location: A study of communications and spatial patterns in central London. *Progress in Planning, Vol. 1, Part 2.* New York: Pergamon Press.

Goddard, J. B. (1980). Technology forecasting in a spatial context. *Futures, 12,* 90–105.

Goddard, J. B., & Morris, D. M. (1976). The communications factor in office decentralization. *Progress in Planning, 6,* 1–80.

Goddard, J., & Pye, R. (1977). Telecommunications and office location. *Regional Studies, 11,* 19–30.

Gottmann, J. (1977a). Megalopolis and antipolis: The telephone and the structure of the city. In I. de Sola Pool (Ed.), *The social impact of the telephone* (pp. 303–317). Cambridge: The MIT Press.

Gottmann, J. (1977b). Urbanisation and employment: Towards a general theory. Paper presented as the Walter Edge Public Lecture at Princeton University, Princeton, N.J., December 8.

Hägerstrand, T. (1970). What about people in regional science? *Papers of the Regional Science Association, 24,* 7–21.

Harkness, R. C. (1973). *Telecommunications substitutes for travel.* Unpublished Ph.D. thesis, University of Washington, Seattle.

Harkness, R. (Ed.). (1977). *Technology assessment of telecommunications/transportation interactions.* Menlo Park, Calif.: Stanford Research Institute.

Hiltz, S. R., & Turoff, M. (1978). *The network nation: Human communication via computer.* Reading, Mass.: Addison-Wesley.

Hupkes, G. (1982). The law of constant travel time and trip-rates. *Futures, 14,* 38–46.

Hymer, S. (1972). The multinational corporation and the law of uneven development. In J. N. Bhagwati (Ed.), *Economics and world order from the 1970s to the 1990s* (pp. 113–140). London: Macmillan.

Illich, I. D. (1974). *Energy and equity.* London: Calder and Boyars Ltd.

Janelle, D. G. (1968). Central place development in a time-space framework. *The Professional Geographer, 20,* 5–10.

Janelle, D. G. (1969). Spatial reorganization: A model and concept. *Annals of the Association of American Geographers, 59,* 348–364.

Kalba, K. (1974). Urban telecommunications: A

new planning context. *Socio-Economic Planning Science, 8*, 37–45.

Knight, R. V. (1982). City development in advanced industrial societies. In G. Gappert & R. V. Knight (Eds.), *Cities in the 21st century* (pp. 47–68). Beverly Hills, Calif.: Sage Publications.

McGraw-Hill Information Systems. (1985). People buying fewer but more expensive personal computers. *London* (Ontario) *Free Press*, 20 June.

Martin, J. (1978). *The wired society*. Englewood Cliffs, N.J.: Prentice-Hall.

Meier, R. L. (1962). *A communications theory of urban growth*. Cambridge: The Joint Center for Urban Studies, MIT and Harvard University.

Meier, R. L. (1973). Urban ecostructures in a cybernetic age: Responses to communications stress. In G. Gerbner, L. P. Gross, & W. H. Melody (Eds.), *Communications technology and social policy* (pp. 353–362). New York: John Wiley.

Nilles, J. M., Carlson, F. R., Jr., Gray, P., & Hanneman, G. J. (1976). *The telecommunications-transportation tradeoff: Options for tomorrow*. New York: John Wiley.

Olander, L. (1979). Office activities as activity systems. In P. W. Daniels (Ed.), *Spatial patterns of office growth and location* (pp. 159–174). Chichester, England: John Wiley.

Palm, R., & Farrington, J. (1976). Telephone use in New Zealand inter-city contacts: The businessman's view. *Area, 8*, 139–142.

Pierce, J. R. (1977). The telephone and society in the past 100 years. In I. de Sola Pool (Ed.), *The social impact of the telephone* (pp. 159–195). Cambridge: The MIT Press.

Pye, R. (1979). Office location: The role of communications technology. In P. W. Daniels (Ed.), *Spatial patterns of office growth and location* (pp. 239–275). Chichester, England: John Wiley.

Reid, A.A.L. (1977). Comparing telephone with face-to-face contact. In I. de Sola Pool (Ed.), *The social impact of the telephone* (pp. 386–414). Cambridge: The MIT Press.

Salomon, I. (1984). Man and his transport behaviour Part la. Telecommuting—promises and reality. *Transport Review, 4*, 103–113.

Schaeffer, K. H., & Sclar, E. (1975). *Access for all: Transportation and urban growth*. Baltimore: Penguin Books.

Short, J., Williams, E., & Christie, B. (1976). *The social psychology of telecommunications*. London: John Wiley.

Spencer, J. E., & Thomas, W. L., Jr. (1969). *Cultural geography: An evolutionary introduction to our humanized earth*. New York: John Wiley.

Svidén, O. (1983). Automobile usage in a future information society. *Futures, 15*, 478–490.

Thompson, W. R. (1968). *A preface to urban economics*. Baltimore: Johns Hopkins University Press.

Thorngren, B. (1970). How do contact systems affect regional development? *Environment and Planning, 2*, 409–427.

Toffler, A. (1980). *The third wave*. New York: William Morrow.

Törnqvist, G. (1970). Contact systems and regional development. *Lund Studies in Geography, Series B, Human Geography*, no. 35. Lund, Sweden: Department of Geography, Royal University of Lund.

Ullman, E. (1954). Amenities as a factor in regional growth. *Geographical Review, 44*, 119–132.

U.S. Bureau of the Census. (1982). *The journey to work in the United States: 1979*. Current Population Reports, series P-23, no. 122. Washington, D.C.: U.S. Government Printing Office.

U.S. Department of Transportation (1973). Home-to-work trips and travel. *National personal transportation study*, Report No. 8.

Westaway, J. (1974). The spatial hierarchy of business organizations and its implications for the British urban system. *Regional Studies, 8*, 145–155.

Webber, M. M. (1963). Order in diversity: Community without propinquity. In L. Wingo, Jr. (Ed.), *Cities and space: The future use of urban land* (pp. 23–54). Baltimore: Johns Hopkins University Press.

Webber, M. M. (1964). The urban place and nonplace urban realm. In M. M. Webber, J. W. Dyckman, D. L. Foley, A. Z. Guttenberg, W. L. C. Wheaton, & C. B. Wurster (Eds.), *Exploration into urban structure* (pp. 79–153). Philadelphia: University of Pennsylvania Press.

Webber, M. M. (1973). Urbanization and communications. In G. Gerbner, L. P. Gross, & W. H. Melody (Eds.), *Communications technology and social policy* (pp. 293–304). New York: John Wiley.

Zelinsky, W. (1975). Personality and self-discovery: The future social geography of the United States. In R. Abler, D. Janelle, A. Philbrick, & J. Sommer (Eds.), *Human geography in a shrinking world* (pp. 108–121). North Scituate, Mass.: Duxbury Press.

16 URBAN TRANSPORTATION: *POLICY ALTERNATIVES*

DAVID A. PLANE
University of Arizona

THE CONTEXT AND GOALS OF URBAN TRANSPORTATION POLICY

From our perspective of the late 1980s, what can we observe about the array of policy alternatives available to those who will design and manage the urban transportation systems of the future? How has the recent academic research and policy experimentation reported in the previous chapters of this book better equipped us to judge the likely success or failure of these policy alternatives? And why is the perspective of the *geographer* particularly useful for understanding urban transportation questions?

Let us begin with the last question. Economist Kenneth Boulding has observed that "of all the disciplines, geography is the one that has caught the vision of the earth as a total system" (1966, p. 108). And, if one were to attempt to distill the major urban transport policy lessons of the past two or three decades into a single thought, one might be tempted to state that we have come to realize the need for a *systemic* approach to designing for the evolution of our metropolitan areas (Jantsch, 1975). In this final chapter we shall discuss in detail how a systemic approach entails the consideration of quite a wide variety of policy options.

We now recognize that a plethora of specific transportation policy alternatives exists. It will be helpful to categorize all policy options into two basic types. One group of alternatives includes methods to manipulate the available *supply* of transportation facilities. Building a new lane on the freeway, opening a new subway line, adding extra buses to existing routes, paving a bicycle path, or instituting a new transport mode such as a paratransit system would all be alternatives of this first type. A second category of alternatives includes those that attempt to manipulate the *demand* for additional transportation facilities. Synchronizing traffic lights so that traffic flows more freely, eliminating downtown parking to encourage fuller use of existing public transit capacity, or promoting carpooling to reduce the number of vehicles on the city street network during peak periods are examples of this second group of policy options.

Although a useful distinction may be made between supply and demand alternatives, a fundamental lesson that must be learned before developing any truly successful urban transportation program is that transportation supply and transportation demand cannot be totally separated. They are, rather, inextricably and complexly interconnected. Moreover, as noted in chapter 1, urban transportation facilities and urban travel demand should be viewed merely as key elements within the much broader context of the nexus of built form and human activity patterns that we now call metropolitan areas. All too often urban transportation planning takes place in isolation from other types of urban planning and policy making. In fact, a third group of nontransportation transportation alternatives exists. Staggering work schedules or better designing cities to integrate work and home locations are examples of policies outside the transportation planner's usual sphere of influence. These policy alternatives could, however, result in substantial transportation benefits.

The geographer, whose quest involves trying to understand how diverse elements interact at various locations on the earth's surface, is ideally prepared to advance knowledge about the extremely complicated systems giving rise to urban transportation problems. The geographer's contribution is a particularly important one at the present time. One result of the increasing awareness of the complexity of urban transportation questions has been rising pessimism about our abilities to articulate successful policy alternatives. This increased awareness, however, also means that we are now in a better position to assess the likely results of proposed transportation projects.

During our current transport and urban development epoch, the auto age, two major shifts have taken place in the direction of U.S. urban transportation policy. Emphasis first shifted away from simply providing more and more highway capacity in an attempt to meet rising urban travel demand. It was realized that increased highway capacity often had the effect of altering land development patterns, thus encouraging longer trips inducing demand for still more new highway construction. Consequently, other supply-side options were explored. Grand-scale policy alternatives involving the building of new public transit systems came to occupy center stage. Soon disenchantment with these policies also began to be expressed. Particularly criticized was the high cost of many of these alternatives. Recently the spotlight has moved again. Now the focus is on lower-cost alternatives designed to manipulate, manage, and maximize the use of existing transportation facilities.

It is interesting that these two major policy shifts have taken place in the context of relatively unchanged urbanization trends. Decentralization of residents, industry, and retailing has continued apace throughout the auto age. The major policy shifts have resulted from changed perceptions about urban transportation problems, not from radically altered circumstances (Altschuler, 1979).

Although the primary concern has been and continues to be summed up in one word, *congestion*, increasingly planners have recognized that other concerns, such as the desire for cleaner air, more social equity, and a higher quality of life in residential areas, imply a multitude of goals that sometimes conflict with one another. A single-objective policy of increasing highway capacity results in smog, neglect of transit (which is often the only form of transportation available to a lower-income segment of the population), and increased noise and traffic safety hazards. No longer can the linkages between transportation policy and other public policy concerns be ignored by the performers in the political circus (Meyer & Gomez-Ibanez, 1981).

A fundamental result of the current disenchantment with the past policy initiatives, which attempted to provide answers to urban transportation problems via massive infusions of public funds to increase transport supply, is a change from a longer- to a shorter-term perspective. Rather than renewed attempts to provide the thus far elusive "final" solutions, efforts now are focusing on less capital-intensive, quick-response methods. Some of the attention formerly directed toward building systems for the future is now being diverted to fine-tuning the systems of the present and to thinking about how those systems can be adapted to meet continually changing conditions.

Transportation policy is also not being viewed now as a simple selection of one program out of a few major candidates. Rather, it is increasingly being thought of as the crafting of an appropriate package of complementary alternatives selected from a much broader spectrum of options. Many of the alternatives now being explored most actively involve managing the demand for transportation rather than increasing the supply of publicly provided facilities.

The present disenchantment with capital-intensive alternatives is part of a pervasive and long-standing predilection of U.S. transport policy for following the market. It has long been accepted wisdom that it is desirable, to the extent possible, for free-enterprise market forces to shape available transportation options. The recent deregulation of the intercity trucking and passenger airline industries is perhaps the strongest expression to date of this predilection, but the current trends in intraurban transportation policy also reflect this conventional wisdom. Many of the policy alternatives now being touted demonstrate a heightened sensitivity to designing incentives in order to harness the spirit of private enterprise to the production of overall public benefit.

A TYPOLOGY OF APPROACHES TO URBAN TRANSPORTATION POLICY

Before studying the specifics of current urban transportation policy alternatives, we should be aware of the ways in which these alternatives are articulated, selected, and ultimately implemented. This will help us understand the popularity of certain alternatives, as well as providing some clues as to why some seemingly potent alternatives rarely get instituted.

Geographer Brian Berry has suggested a useful schema for urban systems planning (Berry, 1981), which is equally relevant for urban transportation planning. He contends that planning generally takes the form of one of four major styles.

First, he observes that frequently what masquerades as planning is, in some sense, the antithesis of planning. It is simply crisis management or a hurried search for answers to current problems. This extremely common style of policy making is not future oriented. Such ameliorative problem-solving planning characterizes well much of the ongoing debate over urban transportation policy. A high degree of interest in urban transportation questions has been engendered in part by the sense that somehow these urban problems have gotten well beyond the reach of existing public control. Most attention is given to coping with the most acute situations that present themselves and to mitigating their short-term consequences. Little thought is given to guiding the evolution of the overall metropolitan system toward some desirable future state—thereby eliminating the causes of the current problems. This ameliorative problem-solving style of planning is akin to the all-too-frequent practice in medicine of treating symptoms and suggesting cures rather than emphasizing ways to promote health.

Berry's second style of planning is essentially a future-oriented version of ameliorative problem solving. The idea here is to attempt to foresee problems before they develop or become serious and then to devise programs to avoid the problems. In the United States such policies have often taken the form of new regulatory mechanisms rather than of direct governmental intervention. The traditional urban transportation planning process, which was discussed at length in chapter 3, also reflects elements of this trend-modifying policy-making style. First, existing trends in population and economic growth are projected and new activity is alloted to zones in the metropolitan area. Then, the amount of traffic likely to be generated in each zone is modeled. Ultimately, forecasted trips are assigned to specific route segments. Thus the anticipated levels of congestion may be studied and the bottlenecks in the existing network can be identified. The implicit presumption is that planning efforts should be allocated on a "worst-first" basis; the goal is simply to avoid the problems that are anticipated in the absence of any other policy.

The trend-modifying type of transportation policy became entrenched during the period in which transport planners were exhibiting their fixations with supply-side-only transportation solutions. Now, however, the interconnections between transportation supply and transportation demand are acknowledged, albeit still not well understood. Though such knowledge is far from uniformly evident in actual transport planning practice, recent academic research has focused on methods for solving several of the traditional steps of the urban transportation modeling process simultaneously. Supply–demand feedbacks will thus become increasingly embedded in the planning process. The larger question posed by the prevalence of trend-modifying approaches to transportation planning is whether the role of planning is simply to respond to what would otherwise take place or to direct the path of events toward some desirable future state of the metropolitan system.

Under the final two styles of planning set forth by Berry policies are not developed simply to avoid present or future problems. Instead, the mission of planning is to achieve some distinguishably different future. These styles are termed the exploitative opportunity-seeking and the normative goal-oriented approaches.

The first of these describes much regional development planning, as practiced by numerous public and quasi-public development agencies, metropolitan chambers of commerce, and a number of other private groups. The overarching (but often implicit) goal of such policy formulation is stated quite concisely: to maximize growth. From economic growth, other social benefits are presumed to accrue. Urban transportation engineering has frequently taken place analogously. The sole (and again often implicit) goal has been to maximize throughput in the system. Ways are sought to handle the greatest possible volume of traffic in the most efficient way. Success or failure of a program is often measured simply by how much additional capacity is achieved. At best, a cost–benefit framework may be utilized.

In exploitative opportunity-seeking economic development planning much attention is devoted to maximizing returns on investment through multiplier effects. An infusion of so-called basic or export-oriented employment is viewed as stimulating additional growth in services and other nonbasic sectors of the economy. The growth of jobs in these derivative industries, in turn, creates the demands for still further nonbasic growth. In traditional transportation planning, the analogous feedback mechanism has been called the *black-hole theory* of highway investment. Initial investments in improved highway facilities result in greater ease of travel and hence altered travel patterns—an increase in average trip length and in the number of trips being made. Over time, as shown in Figure 16-1, this increased demand stimulated by the initial investment in increased transport supply fuels the need for even more facilities, and the feedback process repeats itself.

This commonly observed phenomenon has been called the black-hole theory because some people now claim that money invested in highways has no ultimate impact on reducing congestion. Investing in highways, then, is like throwing money into a black hole. In fact, the only benefits that may accrue from this interactive loop (besides those which pertain to more jobs for engineers, road builders, and auto-industry employees) are the good aspects of living in lower-density environments. Believers in the black-hole theory think that at some point the attendant disbenefits of continued investment outweigh such potential benefits. The willingness to provide ever-expanded highway capacity is now portrayed as the villain causing smog, density levels too low to provide cost-effective transport options other than the private automobile, a pattern of development that makes the provision of

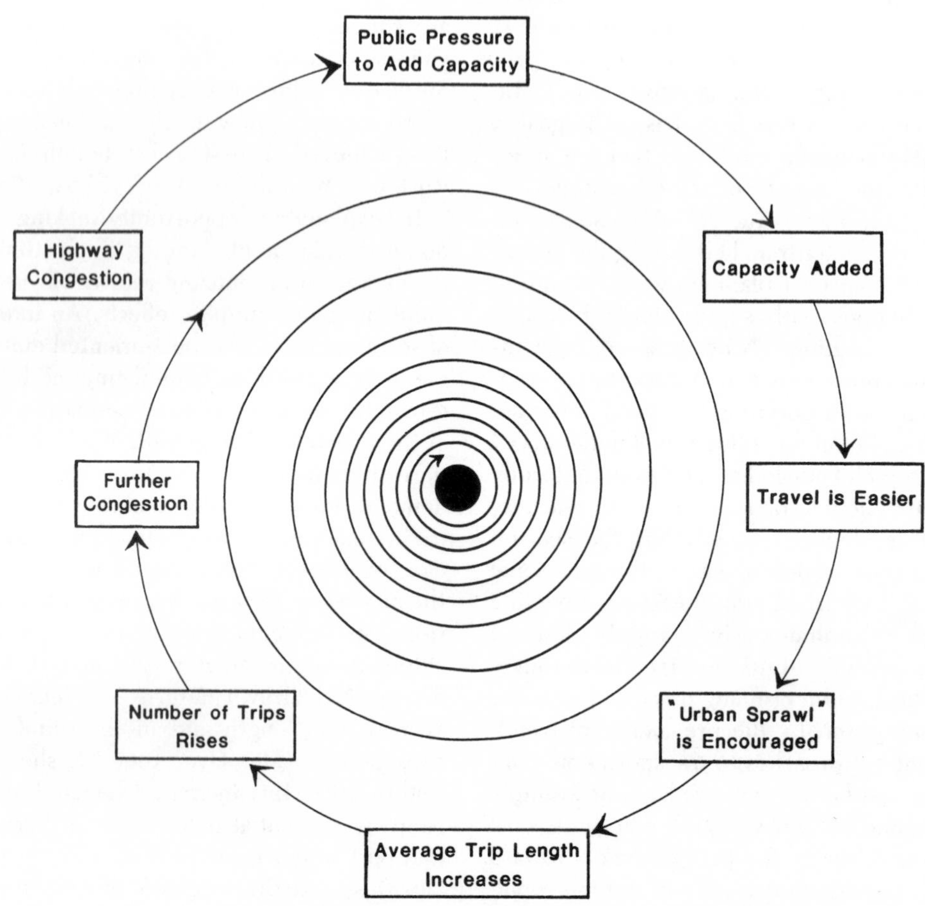

Figure 16-1. The black-hole theory of highway investment.

other public services such as schools, garbage collection, and so on excessively expensive, and an overall reduction in our individually experienced sense of place.

Unlike exploitative opportunity-seeking planning, the *normative goal-oriented* style makes explicit, rather than implicit, the reasons why policy is being formulated. Planning is viewed as a process that identifies both a desired future state of the overall metropolitan system and the means by which the present system could evolve toward it. The word normative reflects the notion that under this style of planning the idea is to attempt to establish some designated standard model or pattern for system evolution. Often urban transportation planning may appear, on paper or in public hearings, to be of this ideal type. Appearances, however, are often illusory. While in form planning is generally posed as involving lofty goal identification, the dearth of effective implementation tools that can be used because of political realities has meant that the preponderance of actual policy making has been of the more primitive problem-alleviation types. The impacts of the decision-making context on the choice of specific urban transportation policy alternatives will be seen as we now examine the spectrum of available options.

TRANSPORT SUPPLY ALTERNATIVES

We first survey the alternative ways in which urban transportation planners can attempt to increase the supply of available transportation facilities. We shall look at past experience and then attempt to assess prospects for new transport supply initiatives—those involving both the installation of new transportation modes and the new packaging of existing modes and options.

Lessons from the Highway-Laying Epoch

As previously mentioned, the spotlight of U.S. urban transportation planning until recently was focused on the provision of facilities to accommodate the inexorable demands of the private automobile (as well as those of big intercity trucking rigs). The concrete lacework of freeways built from the 1950s to the late stages of construction of the nationwide Interstate Highway System in the 1970s is perhaps the most characteristic aspect of the recent pattern of urban growth and development in the U.S. Throughout the prime period of highway building the overarching policy goal was to provide safe, rapid automobile access to virtually any part of the metropolitan area. Concern with congestion was paramount in the thinking of the times. It was believed that easy automobile travel was a necessary ingredient for Americans to reap the pleasures of widely preferred low-density suburban life-styles.

To understand how disillusionment set in with this solitary emphasis on providing more and more lanes of high-speed roadways, we should study the concept of the *carrying capacity* of a street or freeway. A characteristic relationship has been found between the time needed to travel the length of a link in the urban road network and the amount of traffic present on that link (for example, Krueckeberg & Silvers, 1974). Figure 16-2 shows this relationship.

As the amount of traffic on a network link increases, there are initially no delays. At some levels of flow, however, travel times begin to increase as traffic becomes sluggish. The number of vehicles at which traffic just begins to slow down is known as the *free-flow volume*. Free-flow volumes depend on such factors as the width of lanes, posted speed limits, timing of traffic lights, and the mix of vehicle types. At traffic volumes about the free-flow level, congestion sets in rapidly. Eventually, the *carrying capacity* of the link is reached when the maximum possible number of vehicles has been accommodated. Ultimately, if flows in excess of carrying capacity are attempted, traffic may come to a complete halt, and no further vehicles can enter the link. An extreme case of this, when a jamup on one link spreads to surrounding

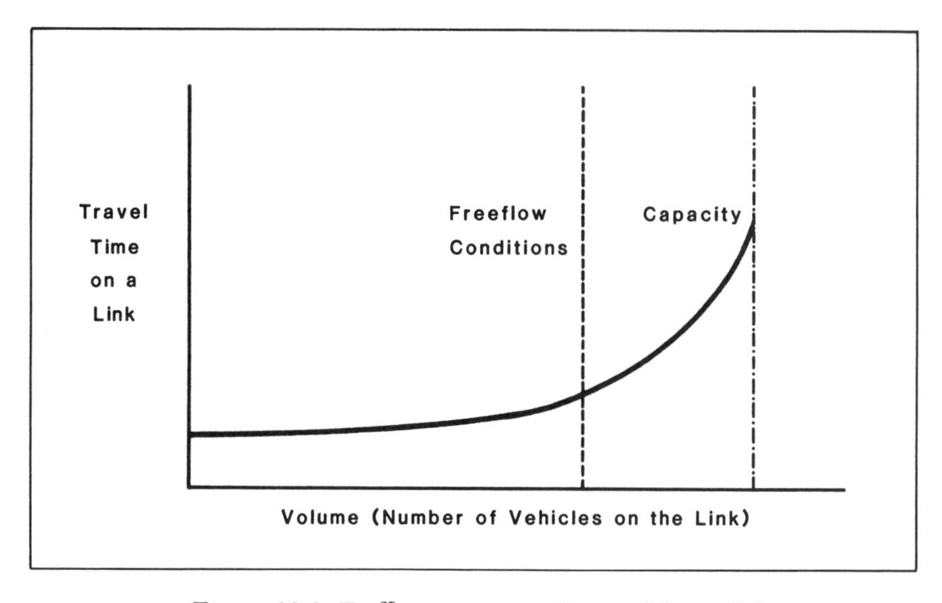

Figure 16-2. Traffic congestion: Hitting "the wall."

links until traffic is tied up in an entire portion of the network, has been given the name *gridlock*. A number of metropolitan areas have suffered gridlock with perhaps the best-documented case in midtown Manhattan, New York City.

Congestion is essentially a peak-period problem. Urban road networks must be designed to cope with rush-hour traffic conditions. A large portion of the carrying capacity of freeways and streets remains unused throughout most of the day. When rush-hour traffic conditions were first acknowledged as a problem, the traditional American response was to add capacity in the hope of providing essentially free-flow travel throughout the day. As supply–demand feedback effects came to be recognized and the black-hole theory of highway investment was formulated, traffic engineers scaled down their expectations. Many urban highway plans today are based upon some acceptable level of trip delay expected under peak-hour traffic conditions. In fact, the carrying capacity of a link is reached at a level of flow corresponding to trip times in excess of those experienced under free-flow conditions.

As it became apparent that free-flow conditions could not always be provided, those who espoused the black-hole theory gained credibility. The dynamic nature of urban transportation and land use systems began to be explored in more detail than previously. Though not everyone reached the conclusion that our metropolitan areas had become much like drug addicts in their need for more and more high-speed highway capacity, the negative side effects of the low-density pattern of development made possible by the freeways were very apparent. The word "smog" had entered the American lexicon, and social commentators were beginning to decry the hectic pace of a workday schedule involving long and tense commuting trips.

The Arab oil embargo of 1973–1974 brought disenchantment with continual road-building solutions to a head. The higher price and scarcity of gasoline during that long cold winter suggested that dependence on a single mode of transportation had its risks. Before turning to examine the alternatives, let us attempt to foresee the future for road-building policy.

It seems evident that in most metropolitan areas (with the exception, perhaps, of the rapidly growing metropolises in such Sunbelt states as Florida, Texas, and Arizona) the highway-laying epoch is past. The days when traffic engineers could sit down and lay out entire new integrated roadway systems to be carved into the urban fabric are long gone. Plans for new links to be added to existing networks and for new lanes to be grafted onto existing roadways must now be accomplished in an atmosphere sensitive to the feedback effects between transportation networks and land use systems.

The previous chapter on transportation and land use linkages cites evidence that the impacts of new highways on urban form are not always as substantial as the public perceives them to be. As a final graphic demonstration of the lessons learned during the highway-laying epoch, however, consider the following case study of two small communities where land use patterns were turned inside out as a result of road-building projects whose impacts on the urban system were neither anticipated nor considered during the planning process.

A Tale of Two 'Bypasses'

In upstate New York on the banks of the St. Lawrence River (the U.S.–Canadian border) lie the city of Ogdensburg and the village of Massena. A state highway, N.Y. Route 37, links the two communities. Historically, the highway passed through the downtown areas of both Ogdensburg and Massena as their main streets. For years New York State transportation officials expressed concern about the slowdowns experienced by through traffic on the highway as a result of the routing through the business centers. Plans for bypass routes were drawn up, debated, and ultimately agreed upon.

As shown in Figure 16-3a, Massena's bypass was built just at the fringe of the existing built-up area of town, with agricultural lands

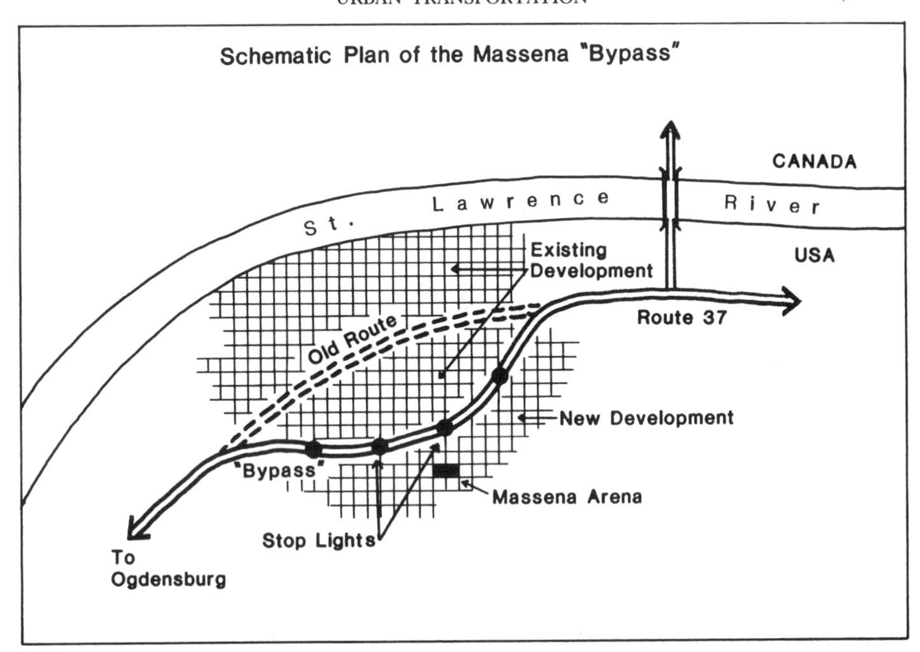

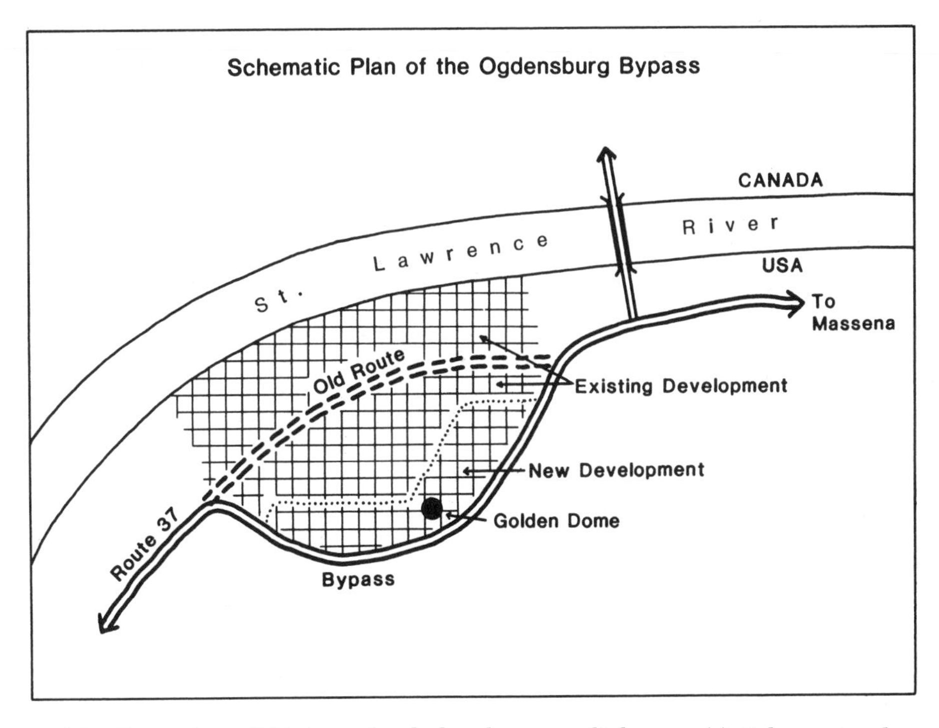

Figure 16-3. Examples of highway/land development linkages. (a) Schematic plan of the Massena bypass. (b) Schematic plan of the Ogdensburg bypass.

lying to its south. For a time after the new route was opened, highway traffic flowed freely around town at the design speed of 55 miles per hour. Soon, however, Massena's land use pattern began to change in response to the altered transportation network. A metamorphosis of urban form common to thousands of smaller communities across the

United States began to take place. A new shopping center was developed on the flat farmlands just across the new highway. The golden arches and other familiar symbols of contemporary U.S. urban growth began to sprout where only farm crops had previously grown. The site chosen for the village hockey rink, the Massena Arena, also lay on the new highway. As the older downtown business district entered a period of declining sales and physical decay, a number of serious accidents began to occur on the bypass as local traffic attempted to cross over to reach this area of new development. Before long, traffic lights had to be installed at the intersections, and, finally, a 30-mile-per-hour speed limit was posted on this stretch of N.Y. 37, built just a few years before to accommodate through travel at highway speeds.

What took place in Massena was by no means unique. At almost the same time similar decay of downtown shopping areas was witnessed in city after city as developers chose shopping center sites on the edges of existing development where land for large parking lots and single-story buildings was available and relatively cheap. In fact, just 35 miles upriver and 35 miles away via Highway 37, similar trends were set in motion by the construction of an Ogdensburg bypass. Here, too, a new shopping center was built and that city's new hockey rink, the Golden Dome, was opened. Unlike the Massena case, however, the bypass route in Ogdensburg left room for new development between the existing built-up area and the highway (see Figure 16-3b). The impairment of the through-travel function that was experienced in Massena has not occurred, at least to date, in Ogdensburg. Perhaps greater sensitivity to the systemic nature of transportation and land use systems was demonstrated in Ogdensburg than in Massena. Perhaps the difference was fortuitous. But the most important lesson is that the linkages between transportation and land development are sometimes quite strong and should always be taken into account when designing urban transportation policy alternatives.

Transit's Turn

During the peak period of urban highway building, public transit systems across the nation were experiencing ever-increasing problems. Years of neglect had resulted in out-of-date, poorly maintained facilities and vehicles. Many municipal transit managers found themselves caught in a vicious circle. The poor physical state of the bus and rapid transit systems, plus the longstanding American love affair with the automobile (and the perceived individual freedom of choice it affords) had resulted in very unfavorable public images of public transportation and in dramatically decreased levels of ridership. Decreased ridership had led inexorably to cutbacks in schedules and in the number of routes served. In turn, the decreased convenience for individual riders brought on by these cutbacks resulted in even lower levels of patronage and thence the pressure for still further cutbacks in service. By the 1960s the situation had deteriorated to the point that governmental subsidies to public transportation came to be almost exclusively justified on equity grounds: transit is now often perceived as an alternative supplied only because certain groups—the very young, the very old, the physically handicapped, and the impoverished—cannot afford or cannot operate their own personal automobiles.

As hopelessness over the ability ever to solve traffic congestion problems and disenchantment with the negative side effects of road building came to the fore, proponents of public transportation seized upon the opportunity to shift funding away from highways. A number of entirely new rapid transit systems were either built (Washington's METRO, Atlanta's MARTA transit system, and BART in the San Francisco Bay Area) or entered the initial planning stages (new subway systems for Los Angeles, Baltimore, and Buffalo). While a number of these new systems represented rather substantial advances in the state-of-the-art of transit technology and offered comfortable travel conditions and thus effective alternatives for suburban commuters (rather than the politically less powerful traditional public transportation users), a barrage of criticism of the rapid transit policy alternative resulted. The extremely high capital investment costs required to build new tracks and stations, to purchase new rolling stock, to acquire land for new rights-of-way and stations, plus the prospect of never-ending government subsidies for operating expenses were seized upon to organize the opposition. In fact, rather than representing a total break with old ways of conceptualizing urban transportation problems, the emphasis on new large-scale transit systems can be viewed simply as an alternative approach to providing a single grand ultimate solution. Unrealistic expectations are perhaps partly to blame for the disillusionment that set in after the new systems entered service; it quickly sank in that rapid rail transit is not a panacea for an entire metropolitan area's transportation woes.

One of the reasons public transportation alternatives are perceived as being especially costly policy options is, as mentioned in chapter 10, that many of the true costs of our reliance on the automobile are not readily apparent. A large portion of the cost of transportation is shouldered directly by the individual, who frequently views it as indirect costs of owning a car, rather than direct travel expenditures. In addition, capital costs for new highway construction have largely not been paid out of general tax revenues. Instead, the federal Highway Trust Fund was established to pay for road building through excise (user) taxes levied on gasoline, tires, new cars, and other items. Auto owners and trucking firms pay rather dearly for their use of the public roadways, but the costs are not nearly as apparent as they would be under a widespread system of road pricing that certain scholars have argued for.

In discussing transportation supply alternatives that involve mode substitution, it is ironic that one of the public transit options now most widely recommended is the mode that was predominant during the urban transport and development epoch immediately preceding the auto age. Light rail transit or LRT systems, which are now being pushed as lower-cost alternatives to subway or rapid rail transit (RRT), are nothing more than modern versions of the old trolley or streetcar networks which, around the turn of the century, had made the development of the vast tracts of American suburbs possible. The rapid replacement of many cities' highly developed streetcar lines with less satisfactory bus routes, which took place during the 1930s and 1940s (as the private automobile began to assert its primary position) is a fascinating vignette of American urban history. As told by Kwitny (1981) the "great transportation conspiracy" is one of the most convincing illustrations of the postulate that free enterprise market forces, rather than explicit public policy decisions, have shaped U.S. metropolitan transportation alternatives up to the present. Before returning to consider the future of urban transportation supply alternatives (a consideration that will reflect the present proclivity for public policy that follows rather than provides substitutes for market forces), let us examine this curious, yet significant episode.

The Great Transportation Conspiracy

Today, as the citizens of Los Angeles ponder the vast expense of an entirely new subway system to be built beneath the smog and sprawling freeways of one of the world's archetypal auto-age cities, few remember the efficient, heavily utilized urban rail system that until the 1940s provided service all the way from Newport Beach in the south, to Pasadena and the San Fernando Valley in the north. Even fewer would think to place direct blame for the tearing up of these tracks, which would make the current renaissance of an alternative transport option so much simpler and cheaper, on the big auto and oil companies. Most of us would think that the auto interests had simply offered a more attractive means of transportation at the time and that natural free market forces and Americans' inherent right to choose to consume as they wish had led to the demise of this and other cities' former streetcar networks. The truth, however, is somewhat different.

In 1949 a federal court jury found General Motors, Mack Trucks, Firestone Tire and Rubber, Standard Oil of California, Phillips Petroleum, and a number of "front" corporations and corporate officials guilty of violating the Sherman Antitrust Act for conspiring to buy up existing trolley systems and replace them with intracity bus routes. Why were these corporations seemingly so keen on promoting bus travel? In city after city when the conspirators ripped up the streetcar tracks and brought in buses, public transportation ridership declined almost immediately. Even though the companies had a stake in supplying the buses or the tires or fuel for the buses, the scheme would appear to have been a money-losing venture. Today we see that buses are neither as clean and comfortable, nor as energy-efficient as light rail transit (LRT, or streetcar) systems. Yet if we examine the impacts of the gutting of the streetcar franchises in more detail, we can see the true motive underlying the scheming, which went on from the depths of the depression in the early 1930s until the trial at the end of the 1940s. The real incentive was to divorce commuters from public transportation as a whole and thus spur the sale of private automobiles, which would prove to be a far more lucrative line of business than supplying buses.

Ironically public policy, rather than opposing this wanton destruction of what had up to then proven to be one of the most effective modes of intraurban transportation yet invented, actually contributed to the cause of the conspirators. Part of the reason why GM and the other colluders were quickly able to buy up streetcar franchises in city after city at bargain prices was a 1935 act of Congress that forced electric utility companies, many of whom had been the original promoters of the systems, to divest themselves of such subsidiary holdings.

As we think today about the millions of dollars it would take to restore even one of the old streetcar networks to provide efficient light rail transit or of the costs imposed by auto exhaust on each and every resident of a major metropolitan area or of the frustration and dangers of fighting rush-hour traffic on a clogged urban expressway, we might keep in mind the penalties that were imposed on the participants in the great transportation conspiracy. The judge at the trial sentenced each of the involved corporations to pay fines of $1,000. And the individuals who were convicted were fined $1 a piece.

Is what's good for General Motors always what's good for America? At a time when U.S. transportation policy is once again proving to be little more than a reassertion of faith in free market forces to supply the most effective alternatives, we should remain vigilant about the need to examine carefully all the implications of eschewing the more direct means of public policy intervention. The experiences of the 1930s and 1940s forcefully illustrate that what appears desirable and inevitable at one time may or may not be what is actually required to shape a desirable long-term future course of evolution for the overall urban system. Private enterprise probably should not be expected to supply anything better than a narrow form of planning for the present, unless truly effective incentives are devised to harness it to broader social tasks.

The substitution of automobile travel for public transportation travel as the primary means of urban movement reflects a number of very compelling advantages for the individual. No longer is one's daily schedule tied to bus or transit timetables. Nor are individuals' activity spaces constrained by the public transportation map. The private automobile is the ultimate demand-responsive transportation mode, providing rapid access between virtually any two points whenever the user desires it.

One of the aphorisms that often gets stated as a truism these days is that current patterns of urban development are predicated upon the near-universal ownership of cars and that residential density levels consequently are simply too low to support cost-effective transit operations, except perhaps in the nation's largest metropolitan areas. A number of the recently constructed high-speed transit systems have sought to counter this conventional wisdom, but, as previously mentioned, have come in for a barrage of criticism because of their expense.

Ironically, providing additional transit and thus relieving auto congestion alters the characteristics of the mode-choice problem that commuters face. It was pointed out soon after the BART system opened that auto travel conditions on the Bay Bridge into downtown San Francisco had improved. The large numbers of commuter buses that had taken up space on the freeway were suddenly removed from service once BART was available. Similarly, if any successful transit system comes into operation, we can expect highway travel conditions to improve as more and more people leave their cars at suburban stations or with spouses who mostly drive in uncongested suburbs or during off-peak hours. Once again the systemic nature of urban transportation is highlighted. All too rarely do the same transportation planners have control over the entire spectrum of modes available to peak-hour travelers. A rare U.S. example of a truly integrated planning approach is found in the Philadelphia metropolitan area.

Case Study: Designing Successful Rapid Rail Alternatives

One of the rapid transit success stories of the 1970s is actually not one of the new, much-discussed metro-wide networks. The Lindenwold High-Speed Line links several suburban communities in southern New Jersey with central-city Philadelphia. Unlike the BART, MARTA, and D.C. METRO systems, this line was built without federal subsidies. It was built by PATCO, the regional authority responsible for the bridges spanning the Delaware River between New Jersey and Pennsylvania. It was justified as an alternative to constructing a new bridge. Bridge toll revenue supplied much of the capital used for the project.

The Lindenwold Line is shrewdly designed to provide high-quality, low-cost rapid transit. The route was built along an existing rail right-of-way, eliminating costly and difficult land acquisition battles. To attract riders, frequent, dependable rush-hour service is emphasized, as is the comfort of riders. Labor costs are held to a minimum through the use of remotely controlled trains and automated stations. Upholstered seating and an aggressive safety and anti-graffiti program are provided. The trains and stations are continuously monitored through a bank of TV screens located in a central facility in downtown Camden. If a "turnstile hopper" is spotted, a tape-recording of barking dogs can be played and the roving security unit (perhaps accompanied by the real dogs!) instantly notified. In addition, the suburban stations are surrounded by large park-and-ride lots, plus kiss-and-ride drop-off areas.

Even though the residential densities in the vicinity of the stations on the Lindenwold Line are far lower than those usually associated with successful rapid transit, the line has attracted and retained a large and loyal ridership. Partly this reflects the difficulty of automobile commuting in the Philadelphia area and the shortage of downtown parking, but it also reflects good planning from the perspective of the overall transport and urban system. Since the managers of the system also operate the bridge crossings on all the major in-commuting routes, use of transit and auto transportation facilities can be optimized by astutely setting both the bridge tolls and speed-line fares. The desirable toll and fare levels can be found through a consideration of all the costs facing commuters using the two modes (Allen & Boyce, 1974). Such integrated facility planning and transportation system management have been achieved all too rarely in the United States.

The Future of Public Transportation Alternatives

Disillusionment has now been experienced over both highway building and grandiose rapid transit alternatives. Neither has proven capable of providing a final solution to the multitude of metropolitan transportation woes now recognized. Some attention continues to be given to the possibilities of new technology to provide such an ultimate solution. Experimentation, for example, has taken place with personal rapid transit (PRT). The idea of PRT systems is to combine the flexibility and demand-responsive characteristics of the private automobile with the congestion-relieving advantages of shared vehicles. Problems with such new-mode solutions do not lie primarily in the realm of technology. The AUTOTRANS system, for example, smoothly distributes people throughout the vast, sprawling Dallas–Ft. Worth Airport complex, and a loop system to shuttle people around the Detroit CBD is now being built. Rather, for broader-scale applications, institutional problems become paramount. Capital construction costs are enormous, and operating costs cannot be so easily disguised (as are those of private auto commuting) in the corona surrounding the free-market system.

Rather than developing new technology and searching for new modes, recent trends have pointed toward reemphasizing older forms of public transportation. Interest now centers on designing public transportation options that involve a variety of modes and on allowing involvement by the private sector (Schofer, 1983). Under the Carter administration renewed attention was given to buses. Buses are, in fact, by far the predominant mode of publicly financed urban transportation, accounting for almost three-quarters of all public transportation trips in the U.S. As we have seen, however, buses have earned a negative image in the minds of U.S. urban residents. Curiously, in Hungary, where buses have only recently been provided as an alternative to the traditional tram (streetcar) routes, just the opposite impression is in vogue. Prestige pricing has been instituted, so that it costs more to make a comparable trip by bus than by tram. Unlike the U.S., the private auto is not nearly so ubiquitous.

U.S. bus ridership is now limited largely to those unable to use the private auto. Though public transit agencies continually mount campaigns to attempt to battle the stereotypes and woo back lost riders, little hope exists (unless gasoline is once again in short supply) that these efforts will succeed in substantially increasing public transportation's over-all share of metropolitan travel. Perhaps more hopeful has been another aspect of the interest in and adoption of marketing techniques by public transit agencies. This is the classic notion of *market segmentation*. Considerable thought has now been given to more satisfactorily serving different groups of existing bus riders and to seeking out certain specific additional groups of people who might be persuaded to use convenient bus transportation.

Because urban congestion is largely a peak-load problem, much experimentation has involved increasing the use of express commuter buses. Some early attention was misdirected toward the construction of very expensive *busways*, exclusive road rights-of-way reserved for buses. The fact is that these facilities lie idle throughout much of the day and that similar results may be obtained by the use of "diamond" lanes on existing freeways, which are reserved for buses during peak commuting hours, or by preferential metering of buses on the ramps allowing access to the freeway. Both of these low-cost alterations to existing transportation facilities have the excellent psychological effect of showing solo drivers, snarled in rush-hour traffic, some tangible benefit to exploring an express-bus alternative.

Recently considerable interest has also been given to expanding the range of traditional public transportation modes available. *Jitneys*, or privately operated minibuses, *dial-a-ride*, and *subscription buses*, plus other demand-responsive services, including the often neglected taxicab, are all being promoted to meet specialized needs. Such policy alternatives have emphasized the need for regulatory reform to make the institutional environment less hostile to those interested in promoting these modes. Tapping private enterprise, with its attendant efficiency, is being touted as an alternative to the continual efforts to keep alive the dinosaur-like metropolitan public transportation systems we have inherited from the past. Under the Reagan administration, former equity arguments have been swept aside, and urban transit subsidies are portrayed as little more than throwing good money onto the tracks and at (though probably not into) the farebox.

DEMAND MANAGEMENT ALTERNATIVES

An appealing alternative to providing costly supply alternatives is to attempt to achieve higher levels of performance from existing urban transportation infrastructure by decreasing demand during peak periods or reorienting demand so that the road and public transport networks are used more effectively. A variety of quick, dirty, and clever responses to the seeming failure to come to grips with overall urban transportation policy issues have evolved. Whereas no one claims that any one of the policy alternatives to be discussed in this section will ultimately solve our current dilemma about urban transportation, recent experience indicates an array of smaller-scale actions that policy makers can take to alleviate the worst aspects of existing problems. These alternatives now in vogue are good examples of Berry's "ameliorative problem-solving" type of planning. If properly packaged, however, and instituted in association with an integrated approach to urban planning in general, *traffic system management options* may, in fact, end up playing a central role in designating satisfactory paths for the gradual evolution of U.S. metropolitan area structure.

Optimizing Network Flow

The notion of traffic demand management is an old one. The lament "if they would only synchronize the lights" is indicative of this approach. Methods of signalization have improved dramatically, as considerable research has gone into network models of urban traffic flow and into studies of urban driver behavior. In addition, under the rubric of traffic demand management there now exist literally hundreds of proven techniques for better use of existing transport facilities. The impetus for more and more cities to adopt transport system management (TSM) plans was provided by the federal government, which made their adoption a prerequisite for cities to obtain transportation funding.

In addition to the relatively low cost of TSM options (at least when compared to the costs of more traditional transport supply options) is the fact that many of these methods can be implemented quite rapidly. Examples of ways in which traffic can be made to flow more smoothly over city streets include (in addition to better signalization) the use of one-way streets, turn prohibitions, parking restrictions, stricter traffic enforcement, and reversible lanes.

One-way streets can sometimes greatly improve throughput in a street network, especially by eliminating the delays at intersections when vehicles attempt to make left turns across the lanes of opposing traffic. Offsetting the potential benefits of slightly enhanced travel speeds, however, is the possibility that vehicles will have to travel longer distances to complete trips on the network than would otherwise be the case. Two arterial streets in relatively close proximity (less than one-half mile) are required if one-way streets are to supply net benefits.

Left-turn prohibitions, particularly during peak traffic periods, may result in benefits similar to those provided by one-way streets. Eliminating parking along major traffic-carrying routes can contribute low-cost "new" lanes for travel, although these policies may run afoul of the merchants or residents along the routes. Stricter enforcement also serves to maximize the carrying capacity of streets, especially by cutting down the number of accidents which can seriously snarl traffic.

Cities where the spacing of arterial routes rules out the use of one-way streets have experimented with reversible travel lanes, like those more commonly used on bridges and in tunnels. During the morning rush hour the center lane of a three- or five-lane avenue is used for traffic entering the downtown area; during the late afternoon period it is used for traffic heading out from the center. At other times it might be used for traffic wanting to make left turns. In fact, a strict policy of no left turns during peak periods is integral to a successful implementation of reversible lanes. In Tucson, Arizona, this approach has been in use for several years on many of the major arterial streets. It was first viewed by planners there as a stopgap measure to keep up with rapidly increasing travel demand until new freeways could be constructed. However, as freeway projects have been consistently voted down over the years and because the spacing of arteries along the one-mile township-and-range section lines precludes a successful use of one-way streets, the city has had to learn to live with this traffic management alternative. The reversible lanes have become known locally as "suicide lanes," because the elderly "snowbirds" who flock to town in the winter are unfamiliar with them and often attempt left turns, rubbing against the grain of local residents who are trying to commute home at their customary high rates of speed. The accident problem persists, despite the best attempts of the city to provide clear signs and strict enforcement of the turn prohibitions.

Although the most common traffic management options involve relatively minor physical alterations to gain additional traffic-carrying capacity on roadways, the term actually is much broader. It encompasses ideas such as road pricing, auto-restraint policies, and the encouragement of increased ride-sharing through car- and vanpooling.

Road Pricing

One of the earliest "radical" policies proposed to help cities cope with traffic congestion problems is known as road pricing. Road pricing per se is nothing new. In the early days of the United States private entrepreneurs built rudimentary roadways through the woods to connect the growing settlements on the East Coast. They adopted the European tradition of charging travelers who sought to travel on these "turnpikes." (The name turnpike was adopted because a large swinging log was often used at toll houses to bar the way until the fee was paid.) When the first limited-access, four-lane divided highways, such as the Massachussetts, New Jersey, and Pennsylvania turnpikes and the New York State Thruway, were constructed in the eastern United States, the practice of charging for travel was continued.

For travel within urban areas, road pricing has not been adopted widely. In the United States, with the exception of certain bridges and tunnels and some isolated examples of expressways with periodic toll booths (such as those in the Chicago, New York, and Richmond, Virginia, metropolitan areas) it is generally thought that the expense of collecting tolls and the possibility of creating traffic jams at the collection points during the rush hour offset any benefits. Two dramatic experiments in Asia, however, are being watched with interest by transportation professionals.

Road Pricing in Hong Kong and Singapore

Hong Kong and Singapore have been called cities that work (Yeung, 1985). The two are frequently compared. Both have similar histories, restricted land areas (1,066 square kilometers in Hong Kong and 602 in Singapore), large populations (5.3 million in Hong Kong and 2.5 million in Singapore), and rapidly expanding economies fueled by exports of manufactured goods. These two tiny nation-states' per capita gross domestic products now rank second and third (behind Japan) in Asia. Similarities exist not only in such statistical terms, but also in the way that success has been achieved in their free-market economies. Both cities have benefited from efficient, and often centralized, planning. A widespread belief underlying their planning is that urban systems can be made to work through incentives to private enterprise. This philosophy is reflected in the equally bold ways that road pricing has been adopted to relieve each city's previously severe congestion problems.

Figures cited by Yeung (1985) show that while Hong Kong's population was doubling in the last two decades, the number of automobiles in the city has increased sevenfold. As a result, in 1982 the government adopted a group of traffic limitation strategies. These included doubling auto registration fees, tripling the annual license fees, and substantially raising the duty on oil. The aim of each of these actions was to limit increases in the number of vehicles attempting to use the crowded streets. Additionally, however, an electronic road-pricing system has been instituted. Special plates have been affixed to the bottom of each automobile. When the vehicle is driven over any of a number of control points distributed throughout the most congested areas of the city, a record of the trip is electronically generated. Bills based on this metering of actual travel patterns are periodically tallied and sent to the car's owner. By moving the locations of the control points, planners can experiment with ways to alter individual trip routes to allow traffic to flow more freely throughout the network. Although the results appear to be encouraging in terms of minimizing congestion, the system has not been uniformly popular.

In evaluating the relevance of the Hong Kong experience for possible adoption in a U.S. context, it should be noted that public transportation is alive and well in Hong Kong; among the most severe public transportation problems is the difficulty of purchasing buses, streetcars, taxis, and ferries fast enough to meet ever-increasing ridership demand!

In Singapore, the recent rise in auto ownership has not been quite so rapid as in Hong Kong. Among the reasons is that a form of road pricing has existed since 1975. The type of pricing adopted in Singapore is less high tech than that now in use in Hong Kong and doesn't allow the same degree of flexibility for manipulating specific patterns of flow. It has, however, produced dramatic results.

The Singapore plan is called the Area Licensing Scheme. The central 620 hectares of the city have been designated as a restricted travel zone. Only specially licensed vehicles are permitted to enter this area between 7:30 and 10:15 A.M. A cordon line was drawn around the zone, with entry restricted to 29 check points where daily or monthly entry passes must be shown. A sliding fee scale exists for these permits. Company-owned vehicles pay the steepest rate: about $100 (U.S.) per month. Carpool vehicles (defined as those carrying four or more persons) are exempted from the licensing requirement.

Shortly after implementation of the scheme, traffic in the restricted area was reduced by 71%. Since then, traffic counts have stabilized at a level slightly above one-third of the preadoption flow. During the same period, carpooling has almost doubled in popularity as a result of the strong financial incentive the plan provides; 44% of all inbound vehicles now contain the four or more persons needed to obtain free entry. Yeung concludes, "Thus, judged from every stated objective of the Scheme, the first road pricing scheme has been immensely successful" (1985, p. 26).

How does road pricing help to control urban travel? The basic idea is to make the social costs of congestion more readily apparent in the form of private costs to be borne by the creators of the problem. If rush-hour travel is perceived, as a result, to be excessively costly, then commuters might be encouraged to use alternative modes or to adjust their residential location decisions accordingly. Road pricing has considerable appeal in that it is an example of using market forces to achieve policy ends, much the same way that tax-code changes have become such a common approach to governmental policy in the United States. Rather than mandating modification of travel behavior directly, the idea is to encourage it through monetary disincentives.

Objections to road pricing have centered on equity issues (it is likely to be lower-income auto commuters whose behavior will be most modified) and on practical aspects. To achieve sufficient control over urban travel patterns to derive the full benefits of road pricing, some sort of electronically sensitive plate, as in Hong Kong, or a meter, would probably have to be placed in private cars to record the charges being accrued as the vehicle travels through the urban highway network. Very strong sentiments about the sanctity of individuals' cars, however, exist in the United States and certain other countries. Large-scale road pricing alternatives are easily painted as heavy-handed "Big Brother" policies. Further experimentation has not been allowed to take place in the United States.

Auto Restraint

Road pricing is, in part, an example of an auto-restraint policy in that one of its goals is to encourage the use of public transportation by making auto travel more difficult. Other examples include banning on-street parking or not actively promoting the construction of additional parking in downtown areas. Just as parking bans to gain additional travel lanes on arterial streets often meet with opposition from business groups, so too downtown auto restraint policies usually provoke considerable resistance. In a number of cases, however, even stronger measures than parking bans have resulted in favorable acceptance by merchants. The conversion of downtown streets into pedestrian malls or transitways (streets open only to buses and perhaps taxis) have become a rather common feature of downtown renewal plans.

The ultimate form of auto restraint—outright prohibition of autos from a major section of town—has rarely been tried. Among the widest-scale applications of auto restraint is the plan tried in Göteborg (Gothenburg), Sweden, where the central portion of the city was divided into a set of five roughly pie-shaped zones (Figure 16-4). Traffic can penetrate within any of the zones to obtain access to the main business area, but no movement is permitted across the zone boundaries. Through traffic is barred from entering the downtown and the use of the city's excellent public transportation system is encouraged.

Somewhat similar auto-restraint policies have been used to discourage through traffic in residential areas and to channel traffic more efficiently over arterial streets. The

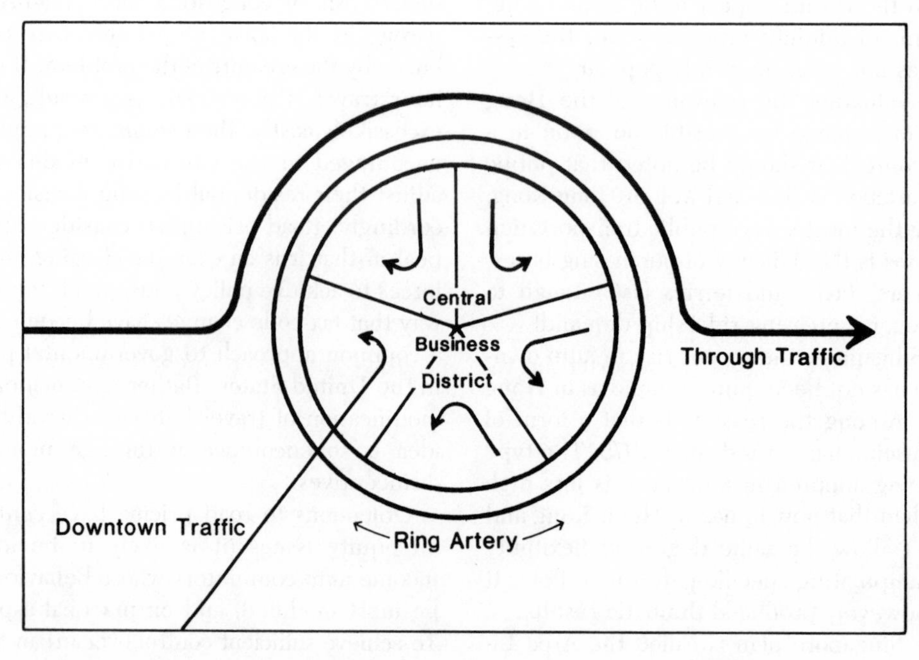

Figure 16-4. Göteborg, Sweden auto-restraint concept.

design of many modern subdivisions, with a series of short cul-de-sacs opening off a few major traffic-carrying "spines" reflects this thinking. Berkeley, California, has tried to create neighborhood environments less hostile to pedestrians, children, and dogs through the use of one-way streets, barricades to create cul-de-sacs, and a series of concrete "bollards" in the center of intersections which, theoretically, cause traffic to slow-down. Residents, however, have become adept at driving the streets much like a ski racer taking a slalom course. "Speed bumps," ubiquitous stop signs, and a policy of not patching potholes can generate similar benefits for residents—and nuisances for drivers. A further consideration that must be kept in mind about such policies is that the traffic thus discouraged from using neighborhood streets is simply transferred to adjacent arterials. To the extent, however, that zoning has been effectively and consistently implemented to channel housing away from the busier thoroughfares, the induced traffic rerouting should result in a community-wide net gain in the quality of life.

Car- and Vanpooling

A form of traffic demand management that is becoming widely adopted is the encouragement by public agencies of greater use of ridesharing—particularly by rush-hour commuters. Auto commuting trips, which are often among the longest of the most frequently made trips in metro areas, are much more commonly made by drivers traveling alone than other types of trips. Table 16-1 shows the American labor force's mode choices for the journey to work, as reported on the 1980 census. Although direct questioning (such as used in the census) does tend to overstate the prevalence of true, daily carpooling, substantial numbers of Americans are now sharing rides to work. According to these census data, of all persons traveling to work in private vehicles, 76.5% are solo drivers. Among those sharing rides, two-person pools predominate. Of all poolers, 70.1% ride with one other person, 17.5% with two others, and 12.4% with three or more. Nationwide, therefore, the average private car, van, or truck transports only 1.15 persons on its trip to the workplace.

Table 16-1. Mode use by the U.S. labor force for the journey-to-work, 1980

Mode	Percent of labor force
Private vehicle	84.1%
Drive Alone	64.4%
Carpool	19.7
Public Transportation	6.4
Walk	5.6
Work at home	2.3
Bicycle	0.5
Motorcycle	0.4
Other	0.7
Total	100.0%

Note. From *1980 Census of Population* (Volume 1, Chapter C, Part 1, Table 101) by Bureau of the Census, U.S. Department of Commerce, 1983, Washington, D.C.: U.S. Government Printing Office.

An appealing alternative to providing additional highway capacity to meet peak-period conditions (which may get rapidly "eaten-up" as travel patterns and residential location decisions adjust to the new conditions on the network) is to increase people-carrying capacity by reducing the number of single-occupant vehicles. Even if only small gains can be made in the number of persons choosing to share rides, either in buses, in carpools using private vehicles, or in vanpools arranged by employers, the critical level at which congestion rapidly occurs on highway links may be avoided.

Metropolitan transportation agencies may now implement effective programs to encourage those who might adopt car- or vanpooling and, equally, to retain existing poolers. Research on the marketing of ridesharing has revealed a number of important points which should be kept in mind. (1) The most effective programs are those arranged through employers. Thus, efforts aimed at helping employers adopt ridesharing as an employee benefit are often far more cost-effective than expensive media campaigns to encourage people to "start pooling around." (2) The likely adopters of ridesharing represent a fairly restricted subset of the overall commuter base. Many people are either attitudinally indisposed to give up the freedom of solo driving or are constrained by their daily activity patterns from participating (for example, the kids must be picked up from child care on the way home, the weekly grocery shopping is usually done as part of the evening commute, or the hours worked or the location of work varies). In addition, it only makes economic sense to car- or vanpool, if certain attributes of the commuting trip, and of the setup of the car- or vanpool are favorable. Thus any successful program should be based on a market segmentation approach.

Designing Effective Ridesharing Incentives

A study of the likely effectiveness of a broad range of incentives to ridesharing (and, equally, disincentives to drive alone) was recently carried out for the Phoenix and Tucson, Arizona, metropolitan areas (Black, Plane, & Westbrook, 1985). First, a nationwide census of ridesharing agencies was conducted to generate a list of available inducements, the extent of their adoption, and the candid opinions of the ridesharing professionals as to the success or failure of each (Table 16-2). From this master list and after discussions with local officials a set of feasible and potentially successful inducements was identified to be tested in surveys of Arizona commuters and, equally important, of their employers.

One of the problems that must be faced when surveying on such issues as the likelihood of ridesharing is that people's direct responses often don't tell the whole story. Since the time of the 1973 Arab oil embargo, ridesharing has been perceived as a patriotic, commendable act that saves energy and reduces smog and congestion. Although a very small fraction of commuters actually car- or vanpool, much larger numbers are in favor of it—at least "for the other guy." A danger in carrying out questionnaire research, therefore, is that the actual increases in ride-

Table 16-2. Most frequently adopted ridesharing inducements: U.S. metro areas

Inducement type, specific inducement	Percentage of ridesharing agencies reporting current usage	Percentage of ridesharing agencies where inducement is used who rate it especially	
		Effective	Ineffective
Sponsored directly by ridesharing agency			
Ridematching service	72	69	7
Public signage	65	70	8
Public service announcements	53	46	10
TV/newspaper publicity	53	38	13
Sponsored by employers			
Preferential parking	65	66	10
Employer-sponsored vanpools	27	86	4
Subsidized transit passes	24	78	4
Ridematching by employer	13	69	23
Sponsored by government			
Park-and-ride lots	31	67	6
Preferential parking	22	61	13
High occupancy vehicle lanes	14	93	0
Transit fare subsidy	14	87	0

Note: From *An Evaluation of Alternative Economic Inducements to Ridesharing for the Arizona Commuter* by W. B. Black, D. A. Plane, and R. A. Westbrook, 1985, Phoenix: Arizona Transportation Research Center, Arizona Department of Transportation.

RIDESHARERS

Likelihood of Continuing to Rideshare

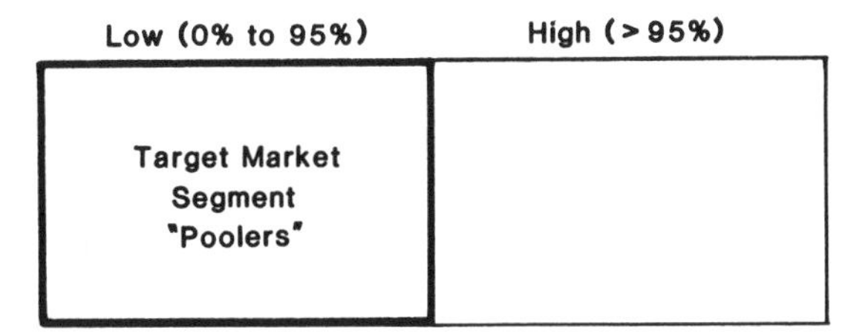

NON-RIDESHARERS

Attitude Toward Ridesharing		Likelihood of Ridesharing: Low (10%)	Moderate (10% to 50%)	High (>50%)
	Mostly Unfavorable			
	Mostly Neutral	Target Market Segment A	Target Market Segment B	
	Mostly Favorable	Target Market Segment C	Target Market Segment D	

Figure 16-5. Ridesharing market segments. From *An evaluation of alternative economic inducements to ridesharing for the Arizona commuter* by W. B. Black, D. A. Plane, and R. A. Westbrook, 1985, Phoenix: Arizona Transportation Research Center, Arizona Department of Transportation.

sharing occasioned by any of the inducements may be overstated.

Before attempting to assess the relative effectiveness of various inducements, the Arizona commuters were segmented on the basis of their overall attitudes toward ridesharing, as well as on assessment of their likely potential to adopt it. As shown in Figure 16-5, the main target market groups that were identified were those that weren't already likely to become poolers in the absence of new inducements and those that weren't unduly biased against the idea. The three attitude groups were identified on the basis of a simple index derived from responses to nine key belief statements, which were chosen from a much larger set of 29. Interestingly, it was found that these nine simple questions could discriminate efficiently between current ridesharers and nonridesharers. Furthermore, the various attitude groups did not vary consistently with either demographic characteristics or the nature of the respondents' journey-to-work trip, thus highlighting the need for true marketing research, rather than simpleminded advertising campaigns aimed at any supposed key groups.

To get around the problem that direct commuter appraisals overstate the actual effectiveness of inducements, a technique known as *conjoint analysis* was employed. Each respondent was asked to rank a set of scenarios involving a combination of different levels of each incentive or disincentive. Thus, a series of forced trade-offs between the inducements was made. The result of the conjoint task was a series of *utility profiles* showing the relative effectiveness of the different levels of the inducements to promote ridesharing among each target market segment.

It was found that the specific inducements appealing to each group varied widely and that they varied between the Phoenix metropolitan area (with about 1.6 million residents) and the somewhat smaller Tucson community (population 600,000). Thus it was suggested that packages of ridesharing inducements needed to be formulated, with implementation of each effectively targeted at its likely adopter group. In addition, an overall finding was the apparent widespread appeal of vanpooling. Governmental programs to assist firms in providing vans for their workers may be far cheaper than building additional highway capacity. They are also particularly effective in that a single vehicle may carry as many as 12 workers. Employees appear to favor vanpools over carpools because their private cars need not be used, they themselves don't have to drive under rush-hour conditions, the arrangements are all made for them, and they have a chance to socialize with their colleagues.

The low cost of car- and vanpool promotion, the ease and speed with which these marketing activities can be carried out, and generally favorable public attitudes (at least among certain segments of the populace) all suggest that imaginative use of ridesharing inducements should be included in all metro areas' near-term transportation plans. Traffic demand management alternatives, in general, would appear to be a fixture of policy for many years to come. The need, however, is not to grasp indiscriminately at any demand-management option that is proposed, but rather to craft a package of self-reinforcing strategies that are integrated into the overall planning process for the metropolitan area.

The last group of policy alternatives we shall look at highlights once again the inseparability of transportation from other aspects of the overall metropolitan system. We shall call these *nontransportation transportation alternatives* because they seek to modify the demand for transportation in less direct ways than do traditional demand management techniques.

NONTRANSPORTATION TRANSPORTATION ALTERNATIVES

Earlier we saw that one of the impacts of increasing the supply of transportation, and thus the ease of traveling in metropolitan areas, was to encourage more movement. Linkages exist between transportation and individuals' daily activity patterns, and through daily activity patterns, linkages are formed between transportation and aggregate patterns of land use and urban development. The final set of alternatives for coping with metropolitan transportation problems takes these linkages into consideration. Alternative work schedules may alleviate transportation problems by shifting the times when a portion of the labor force needs to commute to and from work out of the traditional peak periods. Certain types of alternative work schedules also reduce the number of days per week that employees work, thus cutting down the total number of weekly commuting trips that must use transportation facilities. Technological advances in telecommunications also have an impact on reducing travel, through eliminating the need for certain trips during the business day and also, in some cases, allowing employees to work at home rather than at centralized job sites. Finally, more creative urban design strategies could be formulated that reduce the length of trips that people need to make by more efficiently integrating residential and nonresidential land use.

Alternative Work Schedules

The interrelationship of transportation with the prevailing customs concerning daily patterns of work and discretionary time have long been recognized. When American cities first began to decentralize during the 1800s, the tradition was for business people to go home for the midday meal. Thus the early omnibus and horsecar lines in U.S. cities often benefited from the four trips per day that these early, well-to-do commuters made. Soon, however, as population pressures resulted in the progressive dispersal of more and more income groups and ultimately the mass suburbanization of the rapidly expanding American middle class, custom changed and lunch became a hurried affair consumed at or near the workplace. The change reflected a transformation from the previously dominant agrarian mode of existence to an urban one reflecting the realities of industrialization.

Today life-styles are still changing. Our society has passed from the industrial period of development into postindustrialism. Affluence is widespread, and great emphasis is being placed on leisure-time activities. In addition, a major social change has been the addition of most working-age women to the labor force. The U.S. labor force participation rate for women 18 to 44 years old has increased from just over 40% in 1958 to approximately two-thirds of all such women in the early 1980s. Two-worker households must necessarily budget their time and income differently from the male breadwinner/female breadbaker households of the past.

One result of these social and life-style changes has been increasing pressure to allow variation in the rigid pattern of 40-hour workweeks composed of five, 8-hour days. The federal government, as well as many other public- and private-sector employers, now allows some employees to depart from the usual schedule. A recent survey cited by Weggman and Stokey (1983) disclosed that 13% of all U.S. nongovernmental firms with 50 or more employees allowed some form of flexible working hours. Though becoming more accepted, U.S. companies lag far behind those in the Federal Republic of Germany, where the concept originated at the Messerschmidt-Boerkow-Blohm aerospace plants in 1967. More than half the West German labor force is now permitted to work nonstandard hours, and the idea has spread rapidly throughout other European nations.

As shown in Table 16-3, at least four major types of alternative work schedules have become common. Let us examine the transportation impacts of each. With flex-time employees can tailor their schedules so as to

Table 16-3. Types of alternative work schedules

1. *Flexible work hours ("Flex-time").* Employee chooses his or her own schedule, within some constraints. Employee may be free to vary the schedule daily, vary the lunch break, or "bank" hours from one day to the next or one pay period to the next. Typically all employees are required to be at work five days per week during designated "core" periods (usually 9:30 A.M. to 3:30 P.M.).
2. *Staggered work hours.* Employee works a five-day week, but starting and ending times are spread over a wider time period than usual. Individual employee schedules are assigned by management; employees do not choose them.
3. *Four-day (or compressed) work week.* Employee works the same total number of hours as in a typical five-day workweek, but reports to work only four times per week. The four days may be the same each week, or the extra day off may rotate from week to week.
4. *Job-sharing/part-time work.* Employee works less than the standard 40-hour workweek, either by working fewer than five 8-hour days, or by working less than 8 hours per day. Job-sharing means that two or more people share the same office space and work responsibilities.

Note: "Telecommunications and Alternative Work Schedules: Options for Managing Transit Travel Demand" by P. P. Jovanis, 1983, *Urban Affairs Quarterly, 19,* p. 169. Copyright 1983 by Sage Publications, Inc. Reprinted by permission.

avoid peak-hour auto travel conditions. Staggered work hours have a similar effect, but the individual employee does not have quite the same degree of control to minimize the cost and bother of his or her own commute. Overall, congestion on major commuting routes should be reduced as a result of "spreading" the peaks. An interesting question, however, concerns the impact of staggered and flex-time schedules on the mode-choice decision. Does the possibility of not having to fight rush-hour automobile traffic lead workers to desert public transportation? The evidence (summarized in Jovanis, 1983) is that flex-time does not have negative impacts on transit ridership. In some cases increased use of public transportation has been the result. The biggest impact seems to be that relatively low-income workers are better able to arrange their daily schedules so as to use transit during the off-peak when more seats are available. An added benefit is that transit travel delays are less likely during less crowded periods, so that workers can better rely on transit to get them to work on time.

Four-day workweeks introduce an additional complicating factor, although in some ways the impacts are similar to those of flex-time. The longer (usually 10-hour) workday also results in reduced peak-period flow and increased off-peak travel. The need to commute four rather than five days a week reduces the need for auto travel, but transit operators often fear the results of lost revenues from the fifth day of commuting. Research, however, seems to indicate that those fears may be groundless. Most of the directly lost ridership is during the peak periods, thus freeing seats, which makes transit use more attractive to others who were turned off by the former crowded conditions. The peak-spreading effects of both flex-time and compressed workweeks allow public transportation planners to utilize their vehicle fleet more efficiently.

The increasing prevalence of part-time work also has interesting impacts on transportation, as well as on land use patterns. Many of the impacts will be similar to those of either flex-time or four-day schedules. However, to the extent that part-time workers are people who may not have been working previously (semiretired people or women, for example) the most immediate impact may be *increased* travel demand. In terms of land use patterns, it should be noted that long commuting trips may not be economically rational if, for instance, the employee is only working half days. Indeed, much stronger preferences for jobs close to home have been noted in studies of part-time workers. Firms desiring to hire large numbers of such workers may choose to locate at suburban rather than downtown sites.

If alternative work schedules become common in the United States, as in West Germany, how substantial will the impacts be on reducing peak-period congestion? Jovanis has attempted to provide some clues in a study of work schedules in the San Francisco central business district and commuter travel on the San Francisco–Oakland Bay Bridge.

Impacts of Alternative Work Schedules on Traffic Congestion

Jovanis (1981) sums up past research on the effects of flexible work schedules on traffic congestion by observing that *areawide* impacts appear to be negligible, but that at the specific *corridor* level there may be sizable beneficial aspects. To explore the size and nature of the impacts, he surveyed the actual mix of work schedules chosen by employees of firms in the San Francisco business district already offering flex-time. He then used these work schedule preferences to extrapolate probable shifts in the timing of commuter travel on the San Francisco–Oakland Bay Bridge if flex-time programs were adopted by additional employers. Using a computer model that simulates the queuing process during each 15-minute time slice throughout the morning rush hour, he was able to make a number of interesting observations.

One of the most immediate effects, evident in the choices made by employees of those firms already offering flex-time (all of whom allowed start times between 7:00 and 9:30 A.M.), is that peak congestion would shift to an earlier period. More specifically, he looked at the results from a scenario involving 25,000 additional employees (or about 10% of the central business district work force) being allowed to choose their own starting times. He further assumed that the current 35% share of downtown San Francisco employees living in the East Bay would be maintained, but he did take into account induced mode switching, with a decrease in the percentage of workers driving alone and small increases in those choosing to carpool or ride public transit. The simulation outputs showed a decrease in average commuting time by 8%, a 3% reduction in gasoline consumption, and decreasing emissions of hydrocarbons and carbon monoxide. Nitrous oxide pollution, however, would increase because of faster travel speeds.

One of Jovanis's principal findings was that individuals' time savings would vary greatly depending on their pre- and postadoption work schedules. One group, those changing from an especially congested to a light-travel time slice, could expect to save almost 13 minutes daily. Workers shifting to the latest time periods (slugabeds?) would save time by avoiding the earlier congested periods. Some people, on the other hand, might actually shift to periods where their commuting times would increase. This can be explained as rational behavior if the other benefits of the new time schedule (in terms of being able to arrange before- and after-work personal activities better) outweigh the costs of the extra time spent traveling. Finally, workers without flex-time would generally experience worsened travel conditions. For example, those already working earlier fixed hours would have to contend with sizable numbers of flex-timers choosing to get to the office early.

Probably Jovanis's most intriguing finding is that generally, fairly small increases in the number of flex-time workers can have substantial impacts in alleviating peak-period congestion. Furthermore, his simulation runs do not take into account a feedback effect implicit in flexible work-hour programs. As experience is gained with travel at various hours of the day, workers can make further modifications to their schedules in order to "beat the peak." Thus the results in terms of spreading rush-hour travel more uniformly throughout the day are perhaps understated in this San Francisco study.

Although the transportation benefits of alternative work schedules are now widely recognized, a tougher issue is whether the government should encourage their diffusion. For the most part nontraditional schedules are viewed by employers as employee benefits and consequently as belonging to the realm of private labor negotiations rather than to an area considered more susceptible to government regulation. The American proclivity for leaving the private enterprise system to its own machinations works against policy initiatives to mandate the spreading of peak-hour commuting travel. In addition, those governmental agencies most directly concerned with transportation have rarely been involved in any such nontransportation areas of policy.

Two ways in which government agencies have encouraged this policy option may be noted. One is through setting a good example by experimenting with alternative work schedules for government workers. In many cases, because government employment tends to be highly centralized, this in itself may occasion substantial transportation benefits. A second way is by using the powers of friendly persuasion to encourage other nongovernment employers to explore the possibilities. Appeals can be made to the employer's sense of civic responsibility; the potential employee morale gains can be touted. Traffic congestion is often primarily of concern in central business districts, and the past several decades of public/private cooperation to attempt to revitalize the cores of our metropolitan areas work in favor of government efforts to promote congestion-reducing alterations in daily work schedules.

Telecommunications

Even farther from the sphere of direct policy control but also possessing considerable potential for alleviating urban transportation woes, is the increasing substitution of electronic for face-to-face communications. When videophones first captured the public imagination, the prevalent perception was that personal meetings would soon be relics of the past. In actuality, videophones have been rather disappointing, but the increased use of the telephone and the recent explosion in computer sales are now resulting in some significant transportation impacts. The separation of home and workplace has begun to blur as it has been realized that many routine office operations such as word processing and data management are essentially human/machine operations rather than group activities. For example, this chapter has been written on a home computer looking out over the back patio far from the distractions of students and the telephone of the office. Especially for clerical jobs the potential exists for much more dispersal of work functions to the home environment. Perhaps the central issue, however, is the concern of management over employee control. Are, in fact, the distractions of the refrigerator and the neighbor kids more serious than those of the water cooler and the chance to gossip with one's coworkers? Creative thought is now being given to ways in which work output can be measured electronically, rather than simply through physical presence.

Urban Design

The final group of nontransportation transportation policy alternatives is based on the view that the fundamental problems are not technical but institutional. All too often transportation planners are brought in *after* the developers have put in new subdivisions, new shopping centers, or new office parks. They are charged with attempting to add facilities to the existing urban transportation systems to integrate the new induced travel demands with the existing patterns of flow. It should not be surprising that this form of transportation planning does not always succeed. It can perhaps only "succeed" at the expense of ripping up major portions of the existing urban fabric.

Advocates of urban design options suggest that what is needed are stronger and better forms of land use planning, rather than of post hoc transport planning. The fundamental design of many of our metropolitan areas reflects the realities of the earlier omnibus, horsecar, and streetcar eras rather than those of the automobile age. A better integration of residential and nonresidential land use would have the benefit of reducing the lengths of commuting trips, which are the major cause of traffic congestion problems. Some people have suggested that the focus of public policy should be shifted away from a sole emphasis on propping up the decaying downtowns inherited from the past toward an explicit policy of encouraging and guiding the form of decentralization.

Much of the new development taking place in U.S. urban areas is being carried out by large corporations. In such multiple-use developments the possibility exists for much more effective planning taking into account transportation–land use linkages. Public officials should insist that the transportation impacts on the metropolitan system as a whole be studied carefully and resolved before any new large-scale developments are approved. Furthermore, the almost complete separation of different land uses from one another that is mandated in many cities' zoning plans needs to be rethought. Now that the workplace is no longer, for the most part, a huge factory with belching smokestacks, creative thought should be given to plans to achieve "decentralized centralization." We must escape the mind-set of the monocentric city models and think about how the evolution of polycentric metropolitan area activity patterns can be guided most effectively (Plane, 1981).

Perhaps the most radical proposal for a new form of urban development that would minimize transport requirements, totally eliminating congestion, smog, and most of the time wasted on travel, is *Compact City* (Dantzig & Saaty, 1973). Compact City is visualized as the ultimate urban *megastructure*, a huge three-dimensional unit within which facilities for living, work, shopping, and play are all located within easy walking distance. The plan also has the added advantage of requiring far less space to house a comparable number of people than existing forms of development, thereby providing the possibility of preserving much of the landscape in a pristine, natural state. Largely unanswered, however, is how such a radically different and expensive communal structure could be financed and built.

Some examples of urban design to cut down on the need for transport have already been tried. A polycentric city plan following the *urban village* concept is being tried in Phoenix, Arizona. The urban village idea is to encourage the growth of several high-density, high-rise office building core areas, which are to be surrounded by residential areas housing those working in the core. Integrated with the residential areas will be sufficient retail development, public facilities, schools, recreation, and other services to supply most of the needs of that urban village's residents. Thus, as shown in Figure 16-6, the overall metropolitan area is perceived to be a large sprawling unit composed of several much smaller subsidiary communities. Transportation needs under such a plan would be minimized if in fact the majority of residents worked, lived, and shopped in their own urban village. In addition, it is thought that urban villages would provide a scale for daily living much more compatible with healthy human existence. At the same time, residents can partake of the best things about living in a metropolitan complex of over 1.5 million.

Unfortunately the reality of Phoenix's experience has been rather different from that originally visualized by the planners. The true metropolitan geography consists, in Phoenix, as in most other U.S. metro areas, of a patchwork quilt of (often warring) independent political jurisdictions (Figure 16-7). The degree to which Phoenix can insist on adherence to strict locational criteria is greatly restricted by the real possibility of losing development and its attendant additional tax base, to Mesa, Tempe, Scottsdale, or other surrounding communities offering developers more attractive opportunities. The institutional backdrop has real implications in terms of the physical design (Gale & Moore, 1975) and thus the wasteful transportation needs of our contemporary urbanized areas.

TRANSPORTING THE URBAN SYSTEM INTO THE FUTURE

Throughout this book, particularly in this last chapter, we have studied a plethora of ways in which transportation planning and policy making could proceed. In closing, can we hazard some guesses about how current trends suggest that future planning is, in fact, likely to take place? And can we make some suggestions about how it might take place better?

Perhaps the most fundamental recent change in the way that urban transportation problems are perceived is the realization that no single, grand technological "fix" exists. Considerable disenchantment has resulted from past efforts to provide freeway solutions and later to supply equally grandiose rapid transit solutions. Currently, an array of smaller-scale policy options is receiving the most attention.

The current interest in demand management, in addition to more traditional supply-side alternatives, is an encouraging sign, as are the early tentative steps which have been taken to explore nontransportation transportation policy options. More efficient and effective transportation planning can take place if the systemic nature of metropolitan areas is recognized and the implications carefully studied.

The idea of multiple strategies is also a more sophisticated approach than has previously been evident. If true coordination of all transportation and nontransportation planning could be achieved, we would be in a much better position to cope with present and future transportation challenges.

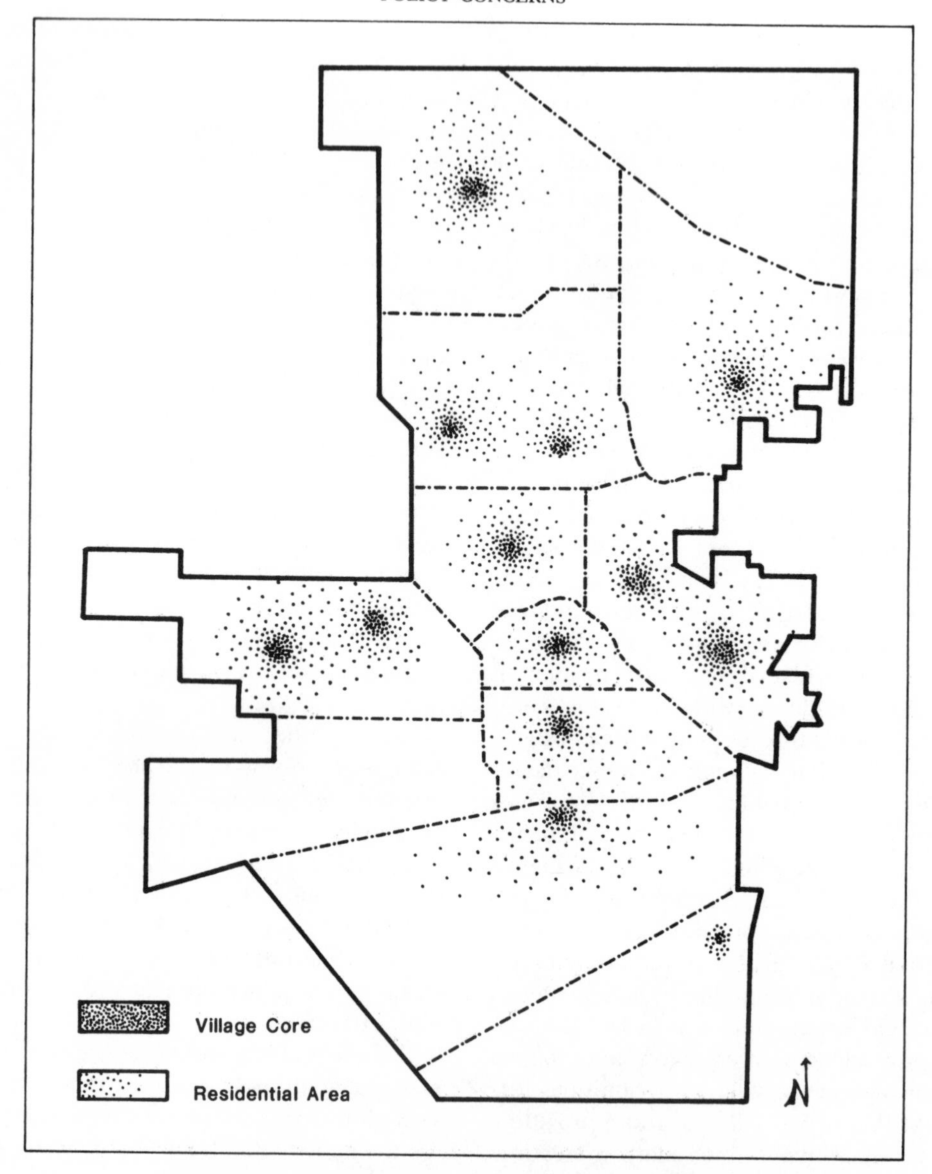

Figure 16-6. The Phoenix "urban village" concept.

The private automobile provides a form of rapid, demand-responsive, random access that no other transportation mode yet invented, or readily foreseeable, is capable of matching. Though future transportation policy will likely continue to emphasize alternative modes for those groups of travelers that can most benefit from them, the fundamental need would appear to be for more creative approaches to accommodate an auto-oriented populace. Some of the most hopeful ways now apparent would appear to be the three forms of nontransportation transportation policy options just described. Americans' daily activity patterns will probably continue to be reoriented as our institutions slowly adjust to the demographic and economic trends concomitant with postindustrial development. Urban form will be altered in response not only to the wider adoption of alternative work schedules and telecommunications, but also to such underlying trends as the increas-

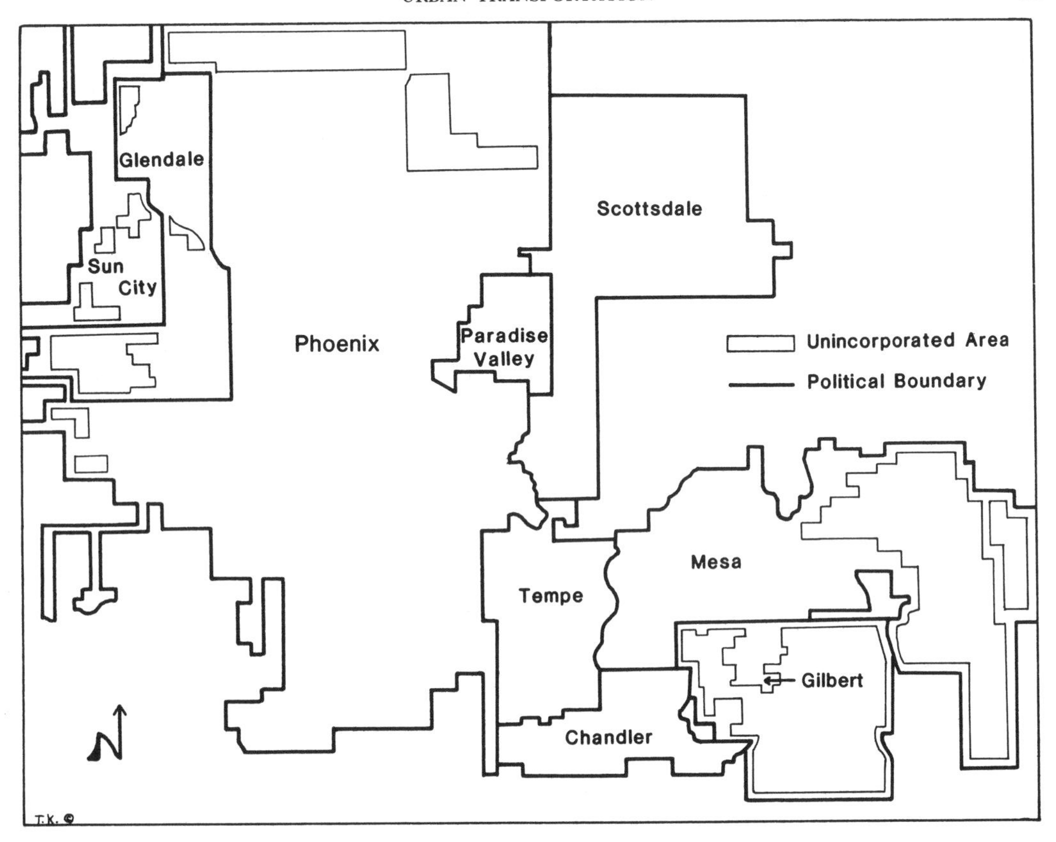

Figure 16-7. The real political geography of metropolitan Phoenix.

ing predominance of two-worker households, smaller families, larger numbers of single people, and the new emphasis on life-style—recreation, fitness, eating out, retirement migration, and so forth. An interesting longer-term question concerns the transportation implications of an aging society. Will the greater proportion of "swinging 50s," empty-nesters, and of the over-65 group spell the doom (or at least de-emphasis) of the Great American Suburbs that were built in response to the needs of the large families begun during the postwar baby boom?

In the 1970s nonmetropolitan areas suddenly began to grow more rapidly than metropolitan areas, reversing a longstanding trend. Among the reasons advanced were the shock of the 1973 Arab oil embargo, the growth of retirement and recreation communities in amenity-rich rural areas, and a desire to return to agrarian, small-town roots. Such trends might suggest hope for alleviating metropolitan transportation problems. In the 1980s, however, we have seen once again a reversal of trends. Now, metropolitan areas are growing more rapidly than nonmetro counties. The fastest growth, however, is being experienced in medium-sized rather than in the largest urban agglomerations. In some ways, such areas face the greatest transportation challenges, as viable competition for the automobile is harder to provide than in the largest metropolises.

Within metropolitan areas themselves, a more compact form of development might ultimately take place as the numbers of elderly persons increase with their preferences for low-care, multifamily housing units. On the other hand, job decentralization seems to be a trend that is here to stay.

U.S. transportation policy has long exhibited a tendency to follow rather than lead the market. To the extent that the energy and creativity of the free enterprise system can be harnessed to meet true transportation needs, then public policy should continue to demonstrate this proclivity. A real danger, however, is that the current de-emphasis on government's role represents an abdication of responsibility to attempt to comprehend and shape the very complex and central role that transportation plays in the evolution of metropolitan systems.

The research and policy experimentation summarized in this book forcefully suggests that policy should be based on careful, rigorous analysis of the indirect as well as the direct linkages that exist between transportation and other components of urban structure. Who is better prepared to make the broad syntheses that are needed to transport the urban system on a route toward a desirable future than those trained to understand the geography of urban transportation?

References

Allen, W. B., & Boyce, D. E. (1974). Impact of a high speed rapid transit facility on residential property values. *High Speed Ground Transport Journal, 8,* 53–60.

Altschuler, A. (1979). *The urban transportation system: Politics and policy innovation.* Cambridge: MIT Press.

Berry, B. J. L. (1981). *Comparative urbanization: Divergent paths in the twentieth century.* New York: St. Martin's Press.

Black, W. B., Plane, D. A., & Westbrook, R. A. (1985). *An evaluation of alternative economic inducements to ridesharing for the Arizona commuter.* Phoenix: Arizona Transportation Research Center, Arizona Department of Transportation.

Boulding, K. (1966). *The impact of the social sciences.* New Brunswick, N.J.: Rutgers University Press.

Dantzig, G. B., & Saaty, T. L. (1973). *Compact city: A plan for a liveable urban environment.* San Francisco: W. H. Freeman.

Gale, S., & Moore, E. G. (1975). *The manipulated city: Perspectives on spatial structure and social issues in urban America.* Chicago: Maaroufa Press.

Jantsch, E. (1975). *Design for evolution: Self-organization and planning in the life of human systems.* New York: Braziller.

Jovanis, P. P. (1981). Assessment of flextime potential to relieve highway facility congestion. *Transportation Research Record,* No. 816, 19–27.

Jovanis, P. P. (1983). Telecommunications and alternative work schedules: Options for managing transit travel demand. *Urban Affairs Quarterly, 19,* 167–189.

Krueckeberg, D. A., & Silvers, A. L. (1974). *Urban planning analysis: Methods and models.* New York: John Wiley.

Kwitny, J. (1981). The Great Transportation Conspiracy: A juggernaut named desire. *Harper's Magazine,* February, 14–21.

Meyer, J. R., & Gomez-Ibanez, J. A. (1981). *Autos, transit and cities.* Cambridge: Harvard University Press.

Plane, D. A. (1981). The geography of urban commuting fields: Some empirical evidence from New England. *Professional Geographer, 32,* 182–188.

Schofer, J. L. (1983). Routes to the future of urban public transportation. *Urban Affairs Quarterly, 19,* 140–166.

U.S. Department of Commerce, Bureau of the Census. (1983). *1980 Census of Population. Volume 1: Characteristics of the Population, Chapter C: General Social and Economic Characteristics, Part 1: United States Summary.* Washington, D.C.: U.S. Government Printing Office.

Wegmann, F. J., & Stokey, S. R. (1983). Impact of flexitime work schedules on an employer-based ridesharing program. *Transportation Research Record,* No. 914, 9–13.

Yeung, Y. (1985). *Cities that work: Hong Kong and Singapore.* Occasional Paper No. 72, Department of Geography, The Chinese University of Hong Kong.

Index